P9-AFZ-463

METHODS IN CELL BIOLOGY

VOLUME VII

Contributors to This Volume

Paul H. Atkinson

C. R. Ball

George Brawerman

Carlo M. Croce

Elaine G. Diacumakos

L. P. Everhart

Richard A. Goldsby

Jack D. Griffith

Peter V. Hauschka

Joseph Kates

J. B. Kirkpatrick

Hilary Koprowski

P. L. Kuempel

Nathan Mandell

C. Moscovici

M. G. Moscovici

Masami Muramatsu

George Poste

R. W. Poynter

D. M. Prescott

Sydney Shall

Jolinda A. Traugh

Robert R. Traut

Elizabeth C. Travaglini

H. W. van den Berg

Woodring E. Wright

Methods in Cell Biology

Edited by

DAVID M. PRESCOTT

DEPARTMENT OF MOLECULAR, CELLULAR AND
DEVELOPMENTAL BIOLOGY
UNIVERSITY OF COLORADO
BOULDER, COLORADO

VOLUME VII

1973

ACADEMIC PRESS • New York and London
A Subsidiary of Harcourt Brace Jovanovich, Publishers

ACADEMIC PRESS, INC.
111 Fifth Avenue, New York, New York 10003

United Kingdom Edition published by
ACADEMIC PRESS, INC. (LONDON) LTD.
24/28 Oval Road, London NW1

LIBRARY OF CONGRESS CATALOG CARD NUMBER: 64-14220

PRINTED IN THE UNITED STATES OF AMERICA

CONTENTS

4. *Recent Advances in the Preparation of Mammalian Ribosomes and Analysis of Their Protein Composition*

Jolinda A. Traugh and Robert R. Traut

5. *Methods for the Extraction and Purification of Deoxyribonucleic Acids from Eukaryote Cells*

Elizabeth C. Travaglini

6. *Electron Microscopic Visualization of DNA in Association with Cellular Components*

Jack D. Griffith

7. *Autoradiography of Individual DNA Molecules*

D. M. Prescott and P. L. Kuempel

8. *HeLa Cell Plasma Membranes*

Paul H. Atkinson

9. *Mass Enucleation of Cultured Animal Cells*

D. M. Prescott and J. B. Kirkpatrick

10. *The Production of Mass Populations of Anucleate Cytoplasms*

Woodring E. Wright

11. *Anucleate Mammalian Cells: Applications in Cell Biology and Virology*
George Poste

12. *Fusion of Somatic and Gametic Cells with Lysolecithin*
Hilary Koprowski and Carlo M. Croce

13. *The Isolation and Replica Plating of Cell Clones*
Richard A. Goldsby and Nathan Mandell

14. *Selection Synchronization by Velocity Sedimentation Separation of Mouse Fibroblast Cells Grown in Suspension Culture*
Sydney Shall

19. *Analysis of Nucleotide Pools in Animal Cells*

Peter V. Hauschka

LIST OF CONTRIBUTORS

Numbers in parentheses indicate the pages on which the authors' contributions begin.

PAUL H. ATKINSON, Department of Pathology and Department of Developmental Biology and Cancer, Albert Einstein College of Medicine, New York, New York (157)

C. R. BALL, Department of Experimental Pathology and Cancer Research, The Medical School, Leeds, England (349)

GEORGE BRAWERMAN, Department of Biochemistry and Pharmacology, Tufts University School of Medicine, Boston, Massachusetts (1)

CARLO M. CROCE, The Wistar Institute of Anatomy and Biology, Philadelphia, Pennsylvania (251)

ELAINE G. DIACUMAKOS, Laboratory of Biochemical Genetics, The Rockefeller University, New York, New York (287)

L. P. EVERHART, Department of Molecular, Cellular and Developmental Biology, University of Colorado, Boulder, Colorado (329)

RICHARD A. GOLDSBY, Department of Chemistry, University of Maryland, College Park, Maryland (261)

JACK D. GRIFFITH, Department of Biochemistry, Stanford University School of Medicine, Stanford, California (129)

PETER V. HAUSCHKA, Department of Molecular, Cellular and Developmental Biology, University of Colorado, Boulder, Colorado (329, 361)

JOSEPH KATES, Department of Chemistry, University of Colorado, Boulder, Colorado (53)

J. B. KIRKPATRICK, Department of Molecular, Cellular and Developmental Biology, University of Colorado, Boulder, Colorado (189)

HILARY KOPROWSKI, The Wistar Institute of Anatomy and Biology, Philadelphia, Pennsylvania (251)

P. L. KUEMPEL, Department of Molecular, Cellular and Developmental Biology, University of Colorado, Boulder, Colorado (147)

NATHAN MANDELL, Yale University, New Haven, Connecticut (261)

C. MOSCOVICI, Tumor Virology Laboratory, Veterans Administration Hospital and Department of Immunology and Medical Microbiology and Department of Pathology, University of Florida College of Medicine, Gainesville, Florida (313)

M. G. MOSCOVICI, Tumor Virology Laboratory, Veterans Administration Hospital and Department of Immunology and Medical Microbiology and Department of Pathology, University of Florida College of Medicine, Gainesville, Florida (313)

MASAMI MURAMATSU, Department of Biochemistry, Tokushima University School of Medicine, Tokushima, Japan (23)

GEORGE POSTE, Department of Experimental Pathology, Roswell Park Memorial Institute, Buffalo, New York (211)

R. W. POYNTER, Department of Experimental Pathology and Cancer Research, The Medical School, Leeds, England (349)

D. M. PRESCOTT, Department of Molecular, Cellular and Developmental Biology, University of Colorado, Boulder, Colorado (147, 189, 329)

SYDNEY SHALL, Biochemistry Laboratory, University of Sussex, Brighton, Sussex, England (269)

JOLINDA A. TRAUGH,* Department of Biological Chemistry, University of California School of Medicine, Davis, California (67)

ROBERT R. TRAUT, Department of Biological Chemistry, University of California School of Medicine, Davis, California (67)

ELIZABETH C. TRAVAGLINI, The Institute for Cancer Research, Fox Chase Center for Cancer and Medical Sciences, Philadelphia, Pennsylvania (105)

H. W. VAN DEN BERG, Department of Experimental Pathology and Cancer Research, The Medical School, Leeds, England (349)

WOODRING E. WRIGHT, Department of Medical Microbiology, Stanford University School of Medicine, Stanford, California (203)

*Present address: Department of Biochemistry, University of California, Riverside, California.

PREFACE

In the ten years since the inception of the multivolume series *Methods in Cell Physiology*, research on the cell has expanded and added major new directions. In contemporary research, analyses of cell structure and function commonly require polytechnic approaches involving methodologies of biochemistry, genetics, cytology, biophysics, as well as physiology. The range of techniques and methods in cell research has expanded steadily, and now the title *Methods in Cell Physiology* no longer seems adequate or accurate. For this reason the series of volumes known as *Methods in Cell Physiology* now continues under the title *Methods in Cell Biology*.

Volume VII of this series continues to present techniques and methods in cell research that have not been published or have been published in sources that are not readily available. Much of the information on experimental techniques in modern cell biology is scattered in a fragmentary fashion throughout the research literature. In addition, the general practice of condensing to the most abbreviated form materials and methods sections of journal articles has led to descriptions that are frequently inadequate guides to techniques. The aim of this volume is to bring together into one compilation complete and detailed treatment of a number of widely useful techniques which have not been published in full detail elsewhere in the literature.

In the absence of firsthand personal instruction, researchers are often reluctant to adopt new techniques. This hesitancy probably stems chiefly from the fact that descriptions in the literature do not contain sufficient detail concerning methodology; in addition, the information given may not be sufficient to estimate the difficulties or practicality of the technique or to judge whether the method can actually provide a suitable solution to the problem under consideration. The presentations in this volume are designed to overcome these drawbacks. They are comprehensive to the extent that they may serve not only as a practical introduction to experimental procedures but also to provide, to some extent, an evaluation of the limitations, potentialities, and current applications of the methods. Only those theoretical considerations needed for proper use of the method are included.

Finally, special emphasis has been placed on inclusion of much reference material in order to guide readers to early and current pertinent literature.

DAVID M. PRESCOTT

CONTENTS OF PREVIOUS VOLUMES

Volume I

Volume II

Volume III

Volume IV

Volume V

Volume VI

Chapter 1

The Isolation of RNA from Mammalian Cells[1]

GEORGE BRAWERMAN

Department of Biochemistry and Pharmacology,
Tufts University School of Medicine,
Boston, Massachusetts

---

[1]Supported by a U.S. Public Health Service research grant (GM 17973) from National Institute of General Medical Sciences.

I. Introduction

Investigators wishing to undertake RNA studies in higher cells are faced with the task of selecting an adequate isolation procedure. For the newcomer in the field, this is no simple task in view of the multitude of techniques described in the literature. The procedures for RNA isolation have been reviewed extensively by Kirby (1967). While the basic isolation procedures are relatively simple and few in number, most investigators have developed their own modifications. Moreover, a procedure that is effective with a given cell extract or cell structure, may prove inadequate when applied to a different type of biological material. This article will attempt to discuss the major approaches to the isolation of RNA. It is hoped that it may serve as a guide for the selection of a procedure best suited for a particular purpose. The discussion will be limited to the isolation of high molecular weight RNA components such as ribosomal RNA, messenger RNA, and nuclear RNA.

The basic initial step in nucleic acid isolation consists of subjecting aqueous cell extracts to agents that will denature proteins and dissociate the nucleic acids from nucleoprotein complexes. In most cases the denatured proteins are either precipitated or extracted in an organic phase. The nucleic acids remain in the aqueous solution, together with other water-soluble components. It is essential to carry out this initial step with great care and rapidity, as the RNA in the cell extracts is exposed to the action of cellular nucleases until these are inactivated and removed along with the rest of the proteins.

For the preparation of defined RNA species, as opposed to total cellular RNA, it is usually necessary to isolate the appropriate subcellular structure. The problem of degradation by ribonucleases can be particularly serious in this case, since the RNA-containing structures remain exposed to these enzymes during the isolation steps. Some cells such as HeLa and rabbit reticulocytes have relatively little RNase. In other cases such as rat liver, a protein component is present that inhibits endogenous RNase action. Some investigators have included RNase inhibitors in the cell homogenates or lysates.

II. The Isolation of Subcellular Components

A. Cell Disruption

The isolation of subcellular components such as nuclei and polysomes requires disruption of the cells under conditions that preserve the integrity of the various cellular structures. One serious difficulty in studies of mRNA metabolism in pulse-labeled cells is the presence in polysome preparations of contaminating highly labeled RNA from damaged nuclei. In cells rich in lyzosomes, disruption of these structures will release nucleases into the cell sap, thus increasing the danger of RNA degradation.

1. Mechanical Disruption

The classical method for cell disruption consists of treating concentrated cell suspensions in a Dounce homogenizer or a motor-driven Potter homogenizer. The cells are squeezed through the space between the glass wall of the homogenizer tube and a tight-fitting pestle. The treatment can be standardized by controlling the clearance between tube and pestle, the rate of rotation of the pestle in the Potter homogenizer, the number of up-and-down strokes, and the thickness of the cell suspension. The procedure is relatively inefficient with many mammalian cell types, and often causes nuclear breakage. Prior swelling of the cells in a hypotonic medium apparently softens the cell envelope, causing the cells to be more readily disrupted. A typical procedure involves suspending the packed cells in RSB (Reticulocyte Standard Buffer; 10 m*M* Tris-HCl pH 7.4, 10 m*M* KCl, and 1.5 m*M* $MgCl_2$). The cells are allowed to swell for 5 minutes, and the suspension is treated in a Dounce homogenizer, using 6 to 10 strokes. The nuclei are removed by centrifugation at 800 *g* (Penman *et al.*, 1963).

2. Cell Lysis with Nonionic Detergents

Mechanical disruption of cells, even after hypotonic swelling, presents various disadvantages. The extent of both cell disruption and nuclear breakage will depend on the severity and length of the shear treatment in the homogenizer. Since the treatment cannot be easily standardized, some uncertainty is introduced in the comparison of homogenates from serial cell samples. Moreover, the absence of nuclear RNA contaminants in the cytoplasmic components cannot be guaranteed.

Mammalian cells can be lyzed by a variety of nonionic detergents such as Nonidet P-40 (Borun *et al.*, 1967), and Triton X-100. The inner nuclear membrane is not solubilized by these agents. This provides a mild, easily standardizable cell disruption procedure, which minimizes damage to nuclei. Lysis in an isotonic medium appears to be effective with mouse L cells (Perry and Kelley, 1968). The cell pellets are suspended in 10 m*M* Tris-HCl pH 7.8, 1 m*M* $MgCl_2$, 0.3 *M* sucrose, and 0.05% Triton X-100, and the suspension is kept in ice for 10 minutes. Polysomes obtained from such lyzates were found to be less contaminated by rapidly labeled heterodisperse RNA (presumed to be of nuclear origin) than those prepared from cells disrupted mechanically after hypotonic swelling.

Detergent lysis in isotonic medium is not effective with all cell types. Mouse sarcoma 180 ascites cells are not lyzed by Triton X-100, unless the cells are first subjected to hypotonic swelling. The cells can be swollen by washing twice in 10 m*M* Tris-HCl pH 7.6, 10 m*M* KCl, and 1 m*M* $MgCl_2$ (Lee *et al.*, 1971a). The packed swollen cells are next suspended in 2.5 volumes of a solution consisting of 50 m*M* Tris-HCl pH 7.6, 130 m*M* KCl, 6.5 m*M* 2-mercaptoethanol, 0.13% Triton X-100, and 13% sucrose. This procedure does not appear to cause any nuclear breakage. A substantial amount of endoplasmic reticulum, however, remains attached to the nuclei (Lee *et al.*, 1971b). Up to 15–20% of the polysomes may remain associated with the nuclear fraction after this treatment. The same procedure, when applied to other types of cells, may leave even greater proportions of cytoplasmic components associated with the nuclei. Brief vigorous shaking of the lyzate on the vortex mixer may reduce the extent of this contamination. Treatment of the lyzate in a homogenizer can also be effective in this respect. Perry and Kelly, however, have reported that such a treatment applied to L cells appears to cause nuclear damage (Perry and Kelley, 1968).

The detergent lysis technique cannot be used when mitochondria or membrane-bound polysomes are to be studied, since these structures are sensitive to the detergents. Also the presence of mitochondrial RNA components in the cytoplasmic extracts of lyzed cells may cause difficulties in some cases. This problem can be circumvented in metabolic studies by preventing labeling of mitochondrial RNA. This can be achieved by incubating the cells in the presence of ethidium bromide (Zylber *et al.*, 1969).

The use of detergents can also lead to disruption of lyzosomes. This constitutes a serious potential hazard, since the cellular nucleic acid components become exposed to the hydrolytic enzymes released from these structures. With some cell types, such as mouse L cells or mouse sarcoma 180 cells, the detergent treatment does not appear to result in polysome fragmentation. It is possible that such cells are either poor in lyzosomes, or contain natural inhibitors of RNase. Rat liver cytoplasm, for instance, which is rich in lyzosomes, can be subjected to detergents with apparent impunity. It has been shown that a RNase inhibitor present in the cell sap protects the RNA components from enzymatic degradation (Blobel and Potter, 1967). HeLa cells have been shown to contain a lyzosomal RNase which remains quite active at 0°C (Penman *et al.*, 1969). This enzyme has its pH optimum around 5, and it is possible to avoid degradation by making the cell extracts at pH 8.5. It is clear that optimal conditions for the disruption of a given cell type cannot be established without some knowledge of the nucleases and of possible inhibitors present in the cell.

3. Lysis of Reticulocytes

Rabbit reticulocytes provide an ideal system as far as cell disruption is concerned. The reticulocyte cell membrane is very fragile, and will burst when the cells are suspended in a hypotonic solution such as RSB. When isotonic conditions are restored immediately after cell lysis, breakage of lymphocytes that contaminate red blood cell pellets can be avoided (Marks *et al.*, 1962).

B. Purification of Nuclei

The numerous procedures available for the isolation of nuclei will not be discussed, as this subject lies outside the area of competence of this writer. This section will be limited to a brief discussion of the approaches to the purification of the crude nuclear pellets obtained from cells disrupted as described above. Mechanical disruption of swollen HeLa cells has been reported to produce nuclei with relatively little attached cytoplasmic material (Holtzman *et al.*, 1966). Treatment of the nuclear pellet with a nonionic detergent does not cause further purification. The cytoplasmic structures that remain associated with the nuclei after lysis of swollen Sarcoma 180 ascites cells with Triton X-100 are also refractory to further treatment with the detergent, and are quite resistant to treatment in the Potter homogenizer (Lee *et al.*, 1971b). Treatment with the ionic detergent sodium deoxycholate leads to lysis of the nuclei. It has been reported by Holtzman *et al.* (1966) that a combination of deoxycholate and the nonionic detergent Tween 40 does not cause lysis of the nuclei and is effective in solubilizing the cytoplasmic contaminants. The washed nuclear pellet is suspended in RSB, sodium deoxycholate and Tween 40 are added to concentrations of 0.5% and 1%, respectively, and the suspension is shaken on the vortex mixer for 3 seconds.

A procedure involving the use of the detergent Triton N-101 has been devised for the preparation of clean nuclei from HeLa cells (Berkowitz *et al.*, 1969). The nuclear pellet from lyzed cells is treated in the Dounce homogenizer in the presence of 0.32 *M* sucrose, 1 m*M* $MgCl_2$, 1 m*M* potassium phosphate pH 6.2 to 6.4, and 0.3% Triton N-101. Two successive treatments yield nuclei apparently free of cytoplasmic contaminants. The purified nuclei obtained by this procedure can be easily resuspended without clumping.

C. Isolation of Polysomes

1. Zone Sedimentation

A variety of techniques are available for the preparation of polysomes. Zone sedimentation of cell lyzates after removal of nuclei and mitochondria is perhaps the simplest method [see Noll (1969) for a detailed description of this procedure]. Deoxycholate or nonionic detergents can be added to the postmitochondrial supernatant to solubilize membrane-bound polysomes. It has been reported, however, that deoxycholate can cause disaggregation of polysomes (Olsnes *et al.*, 1972). Triton X-100, in combination with deoxycholate, appears to prevent this effect, but it seems preferable to avoid deoxycholate altogether when intact polysomes are required.

Labeled RNA associated with the polysomes can be assayed in the fractions recovered from the centrifuge tubes. For more detailed studies, the fractions that contain the polysomes are pooled and subjected to further treatments.

2. Preparative Ultracentrifugation

This is the conventional method for the isolation of ribosomes and polysomes. These particles will usually sediment to the bottom of the centrifuge tubes after 1 to 2 hours at 100,000 *g*. A more efficient separation from soluble proteins is achieved if the cytoplasmic fraction is layered over several milliliters of 20 or 30% sucrose. Centrifugation through 2 *M* sucrose leads to still purer preparations since the dense ribosomal structures will sediment far more rapidly than the proteins (Noll, 1969). This latter procedure, however, requires prolonged centrifugation (4 to 24 hours at 100,000 *g*).

A variation of the centrifugation procedure has been developed to produce a partial separation of polysomes from smaller particles, such as ribosomal subunits. The cytoplasmic fraction is layered over 5 ml of 20% sucrose in a 9-ml centrifuge tube, and sedimented at 80,000 *g* for 40 minutes (Lee and Brawerman, 1971). Nearly all the polysomes are sedimented under these conditions, with little contamination by subunits and by mRNA not associated with polysomes. The ribosomal monomers are distributed between pellet and supernatant. The nonsedimented small particles can be recovered by centrifugation at 150,000 *g* for 2 hours or 220,000 *g* for 1 hour.

3. Magnesium Precipitation

Polysomes and ribosomes can be recovered quantitatively from cell extracts by a simple procedure involving precipitation with Mg^{2+}. The cytoplasmic extract is supplemented with $MgCl_2$ to a concentration of 30 m*M*. This leads to precipitation of the particles. After 30 minutes in ice, they are recovered by centrifugation at 2500 *g* for 10 minutes. The Mg^{2+} concentration is critical, since an excess causes the precipitate to redissolve. KCl interferes with the precipitation, and its concentration must not exceed 10 m*M*. This procedure is effective with Triton X-100 lyzates of mouse Sarcoma 180 ascites cells (Mendecki *et al.*, 1972). For the preparation of rat liver polysomes (see Section II,D,3), the Mg^{2+} concentration must be higher, owing to the presence of deoxycholate added to the cytoplasmic extracts. Polysomes obtained by Mg precipitation appear to be cleaner with respect to contamination by protein than those prepared by centrifugation (Lee and Brawerman, 1971). Glycogen usually present in rat liver extracts is not precipitated by Mg. It has also been shown by Levy and Carter (1968) that polysomes in virus-infected cells can be separated from the virus particles by this procedure.

D. Preparation of Nuclei and Polysomes from Rat Liver

1. Tissue Disruption and Purification of Nuclei

The disruption of cells in tissues is usually performed in a glass homogenizer fitted with a Teflon pestle. The tissue must be first minced into small pieces that can be squeezed around the pestle. The cells of soft tissues such as liver are readily disrupted after a few up-and-down strokes while the pestle rotates slowly. Other tissues, however, are much more resistant to disruption. The presence of lyzosomes in many adult tissues presents a serious problem. The case of liver is particularly favorable in this respect, owing to the presence of a ribonuclease inhibitor among the soluble liver proteins (Roth, 1956).

A variety of procedures have been used for liver homogenization. In all cases, the minced liver is suspended in an isotonic or hypertonic sucrose solution. $CaCl_2$ is usually included, presumably to avoid nuclear damage and clumping. A defined ratio of buffer to liver (usually 6 to 10 ml of buffer per gram of liver) is used in order to keep the conditions of homogenization constant. In the Chauveau procedure (Chauveau *et al.*, 1956), the minced tissue is homogenized in 2.2 *M* sucrose. The high shear forces produced in the highly viscous suspension lead to effective removal of endoplasmic reticulum from the nuclei. In the subsequent centrifugation, the dense nuclei will sediment through the concentrated sucrose solution, while the bulk of the cytoplasmic constituents will float to the top of the centrifuge tube. This procedure yields relatively clean nuclei in a single step, but it is not appropriate for the isolation of cytoplasmic components.

The most commonly used procedure involves homogenization in isotonic sucrose. The procedure used in this laboratory is as follows (Lee and Brawerman, 1971). The excised liver is immersed immediately in ice-cold 0.25 *M* sucrose, 50 m*M* Tris-HCl pH 7.6, and 3 m*M* $CaCl_2$. After weighing, it is minced with scissors, suspended in 6 volumes of the same solution, and disrupted gently in a glass–Teflon homogenizer with 4 up-and-down strokes. The homogenate is filtered through gauze to remove fragments of connective tissue, then centrifuged at 600 *g*. The nuclear pellet obtained in this manner is badly contaminated with cytoplasmic membranous material. Purified nuclei can be obtained by subjecting the pellet to the Chauveau treatment.

The nuclear pellet is suspended in 2.2 *M* sucrose and treated in the homogenizer so as to produce high shear forces; the homogenate is centrifuged at 45,000 *g* for 60 minutes. The nuclei are recovered as a compact colorless pellet. This procedure leads to no obvious nuclear breakage (Hadjivassiliou and Brawerman, 1967). Nuclei can be purified by a simpler procedure that involves resuspending the crude nuclear pellet in an isotonic solution containing 5% Triton X-100. This will solubilize the cytoplasmic membranes. The nuclei are then recovered by low speed centrifugation.

2. Polysome Isolation

For the isolation of polysomes, the cytoplasmic extract is first centrifuged at 17,000 *g*. This serves to remove mitochondria and lyzosomes. Membrane-bound polysomes can be separated from free polysomes as described in Section II, D,3. For the preparation of total polysomes, the post-mitochondrial supernatant is supplemented with Triton X-100 and sodium deoxycholate to concentrations of 5% and 1 %, respectively, to dissolve the membranes of the endoplasmic reticulum. Polysomes can be obtained by the conventional centrifugation procedure as discussed in Section II,C,2, or be precipitated with $MgCl_2$. In the latter case, $MgCl_2$ is added after lysis of the membranes to a concentration of 70 m*M* (Lee and Brawerman, 1971). This high concentration is required because deoxycholate complexes some of the Mg^{2+}. After 30–60 minutes in the cold, the precipitate is collected by centrifugation at 17,000 *g*. Solubilization of the sedimented polysomes requires several extractions of the pellet with 50 m*M* Tris-HCl pH 7.6, 50 m*M* KCl, and 1 m*M* $MgCl_2$. The resuspended polysomes can be separated from the ribosomal subunits by the procedure described in Section II,C,2.

3. Preparation of Membrane-Bound Polysomes

These can be separated from free polysomes on the basis of their lower density. A typical procedure involves layering 4 ml of the postmitochondrial supernatant fraction over a two-layer system that consists of 3 ml of 2 *M* sucrose and 3 ml of 1.38 *M* sucrose. After a 24-hour centrifugation at 100,000 *g*, the free polysomes are at the bottom of the tube and those bound to membranes are in the 1.38 *M* layer (Blobel and Potter, 1967). Free polysomes can be obtained more simply by adding sucrose to the post-mitochondrial supernatant to a concentration of 1.6 *M*, layering over 2 *M* sucrose, and centrifugation at 150,000 *g* for 4 hours (Henshaw, 1968). Membrane-bound polysomes from mouse myeloma cells have been obtained by simpler procedures that involve either centrifuging the cytoplasmic extract at 27,000 *g* for 5 minutes (Baglioni *et al.*, 1971) or at 30,000 rpm for 20 minutes through a 2-ml cushion of 0.8 *M* sucrose (Mach *et al.*, 1973). Polysomes with apparently undegraded mRNA were released from the pellet by 1.2% Nonidet P-40 (Mach *et al.*, 1973).

4. Rat Liver Ribonuclease Inhibitor

The presence of the soluble rat liver ribonuclease inhibitor at all stages of the isolation procedures is essential for the preparation of intact RNA components. The amount of inhibitor normally present in the original tissue homogenate appears to be sufficient to protect the cellular RNA components. Any sedimented material, however, is no longer protected, and its RNA becomes fragmented unless fresh inhibitor is added. The high speed supernatant of rat liver homogenates provides a crude, but adequate, source of inhibitor (Blobel and Potter, 1967). It can be prepared by disrupting liver in 3 volumes of 10% sucrose, 50 m*M* Tris, pH 7.5, and 3 m*M* $CaCl_2$, and centrifuging the extract at 100,000 *g* for 4 hours. The supernatant can be stored at − 15°C after quick-freezing in a Dry Ice-acetone bath. The inhibitor should be included in all solutions used in the isolation steps subsequent to the initial homogenization. It can be omitted from the final suspension of purified polysomes. The inhibitor has been purified extensively (Gribnau *et al.*, 1970).

E. Use of Inhibitors of Ribonuclease

The ubiquitous occurrence of RNases presents a most serious problem for the isolation of intact RNA components. In some cultured cells as well as in reticulocytes, the relatively low levels of RNase do not appear to have any significant effect on the RNA components during isolation. In some cases the lack of effect may be due to the occurrence of natural inhibitors, as in rat liver. In other cells, however, as well as in many tissues, the levels of RNase can be quite high and cause serious damage to the RNA after the cells are disrupted. A variety of agents that are known to inhibit RNase activity, such as bentonite, heparin, polyvinyl sulfate, and dextran sulfate, have been added to cell homogenates in order to avoid RNA degradation. The rat liver RNase inhibitor has also been used with various cells and tissues.

In many cases, no objective evidence for a beneficial effect of the added inhibitor has been presented. One difficulty in the selection of an inhibitor is that optimal conditions for inhibitory action are usually determined with purified RNA in a well-defined ionic environment. Such conditions may not prevail in a cell or tissue homogenate, where a multitude of components may interfere with the action of the inhibitor. Moreover, agents effective against a given RNase may be ineffectual with different enzymes. Heparin and polyvinylsulfate, for instance, have been shown to be effective inhibitors when included in a homogenate of chick oviduct (Palmiter *et al.*, 1970). Their presence permits the preparation of apparently intact polysomes. The same compounds, however, failed to prevent the breakdown of mouse liver polysomes in this laboratory. It is clear that inclusion of RNase inhibitors during cell disruption does not necessarily insure the isolation of intact RNA components.

Regardless of the type of cell used and of the effectiveness of any added RNase inhibitor, it is essential that certain precautions be observed during the preparation of cell fractions, since protection from cellular RNases can never be considered as absolute. The manipulations should be carried out as rapidly as possible and the cell extracts always kept close to 0°C. The use of commercial sucrose not treated to remove RNase that might be present as a contaminant, should be avoided.

III. The Extraction of RNA

A. Phenol Extraction

1. General Procedure

This is the most commonly used procedure for the preparation of RNA. It gained respectability when it was shown to yield infectious RNA from tobacco mosaic virus (Gierer and Schramm, 1956). The procedure consists of mixing a suspension of the biological material with an equal volume of water-saturated phenol. The emulsion is shaken for 5 to 30 minutes. This leads to denaturation of the proteins, some of which are extracted into the organic phase. The remainder of the protein appears as an insoluble gel at the water–phenol interface after the emulsion is centrifuged. Most of the RNA is present in the aqueous phase. A variable amount of RNA, however, remains trapped in the protein gel.

Two types of procedures are used to ensure maximum recovery of RNA. One consists of removing the aqueous phase and reextracting the residue (phenol phase plus protein gel) with a fresh aqueous solution. One reextraction is usually sufficient, but additional reextractions may be required if the protein gel remains bulky. The pooled aqueous phases are extracted several times with fresh aqueous phenol or with a 5:1 chloroform–amyl alcohol mixture to ensure the complete removal of proteins. In the second procedure, the phenol layer is removed after the initial treatment, and the combined aqueous phase plus gel are extracted with fresh phenol. After several such extractions, the size of the protein gel is considerably reduced, and the aqueous phase is then recovered.

The manner in which the initial mixing is done may be quite critical. Mixing of preparations rich in RNases with warm phenol may cause a transient activation of these enzymes and extensive damage to the RNA before the nucleases become denatured. It is preferable, therefore, to use ice-cold aqueous phenol for this initial mixing.

The concentration of cellular materials strongly affects the efficiency of the extraction. The best results are obtained with dilute suspensions. High concentrations of material lead to voluminous protein gels from which it is difficult to recover the trapped RNA. High RNA concentrations also lead to the formation of aggregates between RNA molecules (see Section III,A,4).

2. Agents That Affect the Phenol Treatment

The efficiency of the phenol extraction procedure is affected by a variety of compounds. A multitude of modifications have been applied to the basic procedure with the aim to increase RNA yields and to prevent degradation (Kirby, 1967). Many of the modifications are strictly empirical, and in some cases the beneficial effect of a given agent has not been documented. Moreover, conditions that are favorable for a given type of biological preparation may be damaging for material from another source. Optimal conditions may also be different for different types of RNA.

Sodium dodecyl sulfate (SDS) is commonly included in the phenol mixture. This detergent complexes with proteins and causes their denaturation. It also releases the RNA from ribonucleoproteins. It is used to inhibit nucleases and to ensure a more effective RNA extraction. The SDS is added to the preparations just prior to mixing with phenol. It is usually added from a stock solution kept at room temperature to avoid precipitation. The presence of K^+ in the preparations should be avoided, since this forms a highly insoluble dodecyl sulfate salt. The use of lithium dodecyl sulfate, which remains soluble in the cold, has been recommended (Noll and Stutz, 1968).

Chloroform in addition to phenol is used in some laboratories. $CHCl_3$–phenol mixtures are said to lead to a more compact protein gel, thus promoting the release of trapped RNA into the aqueous phase (Wagner *et al.*, 1967). The $CHCl_3$ also ensures a more effective removal of SDS from the aqueous phase. The effect of chloroform, however, is rather complex. It does not always lead to compact protein gels. Treatment with $CHCl_3$-phenol in the cold tends to produce bulky gels instead. Chloroform is effective in promoting the release of certain types of nucleic acids, but, under certain conditions, the extraction of messenger RNA is prevented in its presence (see Sections III,A,6 and III,A,7).

Ethylenediaminetetraacetic acid (EDTA) is also included in some extraction procedures. This compound is used to complex divalent cations. Removal of Mg^{2+} is believed to promote the disruption of SDS–protein–RNA complexes and thus enhance the release of RNA into the aqueous phase (Wagner *et al.*, 1967). It is also possible that the EDTA functions in preventing RNA degradation caused by some divalent cations (Girard, 1967). The presence of EDTA in the RNA extracts, however, can be troublesome, since it tends to precipitate with the RNA upon addition of ethanol. Its effect on the extraction of nuclear and polysomal RNA is discussed in Sections III,A,6 and III,A,7.

3. The Effect of Monovalent Cations

Monovalent cations such as K^+ and Na^+ are usually included in the phenol extraction mixtures. Tris-HCl buffer at neutral pH provides an additional source of cations. It has been observed, however, that these ions promote the binding of polyadenylate chains to denatured proteins in the phenol mixtures, and the consequent loss in the protein gel of molecules with poly(A) stretches (Brawerman *et al.*, 1972). This effect is particularly serious in studies of messenger RNA, since most mRNA species contain a poly(A) sequence at their 3′ terminus (Mendecki *et al.*, 1972). Maximum yields of mRNA, therefore, are best achieved in the absence of monovalent cations and of Tris-HCl at neutral pH. Tris-HCl buffer at pH 9 is effective for mRNA extraction, probably because most of the Tris molecules are in the basic form at this pH value.

4. Extraction at Elevated Temperature

It was observed by Wecker (1959) that Western equine encephalomyelitis RNA could be extracted from purified viral particles only when the temperature of the phenol mixture was raised substantially. High temperatures were also found to be required for the extraction of DNA-like RNA species, presumably of nuclear origin, from mammalian cells (Georgiev and Mantieva, 1962). Scherrer and Darnell (1962) developed a procedure involving brief treatment with phenol at 50° to 60°C in the presence of 0.01 *M* sodium acetate (pH 5.2) and 0.5% SDS, followed by rapid chilling.

The extent to which heat contributes to the extraction of nuclear RNA is not clear. A considerable amount of nucleoplasmic RNA can be extracted in the cold by the Tris, pH 9, low salt treatment. Poly(A)-containing nuclear RNA molecules can be effectively extracted by both the hot SDS and the cold Tris, pH 9, procedures.

One disadvantage of the hot phenol procedure is that it can lead to aggregation of RNA molecules. Ribosomal RNA will form aggregates when treated with phenol at 65°C (Wagner *et al.*, 1967). Treatment at 55°C reduces the tendency to aggregate, provided that the RNA concentration is low. This tendency to aggregate is probably not an exclusive property of rRNA. It had been observed in this laboratory that phenol extraction of mammalian cytoplasmic fractions at 30°C leads to the presence in the 28 S RNA component of material active in stimulating polypeptide synthesis in *E. coli* extracts. Extraction at 0 to 4°C leads to the presence of most of the active material in the 18 S zone (Brawerman *et al.*, 1965). This old observation may suggest a tendency for mRNA to bind to ribosomal RNA, even at moderate temperatures.

The hot phenol procedure can lead to RNA degradation when applied to certain cell or tissue extracts. It is possible that some RNases may not be completely inactivated in the course of the treatment. The high temperature might then enhance their activity. It appears that the hot phenol procedure is best suited for cell extracts low in RNase.

5. Extraction of Ribosomal RNA

Ribosomal RNA is most easily extracted by the phenol procedure, in spite of its tight association with proteins in the ribosomal particle. This is in contrast to the behavior of cytoplasmic mRNA, which tends to bind to denatured proteins through the poly(A) segment. In view of the relative ease of rRNA extraction, any procedure applied to disrupted cells or cell fractions should be adequate. Dilute suspensions should be used to ensure quantitative yields. It seems preferable to do the phenol extraction in the cold, in order to avoid some of the potential complications associated with the hot phenol treatment. Extraction in the presence of monovalent cations leads to preparations lacking at least part of the mRNA (Brawerman *et al.*, 1972). This should be advantageous in studies of rRNA structure or metabolism.

6. Extraction of Messenger RNA

In view of the interaction of poly(A)-containing RNA molecules with denatured proteins in the presence of monovalent cations, the major critical factor in mRNA extraction is probably the level of these ions in the phenol mixture. The most effective extraction should take place in the absence of Na^+, K^+, or Tris-HCl buffer at neutral pH. Since monovalent cations are usually present in cell extracts, it is useful to reextract the residue after removal of the aqueous phase with a fresh solution lacking these ions. This will cause dissociation of the poly(A)–protein complexes and ensure release of the RNA molecules from the protein gel. Extraction with pH 9 Tris buffer appears to cause a more effective release of poly(A) from the complexes (Brawerman *et al.*, 1972). The final pH in these Tris–phenol mixtures is lower than 9. Exposure of the preparations to the slightly alkaline pH at 0° to 4°C does not appear to cause any damage to the RNA molecules.

An enrichment in mRNA can be achieved sometimes by doing the first phenol treatment in the presence of 0.1 *M* Tris-HCl pH 7.6 and 50 m*M* KCl, and reextracting the residue with 0.1 *M* pH 9 Tris buffer (Brawerman *et al.*, 1972). This fractionation, however, is not always effective, and mRNA molecules with short poly(A) segments are probably lost in the pH 7.6 RNA fraction. It is preferable by far to extract directly the polysomal RNA under conditions optimal for mRNA. The latter can be separated from the rRNA by a variety of procedures using selective adsorption of the poly(A) sequence (see Section V,D).

The extraction procedure used in this laboratory involves mixing an ice-cold aqueous suspension of the material with 0.1 volume of 5% SDS (kept at room temperature) and of 1 *M* Tris-HCl pH 9.0 in rapid succession, and adding ice-cold water-saturated phenol. After 5 to 10 minutes of mixing, the phases are separated by centrifugation, the aqueous phase is removed, and the residue is treated with an equivalent volume of 0.1 *M* Tris pH 9.0 and 0.5% SDS. The combined aqueous phases are extracted three times with fresh phenol to ensure thorough deproteinization as well as removal of the SDS.

Inclusion of $CHCl_3$ in the phenol mixture promotes the release of poly (A)-containing molecules into the aqueous phase under conditions that otherwise would cause binding of poly(A) to proteins. The procedure devised by Perry and co-workers (1972) involves treatment at room temperature with equal volumes of phenol and $CHCl_3$ in the presence of 0.1 *M* NaCl, 10 m*M* Tris pH 7.4, 1 m*M* EDTA and 0.5% SDS. The procedure was found effective in this laboratory, with yields of poly(A)–RNA somewhat lower than those obtained with the pH 9 low salt procedure.

In order to establish the optimal conditions for mRNA extraction from polysomes, the effects of various additives have been compared in this laboratory (see Table I). All treatments were done at 0° to 4°. Phenol treatment in the presence of 0.01 *M* Tris pH 7.6 yielded nearly as much poly(A)-containing RNA as that produced by the alkaline Tris procedure. This is in contrast to the effect of phenol extraction with the more concentrated neutral Tris buffer (Brawerman *et al.*, 1972), and illustrates the monovalent cation effect. EDTA in the presence of 0.01 *M* Tris pH 7.6 led to a reduction in the yield of mRNA. $CHCl_3$ in the presence of neutral Tris completely prevented the extraction of this RNA. Under the slightly different conditions of phenol extraction used by Perry *et al.* (1972), on the other hand, $CHCl_3$ favors the extraction of mRNA. This illustrates the extreme sensitivity of mRNA extraction to variations in the conditions of the phenol treatment. In the presence of Tris pH 9.0, EDTA and $CHCl_3$ did not prevent mRNA extraction, a demonstration of the beneficial effect of the alkaline pH. The various agents had little effect on the recovery of the RNA molecules lacking poly(A).

The detrimental effects of EDTA and chloroform shown in Table I are in sharp contrast to the effects of these agents on the release of RNA molecules at higher temperatures (see following section).

TABLE I

EFFECT OF VARIOUS AGENTS ON THE EXTRACTION OF MESSENGER RNA[a] FROM MOUSE SARCOMA 180 ASCITES CELL POLYSOMES

	Tris, 0.1 M, pH 9.0		Tris, 0.01 M, pH 7.6			
Radioactivity	No addition	EDTA, $CHCl_3$	No addition	EDTA	$CHCl_3$	EDTA, $CHCl_3$
Acid-insoluble	780	740	790	460	290	290
Millipore-bound	470	470	420	130	10	10
Poly(A)	160	150	130	40	3	9

[a] Cells were labeled for 1 hour with (2.8-^{3}H) adenosine in presence of 0.04 μg/ml of actinomycin D. Polysomes were prepared from detergent-treated cells by Mg precipitation and were resuspended in 10 mM Tris-HCl pH 7.6, 10 mM KCl, and 1 mM $MgCl_2$. Aliquots of suspension (0.1 ml) were diluted to 0.9 ml in ice-cold solutions containing the components indicated in the table, and 0.1 ml of 5% SDS was added next. Concentration of ethylenediaminetetraacetetate (EDTA) was 2 mM. Solutions were mixed with 1 ml of aqueous phenol or with 2 ml of a phenol-chloroform mixture (equal volumes of phenol and $CHCl_3$). Extractions were done at 0–4°. After the first treatment, aqueous phases were removed and residues were reextracted with a 1 ml of solution containing the same ingredients as in first extraction. Aqueous phases were pooled and reextracted with phenol. RNA was precipitated with ethanol and redissolved in H_2O. Aliquots were used for determinations of acid-insoluble radioactivity, Millipore-bound radioactivity, and that due to poly(A) (Mendecki *et al.*, 1972).

7. EXTRACTION OF NUCLEAR RNA

The hot SDS–phenol procedure appears well suited for the extraction of nuclear RNA. It offers the distinct advantage of preventing the release of DNA into the aqueous phase. Treatment of nuclei with the pH 9 low salt procedure in the cold, releases large amounts of nuclear RNA, but the presence of DNA in the aqueous phases makes their handling very difficult. Moreover, incubation with DNase must be used to remove the DNA, with the danger that possible RNase contaminants might cause fragmentation of the RNA.

The original procedure of Scherrer and Darnell (1962) has undergone various modifications, such as the inclusion of EDTA and $CHCl_3$ (Wagner *et al.*, 1967), the change from pH 5.2 acetate to neutral Tris buffer (Penman, 1966) and changes in the concentration of sodium acetate up to 0.1 M (Edmonds and Caramela, 1969; Soeiro and Darnell, 1969). It is not clear in what manner the change in pH affects nuclear RNA extraction. The reasons for increasing the sodium acetate concentration are also unclear. In view of the known effect of Na^+ on the extraction of poly(A)-containing molecules, it would seem that the original low Na acetate concentration (0.01 M) should be preferable. A direct comparison of the hot SDS-phenol treatments in the presence of 0.1 *M* Na acetate pH 5.2 and of 0.01 *M* Tris pH 7.6 showed the latter condition to be more effective (Table II). EDTA (2 m*M*) in the presence of Tris buffer led to a still greater yield of nucleoplasmic RNA, and had a preferential effect on the recovery of poly(A)-containing nuclear RNA molecules (Table II). EDTA, however, had a tendency to cause the release of small amounts of DNA into the aqueous phase. Chloroform was also effective in promoting the extraction of nuclear poly(A)-containing molecules, but it caused the release of much DNA. The overall yields of poly(A) were no greater than those obtained with Tris-EDTA. When $CHCl_3$ was included in the initial phenol treatment at 55°C, it led to massive release of DNA into the aqueous phase. When the nuclei were first treated with SDS-phenol at 55°C, and the treatment was next repeated with added $CHCl_3$ (procedure of Wagner *et al.*, 1967), less DNA was released.

A procedure has been devised for the removal of DNA from the nuclei prior to phenol extraction (Penman, 1966). The nuclear pellet, purified by the Tween–deoxycholate procedure (Holtzman *et al.*, 1966) is digested with electrophoretically purified DNase (50 μg/ml) at room temperature in the presence of 0.5 *M* NaCl, 50 m*M* $MgCl_2$ and 10 m*M* Tris-HCl pH 7.4. RNase action during this treatment is apparently prevented by the high ionic strength of the incubation mixture.

B. Treatment with Sodium Dodecyl Sulfate

Purified polysomal and viral preparations can be deproteinized without phenol treatment. The material is dissolved in 0.5% SDS, and subjected to zone centrifugation through a sucrose gradient in which SDS may also be

TABLE II

EFFECT OF DIFFERENT CONDITIONS ON THE EXTRACTION OF NUCLEOPLASMIC RNA[a]

Radioactivity	Acetate, 0.1 *M*, pH 5.2, no addition	Tris, 0.01 *M*, pH 7.6	
		No addition	EDTA, 2 m*M*
Acid-insoluble	2770	3440	3990
Millipore-bound	330	540	970
Poly(A)	53	96	185

[a]Nuclei (from ascites cells treated as in Table I) were washed and resuspended in 10% sucrose, 50 m*M* Tris-HCl pH 7.6 and 3 m*M* $CaCl_2$. Aliquots of suspension (0.2 ml) were diluted with ice-cold solutions as in Table I. Solutions were mixed with 1 ml of aqueous phenol in the cold, and mixtures were heated to 55°C. Chilled mixtures were centrifuged at 12,000 rpm, aqueous phases were removed, and residues were reextracted at room temperature with 1-ml solutions containing buffer used in first treatment (SDS and EDTA not included in second treatment). Other details as in Table I.

included (Girard, 1967; Noll and Stutz, 1968). The RNA components are separated from the slowly sedimenting protein–SDS complexes, and can be recovered by precipitation with ethanol. This procedure leads apparently to quantitative RNA release. A variety of components are usually included in the SDS solution, such as Tris buffer at neutral pH, NaCl (10–100 m*M*), and $MgCl_2$ or EDTA. The function of the added components, other than the buffer, is not clear. K^+ must be avoided, as it makes a highly insoluble dodecyl sulfate salt. The centrifugation is best done at room temperature to avoid precipitation of the SDS. This procedure is rather cumbersome, and not well suited for large-scale preparation.

Zone sedimentation of SDS-treated preparations is a convenient analytical technique for metabolic studies on individual RNA components.

C. Other Isolation Procedures

RNA can be obtained from purified ribosomes by overnight treatment with 2 *M* LiCl in the cold (Kruh, 1967). This procedure takes advantage of the insolubility of macromolecular RNA in concentrated solutions of monovalent cations such as Li^+ or Na^+. The LiCl serves both to dissociate the RNA from the ribosomal proteins and to precipitate it. The precipitate is next washed with cold 2 *M* LiCl. The effectiveness of this procedure for the isolation of polysomal mRNA is not known.

Guanidium chloride has also been used as a means to isolate RNA (Cox, 1968). The proteins are denatured and solubilized in 4 *M* guanidium chloride, and the dissociated RNA is precipitated by 0.5 volume of ethanol.

IV. RNA Purification Procedures

A. Ethanol Precipitation

The RNA extracts obtained by the above procedures are usually free of proteins. Effective deproteinization can be ensured by repeated treatments with phenol or with $CHCl_3$-amyl alcohol, until denatured proteins no longer appear at the water–organic phase interface. The RNA is usually precipitated by 2.5 volumes of ethanol in the presence of 0.1 *M* NaCl or potassium acetate (pH 5.5). Some investigators keep the precipitating RNA at −20°C. Overnight precipitation at 4°C is also effective and reduces the possibility of contamination of the RNA by salt. Phenol, which contaminates the precipitated RNA, can be removed by several washings with a cold 2:1 mixture of ethanol and 0.1 *M* NaCl (v/v). Any phosphate present in the extract will be precipitated with the RNA, and must be removed by other means (see following section).

B. Salt Precipitation

Macromolecules other than proteins may still contaminate the RNA extracts. Polysaccharides are the most frequent contaminants. Overnight precipitation of the RNA with 2 *M* LiCl or NaCl in the cold will eliminate polysaccharides, since the latter are not precipitated under these conditions. Contaminating double-stranded DNA present in moderate amounts can also be removed in this fashion. If large quantities of DNA are present, the high viscosity of the solution will prevent precipitation of the RNA. Both rRNA and mRNA are precipitated in high salt, while tRNA is not. Double-stranded RNA is not precipitated either (Baltimore and Girard, 1966). In the author's experience the precipitation is not effective with very dilute RNA solutions.

C. Partition

Procedures for the purification of RNA by partition between two immiscible phases have been devised (Kirby, 1964). Partition between the two phases created by mixing 2 volumes of 1.25 *M* K phosphate (pH 7.5) and 1 volume of 2-methoxyethanol is effective for the removal of polysaccharides from tRNA. The effectiveness of such procedures for the purification of other types of RNA is not well documented.

D. Dialysis and Gel Filtration

Small molecules such as mono- and oligonucleotides are precipitated by ethanol together with the RNA. They can be removed by gel filtration or dialysis. In order to avoid degradation during dialysis, the tubing should be first boiled with sodium carbonate and EDTA, then washed with distilled water until the washings are no longer alkaline. The treated dialysis tubing can be stored in H_2O in the cold.

V. The Isolation of Messenger RNA

A. Detection of mRNA

The preferential sensitivity of ribosomal RNA synthesis to low concentrations of actinomycin provides a convenient means for labeling mRNA. Incubation of mammalian cells in 0.04 μg/ml of actinomycin D can completely prevent rRNA synthesis. In many cells, nucleoplasmic RNA and polysomal RNA labeling is little affected under these conditions (Perry,

1963; Penman *et al.*, 1968). The resulting radioactivity in polysomes can be attributed mainly to mRNA. In undifferentiated cells, this labeled polysomal RNA consists of a heterogeneous population of molecules with sedimentation values ranging from 10 to 30 S. Release of the labeled RNA by dissociation of the polysomes by EDTA provides an additional criterion for mRNA, by permitting a distinction between polysome-associated RNA and any other cosedimenting RNA species (Penman *et al.*, 1968). The occurrence of poly(A) in most eukaryotic mRNA provides a more precise criterion for mRNA. Poly(A)-containing RNA molecules can be readily recognized by the capacity of the poly(A) segment to bind to Millipore filters at high ionic strength (Lee *et al.*, 1971c). This assay procedure is particularly effective with newly synthesized mRNA molecules, whose poly(A) segment is quite large (150–200 nucleotides) (Mendecki *et al.*, 1972). Molecules with short poly(A) segments are poorly adsorbed on Millipore (Sheiness and Darnell, personal communication). More effective adsorption procedures are available (see Section V,D) but these tend to be more cumbersome. The messenger RNA for histones has been shown to lack the poly(A) sequence, and must be studied by other means (Adesnik and Darnell, 1972).

B. Isolation of mRNA by Zone Sedimentation and Gel Electrophoresis

The mRNA for hemoglobin represents the major messenger species in immature red blood cells. Because of its small sedimentation coefficient (9 to 10 S), it can be easily separated from the rRNA components. Thus zone sedimentation in a sucrose gradient of total polysomal RNA from reticulocytes yields a 10 S peak quite well resolved from the 18 S rRNA. Additional cycles of sedimentation yield a more purified 10 S component (Lockard and Lingrel, 1969). This procedure, however, cannot distinguish between globin mRNA and other RNA species or RNA fragments of similar size.

The histone mRNA, with a 7 to 9 S sedimentation coefficient, can be isolated in a similar fashion. It is present in large amounts in cells synchronized in the S phase, where it appears to represent the major component in this sedimentation range. The histone mRNA has been purified by preparative polyacrylamide gel electrophoresis (Jacobs-Lorena *et al.*, 1972).

C. Messenger Ribonucleoprotein Particles Released from Polysomes

Treatment of polysomes with EDTA leads to dissociation of the ribosomes into 50 S and 30 S subunits. The mRNA is released as complexes with protein. The globin mRNA is released as a 15 S particle when this treatment is applied to reticulocyte polysomes (Huez *et al.*, 1967). The globin mRNP can be separated from the 30 S subunit by zone sedimentation, and deproteinized by treatment with SDS (Labrie, 1969; Housman *et al.*, 1971). Subsequent zone sedimentation yields a 9 to 10 S globin mRNA of greater purity than that obtained by direct sedimentation of total polysomal RNA since the first centrifugation of the mRNP eliminates potential rRNA contaminants.

The EDTA treatment cannot be used as a general method for mRNA isolation, because the mRNP is usually released as a heterogeneous population of particles with sedimentation values overlapping those of the ribosomal subunits (Henshaw, 1968; Lee and Brawerman, 1971).

D. Selective Adsorption of Poly(A)-Containing RNA Molecules

The occurrence of a poly(A) segment on the mRNA molecules permits their isolation from polysomal RNA preparations. Because of unique properties of poly(A), a variety of separation procedures are available. The mRNA can be adsorbed on cellulose nitrate membrane filters (Millipore filters) in the presence of 0.5 *M* KCl (Brawerman *et al.*, 1972). Ribosomal RNA is not retained. The adsorbed RNA can be eluted with 0.5% SDS in 0.1 *M* Tris-HCl pH 9.0 in the cold. Precipitated potassium dodecyl sulfate is removed by centrifugation, and eluted RNA can be obtained by alcohol precipitation. At least 60 μg of RNA can be adsorbed on a single filter. This procedure, however, does not appear to be very effective for the adsorption of molecules with short poly(A) segments.

Several procedures based on the formation of complementary base-paired structures between poly(A) and poly(U) or poly(dT) have been developed. These procedures appear to be more effective than the Millipore binding technique for the recovery of molecules with short poly(A) segments. Poly(dT) coupled to cellulose is an effective adsorbent for the large-scale separation of mRNA from ribosomal RNA (Aviv and Leder, 1972). The RNA is applied to the adsorbent in a small column under conditions that favor base-pairing, and the mRNA is eluted at low ionic strength. Poly(U) coupled to Sepharose also appears to be an effective adsorbent (Adesnik *et al.*, 1972). Poly(U) immobilized on glass fiber by UV irradiation has also been used (Sheldon *et al.*, 1972), but leaching of poly(U) during elution of the mRNA may cause difficulties. Poly(A)-containing molecules have been isolated by chromatography on hydroxyapatite after formation of a triple-stranded hybrid with poly(U) (Greenberg and Perry, 1972). The recovered mRNA, however, is contaminated with poly(U) in this latter procedure.

E. Viral RNA

Poly(A) segments have been found to occur in the RNA of most animal viruses, including the RNA tumor viruses (Johnston and Bose, 1972; Lai and Duesberg, 1972). The isolation of viral RNA, therefore, requires the same conditions as those used for mRNA isolation. It could also be purified using the same techniques for poly(A) adsorption.

References

Adesnik, M., and Darnell, J. E. (1972). *J. Mol. Biol.* **67**, 397.

Adesnik, M., Solditt, M., Thomas, W., and Darnell, J. E. (1972). *J. Mol. Biol.* **71**, 21.

Aviv, H., and Leder, P. (1972). *Proc. Nat. Acad. Sci. U.S.* **69**, 1408.

Baglioni, C., Bleiberg, I., and Zauderer, M. (1971). *Nature (London), New Biol.* **232**, 8.

Baltimore, D., and Girard, M. (1966). *Proc. Nat. Acad. Sci. U.S.* **56**, 741.

Berkowitz, D. M., Kakefuda, T., and Sporn, M. B. (1969). *J. Cell. Biol.* **42**, 851.

Blobel, G., and Potter, V. R. (1967). *J. Mol. Biol.* **28**, 539.

Borun, T. W., Sharff, M. D., and Robbins, E. (1967). *Biochim. Biophys. Acta* **149**, 302.

Brawerman, G., Biezunski, N., and Eisenstadt, J. (1965). *Biochim. Biophys. Acta* **103**, 201.

Brawerman, G., Mendecki, J., and Lee, S. Y. (1972). *Biochemistry* **11**, 637.

Chauveau, J., Moule, Y., and Rouiller, C. (1956). *Exp. Cell Res.* **11**, 317.

Cox, R. A. (1968). *In* "Methods in Enzymology," Vol. 12: Nucleic Acids (L. Grossman and K. Moldave, eds.), Part B, pp. 120–129. Academic Press, New York.

Edmonds, M., and Caramela, M. G. (1969). *J. Biol. Chem.* **244**, 1314.

Georgiev, G. P., and Mantieva, V. L. (1962). *Biochim. Biophys. Acta* **61**, 153.

Gierer, A., and Schramm, G. (1956). *Z. Naturforsch. B.* **11**, 138.

Girard, M. (1967). *In* "Methods in Enzymology," Vol. 12: Nucleic Acids (L. Grossman and K. Moldave, eds.), Part A, pp. 581–588. Academic Press, New York.

Greenberg, J. R., and Perry, R. P. (1972). *J. Mol. Biol.* **72**, 3.

Gribnau, A. A. M., Schoenmakers, J. G. G., Van Kraaikamp, M., and Bloemendal, H. (1970). *Biochem. Biophys. Res. Commun.* **38**, 1064.

Hadjivassiliou, A., and Brawerman, G. (1967). *Biochemistry* **6**, 1934.

Henshaw, E. C. (1968). *J. Mol. Biol.* **36**, 401.

Holtzman, E., Smith, I., and Penman, S. (1966). *J. Mol. Biol.* **17**, 131.

Housman, D., Pemberton, R., and Taber, R. (1971). *Proc. Nat. Acad. Sci. U.S.* **68**, 2716.

Huez, G., Burny, A., Marbaix, G., and Lebleu, B. (1967). *Biochim. Biophys. Acta* **145**, 629.

Jacobs-Lorena, M., Baglioni, C., and Borun, T. W. (1972). *Proc. Nat. Acad. Sci. U.S.* **69**, 2095.

Johnston, R. E., and Bose, H. R. (1972). *Proc. Nat. Acad. Sci. U.S.* **69**, 1514.

Kirby, K. S. (1964). *Progr. Nucl. Acid. Res. Mol. Biol.* **3**, 1.

Kirby, K. S. (1967). *In* "Techniques in Protein Biosynthesis" (P. N. Campbell and J. R. Sargent, eds.), Vol. 1, pp. 265–297. Academic Press, New York.

Kruh, J. (1967). *In* "Methods in Enzymology," Vol. 12: Nucleic Acids (L. Grossman and K. Moldave, eds.), Part A, pp. 609–613. Academic Press, New York.

Labrie, F. (1969). *Nature (London)* **221**, 1217.

Lai, M. M. C., and Duesberg, P. H. (1972). *Nature (London)* **235**, 383.

Lee, S. Y., and Brawerman, G. (1971). *Biochemistry* **10**, 510.

Lee, S. Y., Krsmanovic, B., and Brawerman, G. (1971a). *Biochemistry* **10**, 895.

Lee, S. Y., Krsmanovic, B., and Brawerman, G. (1971b). *J. Cell. Biol.* **49**, 683.

Lee, S. Y., Mendecki, J., and Brawerman, G. (1971c). *Proc. Nat. Acad. Sci. U.S.* **68**, 1331.
Levy, H. B., and Carter, W. A. (1968). *J. Mol. Biol.* **31**, 561.
Lockard, R. E., and Lingrel, J. B. (1969). *Biochem. Biophys. Res. Commun.* **37**, 204.
Mach, B., Faust, C., and Vassali, P. (1973). *Proc. Nat. Acad. Sci. U.S.* **70**, 451.
Marks, P. A., Burka, E. R., and Schlessinger, D. (1962). *Proc. Nat. Acad. Sci. U.S.* **48**, 2163.
Mendecki, J., Lee, S. Y., and Brawerman, G. (1972). *Biochemistry* **11**, 792.
Noll, H. (1969). *In* "Techniques in Protein Biosynthesis" (P. N. Campbell and J. R. Sargent, eds.), Vol. 2, pp. 101–179. Academic Press, New York.
Noll, H., and Stutz, E. (1968). *In* "Methods in Enzymology," Vol. 12: Nucleic Acids (L. Grossman and K. Moldave, eds.), Part B, pp. 129–155. Academic Press, New York.
Olsnes, S., Heiberg, R. and Pihl, A. (1972). *Biochim, Biophys. Acta* **272**, 75.
Palmiter, R. D., Christensen, A. K., and Schimke, R. T. (1970). *J. Biol. Chem.* **245**, 833.
Penman, S. (1966). *J. Mol. Biol.* **17**, 117.
Penman, S., Scherrer, K., Becker, Y., and Darnell, J. E. (1963). *Proc. Nat. Acad. Sci. U.S.* **49**, 654.
Penman, S., Vesco, C., and Penman, M. (1968). *J. Mol. Biol.* **34**, 49.
Penman, S., Greenberg, H., and Willems, M. (1969). *In* "Fundamental Techniques in Virology" (K. Habel and N. P. Salzman, eds.), pp. 49–58. Academic Press, New York.
Perry, R. P. (1963). *Exp. Cell Res.* **29**, 400.
Perry, R. P., and Kelley, D. E. (1968). *J. Mol. Biol.* **35**, 37.
Perry, R. P., LaTorre, J., Kelley, D. E., and Greenberg, J. R. (1972). *Biochim. Biophys. Acta* **262**, 220.
Roth, J. (1956). *Biochim. Biophys. Acta* **21**, 34.
Scherrer, K., and Darnell, J. E. (1962). *Biochem. Biophys. Res. Commun.* **7**, 486.
Sheldon, R., Jurale, C., and Kates, J. (1972). *Proc. Nat. Acad. Sci. U.S.* **69**, 417.
Soeiro, R., and Darnell, J. E. (1969). *J. Mol. Biol.* **44**, 551.
Wagner, E. K., Katz, L., and Penman, S. (1967). *Biochem. Biophys. Res. Commun.* **28**, 152.
Wecker, E. (1959). *Virology* **7**, 241.
Zylber, E., Vesco, C., and Penman, S. (1969). *J. Mol. Biol.* **44**, 195.

Chapter 2

Preparation of RNA from Animal Cells

MASAMI MURAMATSU

Department of Biochemistry, Tokushima University School of Medicine, Tokushima, Japan

I. Introduction

In the history of cell biology and molecular biology the extraction of intact nucleic acids from living organisms has been a subject of prime interest. The functions of deoxyribonucleic acid (DNA) and ribonucleic acid (RNA) mostly require an integrity of the whole macromolecule, so that the isolation of an intact molecule is a prerequisite for the studies of molecular mechanisms of gene action. RNA metabolism, in an older terminology, is now regarded as *transcription* of genetic information from DNA as RNA molecules, and the *processing* and final *degradation* of the latter during the course of their function.

The purification of RNA means the separation of RNA molecules from bound or contaminating proteins, DNA, and polysaccharides. Isolation of individual kinds of RNA from each other is another problem. Older procedures employing chloroform–octanol, guanidine hydrochloride, or high concentration of inorganic salts usually produced low molecular weight RNA fragments, mainly because of ribonuclease (RNase) activity during preparation.

Introduction of aqueous phenol in the extraction of RNA (Kirby, 1956; Gierer and Schramm, 1956) opened up a new era in this field owing to the strong inhibitory effect of this solvent on RNases. The procedure has been sucessfully employed to extract messenger as well as ribosomal RNA from prokaryotic cells, such as *Escherichia coli* (Nomura, Hall, and Spiegelman 1960; Gros *et al.*, 1961). Later the combined use of sodium dodecyl sulfate (SDS) and phenol was found to be even more satisfactory for the purpose of deproteinization and inhibition of RNase activity (Hiatt, 1962).

The preparation of undegraded RNA from eukaryotic cells including animal cells offers special problems. First, certain RNA fractions are tightly associated with cellular proteins and cannot be released into an aqueous phase when the cell homogenate is extracted with phenol alone. Sibatani *et al.* (1960) indicated that this tightly bound RNA was the fraction that was turning over very rapidly in the cell. Georgiev and Mantieva (1962) and Brawerman *et al.* (1963) succeeded in extracting this fraction by raising the temperature and pH, respectively, accomplishing partial fractionation of RNA with respect to its metabolic activity. Scherrer and Darnell (1962), employing SDS and phenol at high temperature successfully extracted whole-cell RNA including the rapidly labeled nuclear RNA in an undegraded form. This is the original basis of the procedures widely used at the present time.

The second problem of eukaryotic cells is the presence of various kinds of organelles with different functions. Some of these organelles have their specific RNA species, and others have nucleolytic enzymes associated with them. Cell fractionation is known to activate these enzymes and often degrade RNA species before they can be extracted. Even RNA extraction from whole cells sometimes cannot prevent partial degradation of RNA, probably because of the strong endogenous RNases. With all these difficulties, many modifications have been attempted to suppress endogenous RNases.

In this chapter, standard procedures for isolation of undegraded RNA from various tissues and cell fractions are described in some detail. Special attention is focused on procedures routinely employed in our laboratory because of their efficacy and reproducibility. In addition, pretreatment of the cell with radioactive isotopes to label RNA is discussed since the procedure is indispensable for the study of some species of RNA and also for some type of analytical procedure.

II. Labeling of RNA with Radioactive Precursors

A. General

Both the radioactive nuclides and the compounds to be employed may vary from one experiment to another depending upon the types of analysis desired, the animal species, type of tissues or cells, the RNA to be analyzed, and the ways of administration.

For the analysis of the primary structure of RNA with, say, Sanger's fingerprinting (Sanger *et al.*, 1965), ^{32}P-labeled inorganic orthophosphate ($^{32}PO_4$) is exclusively used. $^{32}PO_4$ can be injected intravenously or intraperitoneally into whole animals and can also be added directly to the culture medium. $^{32}PO_4$ in dilute HCl is neutralized with a slightly smaller amount of NaOH before injection. Care should be taken not to alkalify the solution beyond pH 6 in order to prevent precipitation.

Theoretically, any radioactive compound can be used to label RNA, provided that it is incorporated into the RNA molecule. However, different tissues or cells take up or utilize different precursors at different rates, making the choice of compound important. For instance, nucleoside triphosphates are poor precursors when employed with whole cells simply because they do not penetrate the cell membrane easily. Nucleosides are commonly employed for culture cells as well as for some tissues in the whole body. ^{3}H- or ^{14}C-labeled uridine has been used most widely, partly because it is not incorporated into DNA unless it is converted to deoxycytidine or thymidine nucleotides. However, labeled cytidine, adenosine, and guanosine can be used equally for the labeling of RNA, especially for relatively short labeling periods. ^{3}H-labeled compounds are preferred for economic reasons as well as to obtain a high specific activity, whereas ^{14}C-compounds provide fewer quenching problems in liquid scintillation counting and also are advantageous in that they can be counted in a low-background gas-flow counter. ^{3}H-labeled compounds with a very high specific activity suffer from high rates of radiolysis during storage.

While uridine can be used to label RNA in such tissues as spleen, bone marrow, and various tumors, orotic acid is preferentially employed to label liver RNA. This is because of the high rate of conversion of orotic acid to UMP in the liver. In most tumors, including many hepatomas, orotic acid is poorly utilized to form RNA when administered *in vivo*.

For the use of methyl-labeled methionine to label ribosomal and transfer RNA, the readers should refer to original papers (Greenberg and Penman, 1966; Zimmerman and Holler, 1967; Muramatsu and Fujisawa, 1968).

B. Rapidly Labeled Nuclear RNA

1. Whole Animal

Intravenous injection is the best procedure for labeling rapidly labeled nuclear RNA, which includes heterogeneous nuclear RNA or DNA-like RNA, and nucleolar precursors of ribosomal RNA. As early as 3 minutes after intravenous injection of labeled orotic acid, radioactivity can be detected in the nuclear and nucleolar RNA of rat liver (Muramatsu, unpublished). The incorporated radioactivity increases almost linearly up to 30 minutes and levels off considerably thereafter (Muramatsu *et al.*, 1964). Figure 1 shows the labeling patterns of nuclear and nucleolar RNA of rat liver labeled with orotic acid ^{14}C. The specific activity (cpm/μg RNA) would be much lower if whole cellular RNA were employed because of the enormous dilution by unlabeled cytoplasmic RNA.

Orthophosphate ^{32}P may be employed intravenously to label nuclear and nucleolar RNA (Muramatsu, Hodnett, and Busch, 1966a). However, extensive purification of RNA, including complete deproteinization by repeated phenol extraction and removal of low molecular phosphate compounds through gel filtration (Sephadex G-25 or G-50), is required especially for labeling times less than 30 minutes.

Nucleolar precursors to ribosomal RNA may be selectively labeled with intravenous injection of methyl-labeled methionine (Muramatsu and Fujisawa, 1968). Natural "chase"-like kinetics is obtained because of the rapid turnover of the methionine pool in the liver *in vivo*. It is important to isolate nuclei, preferably nucleoli, to examine kinetics in order to obtain sufficient specific activity in this type of experiment.

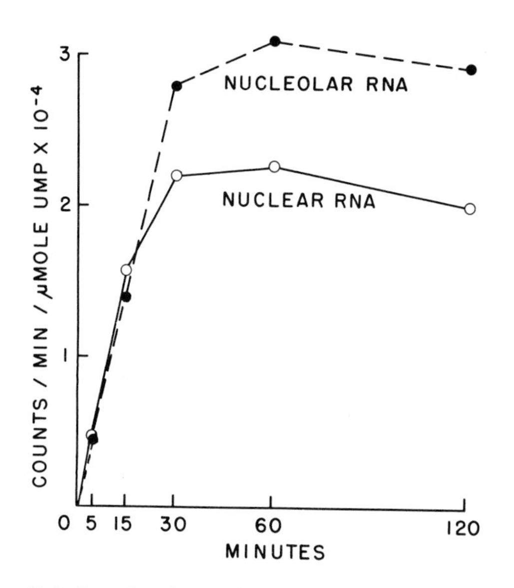

FIG. 1. Kinetics of labeling of nuclear and nucleolar RNA from rat liver. After intravenous injection of 3 μCi of orotic acid-^{14}C into rats weighing 220–250 gm, the animals were sacrificed at indicated intervals, and nuclear (○——○) and nucleolar (●-----●) RNAs were isolated from the liver. Specific radioactivity was normalized with respect to UMP content of each RNA from base composition data from Muramatsu *et al.* (1964). Used by permission of Elsevier, Amsterdam.

2. Cultured Cells

When cells are labeled *in vitro* with radioactive uridine, the specific activity increases for considerably longer than for *in vivo* labeling because of the longer maintenance of the labeled precursors in the surrounding medium. Figure 2 shows the time course of the incorporation of uridine-^{14}C into RNA of a cultured mouse cell line.

When rapidly labeled nuclear RNA with a high specific activity is wanted, e.g., for DNA–RNA hybridization study, the cells must be exposed to the radioactive precursor for relatively longer periods of time. However, the longer the labeling period, the more heavily labeled are the stable RNAs, such as ribosomal RNA and tRNA. This is reflected in the decreasing hybridization efficiency (hybridized RNA: input RNA) of labeled RNA with the increasing labeling period. Therefore, a certain degree of compromise has to be chosen to obtain the best results. When the contamination by ribosomal and other stable RNAs can be ignored, as in the case of the usual DNA-RNA hybridization, a labeling period up to 6 hours may be used without any loss of specificity of hybridization (Levine *et al.*, 1972). How-

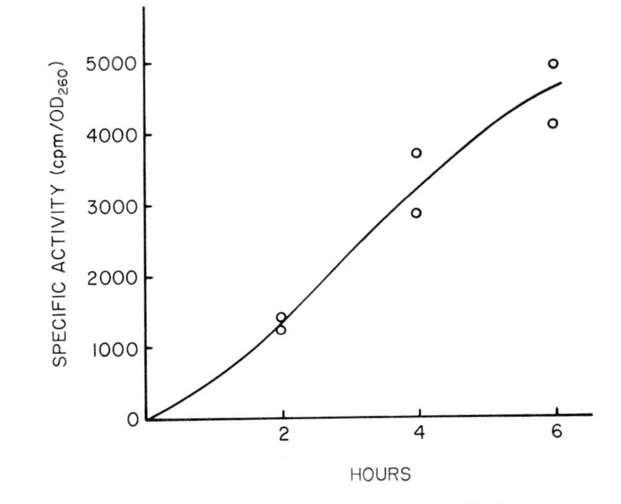

FIG. 2. Kinetics of labeling of nuclear RNA of cultured cells. A monolayer of a mouse cell line, C3H2K (see text) was labeled with 100 μCi/ml of uridine-^{3}H for indicated periods. Nuclei were isolated by lysing the cells with 0.3% of Nonidet P-40 (Muramatsu, 1970). Specific radioactivity was determined for the nuclear RNA extracted by the SDS-hot phenol procedure described in the text. Duplicate experiments are shown.

ever, to obtain size spectra of rapidly labeled nuclear RNA with the aid of sucrose density gradient centrifugation or acrylamide gel electrophoresis, times shorter than 2 hours are recommended so as not to obscure the pattern by labeled ribosomal RNA. At any rate, isolation of nuclei and nucleoli is the best solution to the problem of contamination with cytoplasmic RNA in the analysis of nuclear RNA. For these procedures, the readers are referred to another chapter in this series (Muramatsu, 1970).

3. "CHASE" EXPERIMENTS

To analyze the fate of a labeled RNA species, so-called "chase" experiments have been devised in which an excess of the nonlabeled precursor is added to the culture medium after a certain period of labeling, and the cells are harvested at intervals thereafter. This can be done only for cultured cells, not for the whole animal.

For example, large amounts of cold orotic acid injected into the peritoneal cavity after pulse-labeling do not change the time course of the disappearance of the labeled precursor RNA in the nucleolus (Muramatsu, unpublished). Even in tissue culture, excess cold uridine cannot stop the incorporation of labeled uridine into RNA in a short time. Only gradual dilution of the pyrimidine pool occurs, resulting in a gradual decrease in the incorporation of radioactivity into rapidly labeled nuclear RNA.

An immediate cessation of total RNA synthesis can be produced in tissue culture cells by the addition of 5 μg/ml of actinomycin D. A very small amount of this drug (0.05–0.08 μg/ml) is known to stop specifically nucleolar RNA synthesis, leaving other RNA synthesis almost unaffected (Perry, 1963; Perry and Kelley, 1968). This phenomenon has been utilized for the study of heterogeneous nuclear RNA and messenger RNA (Greenberg and Perry, 1971; Perry *et al.*, 1972). For the whole body, 500–750 μg of actinomycin D per kilogram of body weight can inhibit almost completely nucleolar RNA synthesis in the rat liver within a minute (Muramatsu *et al.*, 1966b; Muramatsu and Fujisawa, 1968). This type of so-called "actinomycin D chase" is especially useful for a short-term chase where immediate cessation of incorporation is essential. However, the effect of the drug in the processing of labeled RNA may not be disregarded, especially for experiments with a relatively long-term chase.

C. Ribosomal RNA and Other Stable RNAs

1. The Whole Animal

Either ^{32}P-labeled orthophosphate, labeled nucleosides, or orotic acid may be used for labeling ribosomal, transfer, and other stable RNAs. Since the half-life of ribosomal RNA is approximately 5 days both for the rat (Loeb, Howell, and Tomkins, 1965; Hiatt *et al.*, 1965) and the mouse (Muramatsu *et al.*, 1972), longer labeling periods are required for the effective labeling of these RNA species. In our laboratory, 20 mCi of ^{32}P-labeled orthophosphate is injected intraperitoneally into a mouse weighing 25 g in two divided doses with an 18- to 24-hour interval between, and the mouse is sacrificed 24 hours after the second injection. The ribosomal RNA prepared from the liver has a specific activity of 4000 cpm per microgram of RNA. When the ascites hepatoma, MH 134 is employed, the specific activity is 12,000 cpm per microgram of RNA under the same labeling conditions (Hashimoto and Muramatsu, 1973). Labeling with orotic acid is also possible for rat liver ribosomal RNA and tRNA (Muramatsu and Fujisawa, 1968), although a considerable amount of labeled precursors is required. Regenerating liver may be employed conveniently to obtain higher specific activities. Labeling of ribosomal RNA with methyl-labeled methionine *in vivo* is not an efficient method because of the very rapid turnover of the methionine pool of the liver (Muramatsu and Fujisawa, 1968). However, we have been able to obtain a specific radioactivity as high as 20,000 cpm per milligram of ribosomal RNA from MH 134 hepatoma, when 1 mCi of methionine-methyl-^3H was injected intraperitoneally into a mouse in four divided doses over a period of 30 hours and the mouse was sacrificed 18 hours after the last injection. Incorporation into the purine ring via 1 C compounds was less than 15% of the total radioactivity of RNA under these conditions.

2. Cultured Cells

Stable RNAs can be labeled in tissue culture cells by addition of radioactive precursors such as ^{32}P-labeled orthophosphate or labeled nucleosides. Higher specific activities than for the whole animal may be obtained because of the longer maintenance of the high specific activity in the precursor pool. Low phosphate medium should not be used for longer labeling periods since it may inhibit cell growth and result in lower synthesis of ribosomal RNA. When 1 mCi of ^{32}P-labeled orthophosphate was added to a Roux bottle containing 20 ml of the minimum essential medium (Eagle) supplemented with 10% calf serum and chased (after a 30-hour pulse) for 18 hours with the same medium containing no radioisotope, the specific activity of the purified ribosomal RNA of a normal cell line, C3H2K (Yoshikura *et al*., 1967), was 20,000 cpm per microgram of RNA. Labeling with radioactive nucleosides is a very common practice and need not be elaborated here. Methyl-labeled methionine is also used to label ribosomal and tRNA. In this type of experiment, 0.3 to 1.0×10^{-4} *M* each of adenosine and guanosine are included in the culture medium to suppress the labeling of purine rings (Iwanami and Brown, 1968; Wagner *et al*., 1967b).

Transfer RNA and other stable RNAs, including a variety of low molecular weight nuclear RNAs, may be labeled in a manner similar to that for ribosomal RNA.

III. Preparation of Rapidly Labeled Nuclear RNA

A. General

Rapidly labeled nuclear RNA includes so-called the heterogeneous nuclear RNA (HnRNA) or DNA-like RNA (dDNA) of chromatin origin and the nucleolar precursors to ribosomal RNA. Both these RNAs bind tightly to nuclear proteins, resisting deproteinization with phenol. The use of SDS (or sodium dodecyl sarcosinate, SDSa) together with high temperature is the best solution at the present time to the extraction of these high molecular weight RNAs in a relatively intact form. Usually, the cells or subcellular fractions are dissolved (by pipetting or homogenization) in a buffer containing appropriate concentrations of SDS and shaken with water-saturated phenol at high temperature. The procedure is especially suitable for tissue culture cells or tumor cells whose cytoplasm is relatively small. For the preparation of rapidly labeled RNA from the liver, where the cytoplasm:

nucleus ratio is very high, the separation of nuclei from cytoplasm is desirable for efficient extraction of nuclear RNA without degradation. Careful and appropriate procedures for isolation of nuclei or nucleoli can prevent rapidly labeled RNA from degradation (Muramatsu, 1970).

B. Procedure of RNA Extraction

The following procedure now in use routinely in this laboratory may be applied to cultured cells as well as to nuclear and nucleolar preparations of any eukaryotic cells (Muramatsu *et al.*, 1970).

1. Dissolve the cell pellet in 20–40 volumes of 0.15 *M* NaCl, 0.05 *M* Na acetate, pH 5.1, 0.3% SDS (or SDSa) by gentle homogenization or by vigorous pipetting.
2. Add an equal volume of phenol–*m*-cresol–H_2O mixture (70:10:20) containing 0.1% 8-hydroxyquinoline.
3. Stir at 65°C for 8 minutes.
4. Shake at room temperature for 15 minutes to complete extraction.
5. Centrifuge at 10,000 rpm for 5 minutes at 4°C.
6. Pipette out the phenol layer leaving the aqueous phase and the interphase.
7. Add $\frac{2}{3}$ volume of phenol–*m*-cresol mixture as described above and shake at room temperature for 5 minutes.
8. Centrifuge as in step 5.
9. Transfer the aqueous phase to another tube and reextract with phenol–*m*-cresol mixture as in 7 and 8.
10. Precipitate the final aqueous phase with 1.5 to 2 volumes of a ethanol–*m*-cresol mixture (9:1) containing 2% K acetate by standing in a deepfreeze for at least 2 hours.
11. Collect the precipitate by centrifugation at 3500 rpm for 30 minutes (or at 10,000 rpm for 10 minutes).
12. Wash the pellet with 75% ethanol containing 1% K acetate, 95% ethanol containing 1% K acetate, and 95% ethanol, successively, by shaking and centrifugation.
13. Dissolve the final pellet in 0.05 *M* Na acetate, pH 5.1, and keep frozen.

The RNA samples can be stored in this form for months without appreciable degradation. The addition of small amounts of SDSa (0.01%) in every solution beyond step 10 may help to protect RNA against contaminating RNase activities.

However, when the RNA is to be used for the substrate for enzymes such as alkaline phosphatase, polynucleotide kinase, various RNase, or DNase to remove contaminating DNA, the RNA pellet must be washed at least three times with 95% ethanol to remove SDSa completely.

C. Comments

1. Extraction Buffer

The pH of the extraction medium is adjusted to 5.1 because of the empirical fact that high molecular weight RNA is better protected at this pH than higher pH's. Neutral or slightly alkaline pH achieved with either phosphate or Tris-HCl buffer may be used for some tissues with better recoveries of nuclear RNA. However, when degradation is encountered, a lower pH buffer should be tested. An ionic strength between 0.05 *M* and 0.2 *M* is recommended to stabilize RNA structure. This should also increase the resistance to RNases. High salt concentrations may cause aggregation of high molecular weight RNA, sometimes leading to precipitation. The SDS concentration is critical. Concentrations higher than 0.5% cause progressively higher contamination of RNA preparations with DNA. When a sufficient amount of buffer is employed, 0.3% SDS is almost as good as 0.5% in the extraction efficiency of nuclear RNA with apparently lower contamination of DNA. Therefore, when DNase treatment is not planned, we use 0.3% SDS rather than 0.5%. Sodium dodecyl sarcosinate (SDSa), which is more soluble than SDS especially at low temperature, may be employed almost in the same way as SDS.

The amount of the solubilizing buffer relative to the tissue weight is an important point of the phenol extraction procedure. At least 20, preferably 40 volumes of the cell or nuclear pellet should be used. Insufficient use of buffer frequently leads to lower recovery of nuclear RNA, higher contamination of DNA, and/or aggregation of high molecular weight RNA.

2. Cell Lysis and Homogenization

Cells or nuclei must be lysed in the SDS-buffer completely by homogenization or pipetting. When the viscous solution containing nuclear DNA is difficult to mix well with the phenol solution by shaking, further homogenization for another minute with a Potter-Elvehjem homogenizer is done after addition of the phenol.

3. Temperature

High temperature not only increases the rate of RNA extraction, but also suppresses the release of DNA into the aqueous phase, resulting in a lower contamination of DNA. At least 55°C is required for efficient extraction of nuclear RNA. Higher temperatures result in a higher rate of extraction of tightly bound nuclear RNA (Georgiev and Mantieva, 1962; Muramatsu, unpublished). Reported aggregation of high molecular weight RNA at temperatures higher than 55°C applies only when relatively small amounts of buffer (as compared to the cell volume) are employed for RNA extraction (Wagner, Katz, and Penman, 1967a; Muramatsu, unpublished).

4. Removal of Contaminating DNA

The removal of DNA contaminating at less than 10% of RNA under standard conditions, can be accomplished by treating the RNA preparation with 50 to 100 μg/ml of DNase I (electrophoretically purified, RNase free) at 37°C for 30 minutes in a buffer containing 5 m*M* Mg^{2+}. For instance, the RNA is dissolved in 0.05 *M* Tris-HCl, pH 7.4, 5 m*M* $MgCl_2$ at a concentration of 1 to 5 mg/ml, and treated with DNase I as described above. The reaction mixture is brought to 0.3% with respect to SDS and shaken with an equal volume of the phenol–*m*-cresol mixture twice to remove DNase, the RNA being precipitated as described above. When the amount of contaminating DNA is exceedingly large, the final RNA precipitate may be further purified by passage through a column (1.5 × 40 cm) of Sephadex G-50 equilibrated with 0.05 *M* Na acetate, pH 5.1, to remove completely the residual deoxyoligonucleotides.

Usually, glycogen and other polysaccharides need not be removed when RNA is prepared from tissue culture cells or from the nuclear fraction.

5. Other RNase Inhibitors

Although SDS (or SDSa) is an excellent RNase inhibitor, which enables one to extract high molecular weight RNA from many cell types without appreciable degradation, there exist situations where this reagent alone cannot completely prevent degradation of RNA. To prevent degradation of RNA in cellular organelles during cell fractionation and/or incubation of isolated organelles (e.g., nuclei, nucleoli, or polysomes), an RNase inhibitor present in the soluble fraction of rat liver cytoplasm has been employed after purification with $(NH_4)_2SO_4$ or DEAE-cellulose column chromatography (Shortman, 1960; Samarina *et al.*, 1968). This factor appears to be effective in suppressing RNase activity which otherwise destroys polysomal messenger RNA or nascent nuclear RNAs.

Degradation of high molecular weight RNA more frequently occurs during the homogenization and phenol extraction. Parish and Kirby (1966) recommended sodium triisopropylnaphthalene sulfonate (Kodak) at 1% concentration in place of SDS, which, however, is still not widely used. Inclusion of 0.5 to 1 mg/ml of either bentonite or Macaloid (Stanley and Bock, 1965) or addition of polyvinyl sulfate (PVS) or dextran sulfate at about 20 μg/ml is sometimes effective in suppressing RNase activity.

In some tissues, nucleolytic activity is so high that it is almost impossible to isolate intact high molecular weight RNA even from fresh whole cells. In this case, diethyl pyrocarbonate, a potent RNase inhibitor (Solymosy *et al.*, 1968; Abadom and Elson, 1970), may be tried. This liquid reagent is dissolved in an equal volume of 95% ethanol and added to the homogenization buffer at the final concentration of 2.5 to 3.0%. The homogenate is kept at 37°C for 5 minutes to inactivate RNase completely. The use of SDS and phenol after this treatment is the same as described above.

6. The Use of a Phenol–Chloroform Mixture Instead of Phenol

Penman (1966) has recommended the use of phenol followed by chloroform to prevent the loss of RNA into the interphase complex. When a sufficient amount of buffer is used for dissolving the material, the loss of nuclear or other RNA into the denatured protein–SDS complex does not seem to be really great. However, recent findings by Perry *et al.* (1972) that poly(A)-containing messenger RNA is specifically lost by extraction with phenol alone, and that this can be prevented by using a 1:1 mixture of phenol and chloroform, indicate that the latter procedure may be the method of choice for extraction of nuclear RNA as well as for extraction of messenger RNA from polysomes (see below).

D. Differential Extraction of RNA

Different species of cellular RNA can be extracted stepwise into the aqueous phase by changing either the temperature (Georgiev and Mantieva, 1962) or the pH of the buffer (Brawerman *et al.*, 1963). Ribosomal and transfer RNAs can be released into the aqueous phase at room temperature upon phenol extraction at neutral pH, but higher temperatures (50 to 65°C) are required for nuclear RNAs to be extracted into the buffer. This could also be accomplished at room temperature at higher pH levels, such as pH 8.0 to 9.0. Usually nucleolar precursors of ribosomal RNA can be extracted at lower temperature or pH than required for HnRNA (or dRNA), thus providing a procedure for partial fractionation of these RNA species. For details, readers are referred to the original papers (Georgiev and Mantieva, 1962; Georgiev *et al.*, 1963; Brawerman *et al.*, 1963; Hadjivassiliou and Brawerman, 1965).

E. Test for Integrity of Nuclear RNA

Both sucrose density gradient centrifugation and acrylamide gel electrophoresis provide good means for the analysis of macromolecular integrity of high molecular weight RNAs. In Fig. 3, a sucrose density gradient profile of total cellular RNA from a monolayer culture of C3H2K cells is presented. The three peaks of optical density represent 28 S ribosomal, 18 S ribosomal,

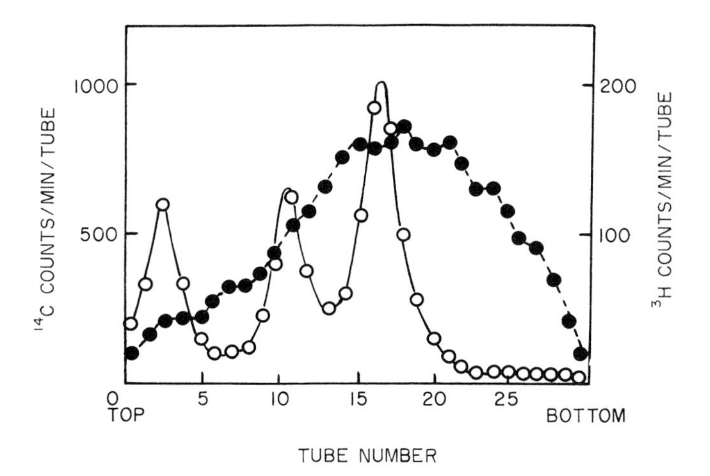

FIG. 3. A typical sedimentation profile of whole-cell RNA from cultured cells. A monolayer culture of C3H2K cells was labeled with uridine-^{14}C for 18 hours followed by pulse labeling with uridine-^{3}H for 25 minutes. Total cellular RNA was extracted by the SDS-hot phenol procedure and analyzed on a 10 to 30% sucrose density gradient as described in the text. ○——○, ^{14}C radioactivity; ●------●, ^{3}H radioactivity.

and low molecular weight RNA containing 5 S and 7 S ribosomal RNA besides transfer RNA. The radioactivity after a 25-minute pulse shows a very polydisperse profile, indicating the overlapping of HnRNA and nucleolar 45 S RNA. When rapidly growing cells such as HeLa or L cells are employed as the starting material, higher radioactivity is usually found in the 45S region because of the higher rate of synthesis of this RNA.

The sucrose density gradient profiles of RNAs extracted from nuclear and subnuclear fractions are somewhat different from those extracted from whole cells, especially in optical density profiles. With shorter pulses, the radioactivity profiles should be similar for nuclear and total cellular RNAs since the major part of cellular RNA synthesis occurs in the nucleus. Figure 4 shows the patterns of RNA extracted from nuclear, nucleolar, and extranucleolar nuclear fractions isolated by our standard procedure (Muramatsu *et al.*, 1966a). When one-third of the absorbance of nucleolar RNA at each point is added to that of the extranucleolar nuclear profile, a pattern is obtained which resembles closely that of total nuclear RNA (Muramatsu *et al.*, 1966b). This fact argues for the maintenance of integrity of high molecular weight RNAs during the course of cellular fractionation and RNA extraction. The radioactivity with a 30-minute pulse did not coincide with the OD peaks. The base compositions of newly synthesized nuclear RNA fractions are presented in Table I. It is apparent that nucleolar RNA has the high GC content characteristic of ribosomal precursor RNAs, but the

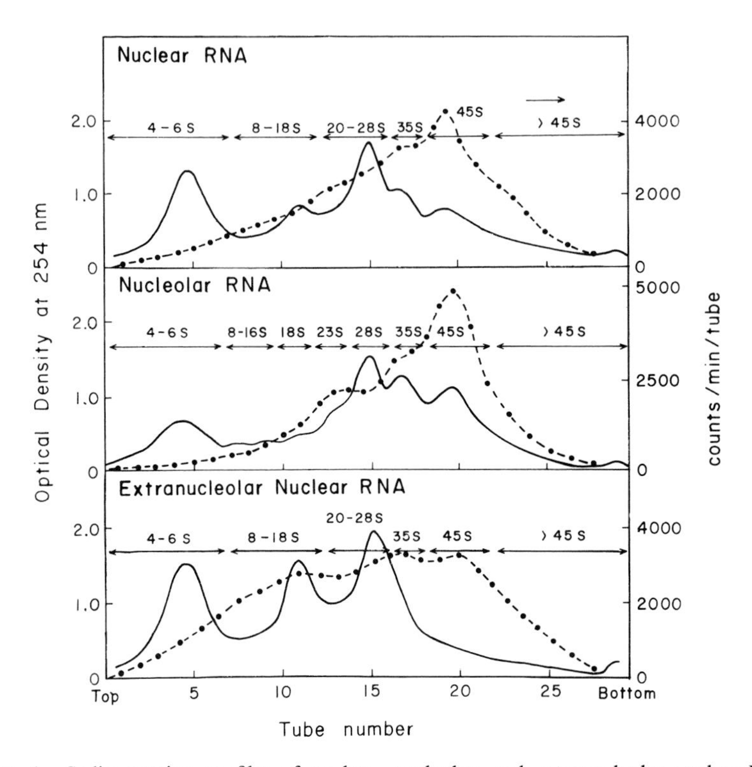

FIG. 4. Sedimentation profiles of nuclear, nucleolar, and extranucleolar nuclear RNA from rat liver. Rats were injected intravenously with 2 μCi/rat of ^{32}P-labeled orthophosphate and sacrificed 30 minutes later. RNAs were extracted from each fraction by the SDS–hot phenol procedure and subjected to sucrose density gradient centrifugation. Pooled fractions were employed to determine base compositions shown in Table I (Muramatsu *et al.*, 1966a). ——, Optical density; ● ----- ●, ^{32}P radioactivity.

extranucleolar nuclear fraction contains RNA with the high AU content characteristic of HnRNA (or dDNA). Reproducible differences in base composition for RNA fractions at different *s* values are noted for both nucleolar and extranucleolar nuclear RNAs.

An example of labeling patterns of nucleolar RNA of rat liver with methionine-methyl-^{3}H is presented in Fig. 5 (Muramatsu and Fujisawa, 1968). Ten minutes after intravenous injection of methionine-methyl-^{3}H the radioactivity is mainly in the peak of 45 S RNA, indicating the methylation of this ribosomal RNA precursor during or immediately after transcription. After 30 minutes the peak moved from 45 S to 35 S with the concomitant appearance of radioactivity on the 28 S peak. After 60 minutes, the major radioactivity peak coincided with the 28 S RNA, with some

TABLE I

BASE COMPOSITIONS (^{32}P) OF SUCROSE DENSITY GRADIENT FRACTIONS AND BASE COMPOSITIONS OF NEWLY SYNTHESIZED RNAs IN SUCROSE DENSITY FRACTIONS[a]

Fraction	Approximate S	Adenylic acid	Uridylic acid	Guanylic acid	Cytidylic acid	$\frac{A + U}{G + C}$
Nuclear RNA						
1–7	4–6	24.3	24.5	26.2	25.0	0.95
8–11	8–18	25.6	23.9	26.2	24.3	0.98
12–16	20–28	25.2	23.5	27.6	23.7	0.95
17–18	35	24.7	22.1	29.1	24.1	0.88
19–22	45	25.1	22.1	28.5	24.3	0.89
23–28	>45	25.7	23.7	26.3	24.3	0.98
Nucleolar RNA						
1–7	4–6	19.8	20.6	34.0	25.6	0.68
8–9	8–16	20.4	19.2	35.2	25.2	0.66
10–11	18	20.7	16.7	35.8	26.8	0.60
12–13	23	19.9	17.5	35.8	26.8	0.60
14–16	28	20.5	17.2	36.3	26.0	0.61
17–18	35	20.3	17.1	39.1	23.5	0.60
19–22	45	19.3	16.6	39.4	24.7	0.56
23–28	>45	20.3	16.6	37.1	26.0	0.58
Extranucleolar nuclear RNA						
1–7	4–6	25.8	27.0	24.2	23.0	1.12
8–11	8–18	26.8	25.8	23.3	24.1	1.11
12–16	20–28	26.2	25.4	23.7	24.7	1.07
17–18	35	26.9	25.1	23.9	24.1	1.08
18–22	45	26.6	24.9	25.0	23.5	1.06
23–28	>45	28.4	25.4	23.1	23.1	1.16

[a]Obtained in Fig. 4 as determined by the distribution of ^{32}P in the four nucleotides. The values are averages of three to four experiments (Muramatsu *et al.*, 1966a).

radioactivity remaining in the 35 S and 45 S region. These results indicate the processing of 45 S RNA into 28 S RNA via a 35 S intermediate, although the production of 18 S RNA from the 45 S precursor cannot be shown in the pattern, most probably because the former leaves the nucleolus quickly after cleavage. Incidentally, virtually no radioactivity was detected in the extranucleolar nuclear RNA at early time periods indicating that HnRNA was not methylated appreciably in the nucleus.

Low molecular weight RNAs in nuclear and nucleolar fractions may be resolved further into many bands with the aid of acrylamide gel electrophoresis (Weinberg and Penman, 1968; Moriyama *et al.*, 1969; Weinberg and Penman, 1969). They are relatively stable but no definite function has been determined so far.

IV. Preparation of Ribosomal RNA

A. General

In animal cells, ribosomal particles contain RNA species with sedimentation coefficients of 28 S, 18 S, 7 S, and 5 S. The large ribosomal subunit contains one molecule of 28 S, 7 S, and 5 S RNA. 7 S RNA is known to be hydrogen-bonded with 28 S RNA (Pene *et al.*, 1968; Sy and McCarty, 1970; King and Gould, 1970), whereas 5 S RNA appears to exist as a mini-subunit in association with a protein (Blobel, 1971). The small ribosomal subunit contains one 18 S RNA molecule. Isolation of these RNAs can be accomplished either from whole cells, postmitochondrial supernatant, ribosomes, or ribosomal subunits. Theoretically, the purest ribosomal RNA

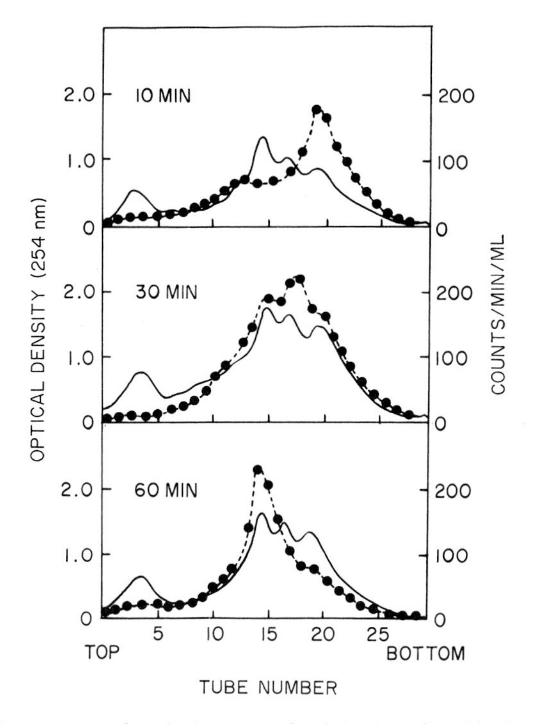

FIG. 5. Labeling patterns of nucleolar RNA after injection of methionine-methyl-^{3}H. Rats were injected intravenously with 0.1 μCi/rat of methionine-methyl-^{3}H and sacrificed at 10, 30, and 60 minutes. Nucleoli were isolated by the sonication technique, and RNA was extracted by the SDS-hot phenol procedure. Approximately 800 μg of RNA was analyzed on a 10 to 40% sucrose density gradient. From Muramatsu and Fujisawa (1968) by permission of Elsevier, Amsterdam. ———, Optical density; ● ------●, ^{3}H radioactivity.

should be obtainable from ribosomal subunits. However, the danger of nucleolytic attack during subunit isolation cannot be overlooked. Therefore, the starting material must be chosen according to the types of analysis to be undertaken on the RNA preparation.

B. Isolation of Ribosomal RNA from Ribosomes

This method is most frequently used for isolation of ribosomal RNA and may be regarded as a standard procedure. Isolation of ribosomes from animal cells may vary from one cell type to another, but in most cases either one of the following two procedures is effective. Pure ribosomes can be obtained by the method of Wettstein *et al.* (1963), in which ribosomes are sedimented through a layer of 2 *M* sucrose. However, the procedure described below requires shorter times and is usually sufficient for isolation of pure ribosomal RNA for most experiments.

1. Isolation of Ribosomes from Rat Liver

The following procedures are performed at 0 to 4°C unless otherwise stated. The rats are fasted for 24 hours to decrease the content of glycogen.

1. Homogenize the liver with 5 volumes of 0.25 *M* sucrose in TKM buffer (0.05 *M* Tris-HCl, 0.025 *M* KCl, 0.005 *M* $MgCl_2$, pH 7.4) in a Potter-Elvehjem type homogenizer with a Teflon pestle, with 7 to 10 strokes. It is recommended that the livers be minced well with scissors and/or passed through a tissue press to remove connective tissue before homogenization.
2. Centrifuge the homogenate at 12,000 *g* for 15 minutes to remove nuclei and mitochondria.
3. Pipette out the upper two-thirds of the supernatant and mix it with one-twentieth volume of freshly prepared 10% sodium deoxycholate.
4. Centrifuge the mixture at 105,000 *g* for 90 minutes. The pellet contains the ribosomes.

2. Isolation of Ribosomes from Tumor Cells or Cultured Cells

Tumor cells, whether solid or ascites type, usually resist the shearing force of the Potter-Elvehjem type homogenizer, and cell breakage cannot be attained under isotonic conditions. The use of the Waring Blender usually causes nuclear breakage as well as cell disruption. The same tendency exists in cultured cells even though they are not really malignant. Under these circumstances, the use of hypotonic shock and/or various detergents become necessary. In this laboratory the following procedure is routinely employed to isolate ribosomes from tumor tissue and tissue culture cells for the purpose of isolating undegraded RNA. The concentration of non-ionic detergent varies depending upon the cells. To avoid breakage of nuclei, which may result in contamination by nuclear RNA, a little lower concentration of detergent should be used than is required for the purification of nuclei (Muramatsu *et al.*, in preparation). For most cases, the following procedure gives satisfactory results.

1. Homogenize the cell in 20 to 40 volumes of RSB (0.01 *M* Tris-HCl, 0.01 *M* NaCl, 0.0015 *M* $MgCl_2$, pH 7.4) and allow to stand in an ice-bath for 10 minutes.
2. Centrifuge the cell at 400 *g* for 5 minutes.
3. Resuspend the pellet in 20 volumes of RSB.
4. Add Nonidet P-40 (a nonionic detergent) to a final concentration of 0.3%. In some normal cells, 0.2% may be sufficient.

This critical concentration should be determined for each case under a light microscope.

5. Homogenize the suspension in a Potter-Elvehjem homogenizer with a Teflon pestle giving 10 strokes.
6. Centrifuge at 12,000 *g* for 15 minutes.
7. Pipette out three-fourths of the supernatant and mix with one-twentieth volume of 10% sodium deoxycholate.
8. Centrifuge at 105,000 *g* for 90 minutes. The pellet contains the ribosomes.

3. Extraction of RNA from Ribosomes

Ribosomal pellets are dissolved in a small volume (2 to 3 volumes of the pellet) of cold TKM buffer by gentle homogenization and then mixed with a relatively large amount of SDS-or SDSa-buffer [0.15 *M* NaCl, 0.05 *M* Na acetate, pH 5.1, 0.3% SDS (or SDSa)]. Approximately 4 to 5 ml of the buffer solution should be employed for the ribosomes obtained from 1 gm of liver tissue. When the amount of SDS-buffer is too small compared with the amount of the ribosomes, both lower recovery and the aggregation of high molecular weight RNA may result.

1. Mix the SDS-lysate of the ribosomes with an equal volume of phenol–*m*-cresol mixture (7:1:2) containing 0.1% 8-hydroxyquinoline, and shake vigorously at room temperature for 15 minutes.
2. Centrifuge at 10,000 rpm for 5 minutes at 4°C.
3. Pipette out the aqueous phase and reextract with two-thirds volume of the phenol–*m*-cresol mixture 3 times. Reextraction involves 5 minutes shaking at room temperature followed by separation of the aqueous phase by centrifugation as in step 2.
4. Add to the final aqueous phase 1.5 to 2 volumes of ethanol–*m*-cresol (9:1) containing 2% K acetate and keep in a freezer at least 2 hours.
5. Collect the precipitate at 10,000 rpm for 10 minutes at 4°C, wash twice with 95% ethanol containing 1% K acetate and dissolve in an appropriate amount of 0.05 *M* Na acetate, pH 5.1.

The RNA solution can be stored at $-70°C$ at least for several months without appreciable degradation. The above method may be applied to the ribosomal subunits in an identical manner.

C. Isolation of Ribosomal RNA from the Postmitochondrial Supernatant

Ribosomal RNA may be extracted from the postmitochondrial supernatant obtained by the above-mentioned procedures. The upper two-thirds to three-fourths of the supernatant of the 12,000 *g* centrifugation is mixed with one-tenth volume of 5% SDS (or SDSa) and shaken with an equal volume of phenol–*m*-cresol mixture (see above) at room temperature for 15 minutes. After centrifugation at 10,000 rpm for 10 minutes at 4°C, the aqueous phase is pipetted out and reextracted 4 to 5 times with two-thirds volume of phenol–*m*-cresol mixture as described in Section IV,B, 3. Reextraction should be repeated until no white material is seen at the interphase. The RNA in the aqueous phase is precipitated as described in the preceding section. The advantage of this procedure lies in the shorter times required for the isolation of postmitochondrial supernatant, which may reduce considerably the chance for nucleases to attack RNA in the ribosomes. Indeed, in some tissues the RNA preparations obtained with this procedure suffer much less from internal nicks than do those obtained from ribosomal particles. However, contamination by small amounts of messenger RNA may occur with this procedure, although this may not cause any real problem when the ribosomal RNA is further purified by sucrose density gradient centrifugation.

D. Isolation of Ribosomal RNA from Whole Cells

When a homogenate of whole cells is shaken with phenol at room temperature without detergent, mainly ribosomal RNA and transfer RNA are extracted into the aqueous phase, leaving tightly bound nuclear RNA in the interphase. After centrifugation, the aqueous phase is transferred to another test tube and one-twentieth volume of 10% SDS is added immediately to suppress residual RNase activity. The mixture is then reextracted with phenol (or phenol–*m*-cresol mixture) repeatedly until no denatured protein is visible at the interphase. This procedure is recommended especially when the starting material is small and loss of ribosomes by cell fractionation must be avoided. Although this procedure is the quickest in preparing ribosomal RNA and the RNA obtained does not suffer appreciably from intramolecular nicks, contamination of small amounts of RNA derived from nuclei may have to be taken into account.

E. Purification and Separation of Ribosomal RNA

The main contaminant of ribosomal RNA prepared with phenol from the liver is polysaccharides, mainly glycogen, especially when the organ has been obtained from a nonfasted animal. Small amounts of glycogen may be removed simply by centrifuging at high speed ($\sim$125,000 *g*) for about 30 minutes (Dingman and Sporn, 1962). However, larger contaminations must be removed by procedures described in detail by Kirby (1964). Since ribosomal RNA tends to degrade at least prartially during the course of these procedures, fasted animals should be used whenever possible. In addition, the following centrifugation procedure is very effective to remove polysaccharides.

In order to purify further ribosomal RNA and also to separate each component, sucrose density gradient centrifugation is performed on ribosomal RNA prepared by the above procedures. In this laboratory, a 5 to 30% linear sucrose gradient containing 0.1 *M* Na acetate, pH 5.1, 0.001 *M* EDTA, 0.01% SDSa is routinely used with excellent results. The slightly acidic pH and the presence of SDSa is expected to suppress possible contaminating RNase activity. SDSa at this concentration is soluble at 4°C. Inclusion of EDTA prevents aggregation of high molecular weight RNA, which may be encountered in the presence of Mg^{2+}.

When ribosomal RNA is extracted at room temperature (or at 4°C), the 28 S RNA carries with it a low molecular weight RNA of about 7 S (Pene *et al.*, 1968; Sy and McCarty, 1970; King and Gould, 1970). Since these RNAs are associated by hydrogen bonding, the use of dimethyl sulfoxide or urea, or heat quenching can dissociate them without breaking phosphodiester linkages. For this purpose, the RNA recovered by ethanol precipitation from the 28 S peak is dissolved in a small amount of 0.01 *M* Tris-HCl, pH 7.6 containing 0.5% SDS (or SDSa) and heated at 60°C for 5 minutes (King and Gould, 1970). The solution is then rapidly chilled in an ice bath and subjected to sucrose density gradient centrifugation as described above.

The procedure is also efficacious for removing "nicked RNA" from intact high molecular weight RNA. Owing to extensive intramolecular hydrogen-bonding, both 28 S and 18 S ribosomal RNA appear to retain overall molecular shape even in the presence of some nicks in the continuous polynucleotide chain. Thus, a 28 S RNA molecule does not change its *s* value even when a few intramolecular scissions are introduced by some nucleolytic attack. That this is indeed the case is revealed by the sucrose density gradient analysis after heat quenching as shown in Fig. 6.

In a number of tissues, the 28 S RNA extracted from ribosomes does release various amounts of partially degraded fragments after heat quenching. The profile of the degradation is frequently reproducible indicating the presence of specific regions particularly susceptible to endogenous (or

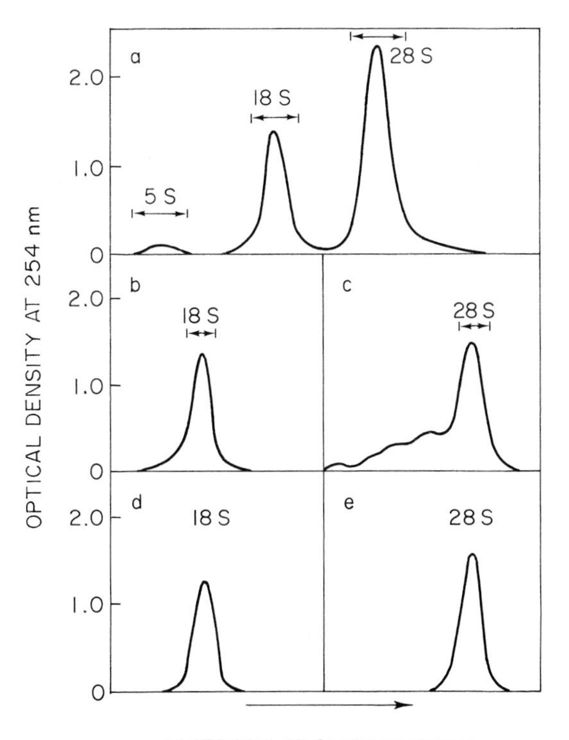

FIG. 6. Sedimentation profiles of ribosomal RNA from a mouse hepatoma, MH 134. Ribosomes were prepared as described in the text. (a) RNA extracted from ribosomes by SDS and phenol at room temperature. (b) 18 S RNA after heat quenching. (c) 28 S RNA after heat quenching. (d) Recentrifugation of the peak portion in (b). (e) Recentrifugation of the peak portion in (c). For details, refer to Hashimoto and Muramatsu (1973).

exogenous) RNases. On the contrary, the 18 S RNA does not usually change its profile even after heat quenching suggesting that this RNA is less susceptible to nucleases during the course of isolation. The second heat quenching applied to the 28 S peak RNA isolated in Fig. 6 does not cause any further degradation, indicating mere heating under these conditions does not cause any scission of phosphodiester linkages in high molecular weight RNA.

F. Purification of 5 S Ribosomal RNA

The RNA in the low molecular weight region of Fig. 6a is precipitated with ethanol and dissolved in a small amount of 0.1 *M* NaCl, 0.01 *M* Tris-HCl, pH 7.4, containing 0.1% SDS and chromatographed on a Sephadex G-100 column (0.5 × 110 cm) equilibrated with the same buffer (Robins and

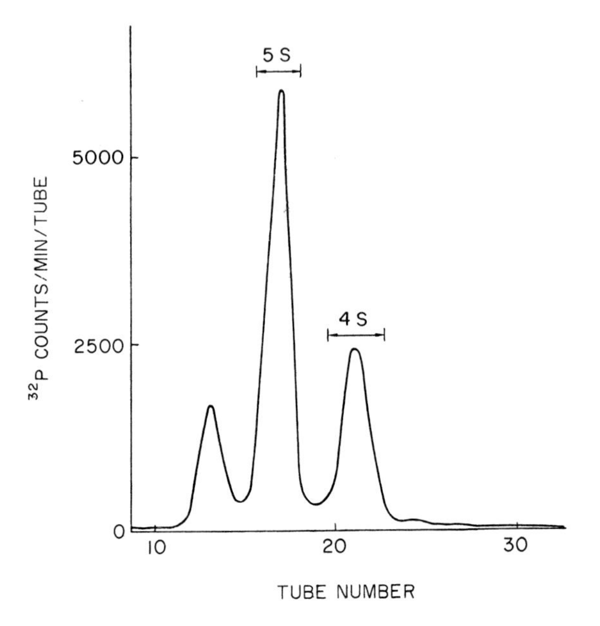

FIG. 7. Purification of 5 S RNA. An elution pattern with a column of Sephadex G-100 of the low molecular weight region obtained in Fig. 6a. See text for detail.

Raacke, 1968). Figure 7 presents the elution pattern of the low molecular weight fraction obtained by sucrose density gradient centrifugation. A small peak in the void volume probably represents fragments of partially degraded ribosomal RNA. Transfer RNA in the 4 S region has been identified by the optical density marker. The middle peak of 5 S is >90% pure as judged by both gel electrophoresis and the oligonucleotide analysis of pancreatic RNase and T1 digests by Sanger's fingerprinting (Sanger *et al.*, 1965).

V. Preparation of Messenger RNA

A. General

Isolation of messenger RNA from animal or eukaryotic cells has been hampered for a few technical reasons. In the first place, the amount of messenger RNA in the cell is supposed to be no more than a few percent of the total RNA present, and because a cell ought to contain many kinds of messenger RNA, the fraction accounted for by one specific messenger must be very small.

Second, these messenger RNAs have a wide range of sizes sedimenting partially at the same *s* values as ribosomal RNA components. Besides, no other specific procedure has yet been devised to separate messenger RNA from ribosomal RNA.

In the third place, messenger RNA is relatively labile and susceptible to nucleolytic attack during the course of isolation. In this connection, messenger RNA cannot be isolated from the nucleus where it is made but has to be isolated from the polyribosomes where it functions. This is so because messenger RNA exists in the nucleus only as a part of the giant heterogeneous nuclear RNA (HnRNA) whose function is poorly understood (Darnell, 1968). However, recent progress in nucleic acid methodology has made it possible to extract and even partially purify messenger RNA from various types of cells.

B. Isolation of Messenger RNA from Special Types of Cells

The first breakthrough came from the system of hemoglobin biosynthesis. Hemoglobin messenger RNA could be isolated from reticulocyte polyribosomes because reticulocytes synthesizing hemoglobin almost exclusively. This messenger, having the sedimentation constant of about 9 S as expected from the molecular weight of hemoglobin, could be banded on a sucrose density gradient between the 18 S and 4 S region (Marbaix and Burny, 1964; Burny and Marbaix, 1965; Evans and Lingrel, 1969a,b).

Other cell types synthesizing one major protein include myeloma cells (Askonas, 1961; Kuff *et al.*, 1964; Namba and Hanaoka, 1968), antigen-stimulated lymph node cells (Kuechler and Rich, 1969), myoblasts (Heywood, 1970), lens cells (Papaconstantinou, 1967), cultured fibroblasts in stationary phase (Goldberg and Green, 1964), and the posterior silk gland cells of the silkworm (Tashiro *et al.*, 1968).

The last system is especially useful for the isolation of pure messenger RNA because during the terminal stage of the fifth larval instar, these cells synthesize almost exclusively silk fibroin, and the messenger RNA can be identified by its unique base ratio as well as base the sequence predicted from the relatively simple structure of fibroin. Indeed, Suzuki and Brown (1972) succeeded in isolating this messenger RNA from the posterior silk gland to a purity greater than 80%.

For other messenger RNAs present in multifunctional cells, isolation of a specific messenger RNA is yet a difficult problem. However, the use of antibody to precipitate polyribosomes (Cowie *et al.*, 1961) is now frequently used with considerable success. When the polyribosomes are treated with the antibody to the protein that is being made on a specific messenger RNA, they are specifically bound to the antibody by way of nascent peptide and can be precipitated by the anti-γ-globulin thereafter. Messenger RNA can be extracted from this complex in a partially purified form (Uenoyama and Ono, 1972). This fraction may be employed *in vitro* for the synthesis of specific protein molecules but, at present, contamination by other messenger RNAs cannot be disregarded. The possibility of partial degradation of messenger RNA during the course of purification is neither completely ruled out. Further refinement of the techniques in every step is necessary for future development.

C. Isolation of Messenger RNA with the Aid of Poly(A) Sequence

Recent findings that most cytoplasmic messenger RNA and heterogeneous nuclear RNA contain polyadenylic acid [poly(A)] sequences 150 to 200 nucleotide long, covalently linked at the 3′ termini of the coding region (Kates, 1970; Darnell *et al.*, 1971a,b; Edmonds *et al.*, 1971; Lee *et al.*, 1971; Mendecki *et al.*, 1972; Sheldon *et al.*, 1972b) provided a unique opportunity to make use of this sequence as a tool in the isolation of messenger RNA. When the poly (A)-containing messenger RNA is passed through a nitrocellulose filter (e.g., Millipore filter, HAWG) under relatively high ionic conditions, it is adsorbed quantitatively to the filter, whereas other RNA species without poly(A) sequence pass through the filter (Lee *et al.*, 1971; Brawerman *et al.*, 1972). Usually, RNA is dissolved in 0.5 *M* KCl, 0.001 *M* $MgCl_2$, 0.01 *M* Tris-HCl, pH 7.6, and gently passed through the Millipore filter.

The poly(A)-containing messenger RNA can be recovered by eluting with 0.5% SDS in 0.1 *M* Tris-HCl, pH 9.0, at 0°C (Mendecki *et al.*, 1972) or with 0.1% SDS in 0.01 *M* Tris-HCl, pH 7.6, at room temperature (Tominaga and Natori, personal communication). Recently, Perry *et al.* (1972) found that the usual extraction with SDS and phenol alone cannot efficiently release poly(A)-containing messenger RNA into the aqueous phase. Sometimes poly(A) sequences are cleaved from the messenger during extraction and remain in the interphase. A mixture of phenol and chloroform (1:1) was found to be useful to extract poly(A)-containing messenger RNA without losing the poly(A) sequence in the interphase. It is now possible with this procedure to obtain an OD amount of total messenger RNA from the polyribosomes of rat liver (Tominaga and Natori, personal communication). The technique of poly(A)-binding with a nitrocellulose filter or with a poly(U) fiber glass filter (Sheldon *et al.*, 1972a) will undoubtedly help in concentrating messenger RNA with a poly(A) sequence, although some smaller messenger RNAs, such as histone messenger RNA, do not seem to contain any poly(A) sequence (Adesnik and Darnell, 1972; Greenberg and Perry, 1972).

VI. Preparation of Transfer RNA

Transfer RNA may be prepared from either whole cells or from the cytoplasmic supernatant fraction. When whole cells are extracted with phenol and fractionated by sucrose density gradient centrifugation, the low molecular weight region should contain, besides transfer RNA, certain amounts of ribosomal 5 S RNA and nuclear low molecular weight RNAs. These can be separated partially by a column of Sephadex G-100 as described in Section IV,F. More rigorous separation may be obtained by slab gel electrophoresis (De Wachter and Fiers, 1971).

If one starts with the soluble cytoplasmic fraction, this sort of complication is much less. The 105,000 *g* supernatant fraction obtained as described in Section IV,B is mixed with one-twentieth volume of 10% SDS and then shaken with an equal volume of phenol–*m*-cresol mixture (see Section III, B) for 15 minutes at room temperature. Reextraction of the aqueous phase is repeated until no denatured protein is visible at the interphase. The precipitated RNA is purified on a column of Sephadex G-100 as described in Section IV,F. Isolation of a single species of transfer RNA involves further fractionation with various kinds of column chromatography, for which the readers are referred to more specialized articles.

VII. Preparation of Mitochondrial RNA

Mitochondrial RNA is the RNA associated with or present in the mitochondria and most probably transcribed from the mitochondrial DNA template. However, part of the RNA synthesized on mitochondrial DNA template may migrate into the cytoplasm, forming specific polyribosomes (Attardi and Attardi, 1968), and the possibility of transport of cytoplasmic RNA into mitochondria cannot be completely ruled out. In view of the extremely small amount of mitochondrial RNA, isolation of pure mitochondria appears to be a prerequisite for isolation of this special class of RNA.

Isolation of mitochondria has been reported by a number of workers. The so-called "mitochondrial fraction" obtained by routine differential centrifugation is grossly contaminated with other components including membranes, polyribosomes etc. A much purer preparation is required for the purpose of isolating mitochondrial RNA. This has been accomplished by the use of sucrose buoyant density centrifugation, in which mitochondria band in a region corresponding to $\rho = 1.19$ to 1.215 (Dawid, 1966; Attardi and Attardi, 1967). The presence of EDTA is apparently necessary to prevent the cobanding with the motochondria of vesicles deriving from endoplasmic reticulum. Eggs of amphibians are good sources of mitochondria because of their extraordinary abundance in this organelle. The following procedure described by Dawid (1966) may be applied to other cell types with appropriate modifications (Attardi and Attardi, 1968; Vesco and Penman, 1969).

1. Homogenize the tissue with 10 volumes of 0.25 *M* sucrose, 0.03 *M* Tris-HCl, pH 7.4, 0.001 *M* EDTA.
2. Centrifuge at 2000 rpm for 15 minutes to sediment nuclei.
3. Centrifuge the supernatant at 10,000 rpm for 20 minutes to sediment mitochondria. In the case of eggs, a black layer of pigment granules is formed at the bottom of the tube, and can be separated from the yellow-brown layer of mitochondria.
4. Suspend the crude mitochondrial fraction in small amounts of homogenizing medium, and layer on a 25-ml linear sucrose gradient, 0.9 to 2.1 *M* containing 0.03 *M* Tris-HCl, pH 7.4, 0.001 *M* EDTA.
5. Centrifuge at 22,000 rpm for 75 minutes. Egg mitochondria band at a density of $\rho = 1.215$ (1.35 *M* sucrose). HeLa cell mitochondria are reported to band at $\rho = 1.19$ (Attardi and Attardi, 1967).

Even the buoyant density-purified mitochondria are still slightly contaminated with other membranous material which is rich in polyribosomal RNA components. Because of the small amount of intramitochondrial RNA, this apparently small contamination may make a great difference as concerns RNA components. Therefore, in some experiments, RNase treatment of purified mitochondria has been attempted to destroy RNA outside the organelle (Attardi and Attardi, 1968). Purified mitochondria were treated with 50 μg/ml pancreatic RNase for 10 minutes at 2°C. Although this treatment does not break down intramitochondrial RNA in an acid-soluble form, the integrity of this RNA after treatment has to be investigated further.

The extraction of RNA from these purified preparations may be accomplished by the usual SDS-hot phenol procedure described in Section III,B.

VIII. Concluding Remarks

Owing to the recent progress in nucleic acid methodology, undegraded RNAs from various animal cells are now obtainable for further analysis of these macromolecules.

In this chapter, a few procedures for isolation of RNA from various tissues, cells, or subcellular fractions have been described in considerable detail. The procedures have been tested in this laboratory and have proved to be effective and reproducible, so that any reader may apply the same procedures to his own experiments. It should be emphasized, however, that any method should be chosen according to the purpose of the experiment, i.e., RNA species to be analyzed, types of analysis to be performed, etc. Moreover, certain modifications are frequently required for different types of cells to obtain the best results. Under these circumstances, insight into the principle and the outcome of the preparation method, especially as regards the intactness of RNA molecules, is of fundamental importance.

Procedures now available for the preparation of intact RNA are yet far from perfect, requiring many future modifications and improvements, especially for special tissues and cells. In any event, suppression of endogenous and exogenous RNases during the isolation and/or incubation of subcellular fractions will undoubtedly be one of the major efforts in future studies.

ACKNOWLEDGMENTS

The author wishes to thank all of his colleagues who cooperated with him in working out many of the procedures described here, and who allowed him to use some of their unpublished observations. Investigations described here have been supported in part by grants from the Ministry of Education of Japan.

REFERENCES

Abadom, P. N., and Elson, K. (1970). *Biochim. Biophys. Acta* **199**, 528.
Adesnik, M., and Darnell, J. E. (1972). *J. Mol. Biol.* **67**, 397.
Askonas, B. A. (1961). *Biochem. J.* **79**, 33.
Attardi, B., and Attardi, G. (1967). *Proc. Nat. Acad. Sci. U.S.* **58**, 1051.
Attardi, B., and Attardi, G. (1968). *Proc. Nat. Acad. Sci. U.S.* **61**, 261.
Blobel, G. (1971). *Proc. Nat. Acad. Sci. U.S.* **68**, 1881.
Brawerman, G., Gold, L., and Eisenstadt, J. (1963). *Proc. Nat. Acad. Sci. U.S.* **50**, 630.
Brawerman, G., Mendecki, Y., and Lee, S. Y. (1972). *Biochemistry* **11**, 637.
Burny, A., and Marbaix, G. (1965). *Biochim. Biophys. Acta* **103**, 409.
Cowie, D. B., Spiegelman, S., Roberts, R. B., and Duerksen, J. D. (1961). *Proc. Nat. Acad. Sci. U.S.* **47**, 114.
Darnell, J. E., Jr. (1968). *Bacteriol. Rev.* **32**, 262.
Darnell, J. E., Phillipson, L., Wall, R., and Adesnik, M. (1971a). *Science* **174**, 507.
Darnell, J. E., Wall, R., and Tushinski, R. J. (1971b). *Proc. Nat. Acad. Sci. U.S.* **68**, 1321.
Dawid, I. B. (1966). *Proc. Nat. Acad. Sci. U.S.* **56**, 269.
De Wachter, R., and Fiers, W. (1971). *In* "Methods in Enzymology," Vol. 21: Nucleic Acids, Part D, (L. Grossman and K. Moldave, eds.), pp. 167. Academic Press, New York.
Dingman, W., and Sporn, M. B. (1962). *Biochim. Biophys. Acta* **61**, 164.
Edmonds, M., Vaughan, M. H., Jr., and Nakazato, H. (1971). *Proc. Nat. Acad. Sci. U.S.* **68**, 1336.
Evans, M. J., and Lingrel, J. B. (1969a). *Biochemistry* **8**, 829.
Evans, M. J., and Lingrel, J. B. (1969b). *Biochemistry* **8**, 3000.
Georgiev, G. P., and Mantieva, V. L. (1962). *Biochim. Biophys. Acta* **61**, 153.
Georgiev, G. P., Samarina, O. P., Lerman, M. I., and Smirnov, M. N. (1963). *Nature* (*London*) **200**, 1291.

Gierer, A., and Schramm, G. (1956). *Nature* (*London*) **172**, 702.
Goldberg, B., and Green, H. (1964). *J. Cell Biol.* **22**, 227.
Greenberg, H., and Penman, S. (1966). *J. Mol. Biol.* **21**, 527.
Greenberg, J. R., and Perry, R. P. (1971). *J. Cell Biol.* **50**, 774.
Greenberg, J. R., and Perry, R. P. (1972). *J. Mol. Biol.* **72**, 91.
Gros, F., Gilbert, W., Hiatt, H. H., Attardi, G., Spahr, P. F., and Watson, J. D. (1961). *Cold Spring Harbor Symp. Quant. Biol.* **16**, 111.
Hadjivassiliou, A., and Brawerman, G. (1965). *Biochim. Biophys. Acta* **103**, 211.
Hashimoto, S., and Muramatsu, M. (1973). *Eur. J. Biochem.* **33**, 446.
Heywood, S. M. (1970). *Proc. Nat. Acad. Sci. U.S.* **67**, 1782.
Hiatt, H. H. (1962). *J. Mol. Biol.* **5**, 217.
Hiatt, H. H., Henshaw, E. C., Hirsch, C. A., Revel, M., and Finkel, R. (1965). *Is. J. Med. Sci.* **1**, 1323.
Iwanami, Y., and Brown, G. (1968). *Arch. Biochem. Biophys.* **124**, 472.
Kates, J. (1970). *Cold Spring Harbor Symp. Quant. Biol.* **35**, 743.
King, H. W. S., and Gould, H. (1970). *J. Mol. Biol.* **51**, 687.
Kirby, K. S. (1956). *Biochem. J.* **64**, 405.
Kirby, K. S. (1964). *Nucleic Acid Res. Mol. Biol.* **3**, 1.
Kuechler, E., and Rich, A. (1969). *Proc. Nat. Acad. Sci. U.S.* **63**, 520.
Kuff, E. L., Potter, M., McIntire, K. R., and Roberts, N. E. (1964). *Biochemistry* **3**, 1707.
Lee, S. Y., Mendecki, J., and Brawerman, G. (1971). *Proc. Nat. Acad. Sci. U.S.* **68**, 1331.
Levine, A. S., Oxman, M. N., Eliot, H. M., and Henry, P. H. (1972). *Cancer Res.* **32**, 506.
Loeb, J. M., Howell, R. R., and Tomkins, G. M. (1965). *Science* **149**, 1093.
Marbaix, G., and Burny, A. (1964). *Biochem. Biophys. Res. Commun.* **16**, 522.
Mendecki, J., Lee, S. Y., and Brawerman, G. (1972). *Biochemistry* **11**, 792.
Moriyama, Y., Hodnett, J. L., Prestayko, A. W., and Busch, H. (1969). *J. Mol. Biol.* **39**, 335.
Muramatsu, M. (1970). *In* "Methods in Cell Physiology" (D. M. Prescott, ed.), Vol. 4, p. 195. Academic Press, New York.
Muramatsu, M., and Fujisawa, T. (1968). *Biochim. Biophys. Acta* **157**, 476.
Muramatsu, M., Hodnett, J. L., and Busch, H. (1964). *Biochim. Biophys. Acta* **91**, 592.
Muramatsu, M., Hodnett, J. L., and Busch, H. (1966a). *J. Biol. Chem.* **241**, 1544.
Muramatsu, M., Hodnett, J. L., Steele, W. J., and Busch, H. (1966b). *Biochim. Biophys. Acta* **123**, 116.
Muramatsu, M., Shimada, N., and Higashinakagawa, T. (1970). *J. Mol. Biol.* **53**, 91.
Muramatsu, M., Azama, Y., Nemoto, N., and Takayama, S. (1972). *Cancer Res.* **32**, 702.
Namba, Y., and Hanaoka, M. (1968). *J. Immunol.* **102**, 1486.
Nomura, M., Hall, B. D., and Spiegelman, S. (1960). *J. Mol. Biol.* **2**, 306.
Papaconstantinou, J. (1967). *Science* **156**, 338.
Parish, J. H., and Kirby, K. S. (1966). *Biochim. Biophys. Acta* **129**, 554.
Pene, J. J., Knight, E., Jr., and Darnell, J. E., Jr. (1968). *J. Mol. Biol.* **33**, 609.
Penman, S. (1966). *J. Mol. Biol.* **17**, 117.
Perry, R. P. (1963). *Exp. Cell. Res.* **29**, 400.
Perry, R. P., and Kelley, D. E. (1968). *J. Cell. Physiol.* **72**, 235.
Perry, R. P., LaTorre, J., Kelley, D. E., and Greenberg, J. R. (1972). *Biochim. Biophys. Acta* **262**, 220.
Robins, H. I., and Raacke, I. D. (1968). *Biochem. Biophys. Res. Commun.* **33**, 240.
Samarina, O. P., Lukanidin, E. M., Molnar, J., and Georgiev, G. P. (1968). *J. Mol. Biol.* **33**, 251.
Sanger, F., Brownlee, G. G., and Barrel, B. G. (1965). *J. Mol. Biol.* **13**, 373.
Scherrer, K., and Darnell, J. E. (1962). *Biochem. Biophys. Res. Commun.* **7**, 486.

Sheldon, R., Jurale, C., and Kates, J. (1972a). *Proc. Nat. Acad. Sci. U.S.* **69**, 417.
Sheldon, R., Kates, J., Kelley, D. E., and Perry, R. P. (1972b). *Biochemistry* **11**, 3829.
Shortman, K. (1960). *Biochim. Biophys. Acta* **51**, 37.
Sibatani, A., Yamana, K., Kimura, K., and Takahashi, T. (1960). *Nature (London)* **186**, 818.
Solymosy, F., Fedoresak, I., Gulyas, A., Farkas, G. L., and Ehrenberg, L. (1968). *Eur. J. Biochem.* **5**, 520.
Stanley, W. M., Jr., and Bock, R. M. (1965). *Biochemistry* **4**, 1302.
Suzuki, Y., and Brown, D. D. (1972). *J. Mol. Biol.* **63**, 409.
Sy, J., and McCarty, K. S. (1970). *Biochim. Biophys. Acta* **199**, 86.
Tashiro, Y., Morimoto, T., Matsuura, S., and Nagata, S. (1968). *J. Cell Biol.* **38**, 574.
Uenoyama, K., and Ono, T. (1972). *J. Mol. Biol.* **65**, 75.
Vesco, C., and Penman, S. (1969). *Proc. Nat. Acad. Sci. U.S.* **62**, 220.
Wagner, E. K., Katz, L., and Penman, S. (1967a). *Biochem. Biophys. Res. Commun.* **28**, 152.
Wagner, E. K., Penman, S., and Ingram, V. (1967b). *J. Mol. Biol.* **29**, 371.
Weinberg, R. A., and Penman, S. (1968). *J. Mol. Biol.* **38**, 289.
Weinberg, R. A., and Penman, S. (1969). *Biochim. Biophys. Acta* **190**, 10.
Wettstein, F. O., Staehelin, T., and Noll, H. (1963). *Nature (London)* **197**, 430.
Yoshikura, H., Hirokawa, Y., and Yamada, M. (1967). *Exp. Cell Res.* **48**, 226.
Zimmerman, E. F., and Holler, R. W. (1967). *J. Mol. Biol.* **23**, 149.

Chapter 3

Detection and Utilization of Poly(A) Sequences in Messenger RNA

JOSEPH KATES

Department of Chemistry, University of Colorado, Boulder, Colorado

I. Introduction

Sequences of polyadenylic acid are attached covalently to the 3′ end of mRNA molecules from a variety of animal viruses and eukaryotic cells (Kates and Beeson, 1970; Lim and Canellakis, 1970; Kates, 1970; Darnell *et al.*, 1971; Edmonds *et al.*, 1971; Lee *et al.*, 1971; Philipson *et al.*, 1971; Molloy *et al.*, 1972; Mendecki *et al.*, 1972; Sheldon *et al.*, 1972a,b). Poly(A) sequences have also been detected on the 3′ end of the heterogeneous nuclear RNA of mammalian cells (Darnell *et al.*, 1971a,b; Sheldon *et al.*, 1972a,b). In general, greater than 85% of the DNA-like RNA associated with the polyribosome fraction of mammalian cells contains polyadenylic acid. The mRNA coding for histones, which is synthesized during the S period of the cell cycle has been shown not to contain polyadenylic acid (Adesnik *et al.*, 1972; Adesnik and Darnell, 1972). Thus, with the known exception of histone message, the occurrence of poly(A) sequences on polyribosomal RNA would appear to be a useful criterion for identification and isolation of mRNA species of both viral and cellular origin.

This chapter will focus on the detection and characterization of poly(A) sequences attached to cellular of viral mRNA, and on the utilization of the poly(A) sequence for isolation of mRNA. Some discussion will also be devoted to the usefulness of poly(A) sequences on RNA for various procedures relating to the structure and origin of various RNA species.

II. Detection and Characterization

A. Purification of Cellular RNA

It has been observed that poly(A) containing DNA-like RNA is easily lost upon purification utilizing conventional phenol extraction of the polyribosome fraction (Lee *et al.*, 1971; Perry *et al.*, 1972). It has also been noted that poly(A) can be cleaved from this RNA during purification (Perry *et al.*, 1972). If the phenol extraction is carried out at pH 9 in the absence of monovalent cations, the loss of poly(A)-containing RNA is greatly reduced (Lee *et al.*, 1971). Phenol extraction at 60°C or release or ribonuclear protein particles from polyribosomes with EDTA prior to extraction of the RNA appears to minimize the loss of poly(A) or poly(A)-containing RNA during purification by the phenol method (Edmonds *et al.*, 1971; Darnell *et al.*, 1971b). More recent work (Perry *et al.*, 1972) has demonstrated that extraction of RNA at neutral pH with a mixture of phenol and chloroform results in efficient extraction of RNA with its associated poly(A) sequences. Since this method is simple and highly effective, it will be described in detail below. Prior to adoption of the chloroform–phenol method, we routinely carried out phenol extraction at room temperature of polyribosomes which had been suspended in a neutral buffer containing sodium dodecyl sulfate. The proportion of RNA extracted by the latter method which contained poly(A) sequences was highly variable but never exceeded 50% of the RNA species, as measured by ability to bind to immobilized poly(U). When the chloroform phenol method was adopted, consistently high yields of polysomal RNA were obtained which showed greater than 85% binding to immobilized poly(U). Thus the method of preparation of RNA from the polyribosomes is extremely important for the demonstration and preservation of poly(A) sequences associated with mRNA. A possible explanation of this phenomenon put forth by Perry suggests that ribosomal proteins associated tightly with poly(A) tails and result in the selective entrapment of mRNAs in the protein after phenol extraction. Use of chloroform–phenol results in a more effective deproteinazation, thus the poly(A) is not trapped in the denatured protein phase. It is further suggested that the poly(A) moiety can be cleaved from mRNA by shear during the phenol extraction, thus resulting in RNA in the aqueous phase which lacks the poly(A) portion.

If RNA purification is from cultured cells, the washed cell pellet is resuspended at a concentration of 5×10^7 to 1×10^8 cells/ml in isotonic buffer containing 0.15 *M* sodium chloride, 0.05 *M* Tris pH 8.5, and 0.0015 *M* magnesium chloride. After the cells are resuspended, Nonidet P-40 detergent (Shell Oil Company) is added to a final concentration 0.5% v/v. The cell suspension is allowed to stand on ice for 5 minutes with occasional pipetting. Examination under a phase contrast microscope should indicate more than 90% cell breakage as evidenced by free nuclei. The lysate is centrifuged at 0°C at approximately 1000 g in a conical centrifuge tube for 3 minutes. The cytoplasmic supernatant is carefully removed with a Pasteur pipette and kept on ice. The nuclear pellet is resuspended in isotonic high pH buffer containing 0.5% Nonidet P-40 and subsequently centrifuged once more to sediment the nuclei. The nuclear wash is combined with the original cytoplasmic supernatant. The combined cytoplasmic fractions are then centrifuged at approximately 20,000 g for 10 minutes. Supernatant solution is removed by Pasteur pipette and further fractionated by sucrose gradient centrification.

Two types of gradients may be used for preparation of the polyribosome fraction. One milliliter of cytoplasmic solution may be layered over a 10-ml gradient of 7–47% w/w sucrose also containing 0.01 *M* Tris pH 7.4, 0.5 *M* sodium chloride, and 0.05 *M* magnesium chloride. This gradient is centrifuged for 110 minutes in the SW 41 rotor at 38,000 rpm and 4°C. Alternatively a 35-ml gradient identical to the one above can be centrifuged in the Spinco SW 27 rotor for 180 minutes at 27,000 rpm at 4°C. The polyribosomal region located by optical density at 260 nm is combined. Under the high salt conditions used in the fractionation, nonribosomal ribonucleoprotein complexes (RNP) are disrupted, and the RNA contained by these particles occurs at sedimentations lower than the 70 S peak of monosomes. Thus, the resulting polyribosomes are not contaminated by cytoplasmic RNP. To the combined polyribosome fractions is added EDTA to a final concentration of 0.01 *M* and sodium dodecyl sulfate (SDS) to a final concentration of 0.5%. The suspension is then precipitated by the addition of two volumes of cold 95% ethanol and allowed to stand at −20°C for at least 2 hours. The suspension is then centrifuged at 20,000 g for 10 minutes. The pellet is resuspended in a buffer containing 0.01 *M* Tris pH 7.5, 0.1 *M* sodium chloride, 0.001 *M* EDTA, and 0.5% SDS. This solution is then extracted with a mixture of chloroform and phenol made up in the following manner. Freshly redistilled phenol is dissolved and saturated with a buffer containing 0.01 *M* sodium acetate pH 6, 0.1 *M* sodium chloride and 0.01 *M* EDTA. The phenol phase is then mixed 1 to 1 with chloroform.

An equal volume of chloroform phenol mixture is added to the RNA suspension. The mixture is shaken at room temperature for 15 minutes. After shaking, the mixture is centrifuged 20,000 *g* for 10 minutes, and the supernatant is removed with a Pasteur pipette. The chloroform–phenol extraction is repeated on this supernatant solution. Usually two chloroform–phenol extractions are required to remove most of the protein which appears as a precipitate at the interphase between the aqueous solution and the organic phase. The supernatant solution is then adjusted to a sodium chloride concentration of 0.5 *M* and precipitated at −20°C with two volumes of 95% ethyl alcohol. At least 2 hours at −20°C are required for precipitation of the RNA. The RNA is collected by centrifugation for 10 minutes at 20,000 *g* and redisolved in 0.01 *M* Tris pH 7.5. At this stage it is often useful to purify the RNA by lithium chloride precipitation. This removes double-stranded RNA structures, DNA, and small RNAs, such as transfer RNA and fragments due to degradation of mRNA. The RNA solution is mixed with an equal volume of 4 *M* lithium chloride. The mixture is allowed to stand for 18 hours at 0°C. The resulting precipitate is centrifuged at 30,000 *g* at 4°C for 45 minutes. This precipitate may be dissolved in any desired buffer. An additional alcohol precipitation is sometimes desirable to remove any traces of lithium chloride present in the lithium chloride pellet.

Heterogeneous nuclear RNA is prepared essentially by the technique of Soero and Darnell (Soero and Darnell, 1969) and will not be described here.

B. Ribonuclease Resistance of Polyadenylic Sequences

In a study by Beers (1960), it was shown that polyadenylic acid was susceptible to degradation by bovine pancreatic ribonuclease. In the latter study it was also shown that moderate salt concentrations can minimize attack by ribonuclease on the poly(A) substrate. Studies in our own laboratory indicate that short incubations with 10 μg/ml of pancreatic ribonuclease in the presence of moderate salt concentration resulted in no detectable degradation of poly(A) to acid-soluble fragments and no significant breakdown in the size of the poly(A) molecules. Routinely, a preparation of RNA which has been labeled *in vivo* with tritiated adenosine is incubated with 10 μg/ml of pancreatic ribonuclease and 1 μg/ml of ribonuclease T1 in 0.01 *M* Tris buffer pH 7.5 containing 0.3 *M* KCl. The reaction mixture is incubated for 20 minutes at 37°C. When the amount of RNA being digested is extremely small (less than 1 μg), approximately 50 μg of pure unlabeled poly(A) per milliliter of reaction mixture was included in order to minimize endonucleolytic cuts in the radioactive poly(A). Under these conditions of incubation most types of single-stranded RNA labeled with uridine are more than 99% degraded.

C. Binding of Poly(A) and Poly(A)-Containing RNAs to Poly(U) Fiber Glass Filters

Immobilized complementary polynucleotide for the isolation and determination of poly(A) sequences was first employed by Mary Edmonds, who demonstrated that poly(A) occurred in ribonuclease hydrolyzates of total nuclear RNA from Ehrlich ascites tumor cells (Edmonds and Caramela, 1969).

In the latter study, oligo dT was immobilized by covalent attachment to cellulose powder (Gilham, 1964). Polyadenylic acid was applied to a column of dT cellulose at 0°C and at moderate salt concentration. The poly(A) was then recovered from the column by elution with a buffer containing low salt at 25°C. Though the latter method is extremely useful for the preparation of larger quantities of poly(A) or poly(A)-containing RNA (see procedure below), a faster method is required for simple quantitative analysis of polyadenylic acid or poly(A)-containing RNA in a mixture of radioactive RNA molecules. Toward this end, we have developed in our laboratory a simple technique utilizing polyuridylic acid immobilized on Whatman fiber glass filters. Using this method a solution containing free poly(A) or poly(A) bound to mRNA can be passed through the filter at room temperature under appropriate conditions and only RNA molecules containing poly(A) sequences will bind to the filter (Sheldon *et al.*, 1972a). This provides a highly specific and extremely rapid method for quantitation of poly(A) content in an RNA population. The principal advantage of this method over the use of immobilized polynucleotide columns or nitrocellulose filters, which also bind preferentially to poly(A) (Lee *et al.*, 1971), is the fact that this method is very simple, rapid and highly specific. We have found that there is good agreement between the binding of poly(A)-containing RNA to poly(U) filters and to nitrocellulose filters under the conditions developed by Lee (Lee *et al.*, 1971). However, since nitrocellulose filters also bind polynucleotides such as poly(U) and poly(G) to very significant extents, it is recommended that the use of nitrocellulose filters on an unknown RNA sample be confirmed by using the immobilized polynucleotide binding assay. This seems desirable since the basis for RNA binding to nitrocellulose filters is unknown at this time.

1. Preparation of Poly(U) Fiber Glass Filters

To each fiber glass filter (Whatman GF/C, 2.4 cm diameter) supported on a stainless steel test tube rack is added 0.15 ml of poly(U) solution (1 mg/ml) in 1 m*M* Tris buffer, pH 7.0. The polyuridylic acid may be purchased from a variety of sources, including Sigma Chemical Co. or Miles Biochemicals. The filters are then dried at 37°C. The dried filters are irradiated with ultraviolet light for 2.5 minutes on each side at a distance of 22 cm from a 30-W Sylvania germicidal lamp. After UV irradiation the filters may be stored dry at room temperature for periods of several weeks or desiccated in a refrigerator for a period of months. Just prior to use, each filter is rinsed with 50 ml of distilled water to remove unbound poly(U). About 67% of the input poly(U) is retained by each filter. If 10 times more poly(U) is applied to the filter, about 10 times more is retained by the filter after UV irradiation. For most purposes, however, we found that approximately 100 μg of bound poly(U) was in great excess of that required to retain the ordinary radioactive samples of poly(A) or poly(A)-containing RNA.

2. Use of Poly(U) Filters for RNA Binding

Radioactive RNA to be tested for binding to poly(U) filters is dissolved in a small volume of binding buffer (0.01 *M* Tris HCl, pH 7.5, containing 0.12 *M* sodium chloride). Usually it is desirable to dissolve the RNA to be applied to a single filter in 50–100 μl of binding buffer. The RNA is spotted onto a filter which has been equilibrated with binding buffer at 25°C. The RNA input is allowed to remain on the filter for 30 seconds, and then the filter is washed with 20 ml of the binding buffer. If larger samples RNA are to be applied, the RNA should be passed through the filter at a rate of approximately 2 ml per minute. This is best with reduced suction. There is no known upper limit of RNA solution which can be passed through the filter with efficient retention of the RNA. After washing of the filters with binding buffer, they are washed with 20 ml of 5% trichloracetic acid at 0°C and 10 ml of 95% ethanol. The filters are dried and counted by liqued scintillation spectroscopy.

The characteristics of binding of RNA to poly(U) filters has been described elsewhere. Basically, binding buffer and temperature has been chosen so that poly(A) sequences greater than 15 to 20 nucleotides in length will bind to the poly(U). Very high salt concentrations result in a slight binding of RNAs which do not contain poly(A) and are, therefore, to be avoided. Elution of RNA from poly(U) filters occurs at a temperature characteristic of melting of poly(A): poly(U) duplexes. The T_m, however, varies somewhat dependent on the length of the RNA chain attached to the poly(A). For a discussion of these properties see Sheldon *et al.* (1972a).

D. Sizing of Poly(A) Sequences on Gels

Purified RNA is dissolved in 0.01 *M* Tris-HCl pH 7.5, 0.3 *M* NaCl, and the amount of RNA present is estimated by measuring the optical density at 260 nm. Pancreatic and T1 ribonucleases are then added at a ratio of 1 μg of each RNase to 10 μg of RNA. When very small amounts of RNA are present (less than 10 μg), cold HeLa ribosomal RNA is added as carrier and the RNase concentration is adjusted accordingly. The RNA is incubated with ribonuclease for 30 minutes at 37°C, then made 0.25% w/v SLS and 0.02 *M* EDTA, extracted with chloroform–phenol, and precipitated with 3 volumes of ethanol at −20°C.

The precipitate is dissolved in 10 m*M* Tris-HCl ph 7.5, 5 m*M* EDTA, 0.125 *M* NaCl, and the poly(A) fragments are purified by chromatography on a dT cellulose column (see below). The purified poly(A) is then precipitated with ethanol.

Gel electrophoresis is carried out essentially according to Peacock and Dingman (1968) and Darnell *et al.* (1971b). The following solutions are used: (1) 20% w/v acrylamide, 0.5% w/v bis-acrylamide (Eastman Kodak) in water; (2) freshly prepared 10% w/v ammonium persulfate in water; (3) 10X gel buffer containing 108 gm of Tris, 9.3 gm of disodium EDTA, 55 gm of boric acid per liter; (4) TEMED (*N,N,N′,N′*,-tetramethylethylenediamine).

The gels are prepared by mixing 10 ml of acrylamide solution, 2 ml of 10X gel buffer, 4 ml of glycerol, and 4 ml of water. To this, 50 μl of ammonium persulfate and 3 μl of TEMED are added. The gels are cast in glass or plastic tubes and polymerize in about an hour. The above mixture suffices for 10–6 cm or 6–10 cm gels.

The tank buffer is prepared by mixing 100 ml of 10X gel buffer, 100 ml of glycerol (final concentration, 10%), 20 gm of sodium lauryl sulfate (SLS; final concentration, 0.2%), and 800 ml of distilled water.

The gels are prerun at least 30 minutes at 5 mA/gel at room temperature.

The sample is then dissolved in about 50 μl of gel buffer which includes 20% v/v glycerol, 0.2% w/v SLS, 5% saturated bromophenol blue. The gels are run at 4 mA/gel at room temperature until the dye reaches the end of the tube. At this time 4 S RNA has migrated about 60% of the distance from the top to the bottom of the tube.

The gels are then removed from their tubes and may be fractionated by several methods. We prefer freezing in a dry ice–ethanol bath and slicing into 2-mm sections with a microtome-like device (Joyce Loebel gel slicer). The fractions are solubilized in 1 ml of a 1:1 mixture of Nuclear Chicago solubilizer and 50% toluene-based fluor at 37°C for 8 hours. (For optimal solubilization it is important to prevent the gel fractions from dehydrating. This may be done by keeping them cold until the NCS is added.) An additional 3 ml of toluene scintillation fluid is added, and the fractions are counted.

E. Purification of Messenger RNA on dT Cellulose Columns

The occurrence of poly(A) sequences on messenger RNA can be exploited for the purification of messenger RNA from the bulk of the cellular stable RNA species. Columns of oligo dT covalently attached to cellulose powder are ideal for the large-scale preparation of messenger RNA. The poly(dT) cellulose synthesized by the method of Gilham (1964) is superior to other types of immobilized polynucleotide columns for preparation of mRNA for several reasons: (a) the oligo dT is firmly bound to the column, and none of it is eluted along with the mRNA; (b) poly(dT) cellulose is resistant to ribonuclease degradation and may be stored for months in the cold room with little loss in efficiency of binding polyadenylic acid sequences. Columns synthesized by immobilizing poly(U) with ultraviolet light onto cellulose powder suffer from the fact that the poly(U) is continually leached out of the column and contaminates RNA samples applied to the column. With a relatively small column of poly(dT) cellulose (10 cm by 1 cm) it is possible to prepare milligram quantities of mRNA from total RNA extracted from any given cell type.

1. Synthesis of dT Cellulose

The pyridine salt of 4 mmoles of 5′-thymidine monophosphate (TMP) is made by dissolving the thymidine monophosphate into a 1:1 mixture of pyridine and water. The solution is then passed through a Dowex 50 column which has been previously equilibrated with the same solution. The column should also have been calibrated to determine the position at which a non-interacting molecule elutes. This is important since pyridine absorbs greatly in the ultraviolet rendering the spectrophotometric detection of TMP impossible. Fractions containing TMP are subjected to repeated evaporation to near dryness *in vacuo* from anhydrous pyridine. This preparation of anhydrous TMP can be carried out in an apparatus similar to that shown in Fig. 1. Pyridine is rendered anhydrous by treatment with molecular sieves for several days. Pyridine is added to the evaporation flask through a side arm from a graduated dropping funnel. Evaporation is carried out *in vacuo* and the solution is prevented from bumping by the emersion of a highly flexible and extremely fine capillary tube into the TMP pyridine solution. Air entering either the capillary or the dropping funnel is first passed through a tube containing Drierite to render it anhydrous prior to entering the system. After the initial pyridine solution is taken nearly to dryness, 10 ml of fresh pyridine is added to the flask and again taken almost to dryness. This

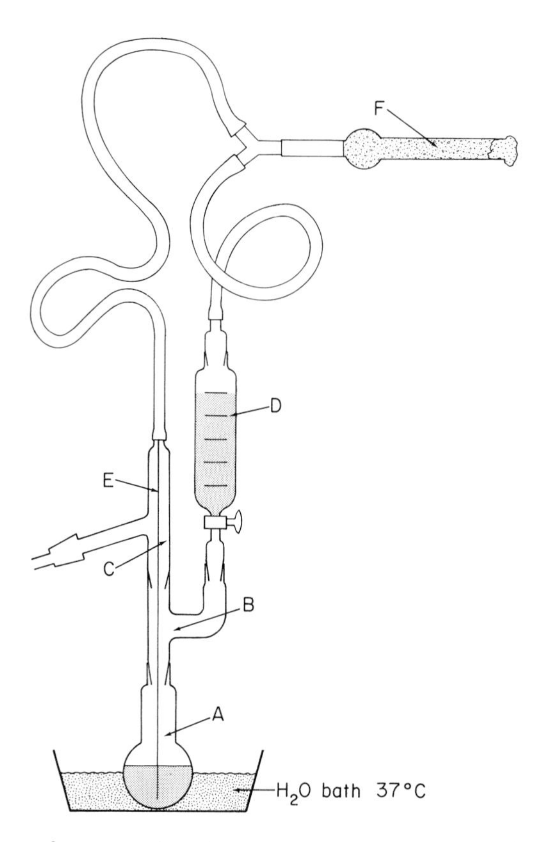

FIG. 1. Apparatus for preparation of anhydrous pyridine solution of thymidine 5′-monophosphoric acid. A, 25-ml round-bottom flask; B, Y-connector; C, distilling head; D, dropping funnel; E, capillary; F, drying tube. All glass fittings are standard taper 19/22.

is repeated a total of 5 times. The TMP solution is finally adjusted to a volume of 6 to 7 ml in pyridine and the flask is carefully removed from the vacuum apparatus and tightly capped. To this is added 8 mmoles of dicyclohexylcarbodiimide. The latter chemical must be stored under anhydrous conditions since it is rapidly degraded in the presence of moisture. The addition of dicyclohexylcarbodiimide results in the formation of a sticky gum. This gum is shaken vigorously in the dark for 5 days. This may be accomplished on a wrist-action shaker. *Great caution should be exercised in the handling of dicyclohexylcarbodiimide since this compound results in severe skin irritation and allergic responses.* During the 5-day shaking in the dark, oligomeric polydeoxythymidylic acid is synthesized. The second step in the reaction involves the linking of this oligomer to cellulose powder. This is accomplished by a similar reaction using dicyclohexylcarbodiimide to link the oligo (dT) via its 5′-phosphate to a hydroxyl group on the cellulose fibers. The cellulose to be used is dried *in vacuo* at 100°C overnight. Whatman standard grade or Munktel cellulose powder is suitable for this procedure. The initial product from the 5-day reaction is mixed with 10 gm of the dried cellulose in 100 ml of dry pyridine, and to this mixture is added 4 gm of dicyclohexylcarbodiimide. This mixture is also shaken for 5 days in the dark. The product is collected by filtration, washed with pyridine and allowed to stand for 4 to 8 hours in 50% v/v pyridine in H_2O. The product is collected again, first by centrifugation, then on a filter and washed extensively with warm ethanol. The product is then washed with deionized water. This material can be stored in 1 *M* NaCl in the refrigerator for a period of at least 6 months.

2. The Running of Poly(dT) Cellulose Columns

The poly(dT) cellulose can be packed into a water-jacketed column. The capacity of this column for binding polyadenylic acid can be gauged by adding a mixture of radioactive and a much larger amount of nonradioactive poly(A) to the column and determining the fraction of this material which binds. For example, if 10 mg of poly(A) are added to the column, and this amount of poly(A) contains 1×10^6 cpm of radioactive poly(A), it is then easy to determine how much of the added poly(A) binds tightly to the column, simply by counting the radioactivity which voids the column as well as the radioactivity which is eluted at low salt and higher temperature. This is a measure of the capacity of the column for poly(A). In general, messenger RNA molecules contain poly(A) tails equal to approximately 0.2 to 0.1 of their length. Thus it can be calculated roughly how much messenger RNA in milligrams will bind to this column.

Figure 2 illustrates a typical elution of RNA from the poly(dT) column. In this experiment tritium-labeled RNA synthesized by vaccinia virus particles *in vitro* was mixed with RNA labeled with ^{14}C which had been prepared from the ribosomes of HeLa cells which had been exposed to ^{14}C-labeled uridine for 36 hours and then chased for an additional 12 hours with excess unlabeled uridine. The mixture of RNA was loaded unto the dT cellulose column in 0.12 *M* sodium chloride in 0.01 *M* Tris pH 7.5 at 25°C. It may be seen that all of the ^{14}C radioactivity voided the column under these conditions while only a small fraction of the tritium-labeled vaccinia RNA eluted at this time. After the column was washed thoroughly, the temperature was raised to 37°C and the column was eluted with Tris buffer 0.01 *M* pH 7.5. It may be seen that a peak of tritium-labeled RNA eluted under these conditions. No ^{14}C labeled ribosomal RNA could be detected in this low salt eluate. The tritium-labeled RNA which eluted from the column in this last peak was bound nearly quantitatively to poly(U) fiber glass filters (see

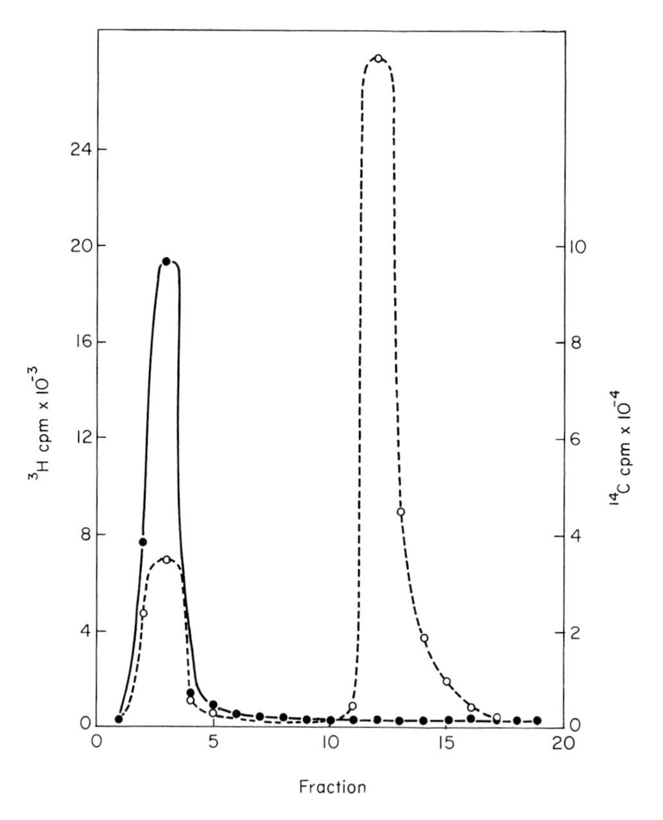

FIG. 2. Fractionation of RNA on dT cellulose column. A mixture of uridine-^{14}C ribosomal RNA and of uridine-^{3}H vaccinia virus messenger RNA were mixed and applied to the column (1 × 4 cm) at 25°C in 0.01 *M* Tris pH 7.5 containing 0.12 *M* NaCl and 0.001 *M* EDTA. After the column was washed with the sample buffer, it was eluted at 37°C with 0.01 *M* Tris pH 7.5 containing 0.001 *M* EDTA. ●——●, ^{14}C-labeled ribosomal RNA; ○------○ vaccinia mRNA.

above). None of the tritium- or ^{14}C-labeled RNA which voided the column in the high salt buffer was capable of binding to poly(U) filters. This illustrates the high specificity of this column for the binding of RNAs which contained polyadenylic acid. In another run of the column, 4.5 mg of total RNA purified from the polyribosomes of HeLa cells was applied to the column. After extensive washing in the higher salt buffer at 25°C, the poly(A)-containing RNA fraction was eluted at 37°C. The fractions containing optical density at 260 nm were precipitated with ethanol and the total yield of poly(A) containing polysomal RNA was 0.31 mg of 6.9% of the total polysomal RNA input. It has been shown in other experiments that this fraction contained the rapidly labeled RNA of DNA-like base composition synthesized under conditions where ribosomal RNA synthesis is blocked with low does of actinomycin D (0.04 μg/ml).

F. Other Applications of Polyadenylic Acid Tails on RNA

Since polyadenylic acid sequences occur on the 3′ end of RNA molecules, it is therefore possible to isolate the fragment of RNA immediately adjacent to the poly(A) sequence. RNA which already contains poly(A) can be endonucleolitically cleaved with specific ribonuclease, such as ribonuclease T1, and the 3′ fragment can be purified by binding to poly(dT) cellulose or poly(U) fiber glass filters. If the RNA has been previously labeled with ^{32}P, the sequence of nucleotides immediately adjacent to the poly(A) can then be determined by conventional sequencing methods. If an RNA does not contain poly(A), polyadenylic acid can be put on specifically to the 3′ end of such and RNA molecule by the action of a terminal nucleotide transferase, such as polynucleotide phosphorylase from *M. lysodeikticus* (Dr. Robert Millitte, Department of Pathology, University of Colorado Medical Center, personal communication). In many cases RNA molecules of high molecular weight have a high degree of secondary and tertiary structure such that their 3′ end is not available for enzymes capable of adding polyadenylic acid to it. If, however, the RNA is first digested with T1 ribonuclease, then only the 3′ fragment will have a free hydroxyl group. This fragment can serve as a primer for polynucleotide phosphorylase for the addition of a polyadenylic acid tail (Millette, personal communication). The 3′ fragment can then be purified by binding to poly(U) fiber glass filters, and the 3′ end of the RNA can then be sequenced. This method has been successfuly employed to determine the sequence of the 3′ end of T7 phage messenger RNA molecules (Millette, personal communication).

Another useful application of poly(A) tails of RNA is their utilization in the synthesis of DNA copies of the RNA using the oncornavirus RNA-dependent DNA polymerases. Since the poly(A) tail occurs on the 3′ end of the RNA molecule, an oligomeric dT molecule added to the RNA solution will complex with the poly(A) tail and serve as a primer for the efficient transcription of the RNA into a DNA copy. This method has been used successfully to make DNA copies of hemoglobin messenger RNA (Verma *et al.*, 1972; Kacian *et al.*, 1972; Ross *et al.*, 1972) and of RNA prepared from vaccinia virus (Zassenhaus and Kates, 1972). It is likely that the ability to synthesize highly radioactive DNA copies of messenger RNA will find diverse uses in the molecular biology of eukaryotic cells. One obvious application is to the problem of the nature of the nuclear precursor of cytoplasmic messenger RNA. This can be achieved by determining which fraction of the nuclear RNA can hybridize with the DNA copy of messenger RNA.

At the present time there are several ideas concerning the role of poly(A) sequences in the RNA of eukaryotic cells and their viruses. However, a definite function remains unknown. In spite of this, the poly(A) tails on messenger RNA are proving to be extremely useful for studies of messenger RNA metabolism, transport, and stability; furthermore, they enable the large-scale purification of messenger RNA from a variety of sources. There is little doubt that, as time goes by, further uses for this mysterious and somewhat amusing phenomenon will be devised. Perhaps even its function will be defined.

References

Adesnik, M., and Darnell, J. E. (1972). *J. Mol. Biol.* **67**, 397.

Adesnik, M., Saldett, M., Thomas, W., and Darnell, J. E. (1972). *J. Mol. Biol.* **68**, 21.

Beers, R. F. (1960). *J. Biol. Chem.* **235**, 2393.

Darnell, J. E., Philipson, L., Wall, R., and Adesnik, M. (1971a). *Science* **174**, 507.

Darnell, J. E., Jr., Wall, R., and Tushinski, R. J. (1971b). *Proc. Nat. Acad. Sci. U.S.* **68**, 1321.

Edmonds, M., and Caramela, M. G. (1969). *J. Biol. Chem.* **244**, 134.

Edmonds, M., Vaughan, M. H., Jr., and Nakagato, H. (1971). *Proc. Nat. Acad. Sci. U.S.* **60**, 1336.

Gilham, P. T. (1964). *J. Amer. Chem. Soc.* **86**, 4982.

Kacian, D. L., Spiegelman, S., Bank, A., Terada, S., Metaford, L., Dow, L., and Marks, P. A. (1972). *Nature (London), New Biol.* **235**, 167.

Kates, J. (1970). *Cold Spring Harbor Symp. Quant. Biol.* **35**, 743.

Kates, J., and Beeson, J. (1970). *J. Mol. Biol.* **50**, 19.

Lee, S. Y., Mendecki, J., and Brawerman, G. (1971). *Proc Nat. Acad. Sci. U.S.* **68**, 1331.

Lim, L., and Canellakis, E. S. (1970). *Nature (London)* **227**, 710.

Mendecki, J., Lee, S. Y., and Brawerman, G. (1972). *Biochemistry* **11**, 792.

Molloy, G. R., Sporn, M. B., Kelley, D. E., and Perry, R. P. (1972). *Biochemistry* **11**, 3256.

Peacock, A. C., and Dingman, C. W. (1968). *Biochemistry* **7**, 668.

Perry, R. P., La Torre, J., Kelley, D. E., and Greenberg, J. R. (1972). *Biochim Biophys. Acta* **262**, 220.

Philipson, L., Wall, R., Glickman, G., and Darnell, J. E, (1971). *Proc. Nat. Acad. Sci. U.S.* **68**, 2806.

Ross, J., Aviv, H., Scolnick, E., and Leder, P. (1972). *Proc. Nat. Acad. Sci. U.S.* **69**, 264.

Sheldon, R., Jurale, C., and Kates, J. (1972a). *Proc. Nat. Acad. Sci. U.S.* **69**, 417.

Sheldon, R., Kates, J., Kelley, D. E., and Perry, R. P. (1972b). *Biochemistry* **11**, 3829.

Soero, R., and Darnell, J. E. (1969). *J. Mol. Biol.* **44**, 551.

Verma, I. M., Temple, G. F., Fan, H., and Baltimore, D. (1972). *Nature (London), New Biol.* **235**, 163.

Zassenhaus, P., and Kates, J. (1972) *Nature (London), New Biol.* **238**, 139.

Chapter 4

Recent Advances in the Preparation of Mammalian Ribosomes and Analysis of Their Protein Composition

JOLINDA A. TRAUGH[1] AND ROBERT R. TRAUT

Department of Biological Chemistry, University of California School of Medicine, Davis, California

---

[1]*Present address*: Department of Biochemistry. University of California, Riverside, California.

I. Introduction

Although many of the early observations on the various basic components required for protein synthesis were made in mammalian systems, the *Escherichia coli* ribosome became the material of choice for investigation of ribosome structure and function (for review see Nomura, 1970; Kurland, 1972). The mammalian ribosome is considerably larger than the bacterial, often attached to membranes, has been less responsive to added synthetic or natural messenger RNA and more difficult to separate into active subunits. Much progress has been made recently toward overcoming the latter difficulties, and mammalian ribosomes can now be manipulated *in vitro* in much the same way as the bacterial. However, one of the major techniques for studying ribosome function now being extensively exploited for studying bacterial ribosomes, reconstitution from RNA and purified proteins, still has not been achieved for the mammalian ribosome.

This article will review some of the recent technical advances in the preparations, manipulation, and characterization of mammalian ribosomes at a time when such investigations are just entering a period of expansion comparable to that undergone with bacterial ribosomes during the last ten years. The article is admittedly selective, and emphasizes investigations from the last 2–3 years. We will review methods for obtaining ribosomal proteins and for the identification and resolution of the individual proteins comprising the 40 S and 60 S subunits. Estimates of the number and molecular weights of mammalian proteins will be presented. The question of the variability of the ribosomal proteins among tissues and among species will be examined as well as the function of membrane-bound ribosomes. Studies on phosphorylation of ribosomal proteins and systems for the translation of purified mRNA into specific protein will be described.

The physicochemical properties of mammalian ribosomes and their comparison with ribosomes from prokaryotes have been described in recent reviews (see Spirin and Gavrilova, 1969; Wittmann, 1970; Maden, 1971) and will not be discussed in detail. Neither will we attempt to review the formation and properties of ribosomal RNA, ribosome assembly, or the mechanism of protein synthesis. Thus, the purpose of this article will be to review current information on the preparation and protein composition of active mammalian ribosomal subunits in the context of structure: function relationships which are being and will be investigated in the years to come.

Ribosomes in mammalian cells are divided into two major classes, those found in the cytoplasm and those found in the mitochondria. Of these two classes of ribosomes, the cytoplasmic ribosome is in the early stages of physical and chemical characterization; even less is known about the mitochondrial ribosomes. Recent reports on the latter indicate they are smaller than bacterial ribosomes with a sedimentation coefficient of about 55 S, and are composed of two dissimilar subunits (O'Brien, 1971). In this review the term mammalian ribosome will refer only to the 80 S cytoplasmic particles and their subunits.

Ribosomes from mammalian tissues have an average sedimentation coefficient $S^{\circ}_{20,w}$ of about 80 S. These 80 S particles are composed of two dissimilar subunits with sedimentation values of approximately 40 S and 60 S. The smaller subunit contains a single species of RNA sedimenting at about 18 S, while the larger subunit includes two species of RNA sedimenting at approximately 28 S and 5 S.

Published values for the protein to RNA ratios of ribosomes from eukaryotes vary greatly, but in general mammalian ribosomes contain roughly 50% protein while bacterial ribosomes are 30–35% protein. The major reason for this discrepancy is the wide variety of methods used in preparation of the ribosomes for analysis. Since different amounts of protein are removed from the ribosome under various ionic conditions, the dividing line between ribosomal proteins and proteins associated with the ribosome is somewhat arbitrary. For the purposes of this review and in the more definitive recent studies on the number of protein species comprising the mammalian ribosome, the proteins associated with the ribosomal particles after sedimentation through a 0.5 *M* concentration of monovalent cations in the presence of magnesium are considered to be ribosomal proteins (see Section II).

Molecular weight values for mammalian 80 S ribosomes obtained by several different methods range from $4.1 \times 4.7 \times 10^{6}$ daltons (Spirin and Gavrilova, 1969). Precise estimates of the molecular weights are not yet available for mammalian ribosomal subunits active in protein synthesis since the procedures for obtaining these subunits have been described only recently. As the mammalian particle is considerably larger than the bacterial and contains a higher proportion of protein, it can be estimated that the former contains roughly twice the amount of protein as the latter. It is important to know whether the mammalian ribosome contains twice as many species of protein as the bacterial, or the same number of proteins having larger molecular weights. The answer to this question represents one of the more important results of recent investigations of mammalian ribosome structure and will be discussed in detail.

II. Preparation of Ribosomes

Cytoplasmic ribosomes of mammalian cells are present in two morphologically distinct states; free in the cytoplasm and attached to the endoplasmic reticulum (ER). Both free and bound forms can exist as native subunits, monosomes, or polysomes. The proportion of free to membrane-bound ribosomes varies in various cell types, and it has been suggested that the proportion of membrane bound ribosomes is related to the degree of specialization of cells for secreting protein products. Thus tissues which secrete a large proportion of their protein products such as pancreas, liver, and plasma cells have a high proportion of their ribosomes attached to membranes, whereas the ribosomes from embryonic cells, tumor cells, and reticulocytes, cells that do not excrete protein, are primarily free in the cytoplasm.

Typical procedures for the preparation of ribosomes with an extensive ER (liver), and from cells containing little or no ER (reticulocytes) are outlined below. In addition, the preparation of purified membrane-bound ribosomes is described.

A. Free Ribosomes

1. Liver

A typical procedure for the isolation of active ribosomes from mouse liver has been described by Falvey and Staehelin (1970; Staehelin and Falvey, 1971). Since this method yields ribosomes with exceptionally high activity in the synthesis of polyphenylalanine, the basic procedure will be outlined here.

Rodents are generally starved for 16 to 20 hours before exsanguination to decrease the glycogen content of the liver. However, it should be noted that Andrews and Tata (1971) found membrane-associated ribosomes from rats starved before death were less active in protein synthesis than ribosomes form normal rats. The livers are quickly removed, rinsed in a buffered salt solution containing sucrose, and homogenized with 8 to 10 strokes in a

Potter-Elvehjem type homogenizer. After centrifugation for 10 minutes at 12,000 rpm to remove intact cells and debris, the upper two-thirds of the supernatant is removed and made 1.3% in sodium deoxycholate. Detergent causes the release of membrane-bound ribosomes, and the resulting preparation contains a mixture of free and released particles. This mixture is immediately layered over discontinuous sucrose gradients consisting of equal volumes of 2 *M* and 0.7 *M* buffered sucrose, and centrifuged. A pellet containing polyribosomes is recovered, resuspended, and frozen at −70°. Most of the ribosome-associated factors involved in protein synthesis can be removed by further centrifugation of the ribosomes through 0.5 *M* salt in buffered sucrose (Skogerson and Moldave, 1967).

2. Rabbit Reticulocytes

In cells such as reticulocytes, which contain a small proportion of membrane-bound ribosomes, detergents are generally omitted from the procedures for preparing ribosomes, and only the major fraction consisting of free ribosomes is isolated. A typical procedure for obtaining ribosomes from rabbit reticulocytes is outlined here (Allen and Schweet, 1962; Collier, 1967; Traugh and Collier, 1971). Reticulocytes from phenylhydrazine-treated rabbits are washed twice in isotonic salt solution containing 5 m*M* KCl and 1.5 m*M* $MgCl_2$. The cells are then lysed by suspending them in 2 m*M* magnesium for 2 minutes followed by the addition of sucrose-KCl. The extract is centrifuged at 15,000 *g* for 15 minutes to remove unlysed cells and membrane fragments, and the supernatant is centrifuged at 60,000 rpm in a Ti 60 Spinco rotor for 3 hours. The ribosomal pellets are washed and allowed to resuspend in buffer overnight at 4°C. Contaminating nonribosomal proteins are removed by pelleting the ribosomes through buffered sucrose containing 25 m*M* KCl, 1 m*M* $MgCl_2$ (Bonanou *et al.*, 1968). In order to remove tightly bound factors involved in protein synthesis, the ribosomes are centrifuged in buffered sucrose containing 0.5 *M* KCl and 0.002 *M* $MgCl_2$ (Traugh *et al.*, 1973).

B. Membrane-Bound Ribosomes

Typical procedures for the preparation of membrane-bound ribosomes from cells containing either a high or low proportion of bound ribosomes involve the following steps: (1) isolation of a microsomal fraction from cell lysates that have been previously centrifuged at low speed (2000 rpm for 2 minutes), in the presence or absence of sucrose, to remove nuclei (Rosbash and Penman, 1971); (2) centrifugation of the supernatant (total cytoplasmic fraction) at 8000 to 20,000 *g* for 10 to 15 minutes. Under these conditions a portion of both the smooth membranes and membrane-bound ribosomes (rough ER) are pelleted (Attardi *et al.*, 1969; Lamaire *et al.*, 1971); (3) separation of the rough ER from the smooth membranes by centrifugation in a Spinco 50 rotor for 3 hours at 90,000 *g* through discontinuous sucrose gradients containing equal volumes of 1.3 *M* and 2.0 *M* sucrose (Blobel and Potter, 1967; Williams *et al.*, 1968). The smooth membrane bands at the interface formed between the supernatant and 1.3 *M* sucrose, the rough ER bands at the boundary of 1.3 *M* and 2.0 *M* sucrose, and the free ribosomes form a pellet. Ribosomes are dissociated from the membranes by suspension in 0.5–1% sodium deoxycholate (DOC) or a combination of this ionic detergent with nonionic detergents.

Adelman *et al.* (1973a) have recently estimated that less than 10% of the total bound ribosome population of rat liver is isolated by procedures like those described above. They have devised an improved scheme involving removal of nuclei and mitochondria in the absence of added salt in order to avoid aggregation and loss of membrane fractions. The methods yields preparations containing approximately 50% of the membrane-bound ribosomes of rat liver. About 85% of these ribosomes are released from the membrane by incubation with 1 m*M* puromycin, 750 m*M* KCl, 5 m*M* $MgCl_2$, and buffered sucrose (Adelman *et al.*, 1973b).

C. Preservation

The general method for preserving ribosomes active in protein synthesis is by rapid freezing and storage at −70°C or under liquid nitrogen. Recently it has been demonstrated that lyophilization is an efficient method for preserving ribosomes functional in poly(U) directed polyphenylalanine synthesis. Lyophilized ribosomes from rabbit reticulocytes have protein synthesizing activities equal to or greater than fresh preparations (95–135%), while aliquots of the fresh ribosomes stored at −70°C for one week were less active (60–85%) (Christman and Goldstein, 1971). The sucrose gradient absorbance profiles of the lyophilized polysomes are the same after freezing as before. Isolated purified subunits can also be lyophilized without loss of activity (Reboud *et al.*, 1972a).

III. Preparation of Ribosomal Subunits

A. Dissociation by Magnesium Depletion

As with prokaryotic ribosomes, the degree of association of mammalian ribosomal subunits is a function of both the ionic strength and the magnesium concentration of the medium. However, depletion of magnesium ions, either with EDTA (Hamilton and Petermann, 1959; Tashiro and Siekevitz, 1965) or with pyrophosphate (Lamfrom and Glowacki, 1962), produces inactive mammalian subunits (Ts'o and Vinograd, 1961; Tashiro and Siekevitz, 1965; Gould *et al.*, 1966). Almost complete removal of bound magnesium is required before the subunits dissociate, and the resulting particles are heterogeneous by sedimentation analysis, with sedimentation values less than 40 S and 60 S due to induced changes in the conformation of the subunits. Therefore, dissociation by magnesium depletion, although widely used in early studies, is of limited value.

B. Dissociation by High Salt

Prior to 1968, there was no reliable procedure for producing subunits which were active in protein synthesis. Martin and Wool (1968; Martin *et al.*, 1971) were the first to prepare active subunits from mammalian ribosomes by using high salt concentrations in the presence of sufficient magnesium to maintain an active subunit conformation.

Ribosomes from rat skeletal muscle suspended in a buffered solution of 1 *M* KCl, 16 m*M* $MgCl_2$, and reducing agent, are separated into subunits by centrifugation through linear sucrose gradients containing 850 m*M* KCl and 15 m*M* $MgCl_2$ in 50 m*M* Tris-HCl, pH 7.8 and 20 m*M* β-mercaptoethanol. When reducing agent is omitted from the preparative sucrose gradients, a component of 25–30 S appears, apparently a derivative of the 40 S subunit, and the large subunit is partially inactivated (Martin *et al.*, 1969). Martin *et al.* (1971) have pointed out the importance of controlling temperature in the formation of ribosomal subunits from different sources; mouse liver ribosomes dissociate into 40 S and 60 S subunits (Falvey and Staehelin, 1970) while rat muscle ribosomes give rise to 90 S and 60 S particles which are apparently dimers of 60 S and 40 S subunits. Centrifugation at 28°C was employed therefore to obtain subunits from rat liver ribosomes. The isolated 60 S and 40 S subunits, when combined in the proportions of 2.5:1 (A_{260}), rapidly form 80 S ribosomes in the absence of nonribosomal cell components such as mRNA or supernatant proteins. These reassociated 80 S ribosomes are active in polyphenylalanine synthesis. Hybridization experiments of subunits derived by high salt treatment (60 S, 40 S) with subunits prepared by EDTA treatment (50 S, 32 S) demonstrate that both ribosomal subunits prepared with EDTA are inactive (Martin *et al.*, 1969; Reboud *et al.*, 1972b).

Ribosomes isolated from rabbit reticulocytes completely dissociate at lower salt concentrations than the muscle ribosomes, 500 m*M* KCl and 2 m*M* $MgCl_2$ (Bonanou *et al.*, 1968; Prichard *et al.*, 1971). This may be due to the high proportion of nonmembrane-associated ribosomes in reticulocytes.

Purified ribosomal subunits are sensitive to the potassium: magnesium ratio, and tend to dimerize at low potassium concentrations or to be inactivated at high potassium concentrations (Nolan and Arnstein, 1969; Faust and Matthaei, 1972). These conditions vary depending upon the source of the ribosomes.

C. Alternative Methods of Dissociation

Dissociation of 80 S ribosomes into subunits at salt concentrations below 0.5 *M* is facilitated by incubation of the ribosomes to remove peptidyl-tRNA and mRNA. Incubation of polysomes from mouse liver with the components required for protein synthesis results in the termination and release of nascent polypeptide chains, and concomitant formation of 80 S ribosomes (Falvey and Staehelin, 1970). Shorter incubation periods with puromycin (Lawford, 1969; Blobel and Sabatini, 1971; Mechler and Mach, 1971) or sodium fluoride (Nolan and Arnstein, 1969) also result in an accumulation of 80 S monosomes. The 80 S ribosomes prepared by any of these methods dissociate into 60 S and 40 S subunits by centrifugation in sucrose gradients containing concentrations of KCl of 0.2 *M* or less in the presence of magnesium. The isolated recombined subunits are active in polyphenylalanine synthesis.

The reconstituted 80 S ribosomes most active in polyphenylalanine synthesis are those prepared from mouse liver by Falvey and Staehelin (1970; Staehelin and Falvey, 1971). The ribosomes are purified as outlined in Section II,A and preincubated in a protein synthesizing system to release 80 S ribosomes from polysomes. The 80 S ribosomes are made 0.5 *M* in KCl and centrifuged through buffered sucrose gradients containing 0.3 *M* KCl, 0.003 *M* magnesium acetate and reducing agent. The fractions corresponding to the 40 S and 60 S subunits are collected, concentrated by pelleting, and frozen (Schreier and Staehelin, 1973). These ribosomes polymerize 15 and 20 phenylalanine residues per reconstituted 80 S particle, all of which are participating in protein synthesis.

An alternative method of dissociating rat liver ribosomes into subunits is depletion of ribosome bound magnesium by dialysis, followed by treatment of the 80 S particles with 2.2 *M* urea (Petermann and Pavlovec, 1971; Petermann, 1971). The resulting subunits are isolated by sucrose density centrifugation and must be used immediately due to the instability of the 40 S subunit. The recombined subunits are 70–80% as active as the controls in polyphenylalanine synthesis.

IV. Preparation of Total Protein from Ribosomes and Ribosomal Subunits

Early methods for the extraction of ribosomal proteins from mammalian ribosomes utilized 2 *M* LiCl to disrupt the ribosome structure and solubilize the protein (Currey and Hersh, 1962). This method is not in general use since other procedures are more effective in extracting the ribosomal proteins from the ribonucleoprotein particles. The methods in common use today employ the following reagents: (1) 4 *M* urea and 2–3 *M* LiCl (Leboy *et al.*, 1964; Spitnik-Elson, 1965); (2) 66% acetic acid alone (Waller and Harris, 1961), or with magnesium acetate (Hardy *et al.*, 1969); (3) 4 *M* guanidine-HCl (Cox, 1968).

The extraction procedures for each of the methods (1–3) are similar; ribosomes or ribosomal subunits are added to the extraction mixture and allowed to stand at 0–4°C for 12–24 hours. The precipitated RNA is removed by low speed centrifugation. The ribosomal proteins remain soluble and are prepared for analysis or further separation by appropriate procedures. The proteins may be dialyzed against 6 *M* urea (pH approximately 5 to minimize carbamylation) to remove the dissociating agents, and frozen in small aliquots. Alternatively, the proteins can be dialyzed against 1 *M* KCl and still remain soluble.

An additional procedure for obtaining total protein consists of subjecting the ribosomes to 0.25 *N* HCl for 30 minutes at 0°C (Welfle *et al.*, 1972). Aggregates are removed by centrifugation, and the proteins are precipitated with acetone, washed with ethanol, and dried under vacuum at 0°C.

Ford (1971) has quantitatively compared several different extraction procedures using ribosomes from *Xenopus*. Radioactive amino acids were incorporated into *Xenopus* ovary proteins *in vivo*. The ribosomes were then purified and subjected to the various extraction reagents. Acetic acid was the least efficient method of removing the total protein, since additional proteins were displaced by further treatment with LiCl: urea. Treatment of the ribosomes with 2 *M* LiCl: 6 *M* urea or with 4 *M* guanidine-HCl were equally efficient, removing 98% and 97% of labeled protein, respectively.

V. Analysis of Ribosomal Proteins

A. One-Dimensional Gel Electrophoresis

Ribosomal proteins from bacteria were initially separated by starch gel electrophoresis in urea at pH 4.5 (Waller and Harris, 1961). Leboy *et al.* (1964) were the first to use one-dimensional gel electrophoresis in polyacrylamide to examine bacterial ribosomal proteins. Since then, polyacrylamide gel electrophoresis has become an indispensable tool in the examination of total ribosomal protein and individual purified proteins from both prokaryotes and eukaryotes. Owing to the basic nature of ribosomal proteins, electrophoresis in urea at pH values less than 5 is an effective method of resolving a large number of protein bands (Leboy *et al.*, 1964). The mobilities of the proteins in this system are affected by both charge and molecular sieving, and when used in conjunction with gel electrophoresis in sodium dodecyl sulfate, it is possible to identify the majority of the individual ribosomal proteins (Traut *et al.*, 1969).

Electrophoresis in polyacrylamide gels containing sodium dodecyl sulfate separates proteins by molecular sieving. A linear relationship exists between the migration distance and the logarithm of the molecular weight (Shapiro *et al.*, 1967; Weber and Osborn, 1969). Acrylamide concentrations of 10–15% with 0.27% bisacrylamide are used for separation of ribosomal proteins (Traut *et al.*, 1969). With the higher concentrations of acrylamide there is better resolution of the smaller molecular weight products. Kabat (1970) and King *et al.* (1971) analyzed ribosomes from rabbit reticulocytes, and Bickle and Traut (1971) studied the ribosomal proteins of mouse plasmacytoma cells using SDS gel electrophoresis. Recently, more sharply defined protein bands were obtained with discontinuous gel electrophoresis in sodium dodecyl sulfate since the sample is concentrated in a spacer gel of different composition than the separating gel (Laemmli, 1970).

B. Two-Dimensional Gel Electrophoresis

The two-dimensional technique of Kaltschmidt and Wittmann (1970), originally designed for separating ribosomal proteins from bacteria, has recently been applied to mammalian ribosomal proteins with individual modifications (Sherton and Wool, 1972; Huynh-van-tan *et al.*, 1971; Welfle *et al.*, 1971; Howard and Traut, 1973). The basic systems which have been employed for analysis of mammalian ribosomal proteins are summarized in Table I. The details of the method employed in this laboratory are given in the legend of Fig. 1.

The two-dimensional systems are of two kinds; the first system employs urea in both dimensions, the first dimension at a high pH with a high-porosity gel, and the second at a low pH with a low-porosity gel. The concentration of the gels and the pH of the buffers are selected to maximize separation of the proteins by charge in the first dimension and size in the second dimension. The second two-dimensional system contains urea at pH 4.5 in the first dimension, and sodium dodecyl sulfate in the second dimension (Martini and Gould, 1971; Hultin and Sjoqvist, 1972). This modification resolves the individual proteins in the second dimension by molecular sieving so that, by appropriate calibration of the gel with standard proteins, molecular weight values for the ribosomal proteins can be obtained directly. However, there appears to be slightly less resolution of the individual proteins

TABLE I

Two-Dimensional Polyacrylamide Gel Systems Used for Separation of Mammalian Ribosomal Proteins

Reference	First-dimension		Second-dimension	
	Buffer	% Acrylamide	Buffer	% Acrylamide
Kaltschmidt and Wittmann (1970)	pH 8.6, urea	4.0; 8.0	pH 4.6, urea	18.0
Howard and Traut (1973)	pH 8.2, urea	4.0	pH 4.0, urea	18.0
Sherton and Wool (1972)	pH 8.6, urea	8.0	pH 4.2, urea	18.0
Welfle *et al.* (1971, 1972)	pH 8.3, urea	5.0	pH 2.0, urea	15.0
Huynh-van-tan *et al.* (1971)	pH 8.6, urea	4.0	pH 4.6, urea	9.0
Martini and Gould (1971)	pH 4.5, urea	4.1	pH 7.1, SDS	10.3
Hultin and Sjoqvist (1972)	pH 4.5, urea	7.5	pH 7.2, SDS	10.0

using the latter method (Table II), due to the fact that a number of proteins have nearly identical molecular weights.

A third analytical procedure following two-dimensional analysis has been described (Howard and Traut, 1973); the stained protein spots from conventional two-dimensional urea slab gels are cut out, crushed in 1% SDS:6 *M* urea, and incubated overnight at room temperature. The supernatant liquid or entire gel slurry is used as sample for electrophoresis in sodium dodecyl sulfate polyacrylamide gels. Electrophoresis in the third step with SDS resolves multiple proteins contained in a single spot and in addition provides molecular weights of the individual proteins.

VI. Molecular Weight Distribution and Number of Ribosomal Proteins

By contrast to the detailed studies on the ribosomal proteins from *Escherichia coli* (in which the number and stoichiometry of ribosomal proteins have been determined, the different proteins have been purified and analyzed for amino acid composition, molecular weights, peptide similarity, and order of assembly, and specific functions have been assigned to certain proteins), the characterization of mammalian ribosomal proteins is still in the preliminary stages. Given the greater size of the mammalian ribosome and its higher protein content, one of the first problems has been to establish the number of different ribosomal proteins.

A. Total Proteins

Heterogeneity of mammalian ribosomal proteins has been shown by resolution of the proteins from the 80 S ribosome and from each subunit by gel electrophoresis. Sixty-nine to seventy-six electrophoretically different components were resolved from the purified subunits of rat liver by two-dimensional electrophoresis (Table II). Two or three additional spots were resolved when protein from the monosomes was subjected to electrophoresis (Sherton and Wool, 1972; Welfle *et al.*, 1972). Two proteins present on the 80 S ribosome, but not on either of the subunits, were also isolated by column chromatography (Westermann *et al.*, 1971a).

It appears, in contrast to results obtained with ribosomal proteins from bacteria (Kaltschmidt and Wittmann, 1970), that all or nearly all of the ribosomal proteins from mammalian cells migrate toward the anode of the first dimension at a basic pH. No major protein species migrates anionically at pH 9.0 in rat liver (Welfle, 1971), rabbit liver, or rabbit reticulocytes (Huynh-van-tan *et al.*, 1971), although 5–7 acidic bands appear with ribosomal protein preparations from rat liver that have not been washed in high salt (Bielka and Welfle, 1968). At pH 8.6, one protein in the 40 S subunit from rat liver migrates in either direction depending upon the individual electrophoretic run, indicating an isoelectric point at approximately 8.6; two proteins of the 60 S subunit migrate anionically (Sherton and Wool, 1972). The three proteins present only with the 80 S ribosomes also migrate toward the anode.

Seventy-five spots were resolved from the monosomes of rabbit liver by the standard Kaltschmidt–Wittmann method of two-dimensional electrophoresis (Huynh-van-tan *et al.*, 1971). In rabbit reticulocytes, 70 spots were identified with a modification of the standard two-dimensional method G. A. Howard, J. A. Traugh, and R. R. Traut, in preparation), and 62–63 spots were resolved when the second-dimension included sodium dodecyl sulfate (Martini and Gould, 1971; see Section V).

There are reproducible differences in the intensity of staining of the resolved proteins (see Fig. 1) that may be due to the intrinsic variability with which individual proteins retain dye, a difference in the relative amounts of each protein, the size of the protein, or the coincidence of more than one protein in the same spot. Examination of well-resolved protein spots from rabbit reticulocyte ribosomal subunits by the third step with SDS mentioned earlier showed that a few of the more heavily stained patterns contained two individual proteins (G. A. Howard, J. A. Traugh, and R. R. Traut, unpublished results). The stoichiometry, or relative molar amounts of mammalian ribosomal proteins, remains an open question at this time; information on both molecular weights and relative amounts of each protein species is required.

B. Subunit Proteins

Initial estimates of the weight average and number average molecular weights for mammalian ribosomal proteins were calculated for the large and small subunits from mouse plasmacytoma (Bickle and Traut, 1971). The proteins from each subunit were resolved by one-dimensional electrophoresis in sodium dodecyl sulfate, stained with Coomassie Brilliant Blue, and scanned spectrophotometrically. Estimates of the number of moles of protein per ribosome showed 18–27 moles of protein per 40 S subunit and 42–52 moles of protein per 60 S subunit. By contrast, bacterial ribosomes analyzed by the same technique were estimated to contain 16–21 moles of protein per 30 S subunit and 29–35 moles per 50 S subunit (Bickle and Traut, 1971). Similar values were found in a variety of bacterial species

TABLE II
NUMBER OF MAMMALIAN RIBOSOMAL PROTEINS FROM 40 S, 60 S, AND 80 S RIBOSOMES

Source of ribosomes	Subunit preparation	Protein preparation	Protein analysis
Mouse plasmacytoma	Puromycin treatment 0.3 *M* KCl	1% SDS	1-Dimensional electrophoresis in SDS
Rabbit reticulocytes	0.5 *M* KCl	2–3 *M* LiCl; 4 *M* urea; SDS Nuclease digestion; SDS	1-Dimensional electrophoresis in SDS
Novikoff hepatoma ascites cells	0.8 *M* KCl	66% acetic acid; 0.1 *M* Mg acetate 3 *M* LiCl; 4 *M* urea	2-Dimensional electrophoresis in urea
Rat liver	0.8 *M* KCl	66% acetic acid; 0.1 *M* Mg acetate 3 *M* LiCl; 4 *M* urea	2-Dimensional electrophoresis in urea
Rat liver	0.88 *M* KCl	66% acetic acid with 0.1 *M* Mg acetate	2-Dimensional electrophoresis in urea
Rat liver	EDTA	66% acetic acid with 0.5 *M* Mg acetate	2-Dimensional electrophoresis in urea
Rat liver	Puromycin treatment 0.55 *M* KCl	0.25 *N* HCl	2-Dimensional electrophoresis in urea
Rabbit liver		2 *M* LiCl	2-Dimensional electrophoresis in urea
Rabbit reticulocytes	0.5 *M* KCl	2–3 *M* LiCl; 4 *M* urea	2-Dimensional electrophoresis in urea and SDS
Rabbit reticulocytes	0.5 *M* KCl	3 *M* LiCl; 4 *M* urea	3-Dimensional electrophoresis in urea and SDS

(Sun, Bickle, and Traut, 1972). The number average molecular weights of the mammalian ribosomal proteins from both subunits were significantly greater than the corresponding bacterial proteins: 60 S mammalian, 27,000; 50 S bacterial, 17,000; 40 S mammalian, 29,000; 30 S bacterial, 16,000. Thus it was concluded that mammalian and bacterial ribosomes differ in that the former have both more proteins and proteins of higher average molecular weight.

A similar study of the molecular weight distribution of ribosomal proteins was conducted with subunits from rabbit reticulocytes (King *et al.*, 1971). The 40 S subunit was estimated to contain 30–40 proteins ranging in molecular weight from 6000 to 39,000, and the 60 S subunit was composed of

Number of individual proteins resolved			Proteins common to both subunits	References
40 S	60 S	80 S		
18–27	42–52	60–79	—	Bickle and Traut (1971)
(moles of protein/ribosome)				
30–40	40–60	70–100	—	King *et al.* (1971)
(moles of protein/ribosome)				
32 ± 2	41 ± 2	72	—	Howard and Traut (in preparation)
30 ± 2	39 ± 2	69	—	Howard and Traut (in preparation
30	39	69–72 (3 add'l.)	1	Sherton and Wool (1972)
42	49	76	16	Welfle *et al.* (1971)
31	39	70–72 (3 add'l.)	2	Welfle *et al.* (1972)
—	—	75	—	Huynh-van-tan *et al.* (1971)
26	36–37	62–63	0	Martini and Gould (1971)
32	39	70	—	Howard, Traugh, and Traut (in preparation)

40–60 proteins ranging in molecular weight from 8000 to 47,000. These figures also agree with estimates from two-dimensional electrophoresis.

The numbers of ribosomal proteins comprising each subunit obtained in several laboratories are summarized in Table II. The methods of preparation of the ribosomal subunits for analysis by gel electrophoresis, and the degree of purity of each subunit fraction vary. Differences in the estimated number of proteins from different laboratories may in part be due to the different techniques used to prepare subunits and to extract ribosomal proteins (see Section IV of this review). For example, Welfle *et al.* (1971) dissociated ribosomal subunits from rat liver by EDTA, a method that also unfolds each subunit to varying degrees, and found 15 proteins common to both subunits. When the same laboratory dissociated the subunits with high salt, only two proteins were common to both subunits (Welfle *et al.*, 1972). Sherton and Wool (1972) observed four common proteins, all but one of which were resolved by varying the degree of cross-linking in the gel and/or the pH of the buffer. The number of individual proteins obtained by two-dimensional electrophoresis in the various laboratories range from 26 to 32 for the 40 S subunit and from 36 to 41 proteins for the 60 S subunit.

VII. Purified Proteins

A. Chemical Studies

Westermann *et al.* (1969, 1971a) obtained a number of electrophoretically pure proteins using ion exchange chromatography to fractionate the ribosomal proteins from rat liver. These were found to differ in amino acid composition, molecular weight, circular dichroic spectra, and cyanogen bromide peptide patterns. In other experiments thirty-one proteins were

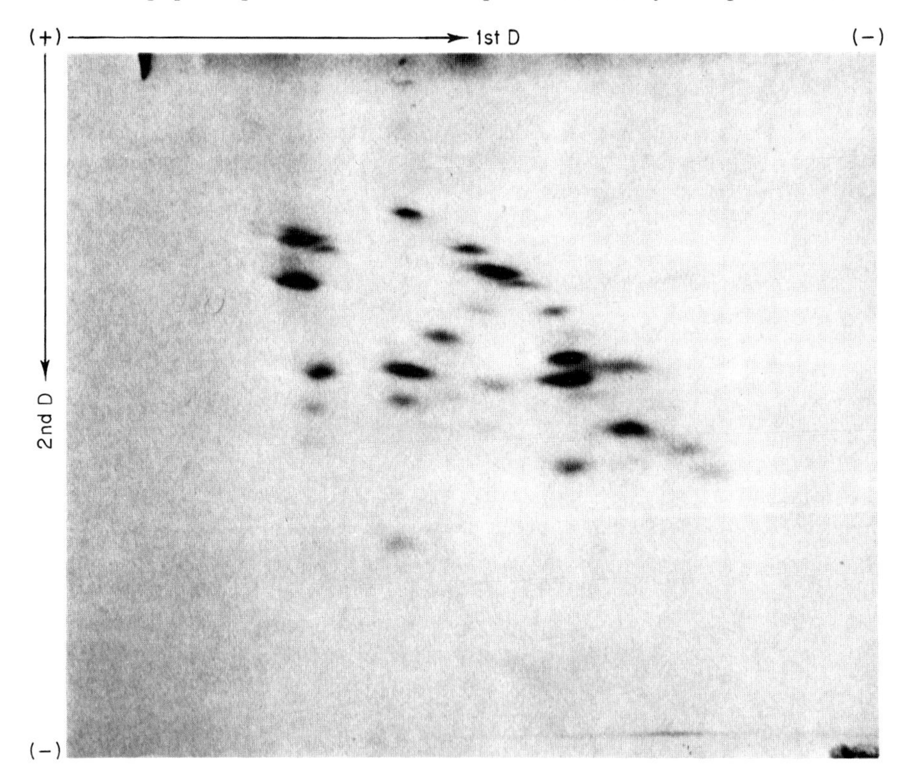

isolated from the total protein mixture in a two-step procedure consisting of extraction of the protein from 80 S ribosomes with 2 *M* LiCl, or with 66% acetic acid and 0.1 *M* $MgCl_2$, followed by chromatography on carboxymethyl cellulose and gel filtration on Biogel P10 or Sephadex G-100. Twelve of the proteins were localized in the 40 S subunit, and 14 in the large subunit. Two additional proteins were found only in the 80 S ribosome. Three proteins were found to be common to both subunits; this may be explained by the method of subunit dissociation employed, since as noted previously, EDTA causes conformational changes in the subunits and may cause proteins to exchange between subunits (King *et al.*, 1971).

The same laboratory also obtained ten purified proteins from rat liver ribosomes by preparative gel electrophoresis, and the molecular weights and amino acid composition of these proteins were determined (Welfle *et al.*, 1969). Physical and chemical analyses of the purified proteins confirm the data obtained by chromatography and show that the 40 S and 60 S subunits are composed of a number of heterogeneous proteins. Thus, the incomplete results on the analysis of mammalian ribosomal proteins are

---

FIG. 1. Two-dimensional separation of ribosomal proteins from 40 S subunits of rabbit reticulocytes.

First dimension apparatus: Standard disc gel apparatus, in glass tubes 0.45 cm × 12.5 cm (Davis, 1964). *Gel composition, 1-D:* Modification of composition described by Kaltschmidt and Wittmann (1970): urea, 360 gm/liter; acrylamide, 40 gm/liter; bisacrylamide, 1.33 gm/liter; EDTA-Na_2, 8 gm/liter; boric acid, 32 gm/liter; TEMED, 0.45 ml/liter; Tris, 48.6 gm/liter, pH 8.7; polymerization was catalyzed with 5 μl of freshly prepared 10% (w/v) ammonium persulfate/1.0 ml of gel solution. *Buffer composition*: EDTA-Na_2, 2.4 gm/liter; boric acid, 4.8 gm/liter; Tris, 7.25 gm/liter, pH 8.2. *Sample applied:* 100 μl of 2.0 mg/ml of protein, or approximately 5 μg per component. *Electrophoresis conditions*: Cathode on top; 6 hours at 6 mA/gel, room temperature; pyronine G (0.5%) as tracking dye. *Apparatus, 2-D:* Described by Reid and Bieleski (1968), as modified by Studier (1972; personal communication) and Howard and Traut (1973). Slab gel size: 13 cm × 15.5 cm × 2 mm. *Treatment of 1-D gel:* Sliced longitudinally; dialyzed 2 × 20 min in "starting buffer" of Kaltschmidt and Wittmann (1970); urea, 480 gm/liter; glacial acetic acid, 0.74 ml/liter; KOH, 0.67 gm/liter, pH 5.2. *Gel composition, 2-D*: Modification of composition described by Kaltschmidt and Wittmann (1970): urea, 360 gm/liter; acrylamide, 180 gm/liter; bisacrylamide, 2.5 gm/liter; glacial acetic acid, 53 ml/l; KOH, 2.7 gm/liter; TEMED, 5.8 gm/liter; pH 4.5. Polymerization was catalyzed with 30 μl of freshly prepared 10% (w/v) ammonium persulfate/1.0 ml of gel solution. *Buffer composition*: Modification of composition reported by Kaltschmidt and Wittmann (1970); glycine, 14 gm/liter; glacial acetic acid, 1.5 ml/liter, pH 4.0. *Electrophoresis conditions*: Cathode on top: 1 hour at 40 V, then 6–12 hours at 80–150 V, room temperature; pyronine G (1% in 20% glycerol) as tracking dye. Electrophoresis stopped when dye is within 1 cm of bottom. *Staining, 2-D*: 1–4 hours in Coomassie Brilliant Blue (R-250 0.1% in 7.5% acetic acid: 50% methanol: H_2O). *Destaining*: Slow shaking of 2-D gel for 2–4 hours with several changes of 7.5% acetic acid:50% methanol:H_2O.

Refer to Howard and Traut (1973) for a complete description of the method. Results of G. A. Howard, J. A. Traugh, and R. R. Traut (article in preparation; see Table II for an outline of the methods of preparation of the ribosomal proteins for electrophoresis).

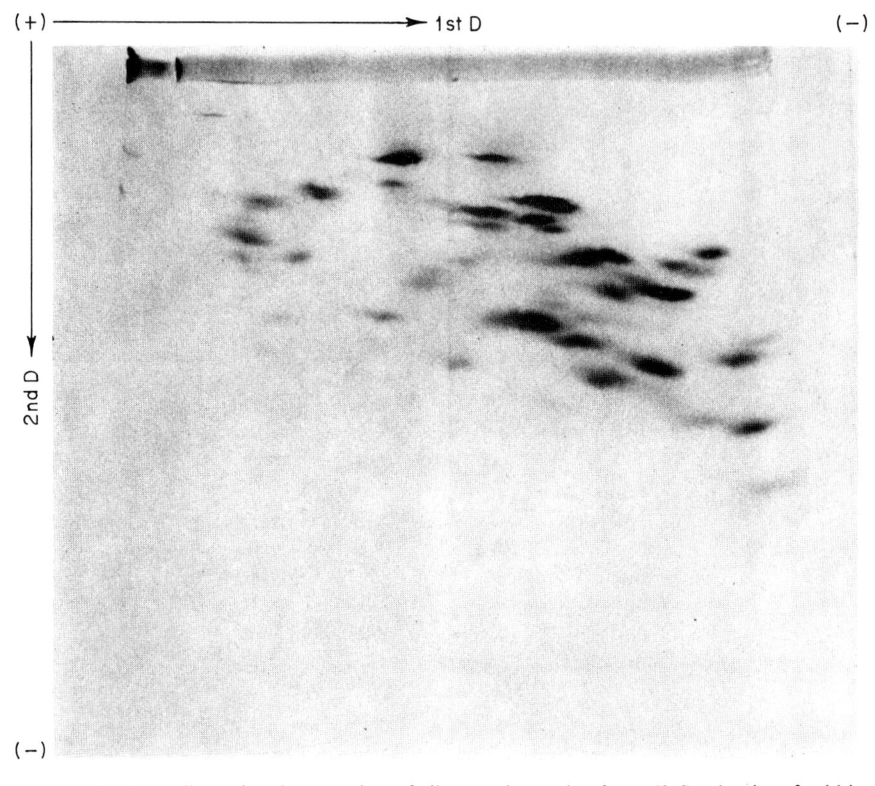

FIG. 2. Two-dimensional separation of ribosomal proteins from 60 S subunits of rabbit reticulocytes. Results of G. A. Howard, J. A. Traugh, and R. R. Traut (article in preparation; see Table II for an outline of the method of preparation of the ribosomal proteins for electrophoresis).

similar to those obtained with bacterial proteins in that each subunit is composed of a number of proteins different in primary structure, and the two subunits have few, if any, proteins in common.

B. Proteins Bound to 5 S RNA

A ribonucleoprotein complex composed of a single protein and 5 S RNA, has been dissociated from the large ribosomal subunits of both rat liver and rabbit reticulocytes by treatment with EDTA (Blobel, 1971; Lebleu *et al.*, 1971) and formamide (Petermann *et al.*, 1972). The isolated ribonucleoprotein complex has a sedimentation coefficient of approximately 7 S, and the protein moiety has a molecular weight between 35,000 and 45,000. In each instance, the large subunit retained a full complement of ribosomal proteins except for those protein species associated with the 5 S RNA. Release of the 5 S RNA is accompanied by inactivation of the 60 S subunit.

C. Proteins Bound to Messenger RNA

Globin messenger RNA released from the salt-washed polysomes of duck and rabbit reticulocytes has been shown to contain two proteins. This ribonucleoprotein complex is dissociated from the polysomes by treatment with EDTA (Lebleu *et al.*, 1971; Morel *et al.*, 1971) or by incubation with puromycin in high salt (Blobel, 1972). The binding of the two proteins to the messenger RNA appears to be specific, since the complex is resistant to dissociation at monovalent salt concentrations less than 1 *M*. In addition, the complex is obtained when either of the two different procedures are used to dissociate the messenger RNA. Both proteins are present in one copy per molecule of mRNA (Blobel, 1972).

By contrast, in rat liver, only a single protein with a molecular weight of 160,000 has been found associated with mRNA, whether the polysomes are dissociated either with EDTA or with puromycin (Olsnes, 1971a,b).

The physiological role of these ribonucleoprotein particles is not known; however, they may have a significant role in the protection of mRNA against attack by RNase, in mRNA transport from nucleus to cytoplasm, in cleavage of large molecular weight nuclear RNA, or in initiation of protein synthesis. Both free globin mRNA and the mRNA: protein complex are highly active in directing the synthesis of globin in purified *in vitro* systems (T. Staehelin, personal communication). The relation of these proteins to ribosomal components is not clear at this time.

VIII. Heterogeneity of Ribosomal Proteins

A. Comparison of Different Tissues from the Same Organism

The ribosomal proteins isolated from various tissues of rat, chicken, and rabbit have been found to be similar by electrophoretic analysis in one-dimensional polyacrylamide gel systems (Low and Wool, 1967; DiGirolamo and Cammarano, 1968; Bielka and Welfle, 1968). In addition, stained protein patterns from ribosomes in both the polysome and monosome region were compared and found to be identical. DiGirolamo and Cammarano (1968) concluded that the "same genes for the synthesis of ribosomal proteins are functioning in the various tissues independently of their embryonic origin, degree of differentiation and functional specialization." They found some differences in the protein patterns of small subunits prepared by dissociation with EDTA, but these results may be due to the methods employed as noted earlier. The conclusion reached by DiGirolamo and Cammarano based on one-dimensional gel analysis in urea at pH 4.5 is strengthened by results from Holland and Tissieres (personal communication) with one-dimensional discontinuous polyacrylamide gel electrophoresis in sodium dodecyl sulfate (Laemmli, 1970). They found no differences in the stained protein patterns of ribosomal subunits of different organs of the same animal; furthermore, no dissimilarities were detected in the protein patterns of cell cultures derived from the various tissues, from virus-infected cell cultures or from interferon-treated cells. Noll and Bielka (1970b) found no evidence for tissue specificity of ribosomal proteins when proteins from a number of tissues were compared by immunological techniques.

In the only study employing two-dimensional electrophoresis, the 80 S proteins from rabbit liver and rabbit reticulocyte ribosomes were compared and found to be similar; however, several specific proteins were present in each tissue (Huynh-van-tan *et al.*, 1971).

These results lead to the conclusion that the ribosomes from different tissues of the same organism are highly similar if not identical. The substantiation of this conclusion will depend upon the application of the newer techniques now available both for the preparation of clean active subunits and protein analysis by two-dimensional gel electrophoresis.

B. Comparison of Different Species

An examination of the ribosomal proteins from rat, hen, tortoise, frog, and bream by one- and two-dimensional electrophoresis indicates that the pattern of ribosomal proteins in vertebrates is conservative (Hultin and Sjoqvist, 1971). The proteins in general have approximately the same size and electrophoretic properties. Two specific proteins in the different species were found to be homologous with respect to their relative susceptibility to attack by trypsin, resistance to chymotrypsin, and in the three-dimensional structural shielding under various conditions.

Approximately 17 of the protein bands from the livers of trout, frog, hen, pig, cow, mouse, and rat resolved by one-dimensional electrophoresis in urea, have common mobilities and may have identical or very similar structures (Bielka and Welfle, 1968). Each species also has two to seven additional protein bands that are specific for that organism. Low and Wool (1967) found species-specific ribosomal proteins from rat and rabbit by one-dimensional electrophoresis in urea at pH 4.7.

It can be concluded from these studies, that distinct differences exist between the ribosomal proteins from different species; however, the size, distribution, and number of proteins from the different species is relatively constant. These results are largely based on one-dimensional gel analysis. More detailed studies of the type undertaken by Hultin and Sjoqvist (1971) may reveal more homology among specific proteins than is apparent from one-dimensional gel electrophoresis at pH 4.5 or even two-dimensional gel analysis. Sun, Bickle, and Traut (1972) have proposed, from a study of prokaryotic ribosomes from several bacterial species by polyacrylamide gel electrophoresis in SDS, that the restrictions imposed in the construction of a ribosomal particle of constant size from a constant number of component proteins infers the existence of size homologies among the component proteins. The same principal may also apply to mammalian ribosomes.

C. Normal vs Tumor Ribosomes

Among the numerous characteristics that distinguish neoplastic and normal cells, the possibility of structural changes at the level of the ribosome has been considered. The stained gel patterns obtained by two-dimensional electrophoresis of the ribosomal proteins from rat liver and several hepatocellular carcinomas have been compared (Delaunay and Schapira, 1972; Bielka *et al.*, 1971). Although more than 60 discrete protein spots were resolved, no major differences were found. No differences were found in the antigenic properties of ribosomes and ribosomal proteins from rat liver and hepatoma as shown by various serological methods (Noll and Bielka, 1970a).

On the other hand, two antigenic differences have been observed between the proteins from 0.5 *M* KCl washed Novikoff hepatoma ascites cells and normal rat liver. These proteins have been isolated from the hepatoma subunits with the use of antiserum to the hepatoma 60 S subunits which has been preabsorbed with normal liver 60 S subunits. The two antigenic determinants have been identified as proteins with molecular weights of 30,000 and 65,000 (Wikman-Coffelt *et al.*, 1972) and have different electrophoretic characteristics when compared to normal liver by two-dimensional electrophoresis (G. A. Howard, J. Wikman-Coffelt, and R. R. Traut, unpublished results).

D. Membrane-Bound vs Free Ribosomes

The ribosomal protein patterns of free and bound ribosomes have been compared by gel electrophoresis in SDS. Fridlender and Wettstein (1970) demonstrated a single protein difference between the free and bound ribosomes from tissue cultures of chick embryo. Preliminary evidence indicated that the 60 S subunit contains the unique protein. Borgese, Blobel, and Sabatini (personal communication) have also resolved a prominent protein band that was associated with the free 60 S subunit, but not with the subunits released from membrane.

IX. Methods for Correlating Structure and Function

Controlled dissociation of complex cellular organelles like ribosomes, coupled with readdition of the individual isolated protein components to produce particles deficient in specific proteins, is a useful tool for examining the role of individual proteins as they relate to the activity of the intact particle. Such methods have been elegantly exploited in the investigations by Nomura and his collaborators (Traub and Nomura, 1969; Traub *et al.*, 1971) on the function of individual *E. coli* 30 S ribosomal proteins. It is likely that similar methods will be applied to the investigation of structure: function relationships in mammalian ribosomes. At present, however, methods for reconstitution of mammalian subunits from individual proteins are not available.

A. Protein-Deficient Particles

Protein-deficient ribosomes have been prepared by controlled dissociation of ribosomal proteins under conditions of increasing concentrations of monovalent salt. LiCl (Reboud *et al.*, 1969; Bielka *et al.*, 1970), CsCl (Grummt and Bielka, 1970), and KCl (Clegg and Arnstein, 1970) have been used to extract specific ribosomal proteins from the intact ribosomal particles. Westermann *et al.*, (1971b) have achieved similar results by binding ribosomal particles to DEAE-Sephadex and then eluting specific proteins with a LiCl gradient ranging from 0 to 1 *M* LiCl. The extracted proteins and the remaining core particles obtained with both procedures were analyzed by gel electrophoresis. A portion of the proteins were completely removed, others were partially extracted, while some protein species remained quantitatively bound to the residual particle.

Discontinous changes were observed in the sedimentation behavior of the large subunit with the concomitant release of certain proteins upon treatment of the subunits with increasing salt concentrations (Clegg and Arnstein, 1970). This result suggests a transition from one discrete stage to another which may be due in part to the release of 5 S RNA (Blobel, 1971). The sedimentation coefficient of the smaller subunit was shown to decrease with increasing salt concentrations (Clegg and Arnstein, 1970).

B. Reconstitution

Partial reconstitution of an active 40 S subunit of rat liver from protein-deficient core particles and split proteins has recently been reported by Terao and Ogata (1971). Split proteins constituting approximately 25% of the total ribosomal proteins were removed from the 40 S subunits with 0.4 *M* LiCl. Reconstitution of 40 S subunits was accomplished by dialyzing the core particles and the split proteins overnight at 0°C in 50 m*M* KCl, 5 m*M* $MgCl_2$, 10 m*M* β-mercaptoethanol and 50 m*M* Tris-HCl buffer (pH 7.6). The putative 40 S particles were examined for activity in polyphenylalanine synthesis with intact 60 S subunits; they were 50% as active as the original subunits. A similar experiment with reconstituted 60 S subunits and intact 40 S subunits showed no activity. The preliminary results obtained with rat liver subunits resemble those with *E. coli.* The small subunit can be easily reconstructed from individual proteins and isolated 16 S RNA, while the large subunit cannot (Traub and Nomura, 1969; Traub *et al.*, 1971). Only with the thermophilic organism, *B. stearothermophilus*, has reconstruction of the 50 S been successful (Nomura and Erdmann, 1970; Nashimoto and Nomura, 1970).

An attempt was made to reconstitute totally the 40 S subunit from 18 S RNA and protein moieties from rat liver (Terao and Ogata, 1971) using the conditions for reconstitution of the small subunit of *E. coli* described by Traub and Nomura (1969). The resulting particles sedimented at 32 S and had considerable activity in binding poly(U); however, their activity in polyphenylalanine synthesis was less then 10% of the activity of the intact 40 S subunits. The reconstructed 32 S subunit had characteristics very similar to those of the 32 S particle obtained by EDTA treatment of the 40 S subunit. The one-dimensional electrophoretic pattern of the reconstituted particles showed a protein pattern qualitatively identical to the native subunit; however, the density of some of the protein bands was reduced.

X. Membrane-Bound Ribosomes

Ribosomes present in the microsomal fraction of cell extracts are localized on the outer membrane surface of the endoplasmic reticulum (ER), giving it a granular appearance as viewed by electron microscopy (Porter, 1961; Siekevitz and Palade, 1960). The site of interaction between the ribosomes and the membrane has been shown to be the large subunit (Sabatini *et al.*, 1966; Baglioni *et al.*, 1971). As described in Section VIII, a single protein difference has been observed between membrane-bound and free ribosomes.

It is an interesting possibility that translational control phenomena may operate at the level of ribosome: membrane interaction.

A. Function

The proportion of free to membrane-bound ribosomes varies in various cell types. It has been postulated that the proportion of free to membrane-bound ribosomes is related to the degree of specialization of these cells for exporting products; thus attached ribosomes would make secretory proteins, and free ribosomes would make nonexportable proteins. Early biochemical evidence for this hypothesis was provided by studies on the guinea pig pancreas, Siekevitz and Palade (1960) demonstrated by radioactive labeling *in vivo* that α-chymotrypsinogen was synthesized by the microsomal fraction rather than the free cytoplasmic ribosomes. The pathway of transport of the completed secreted proteins from the site of synthesis to the condensing vacuoles of the Golgi complex of guinea pig pancreatic cells has been established through electron microscopic autoradiography (Caro and Palade, 1964; Jamieson and Palade, 1967). The proteins are transported from the lumen of the rough ER to the smooth microsomes and finally to the zymogen granule fraction. Evidence for synthesis of other secretory proteins on membrane-bound ribosomes has been obtained from rat liver, where six times more serum albumin proteins are made on the bound ribosomes of rat liver than on the free ribosomes both *in vivo* and *in vitro*, while ferritin, a nonexported protein, is preferentially synthesized on free ribosomes (Redman, 1967, 1969; Takagi and Ogata, 1968). Amylase produced by pigeon pancreas is also synthesized by the microsomal system (Redman *et al*., 1966). Cells which export a large proportion of the synthesized protein, contain a major portion of their ribosomes bound to the ER; whereas in rapidly dividing embryonic cells and tumor cells, and in reticulocytes, only a small fraction of the ribosomes are membrane-bound.

Andrews and Tata (1971) compared the fate of completed protein chains synthesized in the microsomal fraction from liver, where greater than 50% of the ribosomes are membrane-bound and the predominant product is secretory proteins, to those synthesized in the microsomal fraction from cerebral cortex and muscle, where less than 20% of the ribosomes are bound. Puromycin was utilized to induce release of nascent radioactive polypeptide chains from the isolated microsomal fractions and the pathway of the released label was followed. In liver, a greater percentage of the polypeptide chains remained associated with the ER and were found in the interior of the lumen rather than released into the medium, while essentially all the label from both the brain and muscle microsomes was found in the soluble fraction. This suggests that in nonsecretory tissues, a function of ribosome binding may be to compartmentally segregate different systems synthesizing different classes of proteins. Studies comparing the protein composition of membrane-bound and free ribosomes are discussed in Section VIII of this review.

B. Characteristics

Two classes of membrane-bound ribosomes have been described in HeLa cells (Attardi *et al.*, 1969; Rosbash and Penman, 1971) and guinea pig liver (Sabatini *et al.*, 1966). These classes are distinguishable by the tightness of binding to the membrane, and have been descriptively designated as "loose" and "tight." Brief digestion with RNase removes the loosely attached ribosomes; the remainder can be dissociated with DOC. Low concentrations of EDTA completely release the 40 S subunits from microsomes as 32 S components, but only 50% of the heavy subunits are removed. At higher EDTA concentrations, conditions under which the 40 S subunit is almost completely unfolded and partially degraded, 34 to 40% of the 60 S subunits remain bound to the membrane in cells containing both a large and small percentage of the ribosomes in the microsomal fraction. Thus approximately half of the large subunits are removed by RNase digestion or heavy metal chelation, suggesting attachment to the membrane through associated RNA.

Recently Adelman *et al.* (1972b) have shown that up to 40% of the membrane-bound ribosomes of rat liver can be released in high ionic strength media. These appear to consist both of inactive ribosomes and active ribosomes containing relatively short nascent chains. Upon incubation of membrane-bound polysomes with puromycin at high ionic strengths, essentially all the ribosomes are released. One interpretation of these observations is that a portion of the ribosomes are bound to the membrane via the nascent chain and are released by puromycin and high salt, while the noncompleted peptides remain tightly associated with the membrane. The correlation of these ribosomes with the loose and tight designation has not been made.

C. Steroid Control of Endoplasmic Reticulum Formation

The distribution of polysomes between the cytosol and the endoplasmic reticulum can be modified *in vivo* by several drugs and the steroid hormone balance (Rancourt and Litwack, 1968; Cox and Mathias, 1969; Orrenius *et al.*, 1968). An example of hormone control of the rough ER has been shown with alveolar epithelial cells from mouse mammary explants (Mills and Topper, 1969). These cells, cultured in the presence of insulin, contain a small amount of rough ER and are unable to make casein. Addition of hydrocortisone results in formation of an extensive rough ER, and upon addition of prolactin, the secretory protein casein is produced. In the absence of hydrocortisone, there is no increase in polysome attachment to membrane and no prolactin-induced casein formation. Another relationship between the degree of rough ER formation and the amount of secretory product produced has been demonstrated in rats. The ER of the secretory epithelial cells of the seminal vesicle and prostate is almost completely degranulated following castration (Szirmai and van der Linde, 1965).

XI. Phosphorylation of Ribosomal Proteins

A. Evidence for Phosphorylation

Proteins tightly or loosely associated with the ribosome have been found to exist in a phosphorylated form in ribosomes isolated from rat liver (Loeb and Blat, 1970) and rabbit reticulocytes (Kabat, 1970) after administration of radioactive phosphate to the whole animals. Similar findings are reported for heart, liver, and skeletal muscle from chick embryo (Kabat, 1970). Ribosomal proteins are also found to be phosphorylated after incubation of intact rabbit reticulocytes (Kabat, 1970) or rat sarcoma cells (Bitte and Kabat, 1972) with radioactive phosphate *in vitro*. In cell extracts prepared from rat liver (Loeb and Blat, 1970; Eil and Wool, 1971), rabbit reticulocytes (Kabat, 1971; Traugh *et al*., 1973), bovine adrenal glands (Walton *et al*., 1971), and chick embryo fibroblasts (Li and Amos, 1971), ribosomal proteins are phosphorylated by the catalytic action of protein kinase in the presence of (γ-^{32}P) ATP. Kabat (1970, 1971) has shown using one-dimensional disc gel electrophoresis in SDS followed by autoradiography, that the major phosphate-accepting proteins in rabbit reticulocytes, labeled both *in vivo* and *in vitro*, migrate identically. The phosphorylated substrates are proteins, as indicated by their sensitivity to pronase. In addition, phosphoserine and to a lesser extent, phosphothreonine, were identified as the phosphate-accepting groups in several systems (Kabat, 1970; Loeb and Blat, 1970; Eil and Wool, 1971; Walton *et al*., 1971; Traugh *et al*., 1973).

Protein kinases characteristically catalyze the transfer of the terminal phosphate of ATP to the serine or threonine residues of specific receptor proteins [for review see Krebs (1972) and Walsh and Krebs (1973)]. These kinase activities can be divided into two groups, cyclic AMP-activated and cyclic AMP-independent. A portion of the total protein kinase activity in the whole cell extract is attached to ribosomes (Walton *et al*., 1971; Kabat, 1971) and is removed upon dissociation of the ribosomes into subunits (Eil and Wool, 1971; Traugh *et al.*, 1973). However, the existence of a protein kinase integrated specifically in the 80 S ribosome structure remains an open question at this time.

Ribosomal subunits prepared from rabbit reticulocytes and rat liver by dissociation in high salt retain phosphate-accepting activity (Kabat, 1970; Eil and Wool, 1971; Stahl *et al.*, 1972; Traugh *et al.*, 1973) whereas subunits prepared from adrenal cortex are reported to have lost all receptor activity (Walton *et al.*, 1971). A comparison made in this laboratory of the phosphate receptor proteins of ribosomes washed in low salt (0.025 *M* KCl) in high salt (0.5 *M* KCl), and of the proteins removed from the ribosome by a high salt wash is shown in Fig. 3. The majority of the phosphorylated high molecular weight components associated with the ribosome washed only at low salt concentrations are removed by sedimentation through 0.5 *M* KCl. None of the phosphorylated proteins in the high salt wash fraction have been identified at this time; however, it is possible that some of the factors involved in protein synthesis may be phosphate acceptors. It is clear, however, that phosphorylated proteins present on ribosomes washed at high salt concentrations are an integral part of the ribosome structure.

Three protein bands in the 40 S subunit from rat liver are phosphorylated by added cyclic AMP-dependent protein kinase as determined by one-dimensional electrophoresis and autoradiography (Eil and Wool, 1971). Similarly, approximately nine bands are associated with the 60 S subunit. Stahl *et al.* (1972) using bovine skeletal muscle kinase with ribosomes from rat liver, found by analysis with two-dimensional electrophoresis that one protein in the 40 S subunit from rat liver is labeled with high intensity and three other proteins contain lesser amounts of radioactive phosphate. Ten phosphorylated proteins are in the 60 S subunit.

Recent results in this laboratory have shown that two cyclic AMP regulated protein kinases, separated and partially purified from the ribosome-free supernatant of rabbit reticulocytes, have different specificity for the ribosomal proteins of both subunits (Traugh *et al.*, 1973). The enzymes characterized by preferential phosphorylation of the basic histone proteins were found to phosphorylate one protein band in the 40 S subunit having a molecular weight of 35,000 and six protein bands in the 60 S subunit. Another kinase, which preferentially phosphorylates the acidic proteins, casein and phosvitin, was shown to phosphorylate a protein in the 40 S subunit present in less than stoichiometric amounts with a molecular weight of 50,000, and four protein bands in the 60 S subunit, only one of which was also a substrate for the histone specific enzyme. Both (γ-^{32}P) ATP and (γ-^{32}P)6TP serve as substrates for the latter kinase. These cytoplasmic protein kinase activities have also been identified in the 0.5 *M* KCl wash of reticulocyte ribosomes (J. A. Traugh and R. R. Traut, unpublished observations).

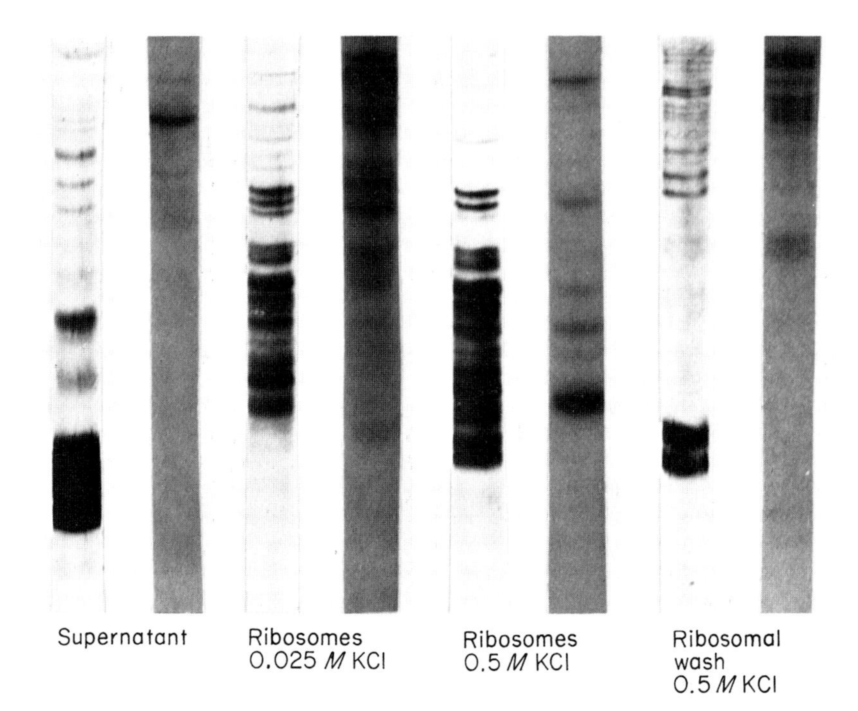

FIG. 3. Protein phosphorylation with endogenous protein kinase and (γ-^{32}P)ATP in cellular fractions from rabbit reticulocytes. Cell extracts of reticulocytes from phenylhydrazine-treated rabbits were fractionated by centrifugation at 50,000 rpm in a Spinco Ti 60 rotor for 5 hours at 4°C. The upper three-fourths of the supernatant was removed, dialyzed, and frozen in small aliquots at −70°C. The ribosome pellet was resuspended in buffer A (10 m*M* Tris-HCl, pH 7.5; 25 m*M* KCl; 1.5 m*M* $MgCl_2$). The ribosomes were washed by sedimentation through 15 ml of 1 *M* sucrose containing either (1) buffer A, or (2) buffer A with 0.5 *M* KCl (the ribosome solution was also made to 0.5 *M* KCl), by centrifugation for 12 hours at 50,000 rpm in a Ti 50 rotor. The 0.5 *M* KCl wash fraction was removed and dialyzed against buffer A. The ribosome pellets were resuspended in buffer A.

The reaction mixtures for phosphorylation of the cellular fractions contained in a final volume of 0.07 ml: 14 m*M* sodium phosphate, pH 7.0; 10 m*M* $MgCl_2$ 1.4 μM cAMP; 0.14 m*M* (γ-^{32}P)ATP; and either 0.5 mg supernatant, 0.15 mg washed or unwashed ribosomes, or 0.1 mg ribosomal wash. The reaction was initiated by addition of each cell fraction, incubated for 30 minutes at 30°C, and terminated by addition of 1% sodium dodecyl sulfate.

The proteins were analyzed by electrophoresis in 10% polyacrylamide gels with sodium dodecyl sulfate. The protein was stained with Coomassie Blue, before the gels were sliced longitudinally, and dried under vacuum; the center slice was exposed to Kodak No-Screen Medical X-ray film for 3 days. The stained protein pattern is on the left; the autoradiogram is on the right.

B. Implications for the Role of Phosphorylation in Translational Control

Since cyclic AMP-dependent protein kinase has been implicated in the control of a number of cellular functions, the phosphorylation of ribosomal proteins suggests their possible involvement in control at the level of translation. This is supported by the nonuniform distribution of the different phosphorylated protein species in the various classes of ribosomes, polysomes, monosomes, and subunits, reported by Kabat (1970). Similar differences in the distribution of phosphorylated proteins in the polysome and monosome fractions of Sarcoma 180 tumor cells supports this idea (Bitte and Kabat, 1972).

Kabat (1972) has measured the specific activity of the ATP pools after incorporation of phosphate into whole cells, and from these values has determined an average figure of 8.4 protein-bound phosphoryl groups per ribosome in the steady state. The single ribosomes contain a higher specific activity than the polysomes. The phosphoryl groups in the different phosphoproteins are exchanged at approximately the same rate, a rate considerably slower than that of protein synthesis. This result places certain limits on the interpretation of possible functions of phosphoproteins in protein synthesis. However, the specificity of proteins phosphorylated in each subunit, the possible existence of two or more protein kinases phosphorylating different ribosomal proteins suggests that there may be functional modifications of the ribosome resulting from phosphorylation. At this time there is no direct evidence for a change in ribosome function resulting from phosphorylation; thus the role of phosphorylation in control of translation is difficult to assess.

XII. Translation of Specific Species of Messenger RNA into Proteins of Proven Structure

Recently, a number of reports on the isolation and accurate translation by mammalian systems of specific species of messenger RNA have appeared. The study of translation and its control in eukaryotic systems had in the past been hampered by difficulties in the preparation of active ribosomal subunits, the lack of purified species of natural mRNA, and in the identification of specific protein products stimulated by the added message. Over the last few years a number of methods have been described for the preparation of active 80 S ribosomes and their subunits (see Sections II and III). Specific species of RNA have been routinely isolated from cells specialized in the production of one or a few major protein products, and these purified messages have been translated and their products identified in systems that do not contain the messenger specific proteins.

The use of heterologous systems has raised the question of the existence of tissue specific or protein specific factors involved in initiation. The number of active protein-synthesizing systems produced by a combination of isolated mRNA from one animal tissue and the translational system from another often unrelated species indicate a lack of tissue specificity, although in most of the cell-free systems the mRNA has been added in saturating amounts. However, cell-free systems prepared from rabbit reticulocytes fail to synthesize virus proteins from the RNA of encephalomyocarditis virus (EMC), whereas extracts from Krebs II ascites cells translate this message (Mathews, 1970). In addition, Heywood (1970) has shown preferential binding and translation of myosin and globin mRNAs with ribosomes containing the respective binding factors which are removed with high salt. At the present time this question of tissue specificity of initiation factors is unresolved.

The various systems reported up to this time for the *in vitro* synthesis of specific proteins programmed by purified mRNAs are summarized in Table III. The protein synthesized, source of mRNA, and the source of the components of cell-free systems are listed. Messenger RNA has, in general, been prepared by SDS treatment of the 10–20 S ribonucleoprotein complex released from polyribosomes by EDTA and isolated by sucrose density gradient centrifugation. The cell-free reaction mixtures contain an energy source, ATP, GTP, radioactive amino acids, tRNA, runoff ribosomes or salt-washed ribosomes, and the dialyzed wash fraction, supernatant, and isolated mRNA in a buffered salt solution. The heterologous protein products are identified by a variety of methods including column chromatography, specific antibody precipitation, peptide analysis and acrylamide gel electrophoresis. In most cases translation has been carried out in lysates or with 80 S ribosomes; however 40 S and 60 S subunits from mouse liver have been shown to be highly active in the synthesis of rabbit or duck hemoglobin (Schreier *et al.*, 1973). Similarly, Prichard *et al.* (1971), have reported the synthesis of rabbit hemoglobin with washed 60 S and 40 S reticulocyte ribosomal subunits. The availability of these systems opens the way for important advances in the study of ribosome structure and function and the control of protein synthesis.

XIII. Concluding Remarks

Despite the fact that mammalian ribosomes contain nearly twice as much protein as bacterial ribosomes, the number of different protein species does not reflect this two-fold increase. Instead, because the average molecular

TABLE III

CELL FREE SYSTEMS USED FOR TRANSLATION OF PURIFIED MESSENGER RNA INTO SPECIFIC PROTEIN PRODUCTS

Protein product	Source of purified mRNA	Cell-free system	Reference
Globin	Rabbit reticulocytes	Human reticulocyte lysate	Nienhuis *et al.* (1971)
Globin	Rabbit reticulocytes	Mouse Krebs II ascites tumor cell lysate	Mathews *et al.* (1971)
Globin	Duck immature erythrocytes	Rabbit reticulocyte lysate	Pemberton *et al.* (1972)
Globin β-chains	Mouse reticulocytes	Rabbit reticulocyte lysate	Lockard and Lingrel (1969)
Immunoglobulin light chain	Mouse plasma cell tumor	Rabbit reticulocyte lysate	Stavnezer and Huang (1971)
Immunoglobulin light chain	Mouse plasma cell tumor	Mouse Krebs II ascites tumor cell lysate	Brownlee *et al.* (1972)
Myosin	Embryonic chick muscle	Chicken reticulocyte lysate	Heywood (1969)
α-Crystalline lens proteins	Calf lens (10 S and 14 S)	Mouse Krebs II ascites tumor cell lysate	Mathews *et al.* (1972)
α-Crystalline lens proteins	Chick lens	Duck reticulocyte lysate	Williamson *et al.* (1972)
α-Crystalline lens proteins	Chick lens	Mouse Landschutz ascites cell lysate	Williamson *et al.* (1972)
α-Crystalline lens proteins	Calf lens (14 S)	Rabbit reticulocyte lysate	Berns *et al.*(1972)
Ovalbumin	Hen oviduct	Rabbit reticulocyte lysate	Rhoads *et al.* (1971)
Ovalbumin	Chick oviduct	Rabbit reticulocyte: salt washed ribosomes; initiation fraction; supernatant	Means *et al.* (1972)
Myosin	Embryonic chick muscle	Rabbit reticulocyte: salt-washed ribosomes; initiation fraction; supernatant; chick muscle initiation factor 1F3	Rourke and Heywood (1972)
Globin	Rabbit reticulocytes	Rabbit reticulocyte: salt-washed ribosomal subunits; purified initiation factors; supernatant	Prichard *et al.* (1971)
Globin	Rabbit reticulocytes	Rabbit liver: salt-washed ribosomes; initiation factors M1 and M2; supernatant; rabbit reticulocyte initiation factor M3	Prichard *et al.* (1971)
EMC virus polypeptides	Encephalomyocarditis virus	Mouse Krebs II ascites tumor cell lysate	Smith *et al.* (1970)
EMC virus polypeptides	Encephalomyocarditis virus	Rabbit reticulocyte lysate; Krebs ascites cell factors	Mathews (1970)

weight of the mammalian proteins is greater than that of the bacterial proteins, there are approximately 70 mammalian ribosomal proteins as compared to the 55 bacterial proteins. Thus the complexity of the mammalian ribosome in terms of the number of constituents is not substantially greater than the better studied bacterial system. It can be foreseen that the 70 mammalian proteins will be isolated, purified and characterized, and perhaps used in reconstitution studies, following much the same pattern as previous investigations with bacteria. At this time a major obstacle would seem to be the absence of a reproducible system for complete reconstruction of mammalian ribosomes from separated RNA and protein. Certainly the elucidation of possible control mechanisms operating at the level of the ribosome would provide a major stimulus for such studies on reconstitution.

In the context of translational control the question of ribosome heterogeneity in the mammalian particles has been raised. The question has four facets: (1) possible differences in the relative molar amounts of the 70 different proteins; (2) possible differences in protein composition corresponding to the tissue or cell line of origin of the ribosome population studied; (3) possible differences in protein composition between membrane-bound and free ribosomes isolated from the same tissue; and (4) differences in the protein composition of the ribosome resulting from exchange between proteins in the ribosome particle with a pool of ribosomal proteins in the cytoplasm.

1. Bacterial ribosomes have been shown to be heterogeneous; there are two classes of ribosomal proteins, those found in amounts equimolar with the RNA and with each other, from which it is inferred that there is one copy per particle, and a smaller number found in amounts less than equimolar with the ribosome. The latter fact means that certain proteins must be absent from a portion of the total ribosome population.

At this time there is insufficient evidence from which to draw any conclusion about this type of heterogeneity in mammalian ribosomes. Molecular weights of all of the 70 protein species are not yet available, and the mass fractions of the different proteins have not been determined by appropriate labeling techniques. Furthermore, physicochemical characterizations of active washed ribosomal subunits are necessary to determine the total protein complement of the particles. All these experiments are now feasible, and an answer to the question of heterogeneity among the ribosomes isolated from a given source should soon be available.

2. The early suggestion that ribosomes from various tissues might differ substantially in their protein composition has not found support. From recent studies it appears that cytoplasmic ribosomes isolated from any part of the organism are nearly, if not completely, identical with respect to both number and type of ribosomal proteins. A more definitive conclusion

will require the application of new high resolution techniques of protein separation combined with quantitation of the individual protein species.

3. Minor protein differences have been found between membrane-bound and free ribosomes. It is an interesting possibility that one aspect of translational control may be in the regulation of the ribosome: membrane interaction. Two or three protein species from hepatoma ribosomes differ with respect to normal rat liver. Since free and membrane-bound ribosomes were not separated in those studies, the differences may be due to the proportion of membrane-bound to free ribosomes in the two kinds of cells.

4. Dice and Schimke (1972) have shown recently that up to 70% of ribosomal proteins in rat liver are readily exchangeable with a pool of ribosomal proteins constituting more than 15% of the total cytoplasmic soluble protein. Garrison *et al.* (1972) have also shown a similar exchange occurs with unfertilized *Xenopus* egg ribosomes. These normally inactive ribosomes can be activated by exchange of proteins with a cytoplasmic fraction from adult liver. Thus, it appears that the mammalian structure may be in a state of active flux with a pool of cytoplasmic ribosomal proteins. Qualitative and quantitative analysis of the proteins present in the ribosomes and of the ribosomal proteins in the cytoplasmic fraction will be required in order to understand fully the significance of this phenomena.

The major conclusion drawn from these studies is that the mammalian ribosome from a given organism is a well defined particle of relatively constant composition, but that small qualitative and quantitative differences in protein composition may exist.

A body of evidence suggests the existence of post-transcriptional control of protein synthesis in a variety of mammalian systems. Such controls may reside at the level of the ribosome and could result from small differences in the amounts of certain rate-limiting ribosomal proteins, or the presence or absence of specific proteins. Another possible mechanism for translational control at the level of the ribosome would be selective phosphorylation or other enzymatic covalent modification of ribosomal proteins. Further quantitative structural analysis correlated with physiological studies as well as with sensitive *in vitro* assays of ribosome function will be required before conclusions can be drawn.

ACKNOWLEDGMENTS

Supported by research grants from the U.S. Public Health Service (GM 17924) and the Damon Runyon Memorial Fund for Medical Research (DRG 1140). J. A. T. is recipient of a U.S. Public Health Service fellowship (GM 505590), and R. R. T. is an Established Investigator of the American Heart Association. We thank Drs. G. Howard, A. Bollen, T. A. Bickle, and J. W. B. Hershey for helpful discussion and critical reading of the manuscript.

References

Adelman, M. R., Blobel, G., and Sabatini, D. D. (1973a). *J. Cell Biol.* **56**, 191.

Adelman, M. R., Sabatini, D. D., and Blobel, G. (1973b). *J. Cell Biol.* **56**, 206.

Allen, E. S., and Schweet, R. S. (1962). *J. Biol. Chem.* **237**, 760.

Andrews, T. M., and Tata, J. R. (1971). *Biochem. J.* **121**, 683.

Attardi, B., Cravioto, B., and Attardi, G. (1969). *J. Mol. Biol.* **44**, 47.

Baglioni, C., Bleiberg, I., and Zauderer, M. (1971). *Nature (London), New Biol.* **232**, 8.

Berns, A. J. M., Straus, G. J. A. M., and Bloemendal, H. (1972). *Nature (London), New Biol.* **236**, 7.

Bickle, T. A., and Traut, R. R. (1971). *J. Biol. Chem.* **246**, 6828.

Bielka, H., and Welfle, H. (1968). *Mol. Gen. Genet.* **102**, 128.

Bielka, H., Grummt, F., and Schneiders, I. (1970). *Acta Biol. Med. Ger.* **24**, 705.

Bielka, H., Stahl, J., and Welfle, H. (1971). *Arch. Geschwulstforsch.* **38**, 109.

Bitte, L., and Kabat, D. (1972). *J. Biol. Chem.* **247**, 5345.

Blobel, G. (1971). *Proc. Nat. Acad. Sci. U.S.* **68**, 1881.

Blobel, G. (1972). *Biochem. Biophys. Res. Commun.* **47**, 88.

Blobel, G., and Potter, K. R. (1967). *J. Mol. Biol.* **26**, 279.

Blobel, G., and Sabatini, D. (1971). *Proc. Nat. Acad. Sci. U.S.* **68**, 390.

Bonanou, S., Cox, R. A., Higginson, B., and Kanagalingam, K. (1968). *Biochem. J.* **110**, 86.

Borgese, D., Blobel, G., and Sabatini, D. D. (1973). *J. Mol. Biol.* **74**, 415.

Brownlee, G. G., Harrison, T. M., Mathews, M. B., and Milstein, C. (1972). *FEBS (Fed. Eur. Biochem. Soc.) Lett.* **23**, 244.

Caro, L. G., and Palade, G. E. (1964). *J. Cell Biol.* **20**, 473.

Christman, J. K., and Goldstein, J. (1971). *Nature (London), New Biol.* **230**, 272.

Clegg, J. C. S., and Arnstein, J. R. V. (1970). *Eur. J. Biochem.* **13**, 149.

Collier, R. J. (1967). *J. Mol. Biol.* **25**, 83.

Cox, R. A. (1968). *In* "Methods in Enzymology," Vol. 12: Nucleic Acids (L. Grossman and K. Moldave, eds.), Part B, p. 120. Academic Press, New York.

Cox, R. A., and Mathias, A. P. (1969). *Biochem. J.* **115**, 777.

Currey, J. B., and Hersh, R. T. (1962). *Biochem. Biophys. Res. Commun.* **6**, 415.

Davis, B. J. (1964). *Ann. N. Y. Acad. Sci.* **121**, 404.

Delaunay, J., and Schapira, G. (1972). *Biochim. Biophys. Acta* **259**, 243.

Dice, J. F., and Schimke, R. T. (1972). *J. Biol. Chem.* **247**, 98.

DiGirolamo, M., and Cammarano, P. (1968). *Biochim. Biophys. Acta* **168**, 181.

Eil, C., and Wool, I. G. (1971). *Biochem. Biophys. Res. Commun.* **43**, 1001.

Falvey, A. K., and Staehelin, T. (1970). *J. Mol. Biol.* **53**, 1.

Faust, C. H., Jr., and Matthaei, H. (1972). *Biochemistry* **11**, 2682.

Ford, P. J. (1971). *Biochem. J.* **125**, 1087.

Fridlender, B. R., and Wettstein, F. O. (1970). *Biochem. Biophys. Res. Commun.* **39**, 247.

Garrison, N. E., Bosselman, R. A., and Kaulenas, M. S. (1972). *Biochem. Biophys. Res. Commun.* **49**, 171.

Gould, H. J., Arnstein, H. R. V., and Cox, R. A. (1966). *J. Mol. Biol.* **15**, 600.

Grummt, F., and Bielka, H. (1970). *Biochim. Biophys. Acta* **199**, 540.

Hamilton, M. G., and Petermann, M. L. (1959). *J. Biol. Chem.* **234**, 1441.

Hardy, S. J. S., Kurland, C. G., Voynow, P., and Mora, G. (1969). *Biochemistry* **8**, 2897.

Heywood, S. M. (1969). *Cold Spring Harbor Symp. Quant. Biol.* **34**, 799.

Heywood, S. M. (1970). *Proc. Nat. Acad. Sci. U.S.* **67**, 1782.

Howard, G. A., and Traut, R. R. (1973). *FEBS (Fed. Eur. Biochem. Soc.) Lett.* **29**, 177.

Hultin, T., and Sjoqvist, A. (1971). *Comp. Biochem. Physiol. B* **40**, 1011.

Hultin, T., and Sjoqvist, A. (1972). *Anal. Biochem.* **46**, 342.

Huynh-van-tan, Delaunay, J., and Schapira, G. (1971). *FEBS (Fed. Eur. Biochem. Soc.) Lett.* **17**, 163.
Jamieson, J. D., and Palade, G. E. (1967). *J. Cell Biol.* **34**, 577.
Kabat, D. (1970). *Biochemistry* **9**, 4160.
Kabat, D. (1971). *Biochemistry* **10**, 197.
Kabat, D. (1972). *J. Biol. Chem.* **247**, 5338.
Kaltschmidt, E., and Wittmann, H. G. (1970). *Anal. Biochem.* **36**, 401.
King. J. S. W., Gould, J. J., and Shearman, J. J. (1971). *J. Mol. Biol.* **61**, 143.
Krebs, E. G. (1972). *Curr. Top. Cell. Regul.* **5**, 99.
Kurland, C. G. (1972). *Annu. Rev. Biochem.* **41**, 377.
Laemmli, U. K. (1970). *Nature (London)* **227**, 680.
Lamaire, S., Pelletier, G., and Labrie, F. (1971). *J. Biol. Chem.* **23**, 7303.
Lamfrom, H., and Glowacki, E. R. (1962). *J. Mol. Biol.* **5**, 97.
Lawford, G. R. (1969). *Biochem. Biophys. Res. Commun.* **37**, 143.
Lebleu, B., Marbaix, G., Huez, G., Temmerman, J., Burny, A., and Chantrenne, H. (1971). *Eur. J. Biochem.* **19**, 264.
Leboy, P. S., Cox, E. C., and Flaks, J. G. (1964). *Proc. Nat. Acad. Sci. U.S.* **52**, 1367.
Li, C.-c., and Amos, H. (1971). *Biochem. Biophys. Res. Commun.* **45**, 1398.
Lockhard, R. E., and Lingrel, J. B. (1969). *Biochem. Biophys. Res. Commun.* **37**, 204.
Loeb, J. E., and Blat, C. (1970). *FEBS (Fed. Eur. Biochem. Soc.) Lett.* **10**, 105.
Low, R. B., and Wool, I. G. (1967). *Science* **155**, 330.
Maden, B. E. H. (1971). *Progr. Biophys. Mol. Biol.* **22**, 127.
Martin, T. E., and Wool, I. G. (1968). *Proc. Nat. Acad. Sci. U.S.* **60**, 569.
Martin, T. E., Rolleston, F. S., Low, R. B., and Wool, I. G. (1969). *J. Mol. Biol.* **43**, 135.
Martin, T. E., Wool, I. G., and Castles, J. J. (1971). *In* "Methods in Enzymology," Vol. 20: Nucleic Acids and Protein Synthesis, Part, C. (K. Moldave and L. Grossman, eds.), p. 417. Academic Press, New York.
Martini, O. H. W., and Gould, H. J. (1971). *J. Mol. Biol.* **62**, 403.
Mathews, M. B. (1970). *Nature (London)* **228**, 661.
Mathews, M. B., Osborn, M., and Lengrel, J. B. (1971). *Nature (London), New Biol.* **233**, 206.
Mathews, M. B., Osborn, M., Berns, A. J. M., and Bloemendal, H. (1972). *Nature (London), New Biol.* **236**, 5.
Means, A. R., Comstock, J. P., Rosenfeld, G. C., and O'Malley, B. W. (1972). *Proc. Nat. Acad. Sci. U.S.* **69**, 1146.
Mechler, B., and Mach, B. (1971). *Eur. J. Biochem.* **21**, 552.
Mills, E. S., and Topper, Y. J. (1969). *Science* **165**, 1127.
Morel, C., Kayibanda, B., and Scherrer, K. (1971). *FEBS (Fed. Eur. Biochem. Soc.) Lett.* **18**, 84.
Nashimoto, H., and Nomura, M. (1970). *Proc. Nat. Acad. Sci. U.S.* **67**, 1440.
Nienhuis, A. W., Laycock, D. G., and Anderson, W. F. (1971). *Nature (London), New Biol.* **231**, 205.
Nolan, R. D., and Arnstein, H. R. V. (1969). *Eur. J. Biochem.* **10**, 96.
Noll, V. G., and Bielka, H. (1970a). *Arch. Geschwulstforsch.* **35**, 338.
Noll, V. G., and Bielka, H. (1970b). *Mol. Gen. Genet.* **106**, 106.
Nomura, M. (1970). *Bacterial. Rev.* **34**, 228.
Nomura, M., and Erdmann, V. A. (1970). *Nature (London)*, **228**, 744.
O'Brien, T. W. (1971). *J. Biol. Chem.* **246**, 3409.
Olsnes, S. (1971a). *Eur. J. Biochem.* **23**, 248.
Olsnes, S. (1971b). *Eur. J. Biochem.* **23**, 557.
Orrenius, S., Gnosspelius, Y., Das, M. L., and Ernster, L. (1968). *In* "Structure and Function of the Endoplasmic Reticulum in Animal Cells," (F. C. Gran, ed.), p. 81. Academic Press, New York.

Pemberton, R. E., Houseman, D., Lodish, H. F., and Baglioni, C. (1972). *Nature (London), New Biol.* **235**, 99.

Petermann, M. L. (1971). *In* "Methods in Enzymology," Vol. 20: Nucleic Acids and Protein Synthesis, Part, C. (K. Moldave and L. Grossman, eds.), p. 429. Academic Press, New York.

Petermann, M. L., and Pavlovec, A. (1971). *Biochemistry* **10**, 2770.

Petermann, M. L., Hamilton, M. G., and Pavlovec, A. (1972). *Biochemistry* **11**, 2323.

Porter, K. R. (1961). *In* "The Cell" (J. Brachet and A. E. Mirsky, eds.), Vol. 2, p. 621. Academic Press, New York.

Prichard, P. M., Picciano, D. J., Laycock, D. G., and Anderson, W. F. (1971). *Proc. Nat. Acad. Sci. U.S.* **68**, 2752.

Rancourt, M. W., and Litwack, G. (1968). *Exp. Cell Res.* **51**, 413.

Reboud, A.-M., Hamilton, M. G., and Petermann, M. L. (1969). *Biochemistry* **8**, 843.

Reboud, A.-M., Arpin, M., and Reboud, J.-P. (1972a). *Eur. J. Biochem.* **26**, 347.

Reboud, A.-M., Buisson, M., and Reboud, J.-P. (1972b). *Eur. J. Biochem.* **26**, 354.

Redman, C. M. (1967). *J. Biol. Chem.* **242**, 761.

Redman, C. M. (1969). *J. Biol. Chem.* **244**, 4308.

Redman, C. M., Siekevitz, P., and Palade, G. E. (1966). *J. Biol. Chem.* **241**, 1150.

Reid, M. S., and Bieleski, R. L. (1968). *Anal. Biochem.* **22**, 374.

Rhoads, R. E., McKnight, G. S., and Schimke, R. T. (1971). *J. Biol. Chem.* **246**, 7407.

Rosbash, M., and Penman, S. (1971). *J. Mol. Biol.* **59**, 227.

Rourke, A. W., and Heywood, S. M. (1972). *Biochemistry* **11**, 2061.

Sabatini, D. D., Tashiro, Y., and Palade, G. E. (1966). *J. Mol. Biol.* **19**, 503.

Schreier, M. H., and Staehelin, T. (1973). *J. Mol. Biol.* **73**, 329.

Schreier, M. H., Staehelin, T., Stewart, A., Gander, E., and Scherrer, K. (1973). *Eur. J. Biochem.* **34**, 213.

Shapiro, A. L., Vienuela, E., and Maizel, J. V. (1967). *Biochem. Biophys. Res. Commun.* **28**, 815.

Sherton, C. C., and Wool, I. G. (1972). *J. Biol. Chem.* **247**, 4460.

Siekevitz, P., and Palade, G. E. (1960). *J. Biophys. Biochem. Cytol.* **7**, 619.

Skogerson, L., and Moldave, K. (1967). *Biochem. Biophys. Res. Commun.* **27**, 568.

Smith, A. E., Marcker, K. A., and Mathews, M. B. (1970). *Nature (London)* **225**, 184.

Spirin, A. S., and Gavrilova, L. P. (1969). "The Ribosome." Springer-Verlag, Berlin and New York.

Spitnik-Elson, P. (1965). *Biochem. Biophys. Res. Commun.* **18**, 557.

Staehelin, T., and Falvey, A. K. (1971). *In* "Methods in Enzymology," Vol. 20: Nucleic Acids and Protein Synthesis, Part. C, (K. Moldave and L. Grossman, eds.), p. 433. Academic Press, New York.

Stahl, J., Welfle, H., and Bielka, H. (1972). *FEBS (Fed. Eur. Biochem. Soc.) Lett* **26**, 233.

Stavnezer, J., and Huang, R. C. C. (1971). *Nature (London), New Biol.* **230**, 172.

Studier, F. W. (1972). *Science* **176**, 367.

Sun, T. T., Bickle, T. A., and Traut, R. R. (1972). *J. Bacteriol.* **111**, 474.

Szirmai, J. A., and van der Linde, P. C. (1965). *J. Ultrastruct. Res.* **12**, 380.

Takagi, M., and Ogata, K. (1968). *Biochem. Biophys. Res. Commun.* **33**, 55.

Tashiro, Y., and Siekevitz, P. (1965). *J. Mol. Biol.* **11**, 149.

Terao, K., and Ogata, K. (1971). *Biochim. Biophys. Acta* **254**, 278.

Traub, P., and Nomura, M. (1969). *J. Mol. Biol.* **40**, 391.

Traub, P., Mizushima, S., Lowry, C. V., and Nomura, M. (1971). *In* "Methods in Enzymology," Vol. 20: Nucleic Acids and Protein Synthesis, Part. C, (K. Moldave and L. Grossman, eds.), p. 391. Academic Press, New York.

Traugh, J. A., and Collier, R. J. (1971). *Biochemistry* **10**, 2357.

Traugh, J. A., Mumby, M., and Traut, R. R. (1973). *Proc. Nat. Acad. Sci. U.S.* **70**, 373.

Traut, R. R., Delius, H., Ahmad-Zadeh, C., Bickle, T. A., Pearson, P., and Tissieres, A. (1969). *Cold Spring Harbor Symp. Quant. Biol.* **34**, 25.

Ts'o, P. O. P., and Vinograd, J. (1961). *Biochim. Biophys. Acta* **49**, 113.

Waller, J. P., and Harris, J. I. (1961). *Proc. Nat. Acad. Sci. U.S.* **47**, 18.

Walsh, D. A., and Krebs, E. G. (1973). *In* "The Enzymes." (P. D. Boyer, ed.), Vol. 8, p. 555. Academic Press, New York.

Walton, G. M., Gill, G. N., Abrass, I. B., and Garren, L. D. (1971). *Proc. Nat. Acad. Sci. U.S.* **68**, 880.

Weber, K., and Osborn, M. (1969). *J. Biol. Chem.* **244**, 4406.

Welfle, H. (1971). *Acta Biol. Med. Ger.* **27**, 547.

Welfle, H., Bielka, H., and Bottger, M. (1969). *Mol. Gen. Genet.* **104**, 165.

Welfle, H., Stahl, J., and Bielka, H. (1971). *Biochim. Biophys. Acta* **243**, 418.

Welfle, H., Stahl, J., and Bielka, H. (1972). *FEBS (Fed. Eur. Biochem. Soc.) Lett.* **26**, 228.

Westermann, P., Bielka, H., and Bottger, M. (1969). *Mol. Gen Genet.* **104**, 157.

Westermann, P., Bielka, H., and Bottger, M. (1971a). *Mol. Gen. Genet.* **111**, 224.

Westermann, P., Koppitz, D., and Bielka, H. (1971b). *Acta Biol. Med. Ger.* **26**, 611.

Wikman-Coffelt, J., Howard, G. A., and Traut, R. R. (1972). *Biochim. Biophys. Acta* **277**, 671.

Williams, D. J., Gurari, D., and Rabin, B. R. (1968). *FEBS (Fed. Eur. Biophys. Soc.) Lett.* **2**, 133.

Williamson, R., Clayton, R., and Truman, D. E. S. (1972). *Biochem. Biophys. Res. Commun.* **46**, 1936.

Wittmann, H. G. (1970). *Symp. Soc. Gen. Microbiol.* **20**, 55.

Chapter 5

Methods for the Extraction and Purification of Deoxyribonucleic Acids from Eukaryote Cells[1,2]

ELIZABETH C. TRAVAGLINI

The Institute for Cancer Research,
Fox Chase Center for Cancer and Medical Sciences,
Philadelphia, Pennsylvania

I. Introduction

A cursory survey of the methods available for the extraction and purification of deoxyribonucleic acid (DNA) might lead a novice in the field of nucleic acid chemistry to conclude that for every cell whose DNA is to be

[1]This work was supported by Grants CA-06927 and RR-05539 from the National Cancer Institute, U.S. Public Health Services and by an appropriation from the Commonwealth of Pennsylvania.

[2]This chapter is dedicated to the memory of Jack Schultz, who so early and so clearly saw that the isolation of pure DNA would be required for understanding the role of DNA in the cell.

studied, a new or modified method has to be devised for extracting and purifying that DNA. A slightly closer look at the literature might even suggest that, depending on what aspect of a particular DNA is to be studied, several methods of extraction and purification may be required. In practice, the choice of a method is usually determined by the type and amount of tissue available, whether a quantitative recovery of the DNA per cell is mandatory, and in what molecular state the DNA is desired.

It is the purpose of the present communication to acquaint the reader with the problems in the extraction of DNA, describe the methods which have been designed to solve these problems, and show how these methods have been adapted by modern investigators.

II. Physical-Chemical Characteristics of DNA

In order to extract DNA from whole tissues or cells, it is necessary to have a basic understanding of the physical and chemical properties of DNA as a macromolecule, where it is localized in the cell, and how it is associated with other molecules in the cell. The following is a brief description of the properties of DNA, both *in vitro* and *in vivo*, and is designed to help the reader choose a method or a combination of methods to isolate the DNA from a particular tissue.

A. DNA *in Vivo*

The genetic material of eukaryotes consists of DNA. It is usually extracted from the nuclei of eukaryotic cells as a linear double-stranded polymer which consists of two single chains, whose links are purine and pyrimidine deoxyribonucleotides, held together by hydrogen bonds between the purine and pyrimidine subunits to form a double helix. However, in certain prokaryotes and in the mitochondria and chloroplasts of most eukaryotes, DNA has been found in the form of covalently closed circular double-stranded molecules. In prokaryotes and the mitochondria and chloroplasts of eukaryotes, DNA is usually free of proteins, but in the nuclei of eukaryotes, it usually is found in tight association with basic proteins, such as histones. When preparations of pure organelles, such as nuclei and mitochondria can be obtained from eukaryotes, the methods of lysis and DNA extraction that have been developed for prokaryotes can be used with very little modification. However, when DNA is to be extracted from whole tissue, the methodology becomes more complex insofar as the DNA has to be removed from large amounts of proteins, lipoproteins, RNA, and polysaccharides, some of which have solubility characteristics similar to those of DNA.

B. DNA *in Vitro*

DNA molecules in their native state can be quite large. DNAs with molecular weights of several hundred million daltons have been isolated; the DNA from T2 bacteriophage has been found to be a single linear duplex molecule with a molecular weight of 130×10^6 (Levinthal and Davidson, 1961), while a chromosome from yeast contains DNA which is a single molecule of molecular weight 800×10^6 (Petes and Fangman, 1972; Blamire *et al.*, 1972a). These DNA molecules, which *in vivo* must be compactly coiled and folded, are, in solution, flexible and, because of their massive size and semirigid confirmation, exquisitely sensitive to shear degradation. Shear forces are easily set up when one surface slides relative to another parallel to their plane of contact as they do in the processes of tissue grinding, shaking of immiscible solutions, pipetting, etc., that is, in the processes involved in the extraction and purification of DNA. A single shear-induced double-stranded break per molecule will halve the mean molecular weight of a sample of linear DNA since the molecule tends to break near its middle. As the DNA decreases in molecular weight, it tends to become more rodlike and less prone to shear [for a detailed review, see Josse and Eigner (1966)].

The circular DNAs in eukaryotes are usually of low molecular weight, approximately 10×10^6 for mitochondria (Borst and Kroon, 1969) and approximately 100×10^6 for chloroplasts (Kolodner and Tewari, 1972), and can be isolated easily as covalently closed circular molecules whereas the circular DNA of the *Escherichia coli* genome, whose molecular weight is 2 to 4×10^9 (Davern, 1966), is difficult to isolate as an intact circle because of its size. As we shall see below, it is possible to separate the circular mitochondrial DNA from the linear DNA of eukaryotes because the circular DNA has a lower capacity for intercalating certain dye molecules, such as ethidium bromide (EBr) (see Section V, Glossary of Reagents, for explanation of reagent abbreviations and full names).

The fact that DNA is a double helix makes it appear to be more stable to certain types of chemical and enzymatic degradation than it really is. At low pH (i.e., pH 5, 100°C, 5 minutes, or pH 2, 37°C, 26 hours), the molecule becomes depurinated although it retains its deoxyribose-phosphate backbone. When this happens, it is easily broken at the depurinated sites and its average molecular weight is lowered greatly [see Chargaff (1955)]. At pH > 9.5, the two strands of the DNA begin to separate; at pH > 11, the hydrogen bonds which connect the base pairs are completely broken and the DNA molecules become single-stranded (Vinograd *et al.*, 1963). High temperatures, depending on the base composition of the DNA and the ionic strength of the solution in which it is dissolved, will also cause the molecules to become single-stranded. The melting temperature, T_m, at which a DNA containing 40% guanine, plus cytosine, becomes single-stranded is 85°C in 1 × SSC (0.15 *M* NaCl, 0.015 sodium citrate) and 65°C in 0.1 × SSC (Marmur and Doty, 1962). It is possible that, during the extraction procedure, endonucleases, often called "nickases," will introduce breaks in one of the strands of the DNA. Such a molecule will appear intact until it is melted, and only then will it be apparent that the DNA has been degraded.

These kinds of lability, plus the fact that the DNA double helix can be hydrolyzed to oligonucleotides by endonucleases (DNases), put severe limitations on the way tissues can be handled for DNA extraction and purification. The pH and temperatures limits have to be kept relatively narrow; at the same time, nucleases have to be inhibited and shear forces kept at a minimum. In the following paragraphs, the basic methods which have been devised to solve these problems are elucidated.

III. General Methodology

A. History

Prior to the experiments of Avery, McLeod, and McCarty in 1944 on bacterial transformation (Avery *et al.*, 1944), which identified DNA as the probable carrier of hereditary information, the studies on nucleic acids were limited to defining its chemical constituents. With the elucidation of DNA's double helical structure (Watson and Crick, 1953a,b), which showed that the molecule could transmit information, it became important to extract DNA in such a way that its innate physical structure be retained.

The early investigators used high salt concentrations to dissociate DNA from the protein of eukaryote nuclei, and chloroform–isoamyl alcohol to separate DNA from the protein (Sevag *et al.*, 1938). Usually, tissues were used that were rich in DNA and contained very little RNA or polysaccharide, such as calf thymus nuclei or fish spermatozoa and testes. Little attention was paid to shear forces or to the action of nucleases during either cell fractionation or extraction [for a review, see Chargaff (1955)]. Such a method is described as follows (Gulland *et al.*, 1947): Fresh calf thymus glands were minced and suspended in an equal volume of 0.9% sodium chloride and milled to form a fine suspension, which was centrifuged at approximately 6000 *g*. The pellet was resuspended in an equal volume of 0.9% NaCl, and the above procedure was repeated. The pellet was then suspended in 5 volumes of 10% NaCl with vigorous stirring at 0°C for 48 hours. The insoluble material was removed by centrifugation at approximately 6000 *g* and the supernatant was precipitated in an equal volume of methanol. The precipitant was washed with methanol and dried in a vacuum at 25°C. The nucleoprotein was powdered to assist solution and dissolved in 10 volumes of 10% NaCl with virgorous stirring. The solution was clarified by centrifugation at 6000 *g*, an equal volume of chloroform–isoamyl alcohol added and the mixture emulsified by rapid stirring. The emulsion was broken by low speed centrifugation, and the aqueous layer was reextracted with chloroform–isoamyl alcohol at least 9 times to remove all the protein. The DNA was precipitated in an equal volume of methanol, washed with varying concentrations of methanol and dried. When used for large amounts of tissue, this method extracted DNA more or less quantitatively and sufficiently pure and intact for qualitative chemical analysis.

However, when it was found that DNA is the molecule which contains the genetic information of the cell, it became interesting to extract it from a variety of organisms in its native state. Since these organisms differed widely in their nuclear and cytoplasmic components, as well as in their membrane properties, modifications of the known extraction methods became imperative. In 1948, Chargaff and Zamenhof introduced the use of 0.05 *M* sodium citrate in the initial cell brei to bind Mg^{2+} and thus inhibit deoxyribonuclease. They also devised a method for removing polysaccharides by precipitating them in CsCl; RNA was removed by enzymatic digestion with ribonuclease. A variety of such modifications were introduced as investigators sought methods of retaining DNA in its native highly polymerized state during the extraction procedure. In 1940, Bowden and Pirie had made a survey of reagents that would inactivate viruses by separating the viral coat protein from its nucleic acid. Among these reagents were the anionic detergents sodium dodecyl sulfate and sodium deoxycholate and various organic agents, such as phenol, urea, and guanidine-HCl. It was among these reagents that subsequent investigators found more gentle yet effective methods to extract DNA from eukaryotes.

Most of the methods presently employed to extract DNA from eukaryotic cells can be outlined as follows: (1) the disruption or lysis and sometimes fractionation of the tissue to be extracted; (2) the use of a detergent or salt solution to dissociate the DNA from protein in cells or cell organelles; (3) the addition of an extractant to separate the bulk of the protein from the DNA; and (4) the use of enzymes or differential precipitation to remove the RNA and polysaccharides that are coextracted with the DNA.

The most popular of these methods are used either singly or in combination with each other and are known in laboratory jargon as "the detergent method" (Marko and Butler, 1951; Kay *et al.*, 1952; Marmur, 1961) and "the phenol method" (Kirby, 1957, 1961, 1964). In our laboratory, there is yet another, "the CsCl method," which is based on the CsCl density gradient technique developed by Meselson *et al.* (1957) and modified by us to extract DNA quantitatively from whole tissues under extremely mild conditions (Travaglini and Meloni, 1962).

All these methods are equally effective provided the organelles that contain the DNA can be separated from excessive cytoplasm and cell debris, then concentrated and disrupted in a gentle manner with minimal shear and nuclease activation. Often this is not the case, particularly with eukaryotic cells, and thus the basic extraction methods appear to be modified when essentially the changes are in the initial cell disruption procedures.

B. Methods for Lysing and Disrupting Cells

Many eukaryote cells and cell organelles, including mouse embryos (Williamson, 1969), certain mammalian organs, e.g., liver, spleen, testis (Kirby, 1964; Irving and Veazey, 1968; Okuhara, 1970), insect eggs (Travaglini *et al.*, 1972) and larvae (Papaconstantinou *et al.*, 1972), starfish oocytes (Huez *et al.*, 1972), *Xenopus* eggs and ovaries (Dawid, 1965; Dawid *et al.*, 1970), Ascaris embryos (Bielka *et al.*, 1968), slime mold (Sussman and Rayner, 1971), ascites tumor cells (Harrison, 1971a), and most mammalian tissue culture cells (for example, McCallum and Walker, 1967; Cummings, 1972; Terasima and Tsuboi, 1969), can be lysed by homogenization in the extraction medium of choice—2% SDS, 6% *p*-aminosalicylate, cesium chloride, etc. Several investigators (Kirby, 1964; Kay *et al.*, 1952; Savitsky and Stand, 1966) claim that in order to obtain complete mammalian cell lysis, it is necessary to freeze-thaw the cells before homogenization. Savitsky and Stand (1966) have also found that the recovery of DNA from mammalian cells was more reproducible when 0.04% deoxycholate was added to the homogenization medium.

Often, however, it is necessary to fractionate cells before extracting DNA. The cells of certain eukaryote tissues, such as those of green plants and insects, are difficult to break open because of their tough exterior coating. Care must be taken to break all the cells in such tissues, otherwise some types of cells may be broken in preference to others. In extreme cases of this type, it may be important to isolate particular organs before attempting cell disruption. Stern (1968) has reviewed the methods developed for the mechanical breakage of plant tissue. These methods involve the use of a blender, a mortar and pestle, or a glass homogenizer in media containing various concentrations of glycerol, sucrose, $CaCl_2$ or $MgCl_2$ and Tris buffer pH 6.8–7.0. The disrupted cells are then filtered through a screen or nylon mesh to remove cell debris and unbroken tissue. Pitout and Potgieter

(1968) have shown that the mortar and pestle is to be preferred in the case of maize kernels because the blender tends to shear the DNA in the final product. Ritossa and Spiegelman (1965) use a mortar and pestle to grind up adult flies. When these methods are used, care should be taken that cell nuclei, mitochondria, and chloroplasts are not broken in the process of cell disruption. In the case of insect larvae, Zweidler and Cohen (1971) found it important to break the larvae without disrupting whole organs; otherwise, the large amount of nuclease activity present at this stage of insect development degrades the nucleic acids. Sometimes, in an organism such as yeast, it is advisable to obtain spheroplasts before lysing the cells (Blamire *et al.*, 1972a).

In many tissues, it is often desirable to separate the nuclei, mitochondria, and chloroplasts from the bulk of cytoplasmic constituents. This is true particularly in the case of plants and insects, where large amounts of polysaccharide particles and/or starch granules or excessive protein are present. Methods for doing this have been devised by several investigators: for yeast (Blamire *et al.*, 1972a; Wintersberger and Tuppy, 1964); for Euglena (Nass and Ben-Shaul, 1972); for slime mold (Sussman and Rayner, 1971); for green leaves and seeds (Stern, 1968; Quetier and Guille, 1968; Manning *et al.*, 1972; Tautvydas, 1971; Thompson and Cleland, 1971; Chun *et al.*, 1963); for insects (Ritossa and Spiegelman, 1965; Zweidler and Cohen, 1971). Several reviews have been written on the methods devised for extracting nuclei from mammalian cells [see Muramatsu (1970)]. Always in these fractionation procedures, the necessity to avoid nuclease activity and shear forces must be kept in mind.

Once the cells have been disrupted, the methods of DNA extraction are relatively few in number and are described as follows.

C. Basic Methods for DNA Extraction

1. The Detergent Method

The most used version of the detergent method is that devised by Marmur (1961) for extracting DNA from bacteria. This procedure involves the use of sodium dodecyl sulfate (SDS) to release nucleic acids from proteins (see Glossary for abbreviations, etc.). The procedure is essentially as follows: 2–3 gm of wet packed cells are suspended in 25 ml of STE (0.15 *M* NaCl, 0.05 *M* EDTA, 0.05 *M* Tris, pH 8) to which 2 ml 25% SDS is added, and the mixture is incubated at 60°C for 10 minutes and cooled to room temperature; 5 *M* sodium perchlorate is added at a final concentration of 1 *M* to the lysed suspension, and the whole mixture is shaken with an equal volume of chloroform–isoamyl alcohol for 30 minutes. The resulting emulsion is separated into 3 layers by a 5-minute low speed centrifugation (3000–12,000 g in the Sorvall centrifuge). The upper aqueous phase is pipetted into a flask with a wide-mouth pipette. The nucleic acids are precipitated from the aqueous phase by layering two volumes of 95% ethanol onto it and gently mixing the two solutions. The nucleic acids form a fibrous precipitate that can be removed by spooling the fibers on a stirring rod. The precipitate is dissolved in 10–15 ml of 1 × SSC (0.15 M NaCl, 0.015 M Na citrate), and the deproteinization step using chloroform–isoamyl alcohol is repeated several times until very little protein is seen at the interface. The supernatant is then precipitated with ethanol and dissolved in 1 × SSC (approximately 0.7 times the supernatant volume). RNase is added to a final concentration of 50 μg/ml, and the mixture is incubated for 30 minutes at 37°C (see Glossary for sources of enzymes). The digest is then deproteinized with chloroform–isoamyl alcohol as described above, the supernatant is precipitated with an equal volume of ethanol, and the nucleic acids are dissolved in 9 ml of 1 × SSC. To the dissolved nucleic acids is added 1.0 ml (3.0 M sodium acetate, 0.001 M EDTA, pH 7.0) and while the solution is stirred, 0.54 volume isopropanol is added dropwise to the vortex. The DNA precipitates in fibrous form.

This method is very mild and has been successfully used for the extraction of DNA from eukaryotic cell nuclei (Marko and Butler, 1951; Kay *et al.*, 1952; Williamson, 1969; Klett and Smith, 1968; Quetier and Guille, 1968; Chum *et al.*, 1963; Weintraub and Holtzer, 1972). However, when the tissues to be extracted contain a large amount of protein or polysaccharide, this method is usually combined with the phenol or CsCl extraction methods (see below) so as to avoid the tedious number of deproteinizations with chloroform–isoamyl alcohol (and thus, the shear forces set up by shaking) which are required.

2. The Phenol Method

Kirby used phenol as an alternative reagent to chloroform–isoamyl alcohol to separate nucleic acids from protein. Phenol inactivates nucleases and when it is saturated with a lipophilic salt such as *p*-aminosalicylate, it is a much more effective deproteinizer than an SDS–chloroform–octanol solution. Kirby (1964) has used a variety of phenol–salt combinations; the one most commonly used to extract DNA from mammalian tissues is essentially the following (Kirby, 1957):

Tissues (75 gm) are broken down in a high-speed mixer (45 seconds) with a solution of 6% sodium *p*-aminosalicylate (600 ml). The mixture is poured through a Büchner funnel to remove fibers and debris, and the filtrate is stirred while 600 ml of 90% (w/w) phenol is added quickly. Stirring is continued for 1 hour, after which the mixture is centrifuged in an International centrifuge at 0°C (200 *g* for 1 hour). The aqueous layer, which contains the nucleic acids, is removed by suction. The phenolic layer and the insoluble material are washed once with a small quantity of 6% (w/v) sodium *p*-aminosalicylate solution, and the aqueous layer separated by centrifuging. The combined aqueous layers (400 ml) are stirred, and 2-ethoxyethanol (400 ml) is added. The fibrous precipitate is removed with a glass rod and dissolved immediately in water (100 ml). Sodium *p*-aminosalicylate (6 gm) is added to the dissolved DNA, and the DNA is precipitated again with 2-ethoxyethanol (100 ml). In each case a flocculent precipitate of RNA remains in the water–ethoxyethanol mixture. The DNA precipitate is dissolved quickly in water (100 ml); sodium acetate (NaAc, H_2O, 4 gm) is added and the DNA precipitated again with 2-ethoxyethanol (100 ml). The DNA is then dissolved in 50 ml of water, to which sodium acetate (2 gm) and ribonuclease (1–5 mg in 1 ml of water) are added and the mixture is placed at 2°C for 16 hours. The DNA is precipitated with 2-ethoxyethanol (50 ml), and as much as possible of the solvent is removed from the precipitate before it is dissolved in water (33 ml), which takes 15–30 minutes. Then potassium phosphate [33 ml of 2.5 *M* K_2HPO_4 and 1.65 ml of 33% (v/v) H_3PO_4] and 2-methoxyethanol (33 ml) is added; after shaking the mixture is allowed to stand until the layers separate. The top layer is removed and centrifuged in a Sorvall centrifuge at 10,000 *g* for 1 hour. The clear organic layer is carefully poured from any insoluble sediment, a few drops of toluene are added, and the mixture is dialyzed twice against water (2 liters each time), and twice against 1% sodium acetate (2 liters each time). The contents of the bag are then removed, centrifuged, made up to 4% (w/v) with respect to sodium acetate (the volume is usually about 100 ml), and the DNA is precipitated with an equal volume of 2-ethoxyethanol. The fibrous precipitate is removed, washed twice with ethanol–water (3:1) and once with ethanol, and then dried over $CaCl_2$ in a vacuum desiccator.

The phenol extraction method is to be preferred when large amounts of bulky tissue are to be extracted. However, care should be taken that all the cells are disrupted, else the DNA extracted will be representative only of the cells which are most easily homogenized. Reagent volumes should be kept in relative ratio to the amount of tissue extracted; if the reagent volumes are too low, the large amounts of denatured protein will trap the DNA at the phenol-*p*-aminosalicylate interface and the yield of DNA will decrease markedly. The part of the methodology which separates DNA from polysaccharides by extraction with 2-methoxyethanol from potassium phosphate solution is cumbersome and causes some denaturation of the DNA (Kirby, 1968). The DNA is more easily purified with RNase and α-amylase (see below). Again, there is the problem of shear forces which arise from the use of a blender and excessive shaking. These problems have in large part been solved by combining the phenol method with the detergent method (see below).

Phenol extractions should not be done at temperatures above 25°C. Massie and Zimm (1965) have shown that even the presence of small amounts of phenol in the extracted DNA will denature the DNA by lowering its T_m. Smith *et al.* (1970) have shown that poly (dAT) and other repeating DNA-like polymers are solubilized in phenol saturated with molar salt solutions, so this method should be avoided when these polymers are to be extracted.

3. The Detergent–Phenol Method

Many investigators have used the detergent–phenol method: for slime mold (Sussman and Rayner, 1971), for mammalian tissue (Okuhara, 1970; Harrison, 1971a,b; McCallum and Walker, 1967; Thomas *et al.*, 1968), plants (Pitout and Potgieter, 1968; Manning *et al.*, 1972), amphibians (Dawid, 1965; Dawid *et al.*, 1970; Brown *et al.*, 1971; Baker, 1972), insects (Ritossa and Spiegelman, 1965; Kram *et al.*, 1972), etc. An example of such a method is the one devised by Ritossa and Spiegelman (1965) for adult *Drosophila*: *Drosophila* are suspended in 10 volumes (w/v) of buffer at pH 7.6, Tris (0.05 *M*), KCl (0.025 *M*), Mg acetate (0.005 *M*), sucrose (0.35 *M*), and homogenized in a mortar at 0°C. The homogenate is then filtered through 8 layers of gauze. The filtrate is centrifuged for 10 minutes at 700 *g* in the International centrifuge and the resulting pellet is suspended in 0.15 *M* NaCl, 2% SDS, and 0.1 *M* EDTA, and adjusted to pH 8. For each 10 gm of flies, about 35 ml of this medium is used and the resulting suspension shaken for 10 minutes at 60°C. All subsequent steps follow the procedure detailed by Marmur (1961), with the exception that at the first precipitation 1 volume (rather than 2) of cold ethanol is added. This avoids interference by salt precipitation and increases the yield. Fibers are collected on a glass rod, and any flocculent DNA precipitate remaining behind is centrifuged, resuspended in 1 × SSC to which 1 volume of ethanol is added yielding fibrous precipitates. To remove RNA, the fibrous DNA, dissolved in SSC, is treated with RNase (150 μg per milliliter of sample) for 4 hours at 37°C. To remove polysaccharide, the preparation is digested for 45 minutes at 37°C with 250 μg of α-amylase per milliliter of sample, and finally pronase (50 μg per milliliter of sample) for 30 minutes at 37°C (see Glossary for sources of enzymes). SDS (1%) is added to the digest, which is then treated twice with phenol at room temperature and by two successive deproteinizations with chloroform–isoamyl alcohol for 10 minutes each. To remove traces of chloroform and/or phenol, the extract is shaken with ether several times. The DNA is precipitated by the addition of 2 volumes of ethyl alcohol, and the fibers are collected and dissolved in 0.01 × SSC.

Thomas *et al.* (1968) have observed that treatment of the original homogenate with pronase increases the yield of DNA by decreasing the amount of interfacial protein which tends to trap DNA in the subsequent phenol extraction. They also recommend gentle agitation instead of shaking during the extraction procedure to avoid shearing the DNA.

4. The CsCl Method

The methods hitherto discussed are not easily applicable to the problem of isolating DNA from small quantities of tissues which contain large amounts of cytoplasmic constituents. To do this, we developed a method which adapts CsCl equilibrium centrifugation (Meselson *et al.*, 1957) to separate nucleic acids from the bulk of cytoplasmic protein in the cell (Travaglini and Meloni, 1962; Travaglini *et al.*, 1972). In a CsCl gradient, RNA, DNA, proteins, lipoproteins, and polysaccharides separate at equilibrium by virtue of their different buoyant densities. If the loosely packed cells are directly homogenized in 4.0 *M* CsCl and the mean density of the CsCl is adjusted to 1.40 gm cm^{-3}, the nucleic acids and polysaccharides pellet when the homogenate is centrifuged to equilibrium. The high CsCl concentration inhibits enzymatic degradation of the DNA. The method has been successfully used for both plant and animal material when the sample sizes are small. The method is described as follows for obtaining DNA from *Drosophila* embryos.

The embryos (2 ml per 10 ml of CsCl) are homogenized in 4 *M* CsCl (ρ = 1.40 gm cm^{-3}) with a Potter-Elvehjem glass homogenizer at 4°C. The homogenate is placed in a Spinco type 50 rotor and spun at 40,000 rpm for 20 hours at 20°C. As equilibrium is approached, the total nucleic acids and polysaccharides of the egg are found in the pellet. This pellet is dissolved in a minimal amount of 1 × SSC (usually 9 ml of 1× SSC for a pellet from 2 ml of eggs) and digested with the following enzymes successively for 1 hour each at 37°C: 0.01 volume of α-amylase (10 mg/ml); 0.1 volume of ribonuclease (0.1 mg/ml); and finally, 0.01 volume of pronase (50 μg/ml) (see Glossary for sources of enzymes). Insoluble materials in the digest are removed by low speed centrifugation and the DNA is pelleted out of solution by centrifugation (Spinco type 50 rotor at 40,000 rpm for 1 hour at 4°C). The DNA is dissolved in a minimal amount of 1 × SSC and stored at −30°C until analyzed. In the case of some tissues, such as adult *Drosophila*, certain pigments are pelleted with the nucleic acids; it is then necessary to extract the digest with an organic solvent, such as water-saturated phenol, to remove the pigments.

The virtue of this method is that it effectively extracts undegraded DNA quantitatively from disrupted cells. In many tissues, it is sufficient to expose the cellular nucleases to 4 *M* CsCl to inhibit them completely. However, in certain tissues, such as sea urchin embryos, it is important to shake the solubilized nucleic acid–polysaccharide pellet with chloroform–isoamyl alcohol prior to the enzymatic digestion in order to remove any trace of nucleases that remain in the pellet. The quantitative aspects of this method are to be emphasized.

D. Purification and Fractionation of DNA

Simple enzyme hydrolysis of the macromolecules which coprecipitate with DNA followed by precipitation of the DNA in alcohol is usually not sufficient treatment to yield DNA completely free of trace contaminants. Often, as a result, other fractionation methods, such as column chromatography, equilibrium density gradient centrifugation, rate sedimentation centrifugation, are used to remove any trace impurities from DNA and at the same time to fractionate the DNA analytically. Some investigators [for example, Britten *et al.* (1970), Petes and Fangman (1972), Nass and Ben-Shaul (1972), Blamire *et al.* (1972a,b), McBurney and Whitmore (1972) etc.] have used such fractionation procedures as a direct method for extracting DNA from lysed cell organelles. To better understand these methods of DNA extraction, the most frequently used fractionation methods will be described.

1. Selective Precipitation

It is common practice to recover DNA after extraction by precipitation in ethanol, isopropanol, or other reagents in which the DNA is insoluble. For a quantitative recovery of DNA precipitated in this manner, it is important that the concentration of DNA be approximately 5 OD units/ml at 260 nm; only if the DNA is of extremely high molecular weight can lower concentrations be used. At low concentrations of DNA, the recovery is not quantitative; and at extremely low concentrations, the DNA is not recoverable unless carrier molecules are added to aid the precipitation. Unfortunately, the presence of such carriers is exactly what one is trying to avoid.

When DNA, which has been extracted and treated with amylase, RNase, and pronase, is precipitated with an equal volume of ethanol, two types of precipitate result: a fibrous precipitate of relatively pure DNA and a flocculent precipitate which is usually contaminated with oligoribonucleotides. To recover the DNA in the flocculent precipitate in fibrous form, it is necessary to solubilize the DNA in an appropriate volume of SSC and repeatedly reprecipitate it in an equal volume of ethanol until no fibers can be recovered.

A more selective precipitation of DNA can be obtained with isopropanol

(Marmur, 1961), ethoxyethanol (Kirby, 1957), CTAB (Jones, 1953; Ralph and Bellamy, 1964; Bellamy and Ralph, 1968; Thompson and Cleland, 1971) with diphenolic diphosphates (Kirby, 1964) or with varying concentrations of ethanol in the presence of potassium acetate (Harrison, 1971a). These methods have the advantage of being simple and reproducible and are fairly effective for obtaining pure DNA. However, the risk of losing a large amount of product in the precipitation step is great, particularly when DNA in low concentration is being precipitated.

2. Column Chromatography

Two of the most commonly used adsorbing materials for fractionating DNA are methylated albumin-kieselguhr (MAK), and hydroxyapatite (HAP). The MAK column can be used to fractionate DNA and RNA by molecular size when a continuous salt gradient is used to elute the nucleic acids (Mandell and Hershey, 1960). The nucleic acids are adsorbed onto the column at a NaCl concentration less than 0.4 *M* and are eluted at higher concentrations. Sueoka and Cheng (1962) showed that by using stepwise salt elution, DNAs can be separated on MAK not only on the basis of molecular weight, but also by their base composition and extent of hydrogen bonding; i.e., double-stranded DNA can be separated from single-stranded DNA as well as RNA. Analysis of each fraction is usually required to interpret the elution pattern of a complicated mixture of nucleic acids is applied to the columns. However, the column is a useful tool if one is concerned with analyzing the heterogeneity of a DNA preparation and obtaining a pure fraction of DNA of a particular molecular weight and base composition which will be free of RNA and protein [for a review of MAK methodology, see Ishida *et al.* (1971)].

HAP columns can be used to purify DNA by applying the DNA onto the column, and eluting it in the double-stranded state with a linear phosphate buffer (PB) gradient (Bernardi, 1962) at 25°C, or eluting it as single strands with 0.08 or 0.12 *M* PB at increasing temperatures (Miyazawa and Thomas, 1965). Up to a temperature of 70°C, unbroken, undenatured DNA molecules can be eluted from a HAP column at 0.27 *M* PB, whereas denatured DNA is eluted at 0.1 *M* PB at 25°C. RNA elutes with denatured DNA, oligonucleotides elute at even lower molarities (0.01–0.015 *M* PB), and proteins and polysaccharides are adsorbed much more weakly than native DNA (Bernardi, 1969). The molecular weight of the DNA does not appear to affect greatly the elution from HAP. However, when DNA is eluted from a HAP column at stepwise increments in temperature, the molecular weight has some effect on the elution pattern although it is the base composition of the DNA which is the predominant factor in the fractionation of the DNA. DNAs having a high adenine, plus thymine, composition will come off first because their T_m's are low [for a review of HAP chromatography, see Bernardi (1971)]. McCallum and Walker (1967) have made a detailed study of the fractionation of mammalian DNA on HAP and show that information about both the base composition and secondary structure of the DNA can be obtained.

3. Equilibrium Density Gradient Centrifugation

In 1957, Meselson and co-workers demonstrated that macromolecules could be fractionated in a self-generating CsCl gradient if their buoyant (isopycnic) densities were different. It is possible in the preparative ultracentrifuge to separate DNA from RNA and protein on a gradient when the mean density of the CsCl is 1.70 gm cm^{-3} before centrifugation. The buoyant density of DNA in CsCl is approximately 1.70 gm cm^{-3}; it will segregate as a band on the gradient, whereas proteins with a density less than 1.30 gm cm^{-3} will float, and RNA, with a density greater than 1.85 gm cm^{-3}, will pellet. Schildkraut *et al.* (1962) have shown that it is possible to determine the base composition of DNA on a CsCl gradient in the analytical ultracentrifuge.

CsCl equilibrium gradients can also be used to separate linear from circular DNAs if the DNA is combined with an intercalating dye, such as ethidium bromide (EBr). EBr intercalates to a greater extent with linear double-stranded DNA than with circular DNA, thereby decreasing the buoyant densities of the DNAs disproportionately (Radloff *et al.*, 1967). This method is of extreme importance when it is found necessary to separate mitochondrial DNA from nuclear DNA in a sample of whole-tissue DNA (Travaglini and Schultz, 1972). EBr can also be used to separate dAT polymers from other DNAs because it intercalates preferably between adenine and thymine base pairs (Bauer and Vinograd, 1970). The CsCl–EBr gradient has the added advantage of making the recovery of the DNA easy since it is possible to see the banded DNA in the near ultraviolet light (the EBr changes color from red to yellow when it intercalates with DNA). After fractionation, the dye can be extracted by shaking the DNA–CsCl–EBr mixture with an equal volume of isopropanol saturated with CsCl (Travaglini and Schultz, 1972).

Cs_2SO_4 has also been used in equilibrium density gradients to fractionate DNA (Szybalski and Szybalski, 1971). DNA has a lower buoyant density in Cs_2SO_4; this makes it possible to bind DNA with heavy metals which increase its buoyant density. For example, Hg^{2+} ions preferentially bind to AT rich DNAs, and Ag^{2+} ions to GC rich DNAs, and these metals afford a way to separate DNAs on the basis of their base composition in the preparative ultracentrifuge (Davidson *et al.*, 1965).

4. Density Gradient Sedimentation (Band Centrifugation)

When a thin lamella of a solution of macromolecules is layered onto a preformed density gradient, such as a sucrose gradient (Britten and Roberts, 1960), or onto a denser liquid in a centrifugal field, i.e., NaCl or CsCl (Vinograd *et al.*, 1963), the macromolecules can be separated on the basis of molecular weight when sedimented in the ultracentrifuge. This can be done with both native and denatured DNA. If native DNA is sedimented through an alkaline sucrose gradient, it becomes single-stranded, and one can, by determining the molecular weight of the single strands, learn whether the strands are nicked or intact. The advantage of a preformed sucrose gradient is that it can be used in the preparative ultracentrifuge whereas the Vinograd technique is more easily performed in the analytical centrifuge.

E. Methods for DNA Extraction Using DNA Fractionation Techniques

Recently, many investigators have devised DNA extraction procedures that involve simple cell lysis and immediate layering of the crude lysate onto a column or gradient that will separate the DNA with a high degree of purity from the other cell constituents. The most useful of these methods are described below.

1. The MUP Method

Britten *et al.* (1970) devised a method for separating DNA from tissue lysates by direct absorption chromatography on hydroxyapatite (HAP). The method is as follows: tissue is suspended in 8 *M* urea, 0.24 *M* PB, 1% SLS, 0.01 *M* EDTA and homogenized in a blender. The homogenate is passed over HAP (stir to prevent channeling) and the HAP is washed with several volumes of buffer (8 *M* urea, 0.24 *M* PB) and then washed with 0.14 *M* PB to remove the urea. RNA is not absorbed under these conditions. The DNA is then eluted with 0.4 *M* PB. This method can be carried out by a column or batch procedure; its only limitation is the capacity of the HAP (19 mg of DNA were recovered on 20 ml of HAP from 20 gm of human liver). Plant materials require that molar $NaClO_4$ be added to the lysing mixture and the lysate be extracted with an equal volume of chloroform–isoamyl alcohol and centrifuged before the supernatant is passed over HAP. Using the blender reduces the DNA size, but according to the authors, it should be possible to recover 6×10^7 dalton DNA if the tissue is homogenized in another manner. DNA has been extracted from many eukaryote tissues successfully using this technique. It is routinely used in our Institute, by Greenberg and Perry (1971), for tissue culture cells. It is a convenient method for the preparation of DNA from small quantities of tissue.

2. The PLK Method

A very similar procedure to the MUP method is that described by Blamire *et al.* (1972b), who use kieselguhr columns coated with poly(L-lysine) to separate DNA from a crude lysate of yeast cells which have been treated extensively with RNase. The method is essentially the following: 10 gm of yeast cells are suspended in 10 ml of 0.4 *M* NaCl–0.02 *M* KH_2PO_4 (pH 6.8), 0.1 *M* EDTA and disrupted in a Bräun MSK homogenizer. The lysate is centrifuged at 27,000 *g* for 30 minutes at 4°C to remove cellular debris; the supernatant, after being treated for 30 minutes at 30°C with 50 μg/ml RNase, is diluted with 0.4 *M* buffered saline and then loaded immediately onto a standard PLK column (Ayad and Blamire, 1968). The column is washed with 1.4 *M* NaCl–0.02 *M* KH_2PO_4 and eluted with a linear salt gradient of 1.4–3.0 *M* NaCl. Nuclear DNA from yeast is eluted at 1.6 *M* NaCl while yeast mitochondrial DNA is eluted at 1.7 *M* NaCl (Finkelstein *et al.*, 1972). The method can be carried out by means of a batch procedure if desired. It is important that the crude lysate contain no trace of detergent and that the ratio of polylysine: DNA be greater than 10:1 w/w, although the ratio of polylysine to kieselguhr may range from 0.5 to 10.0 mg per gram of kieselguhr. A good ratio is the following: 500 μg DNA: 3 mg poly(L-lysine): 2.5 gm kieselguhr. The DNA obtained in this manner is in good yield and suitable for hybridization experiments without further purification.

3. The Sucrose Sedimentation Gradient Method

Several investigators (Terasima and Tsuboi, 1969; Petes and Fangman, 1972; Blamire *et al.*, 1972a; Wray *et al.*, 1972) have been interested in studying high molecular weight DNA from eukaryote cells. To do this, they have layered lysates of whole cells, spheroplasts, nuclei, intact chromosomes, mitochondria, or chloroplasts directly onto sucrose gradients and separated out the DNA by sedimentation. One of the more elegant descriptions of this methodology is that given by Blamire *et al.* (1972a) for yeast spheroplasts: 1 ml of spheroplasts, suspended in a buffer (0.1 *M* NaCl, 0.02 *M* EDTA) which causes them to collapse, are layered onto 1 ml of detergent mixture [2% sarkosyl, 3% sodium deoxycholate, 5% SDS in 0.02 *M* EDTA, 0.01 *M* Tris (pH 8.0), sterile filtered]. The detergent mixture in turn had been layered onto 50 ml 5 to 25% sucrose (w/v) gradient containing 0.1 *M* NaCl and 0.02 *M* EDTA (pH 7) in a Spinco SW 25.2 rotor at 18°C. Nuclear DNA is centrifuged 16 hours at 11,000 rpm, and mitochondrial DNA 16 hours at 14,000 rpm. (When high molecular weight DNA is to be analyzed, rotor speeds should be kept low. High rotor speeds have an anomalous effect on the sedimentation coefficient of the DNA). Forty-five 1.1-ml fractions are collected from the gradient and analyzed. DNA of molecular weight 500×10^6 daltons is usually found in the tenth fraction from the bottom of the tube.

As a marker, ^{14}C-labeled phage T2 DNA, whose molecular weight is 130 $\times 10^6$ daltons, is added. This procedure minimizes degradation by protecting the DNA in a collapsed spheroplast until just before sedimentation commences, thus reducing breakdown from shear forces and/or enzyme action. To ensure that no protein (the bulk of which stays at the top of the gradient) remains with the DNA in trace amounts, pronase (50 μg/ml) can be added to the sucrose gradients. If single-stranded DNA is to be studied, alkaline sucrose is used for the gradients (McGrath and Williams, 1966; Wray *et al.*, 1972; McBurney and Whitmore, 1972).

4. CsCl Equilibrium Density Gradients

Both the CsCl and CsCl-EBr gradient techniques have been used to extract DNA from lysates of whole cells, cell nuclei, mitochondria, and chloroplasts (Sussman and Rayner, 1971; Williamson, 1969; Nass and Ben-Shaul, 1972; Ittel *et al.*, 1970; Travaglini and Schultz 1972; Koch, 1972). When the CsCl gradient is used, the lysate is mixed with CsCl so that the final density of the solution is 1.700 gm cm^{-2}; the mixture is centrifuged to equilibrium and fractionated (5-ml samples in the SW 50.1 rotor are centrifuged for 48 hours at 20°C at 40,000 rpm). The density of each fraction is determined with a refractometer, and the fractions are dialyzed against 1 $\times$ SSC. A simple method for the microdialysis of a large number of fractions is that described by Travaglini and Meloni (1962). If the sample size is small, and the DNA can be analyzed only by its ultraviolet absorption, it is desirable to use the analytical ultracentrifuge both for CsCl equilibrium gradients (Meselson *et al.*, 1957) or for band sedimentation gradients of the kind described by Vinograd *et al.* (1963).

CsCl–EBr or CsCl–propidium diiodide equilibrium gradients have also been used to extract DNA from lysates. The method described for mitochondria (Smith *et al.*, 1971) is essentially the following: Mitochondria are suspended for 15 minutes in 1% SDS at 20°C. The mixture is then made 1.0 *M* in CsCl and chilled for 15 minutes. The density of the supernatant remaining after centrifugation for 15 minutes at 13,000 rpm was adjusted to 1.57 gm/ml with solid CsCl; propidium diiodide is added to approximately 400 μg/ml or EBr to 100 μg/ml, and 3-ml samples are centrifuged for 24 hours at 38,000 rpm at 20°C in an SW 50.1 rotor. The tubes are fractionated, and the DNA bands are collected. The dye is removed with a Dowex-50 resin column and the DNA is dialyzed against 2 $\times$ SSC. This method has also been used by Smith *et al.* (1971) for cell nuclei and whole HeLa cells.

The advantages of the gradient equilibrium methods are that they not only extract the DNA, but make it possible to fractionate it on the basis of base composition and/or secondary structure at the same time.

IV. Concluding Remarks

The methods described in this chapter are routinely used in the extraction and purification of DNA. The choice of a method is usually determined by the type and amount of material available, whether a quantitative recovery of the DNA per cell is mandatory and in what molecular state the DNA is desired. For many analytical studies (molecular hybridization, renaturation kinetics, base sequencing, etc.), the molecular weight of the DNA need not be extremely high; however, the DNA should be in a native state, and it must be pure. If the material to be extracted is available in large quantities, either the detergent or the phenol methods, or a combination of the two, are to be recommended. However, when tissue is in short supply and/or a quantitative recovery of the DNA is desired, the CsCl, the MUP, or the PLK method is preferable. The methods which involve direct layering of lysates onto sedimentation gradients are of importance when the molecular weight of the DNA is to be analyzed.

Often the investigator is interested in measuring the amount of DNA per cell. Usually, he resorts to the Schmidt and Thannhauser or a similar procedure [for a review, see Leslie (1955)] wherein DNA is precipitated by acid from an alkaline digest of a tissue homogenate and measured colorimetrically. This method is easy and very reproducible for most tissue culture cells and other eukaryote cells with a minimal amount of cytoplasm. Unfortunately, it is not good for plant or insect cells or other cells where pigments and/or starch and glycogen interfere with the colorometric assays. In these cases, it is important to free the DNA from these interfering substances before measuring the DNA colorimetrically (Travaglini *et al.*, 1972); the CsCl method is ideal for this purpose.

The gradient extraction techniques are the best suited for microextractions of DNA, particularly the DNA from pure organelles. In these cases, a consideration of the manner in which the DNA is to be detected determines the methodology. If the DNA can be detected by its radioactive label, gradients in the preparative ultracentrifuge are to be preferred because the DNA can be recovered. However, if the DNA has to be analyzed spectrometrically and is free of ultraviolet-absorbing low molecular weight substances, then gradients in the analytical ultracentrifuge should be considered, particularly if the amount of DNA is critical.

Certain precautionary measures should be taken prior to extracting DNA from any material. It is recommended that DNA be extracted from fresh material; this is not always possible. When tissue has to be stored, it should be kept at $-70°C$. In our experience, when tissue is stored at $-30°C$ for any period of time, its DNA is slowly degraded and, upon extraction, recovery of the DNA is very poor. Also, the investigator should be reminded that when DNA is to be extracted from the tissue of a developing organism, the DNA from one stage of development of that organism might be more easily obtained in a native state than the DNA from another stage of development, i.e., the DNA from a prelarval *Drosophila* embryo is usually obtained in much higher molecular weight than the DNA from a third-instar larvae. Finally, a word about storing the extracted DNA: if the DNA is precipitated in alcohol, it is possible to lyophilize the DNA fibers and store them *in vacuo*. Many investigators prefer dialyzing the DNA versus 1 × SSC and freezing it in solution at −30°C. However, freezing should be avoided if extremely high molecular weight DNA is to be stored because the shear forces set up by crystal formation might break the molecules.

V. Glossary of Reagents

p-Aminosalicylate. A lipophilic salt with chelating properties which has been shown to be effective in separating DNA from protein (Kirby, 1957).

α-Amylase (pancreas, Worthington). An enzyme that catalyzes the hydrolysis of internal α-1,4 glucan links in polysaccharides containing 3 or more α-1,4-linked D-glucose units yielding a mixture of maltose and glucose. Needs Cl^- ions for activity. Usually 0.01 ml of α-amylase (10 mg/ml) is used per milliliter of reaction mixture.

Cesium chloride (*CsCl*). A salt which in solution is capable of forming a density gradient when sedimented to equilibrium in a constant centrifugal field. Since it is used in high molar concentrations, it also inhibits the action of nucleases and separates proteins from nucleic acids (obtainable from Gallard and Schlessinger, New York, or American Chemical and Potash Co., Trona, California, for approximately $40.00 per pound).

Cesium sulfate (Cs_2SO_4). A salt with gradient forming capabilities similar to CsCl. DNA has a lower buoyant density in Cs_2SO_4 than in CsCl.

Cetyl-trimethyl ammonium bromide (*CTAB*) (obtainable, Eastman Organic Chemicals). A cationic detergent which is used to precipitate extracted DNA to remove traces of polysaccharides, polynucleotides, traces of nucleases and other basic proteins.

Chloroform–isoamyl alcohol (24:1 v/v). Used to deproteinize nucleic acids; the chloroform causes surface denaturation of proteins; the isoamyl alcohol reduces foaming, aids the separation, and maintains the stability of the layers of the centrifuged, deproteinized solution (Marmur, 1961).

Deoxycholate. An anionic detergent used to break lipoprotein links in nuclear matrices and membranes.

Ethanol. Used to recover extracted DNA by precipitation.

Ethidium bromide (*EBr*) (2,7-diamino-10-ethyl-9-phenyl-phenanthidium bromide). A dye which intercalates with double-stranded DNA and thereby causes a lowering of the buoyant density of DNA in a CsCl equilibrium density gradient (Radloff *et al.*, 1967).

Ethoxyethanol. An alcohol which preferentially precipitates DNA in the presence of RNA (Kirby, 1961).

Ethylene diaminetetraacetate (*EDTA*). A wide-spectrum chelating agent which binds divalent and polyvalent metal ions.

Hydroxyapatite (*HAP*) (purchasable, Bio-Rad, Richmond, California; or Clarkson Chemical Co., Williamsport, Pennsylvania). An alkalized mixture of Na_2HPO_4 and $CaCl_2$ which is used to purify and fractionate DNA.

Isopropanol. An alcohol which preferentially precipitates DNA in the presence of RNA (Marmur, 1961).

Lithium dodecyl sulfate (*LDS*) ($CH_3(CH_2)_{11}SO_4Li$). A detergent which has the same lysing and deproteinizing properties as SDS, it has the advantage of not solidifying at 4°C (Noll and Stutz, 1968).

Methylated albumin-kieselguhr (*MAK*). Used to purify and fractionate DNA (Mandell and Hershey, 1960).

Phenol (Mallinckrodt, analytical grade, loose crystals). Extracts protein from DNA after dissociation of the protein by salt or detergent treatment. Usually used as a water-saturated solution (500 parts phenol: 50 parts water w/v). To prevent oxidation of the phenol, it is desirable to add 8-hydroxyquinoline (Amalar) to a final concentration of 0.1%.

Phosphate buffer (*PB*). A mixture of Na_2HPO_4 and NaH_2PO_4 solutions used in the pH range 6–8.

Poly-L-lysine-kieselguhr (*PLK*). Used to purify and fractionate DNA (Ayad and Blamire, 1968).

Pronase (Calbiochem, Grade B). An enzyme which catalyzes the hydrolysis of almost any protein to free amino acids. To free it of contaminating nucleases, it is predigested for 18 hours at 37°C before use. It is usually used in concentrations of 50 μg/ml of reaction mixture.

Propidium-diiodide. An analog of ethidium bromide which intercalates with DNA and is also used for the detection and isolation of DNA (Hudson *et al.*, 1969).

Ribonuclease A (*RNase*) (Sigma, 5 × crystallized). An enzyme which catalyzes the cleavage of the phosphodiester bond between the 3′ and 5′ positions of the ribose moieties in RNA with the formation of oligo-nucleotides terminating in 2′,3′ cyclic phosphate derivatives. Prior to use, RNase is heated at 100°C for 10 minutes to destroy any possible contaminating DNase activity; it is usually used in the range of concentration of 25–100 μg/ml of reaction mixture.

Sodium chloride–sodium citrate (*SSC*). A buffer used in various concentrations to stabilize DNA in solution and chelate Mg^{2+} (1 × SSC is equivalent to 0.15 *M* NaCl, 0.015 *M* Na citrate).

Sodium chloride-Tris-EDTA (*STE*). A buffer used for the same purpose as SSC. The most commonly used concentration is 0.15 *M* NaCl, 0.05 *M* EDTA, 0.05 *M* Tris, pH 8.

Sodium citrate (*Na citrate*). A chelating agent used in preference to EDTA to specifically chelate Mg^{2+} ions which are necessary for the activation of some nucleases.

Sodium dodecyl sulfate (*SDS*) (Dupanol, M.E.), $CH_3(CH_2)_{11}SO_4$ Na, also known as sodium lauryl sulfate (*SLS*). A detergent which is an effective cell-lysing agent, a nuclease inhibitor and a deproteinizing agent (Noll and Stutz, 1968). It is purified by recrystallizing from hot ethanol and dried by washing with ether. It is usually stored as a 25% w/v stock solution at 4°C, at which temperature it solidifies.

Sodium dodecyl sarcosinate, Sarkoysyl, $CH_3(CH_2)_{10}CON(CH_3)CH_3COO$ Na. An amide derivative of SDS which has the lysing and deproteinizing properties of SDS but is soluble in solutions of high salt concentrations, such as 4 *M* CsCl.

Sodium perchlorate, $NaClO_4$. At high concentrations this salt helps dissociate protein from nucleic acids.

Tris. 2-Amino-2-(hydroxymethyl)-1, 3-propandiol, in combination with HCl, is an excellent buffer in the physiological pH range. Divalent cations are more soluble in it than in most buffers in this pH range. Since this buffer has a very significant temperature coefficient (pH decreases as temperature increases), care should be taken to measure the pH of the buffer at the temperature at which the experiment is to be carried out.

Urea, NH_2CONH_2. Used to denature proteins and thus inactivate enzymes. It may also help disrupt the cell and chromatin structure (Britten *et al.*, 1970).

References

Avery, O. T., McLeod, C. M., and McCarty, M. (1944). *J. Exp. Med.* **79**, 137–158.

Ayad, S. R., and Blamire, J. (1968). *Biochem. Biophys. Res. Commun.* **30**, 207–218.

Baker, R. F. (1972). *J. Cell. Sci.* **11**, 153–171.

Bauer, W., and Vinograd, J. (1970). *J. Mol. Biol.* **54**, 281–298.

Bellamy, A. R., and Ralph, R. K. (1968). *In* "Methods in Enzymology," Vol 12: Nucleic Acids (L. Grossman and K. Moldave, eds.), Part B, pp. 156–160. Academic Press, New York.

Bernardi, G. (1962). *Biochem. J.* **83**, 32p–33p.

Bernardi, G. (1969). *Biochem. Biophys. Acta* **174**, 423–433.

Bernardi, G. (1971). *In* "Procedures in Nucleic Acid Research" (G. L. Cantoni and D. R. Davies, eds.), Vol. II, pp. 455–499. Harper, New York.

Bielka, H., Schultz, I., and Böttger, M. (1968). *Biochim. Biophys. Acta* **157**, 209–212.

Blamire, J., Cryer, D. R., Finkelstein, D. B., and Marmur, J. (1972a). *J. Mol. Biol.* **67**, 11–24.
Blamire, J., Finkelstein, D. B., and Marmur, J. (1972b). *Biochemistry* **11**, 4848–4853.
Borst, P., and Kroon, A. M. (1969). *Int. Rev. Cytol.* **26**, 107–190.
Bowden, F. C., and Pirie, N. W. (1940). *Biochem. J.* **34**, 1278–1292.
Britten, R. J., and Roberts, R. B. (1960). *Science* **131**, 32–33.
Britten, R. J., Pavich, M., and Smith, J. (1970). *Carnegie Inst. Washington, Year b.* **68**, 400–402.
Brown, D. D., Wensink, P. C., and Jordan, E. (1971). *Proc. Nat. Acad. Sci. U.S.* **68**, 3175–3179.
Chargaff, E. (1955). *In* "Nucleic Acids" (E. Chargaff and J. N. Davidson, eds.), Vol. I, pp. 308–371. Academic Press, New York.
Chargaff, E., and Zamenhof, S. (1948). *J. Biol. Chem.* **173**, 327–335.
Chun, E. H. L., Vaughan, M. H., and Rich, A. (1963). *J. Mol. Biol.* **7**, 130–141.
Cummings, D. E. (1972). *Exp. Cell Res.* **71**, 106–112.
Davern, G. I. (1966). *Proc. Nat. Acad. Sci. U.S.* **55**, 792–797.
Davidson, N., Widholm, J., Nandi, U. S., Jensen, R., Olivera, B. M., and Wang, J. C. (1965). *Proc. Nat. Acad. Sci. U.S.* **53**, 111–118.
Dawid, I. B. (1965). *J. Mol. Biol.* **12**, 581–599.
Dawid, I. B., Brown, D. D., and Reeder, R. H. (1970). *J. Mol. Biol.* **51**, 341–360.
Finkelstein, D. B., Blamire, J., and Marmur, J. (1972). *Biochemistry* **11**, 4853–4858.
Greenberg, J. R., and Perry, R. P. (1971). *J. Cell Biol.* **50**, 774–786.
Gulland, J. M., Jordan, D. O., and Threlfall, C. J. (1947). *J. Chem. Soc.* 1129–1130.
Harrison, P. R. (1971a). *Biochem. J.* **121**, 27–31.
Harrison, P. R. (1971b). *Eur. J. Biochem.* **19**, 309–310.
Hudson, B., Upholt, W. B., Devinny, J., and Vinograd, J. (1969). *Proc. Nat. Acad. Sci. U.S.* **62**, 813–820.
Huez, G., Zampetti-Bosseler, F., and Brachet, J. (1972). *Nature (London) New Biol.* **237**, 155–157.
Irving, C., and Veazey, R. A. (1968). *Biochim. Biophys. Acta* **166**, 246–248.
Ishida, T., Kan, J., and Kano-Sueoka, T. (1971). *In* "Nucleic Acid Research" (G. L. Cantoni and D. R. Davies, eds.), Vol. II, pp. 608–617. Harper, New York.
Ittel, M. E., Wintzerith, M., Zahnd, J. P., and Mandell, P. (1970). *Eur. J. Biochem.* **17**, 415–424.
Jones, A. S. (1953). *Biochim. Biophys. Acta* **10**, 607–612.
Josse, J., and Eigner, J. (1966). *Annu. Rev. Biochem.* **35**, 789–834.
Kay, E. R. M., Simmons, N. S., and Dounce, A. L. (1952). *J. Amer. Chem. Soc.* **74**, 1724–1726.
Kirby, K. S. (1957). *Biochem. J.* **66**, 495–504.
Kirby, K. S. (1961). *Biochim. Biophys. Acta* **47**, 18–26.
Kirby, K. S. (1964). *Progr. Nucl. Acid. Res. Mol. Biol.* **3**, 1–32.
Kirby, K. S. (1968). *In* "Methods in Enzymology," Vol. 12: Nucleic Acids (L. Grossman and K. Moldave, eds.), Part B, pp. 87–98. Academic Press, New York.
Klett, R. P., and Smith, M. (1968). *In* "Methods in Enzymology," Vol. 12: Nucleic Acids (L. Grossman and K. Moldave, eds.), Part B, pp. 112–115. Academic Press, New York.
Koch, J. (1972). *Eur. J. Biochem.* **26**, 259–266.
Kolodner, R., and Tewari, K. K. (1972). *J. Biol. Chem.* **247**, 6355–6364.
Kram, R., Botchan, M., and Hearst, J. E. (1972). *J. Mol. Biol.* **64**, 103–117.
Leslie, I. (1955). *In* "Nucleic Acids" (E. Chargaff and J. N. Davidson, eds.), Vol. II, pp. 1–50. Academic Press, New York.
Levinthal, C., and Davidson, P. F. (1961). *J. Mol. Biol.* **3**, 674–683.
McBurney, M. W., and Whitmore, G. F. (1972). *Biochem. Biophys. Res. Commun.* **46**, 898–904.
McCallum, M., and Walker, P. M. B. (1967). *Biochem. J.* **105**, 163–169.
McGrath, R. A., and Williams, R. W. (1966). *Nature (London)* **212**, 534–535.
Mandell, J. D., and Hershey, A. D. (1960). *Anal. Biochem.* **1**, 66–77.

Manning, J. E., Wolstenholme, D. R., and Richards, O. C. (1972). *J. Cell Biol.* **53**, 594–601.
Marko, A. M., and Butler, G. C. (1951). *J. Biol. Chem.* **190**, 165–176.
Marmur, J. (1961). *J. Mol. Biol.* **3**, 208–218.
Marmur, J., and Doty, P. (1962). *J. Mol. Biol.* **5**, 109–118.
Massie, H. R., and Zimm, B. H. (1965). *Proc. Nat. Acad. Sci. U.S.* **54**, 1641–1643.
Meselson, M., Stahl, F. W., and Vinograd, J. (1957). *Proc. Nat. Acad. Sci. U.S.* **43**, 581–588.
Miyazawa, Y., and Thomas, C. A. (1965). *J. Mol. Biol.* **11**, 223–236.
Muramatsu, M. (1970). *In* "Methods in Cell Physiology" (D. Prescott, ed.), Vol. IV, pp. 195–228. Academic Press, New York.
Nass, N. M. K., and Ben-Shaul, Y. (1972). *Biochim. Biophys. Acta* **272**, 130–136.
Noll, H., and Stutz, E. (1968). *In* "Methods in Enzymology," Vol. 12: Nucleic Acids (L. Grossman and K. Moldave, eds.), Part B, pp. 129–155. Academic Press, New York.
Okuhara, E. (1970). *Anal. Biochem.* **37**, 175–178.
Papaconstantinou, J., Bradshaw, W. S., Chin, E. T., and Julku, E. M. (1972). *Develop. Biol.* **28**, 649–661.
Petes, T. D., and Fangman, W. L. (1972). *Proc. Nat. Acad. Sci. U.S.* **69**, 1188–1191.
Pitout, M. J., and Potgieter, D. J. J. (1968). *Biochim. Biophys. Acta* **161**, 188–196.
Quetier, F., and Guille, E. (1968). *Arch. Biochem. Biophys.* **124**, 1–11.
Radloff, R., Bauer, W., and Vinograd, J. (1967). *Proc. Nat. Acad. Sci. U.S.* **57**, 1514–1521.
Ralph, R. K., and Bellamy, A. R. (1964). *Biochim. Biophys. Acta* **87**, 9–16.
Ritossa, F., and Spiegelman, S. (1965). *Proc. Nat. Acad. Sci. U.S.* **53**, 737–745.
Savitsky, J. P., and Stand, F. (1966). *Biochim. Biophys. Acta* **114**, 419–422.
Schildkraut, C. L., Marmur, J., and Doty, P. (1962). *J. Mol. Biol.* **4**, 430–443.
Sevag, M. G., Lackman, D. B., and Smolens, J. (1938). *J. Biol. Chem.* **124**, 425–436.
Smith, C. A., Jordan, J. M., and Vinograd, J. (1971). *J. Mol. Biol.* **59**, 255–272.
Smith, D. A., Martinez, A. M., and Ratliff, R. L. (1970). *Anal. Biochem.* **38**, 85–89.
Stern, H. (1968). *In* "Methods in Enzymology," Vol. 12: Nucleic Acids (L. Grossman and K. Moldave, eds.), Part B, pp. 100–112. Academic Press, New York.
Sueoka, N., and Cheng, T. Y. (1962). *J. Mol. Biol.* **4**, 161–172.
Sussman, R., and Rayner, E. F. (1971). *Arch. Biochem. Biophys.* **144**, 127–137.
Szybalski, W., and Szybalski, E. H. (1971). *In* "Procedures in Nucleic Acid Research" (G. L. Cantoni and D. R. Davies, eds.), Vol. II, pp. 311–354. Harper, New York.
Tautvydas, K. J. (1971). *Plant Physiol.* **47**, 499–503.
Terasima, T., and Tsuboi, A. (1969). *Biochim. Biophys. Acta* **174**, 309–314.
Thomas, C. A., Berns, K. I., and Kelley, T. J. (1968). *In* "Methods in Enzymology," Vol. 12: Nucleic Acids (L. Grossman and K. Moldave, eds.), Part B, pp. 535–540. Academic Press, New York.
Thompson, W. F., and Cleland, R. (1971). *Plant Physiol.* **48**, 663–670.
Travaglini, E. C., and Meloni, M. L. (1962). *Biochem. Biophys. Res. Commun.* **7**, 162–166.
Travaglini, E. C., and Schultz, J. (1972). *Genetics* **72**, 441–450.
Travaglini, E. C., Petrovic, J., and Schultz, J. (1972). *Genetics* **72**, 419–430.
Vinograd, J., Morris, J., Davidson, N., and Dove, W. F., Jr. (1963). *Proc. Nat. Acad. Sci. U.S.* **49**, 12–17.
Watson, J. D., and Crick, F. H. C. (1953a). *Nature* (*London*) **171**, 737–738.
Watson, J. D., and Crick, F. H. C. (1953b). *Nature* (*London*) **171**, 964–967.
Weintraub, H., and Holtzer, H. (1972). *J. Mol. Biol.* **66**, 13–35.
Williamson, R. (1969). *Anal. Biochem.* **32**, 158–160.
Wintersberger, E., and Tuppy, H. (1964). *Biochem. Z.* **341**, 399–408.
Wray, W., Stubblefield, E., and Humphrey, R. (1972). *Nature* (*London*), *New Biol.* **238**, 237–238.
Zweidler, F., and Cohen, L. (1971). *J. Cell Biol.* **51**, 240–248.

Chapter 6

Electron Microscopic Visualization of DNA in Association with Cellular Components

JACK D. GRIFFITH

Department of Biochemistry, Stanford University, School of Medicine, Stanford, California

I. Introduction

The increasing interest in the study of DNA–protein complexes of both prokaryotic and eukaryotic cells has underscored the need for simple electron microscopic methods for visualizing them. Electron microscopy can provide both a means of monitoring the purification and a tool for delineating the structure of DNA–protein complexes. The object of this chapter is to describe two simple microscopic techniques, one standard and one new, for this purpose. Examples from the author's work will be discussed to illustrate the variety of ways these methods can be used.

II. Isolation of DNA–Protein Complexes

The purification of eukaryotic chromatin has been recently reviewed by Bonner *et al.* (1968). Chromatins are best prepared from lysates of purified nuclei. Triton X-100 treatment aids in the removal of contaminating cytoplasm but appears to release nuclease activity (J. Huberman, personal communication, 1971). Crude or purified chromatins precipitate in 0.15 *M* NaCl, a useful washing step. High speed Waring blending allows the preparation of a soluble relatively homogeneous material termed nucleohistone (Bonner *et al.*, 1968). This material may, however, have lost some of the structural properties characteristic of native chromatin.

In some instances, chromatins must be prepared from whole-cell homogenates—if, for example, it is difficult to obtain a good yield of nuclei. An example of such a procedure is the isolation of pea plant chromatin described by Fambrough and Bonner (1966).

Specific DNA–protein complexes can be isolated from virus-infected nuclei or from phage-infected bacteria. A method for purifying a polyoma DNA–protein complex from nuclei of infected cells has been developed by Green *et al.* (1971). Pratt *et al.* (1972) have described the purification of a small phage DNA–protein complex from spheroplasts of *E. coli*.

To be useful as an aid in purification, a microscopic technique should be rapid and should visualize DNA–protein complexes and contaminants equally well. Heavy metal staining, though rapid, can introduce troublesome artifacts and has proved too selective (Griffith, 1969). The two methods described here are quick and can be used in a complementary manner to provide a representative picture of the sample.

III. Electron Microscopy

A. Protein Surface Spreading

Denatured protein films were first used by Kleinschmidt and Zahn (1959) to spread and visualize DNA.

Many cellular structures in addition to DNA can be visualized by this technique. Controlled changes in the ionic strength or temperature during spreading add useful parameters for the study of the structure and associations of one or more cellular components.

Surface spreading to visualize components other than DNA has not been emphasized. Several examples, including the penetration of the filamentous phage M13, will help illustrate the more general use of this technique.

The recent article of Davis, Simon, and Davidson (1971) discusses variations of the Kleinschmidt procedure now in use. These differ in the presence or in the absence of the reagent formamide included to spread single-stranded DNA. The conditions for these two procedures are summarized here for the reader's convenience.

AQUEOUS SPREADING

Spreading vessel	A teflon dish 1 inch deep by 2 inches in diameter and a dry, acid-cleaned glass slide
Sample phase	Tris or ammonium acetate pH 8, 0,5 *M*, cytochrome *c* 100 μg/ml, sample 1 μg/ml
Vessel phase	Tris or ammonium acetate 0.01 to 0.1 *M*, pH 8
Supporting grids	200-mesh grids covered with a film of 3% Parlodion in amyl acetate formed on a water surface
Permissible conditions	Both sample and vessel phase can be varied over a wide range of ionic strengths; the sample in 0.4 *M* to 4.0 *M* ammonium acetate may be spread over a 0.01 to 0.2 *M* lower phase

A 30 μl to 50 μl quantity of the sample phase is spread slowly down the glass slide onto the vessel solution. The sample-containing film is picked up with a Parlodion-coated grid and dehydrated in 90% ethanol for 30 seconds.

Best contrast enhancement is achieved by platinum shadowing at an angle of 1:8 with rotation.

This procedure is much less sensitive to the presence of detergents than the formamide method, and spreading can be carried out at much higher ionic strength.

FORMAMIDE SPREADING[a]

Sample phase	Tris 0.1 *M* pH 8.5, EDTA 0.01 *M*, cytochrome *c* 100 μg/ml, 40% formamide, sample 1 μg/ml
Vessel phase	Tris 0.01 *M* pH 8.5, EDTA 0.001 *M*, 10% formamide
Permissible conditions	Best results are obtained when the sample is in a low ionic strength buffer. This procedure is required for the visualization of single-stranded DNAs. It is very sensitive to the presence of detergents

[a]From Davis *et al.* (1971).

B. Direct Visualization: Tungsten Contrast Enhancement

Kleinschmidt's technique solves two problems encountered in visualizing the 20 Å DNA duplex. The denatured protein increases the contour width from 20 Å to over 250 Å. This complex is easy to visualize by the simplest shadowing methods. In addition, surface forces and bound protein spread the DNA into a much more open conformation. The 250 Å protein coat, however, masks any other protein bound to the DNA backbone. Repressors or polymerases would be hidden and DNA uniformly complexed with histone will not bind cytochrome to form the Kleinschmidt complex.

The procedures described below provide a rapid and reproducible method for visualizing bare DNA and DNA–protein complexes, as well as many other cellular structures. Visualization is accomplished by binding the sample to a reactive supporting film, and following dehydration, contrast enhancement by vacuum deposition of a staining amount of tungsten.

1. Supporting Films: Activation

Thin carbon films formed in the manner illustrated in Fig. 1 (a–d) are both smoother and, for a given tensile strength, less electron opaque than colllodion or Formvar films. A 1% solution of polybutene_{6000} in xylene has been useful as a glue to bind the film to the copper mesh.

DNA and proteins bind poorly to carbon films formed in the manner described above. The sample must be linked in some way to the hydrophobic carbon film. This may be done by using a large molecule having both a hydrophobic portion and a DNA or protein binding site.

The method used in the author's work and in other laboratories is to form such amphiphilic molecules in the vacuum system by high voltage discharge. The grids are placed in a system rough pumped with Kinney Super X oil at a vacuum not above 300 μ or below 100 μ. A 10,000 to 20,000 V potential placed across two separated plates in the system for 5 minutes is sufficient to produce grids which will bind DNA for 20 minutes. This is presumably producing amphiphilic molecules from the oil vapor.

Koller *et al.* (1969) have described the use of a dilute solution of alkyl dimethyl benzyl-ammonium chloride for the same purpose. We have used the charged polymer trimethyl ammonium polyethylene glycol 6000 described by Albertsson (1971) in a similar way.

2. Mounting and Dehydration

A drop of the sample is placed on an activated grid for 1 to 5 minutes.

Dehydration may be accomplished rapidly by direct immersion into 100% ethanol, or more slowly through a graded series of ethanol solutions. Anderson's critical point drying can be used, but in practice the requirement for cleanliness has made it difficult to obtain good results on DNA–protein samples with this method.

3. Contrast Enhancement

Biological samples are normally shadowed with heavy metals, such as gold or platinum, to provide the greatest electron contrast. This simple shadowing relies on the differential deposition of the metal on objects exposed to or protected from the metal source.

A different phenomenon, selective nucleation, has turned out, however, to be useful. If a small amount of a reactive metal such as tungsten is evaporated under conditions similar to those used for gold or platinum shadowing, it might be expected to bind selectively to DNA (e.g., form phosphotungstate) or to protein rather than to the inert carbon film, i.e., result in vapor-

FIG. 1. Grid preparation. (a) With a razor blade, a sheet of mica is cleaved to obtain a fresh surface. (b) A thin layer of carbon is deposited by vacuum evaporation. (c) After 1 hour in a humid atmoshpere, the carbon film is stripped off onto a water trough. (d) Copper screens dipped in adhesive are dropped onto the floating carbon film and rescued with a Cellophane–petri dish drum. (e) Vacuum evaporation apparatus: rotating table (T), tungsten filament (W), and evaporation monitor (M). The evaporation monitor consists of two Teflon rods slotted to hold a 22 mm × 22 mm glass microscope coverslip. The edges of the coverslip are painted with silver conductive paint. Leads from a resistance meter run through a high vacuum feedthrough to the slots. A resistance change across the coverslip of greater than 5×10^7 ohms to 5×10^4 ohms in this apparatus corresponds to the evaporations used in Fig. 3.

phase positive staining. This effect would be expected to dominate only as long as the sample is relatively free of the metal. Once a monolayer of metal is deposited on it, then only simple shadowing will occur.

This effect has been studied using tungsten. When small amounts are evaporated onto a DNA–protein sample, selective binding or nucleation does appear to occur such that the buildup is greatest on protein, somewhat less on DNA, and much less on the carbon support.

The following conditions are used in the shadowing apparatus illustrated in Fig. 1e.

A straight 20 mil tungsten wire is clamped between two electrodes with a 3.0 cm separation. A current is determined such that, when set, the wire will heat, evaporate, and break in 5 to 7 minutes. The wire is placed 8 cm from the sample. Because selective nucleation rather than shadowing is being used, wire-sample angles between 8° and 45° appear to be equally effective. The sample is rotated during the evaporation. An evaporation monitor is valuable for reproducing optimal levels. The monitor illustrated in Fig. 1e has worked well for this purpose.

IV. Direct Visualization of Defined Complexes

Complexes of protein with purified DNAs carrying a known number and distribution of binding sites for that protein have been useful in studying the fidelity of the direct visualization technique. These studies have also uncovered new details of certain DNA–protein interactions.

Purified DNAs prepared for visualization by the two methods described above are contrasted in Fig. 2. The contour widths measured on the micrographs are 30 Å (a) and 250 Å (b). This helical DNA, indistinguishable by these methods from DNA purified from natural sources, was formed from the polynucleotides $d(A)_{4000}$ and $d(T)_{200}$ (T to A ratio of 0.75:1).

This defined DNA molecule carries 20 3′-hydroxyl termini at spacings of roughly 700 Å. It has been shown that under conditions of DNA synthesis, *Escherichia coli* DNA polymerase I, a molecule of 109,000 daltons and 65 Å diameter binds to these 3′ termini and that the complexes formed are stable

FIG. 2. Visualization of $d(A)_{4000}$:$d(T)_{200}$. The ratio of T residues to A residues was 0.75:1. (a) Direct visualization with tungsten. (b) Prepared by the aqueous surface spreading method, platinum shadowed. (a) and (b) are shown here at the same magnification for comparison of the two techniques.

These and the following micrographs were printed on Kodak Ortho-A-Litho sheet film followed by printing on a soft grade paper to obtain the optimum balance of contrast and detail.

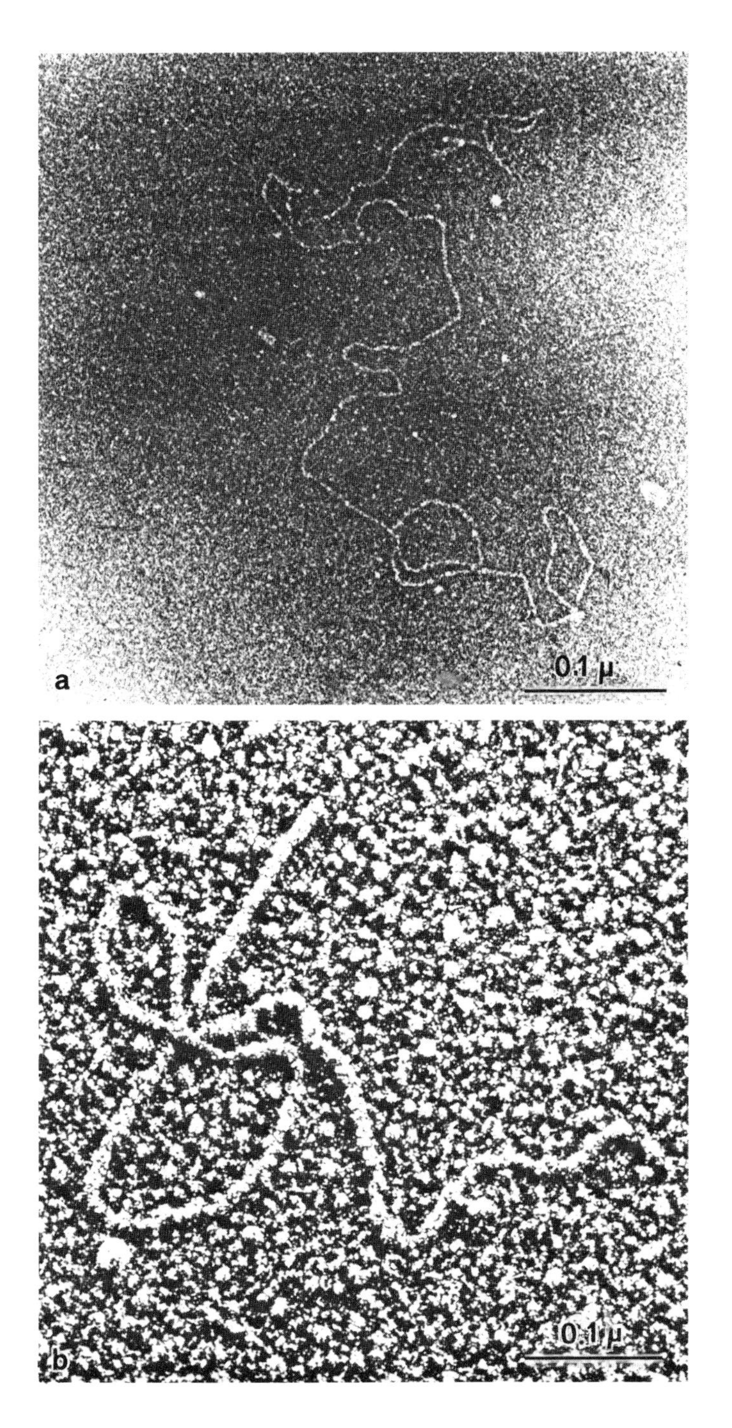
a
0.1 μ
b
0.1 μ

at low ionic strengths (Englund *et al.*, 1969). Such complexes were formed at polymerase to termini ratios of from 0.5 to 2.0 and were visualized by tungsten evaporation (Fig. 3a). Dimers of DNA polymerase I are formed in the presence of mercury and retain their specificity of binding (Fig. 3b) (Griffith *et al.*, 1971; Jovin *et al.*, 1969).

The correspondence of the regular polymerase spacing (mean of 680 Å) with the termini separation shows that the specific DNA–protein interaction is preserved during sample preparation.

Aggregation of DNA or DNA complexes is commonly encountered during preparation of samples for electron microscopy. This may be reduced by a variety of methods: (1) The DNA may be diluted to a concentration of 10 μg/ml or less, heated to 60°C, a nondenaturing temperature, for 5 minutes and cooled just prior to forming the complex. (2) Aggregation is reduced in the presence of EDTA and is reduced at low ionic strength. (3) Formamide may be added to the complexes (to 30%) just prior to mounting. The latter treatment has been shown to have no effect on the specificity of DNA polymerase I binding (Griffith *et al.*, 1971).

In many cases, unlike DNA polymerase I, the specific interaction of a protein with DNA is favored at moderate ionic strengths in the presence of Mg^{2+}. Because high concentrations of salt may obscure the complex and magnesium causes DNA to appear condensed, the samples should be put into a dilute, magnesium-free buffer just prior to mounting.

The interaction of the lambda repressor, a protein of 160,000 daltons with 4 identical subunits, with lambda DNA carrying the lambda operator is well known (Ptashne and Hopkins, 1968). In collaboration with Reichart and Kaiser, complexes of lambda DNA and lambda repressor were formed under conditions of high DNA concentration (10 μg/ml), high salt (0.2 *M* KCl) and high Mg^{2+} (5 m*M*) at 37°C for specific binding and then diluted 1:100 in a dilute Mg^{2+}-free buffer for mounting (Griffith *et al.*, 1972).

Under these conditions, lambda repressor was observed bound to the operator site as determined by its distance from the nearest end of the DNA molecule (Fig. 3c). Similar specific binding has been visualized with the galactose repressor and a ϕ80dlac DNA (Ptashne and Gilbert, 1970) (Fig. 3d) and with *E. coli* RNA polymerase and the phage T7 DNA (Griffith and Chamberlin, 1972).

Murti, Prescott, and Pene (1972) have used similar methods to locate RNA polymerase binding sites on the DNA of the protozoan *Stylonychia*.

These results demonstrate the capability of the direct visualization technique for locating even a single site of interaction of a protein and a large DNA. Because of the opportunity for nonspecific binding, the use of this technique for mapping must be carefully controlled.

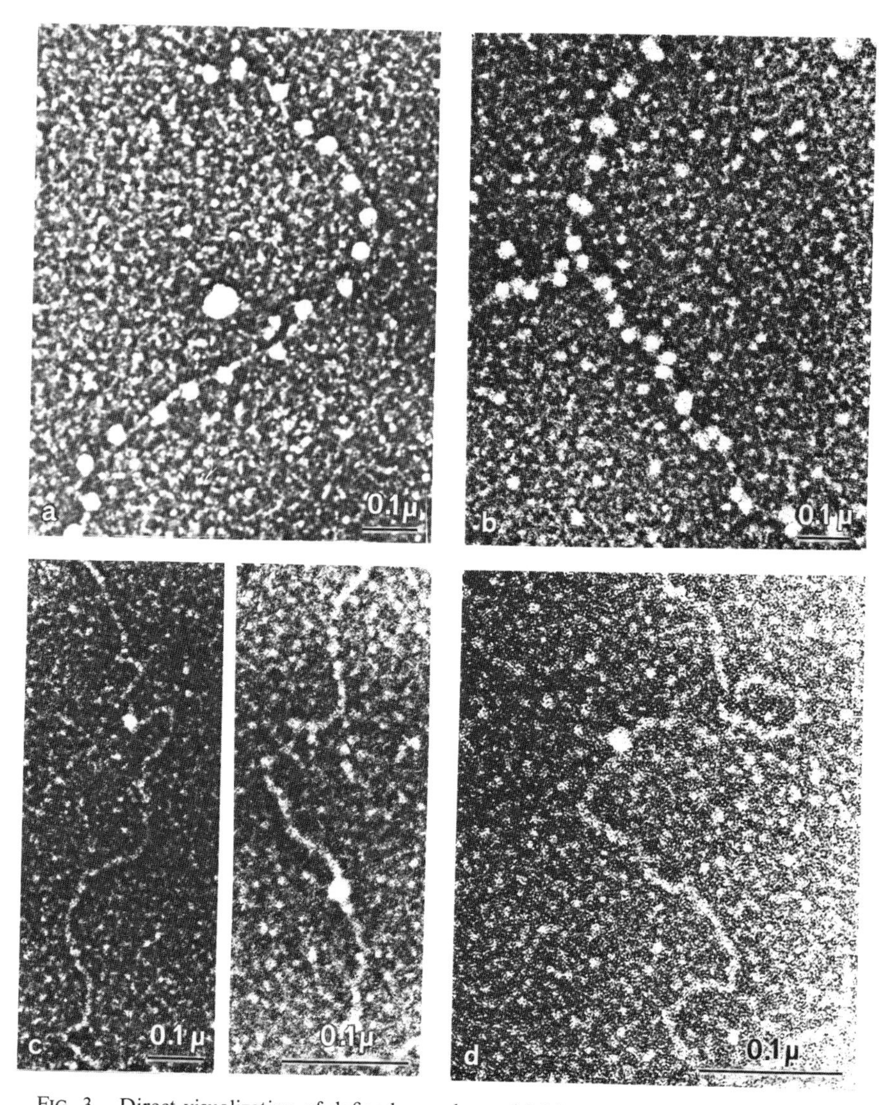

FIG. 3. Direct visualization of defined complexes. (a) Monomers of *Escherichia coli* DNA polymerase I bound to the synthetic DNA poly (dA_{4000}): poly (dT_{200}). (b) Dimers of DNA polymerase I bound to the same DNA. The uniform spacing of the polymerase I molecules is due to the even spacing of polymerase I binding sites along the DNA. (c) Two examples of lambda repressor bound to the operator site of lambda DNA (d) Lac repressor bound to the operator region of a ϕ80dlac DNA.

Samples (a) and (b) were brought to 30% formamide just prior to spreading. Dehydration was accomplished by direct immersion into 100% ethanol (a) and (b), and by graded ethanol series (c) and (d).

V. Phage-Membrane Association: A Combination of Two Methods

The filamentous phage M13 is a single-stranded DNA phage whose host is male strains of *E. coli*. The entry and extrusion of M13 is intimately associated with the bacterial envelope. The life cycle has been recently reviewed by Marvin and Hohn (1969).

The structure of the phage is that of a filament, 9750 Å in length and 60 Å in diameter, consisting of a circular single-stranded DNA (2.3 μ in length) and one major capsid protein. The phage can be visualized both by surface spreading and by tungsten evaporation (Fig. 4). Chloroform is known to inactivate M13 without disassociating any of the major phage components (Williams and Fenwick, 1967). Chloroform-treated phage visualized by tungsten evaporation appear collapsed. However, when visualized by formamide surface spreading, the treated phage open into circles roughly half the width of untreated phage (Fig. 4b). Thus the strong surface forces involved provide a means of dissecting and identifying the basic structure of M13, possibly that of a thin DNA–protein torus (Griffith and Kornberg, 1972).

M13 associates with the bacterial F pilus in its first step of entry into the cell. Purified M13 and F pilus fragments mixed and visualized by tungsten evaporation are shown in Fig. 4d. The thin phage and thicker pilus can be differentiated, and a wrapping of the phage about the pilus is observed. This wrapping is not resolved by surface spreading or by simple platinum shadowing.

Methods have been recently described for separating inner and outer membranes of *E. coli* by sucrose density centrifugation (Schnaitman, 1970; Osborn *et al.*, 1972). These layers can be distinguished by the two microscopic techniques (Fig. 5). Intermediates in M13 penetration, replication, and extrusion have been visualized in association with inner or outer membrane by these methods. Two examples are shown in Fig. 5. Rifampicin pretreatment of cells allows M13 adsorption but blocks its uncoating (Brutlag *et al.*, 1971). Such phage are seen bound to a portion of the cell outer membrane. Partially extruded phage can be purified bound to a fragment of the cell envelope.

Using one or a combination of the two methods, one can distinguish single-stranded DNA, duplex DNA, or inner membrane and outer membrane in association with each other. Another dimension could be added by coupling these methods with specific antibody labeling.

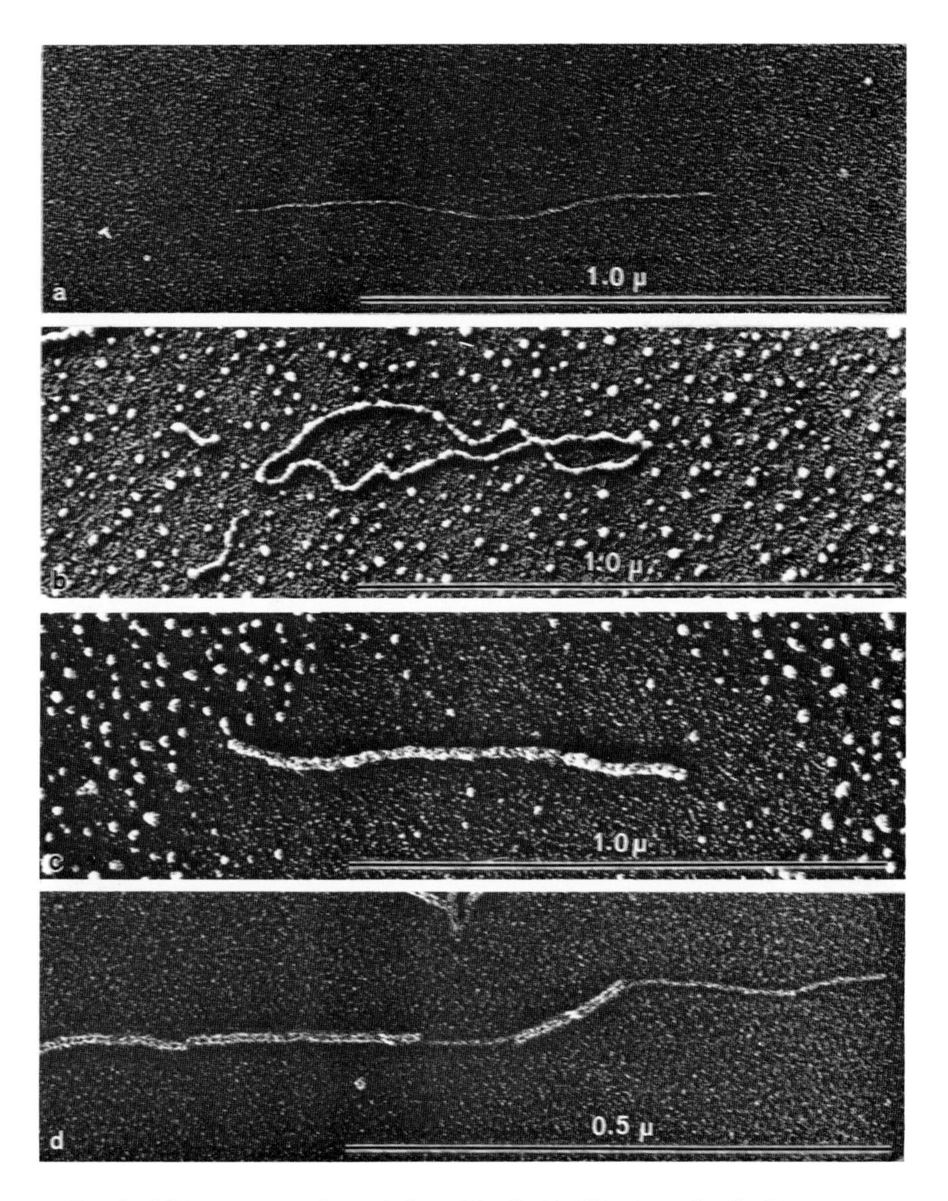

FIG. 4. M13 structure and association with pili. (a) M13 phage visualized by tungsten evaporation. (b) Treated with chloroform and surface spread (formamide technique). (c) Formamide surface spread without chloroform treatment. (d) M13 in association with F pili fragments. Tungsten evaporation (a) and (d); plantinum shadowing (b) and (c).

VI. Eukaryote Chromatin Fibers

The two techniques outlined above can be valuable in studies of isolated chromatins. The purity of the preparations, extent of aggregation, and degree of DNA degradation can be monitored. In addition, they make possible direct studies of chromatin structure. A number of examples from recent work of the author in collaboration with Dr. James Bonner illustrate a variety of ways in which these methods can be used (Griffith, 1969).

Purification of chromatins follows the procedures cited above (Bonner *et al.*, 1968). Membrane fragments comprise a major contaminant of crude nuclear lysates and are easily visualized by surface spreading. Chemical

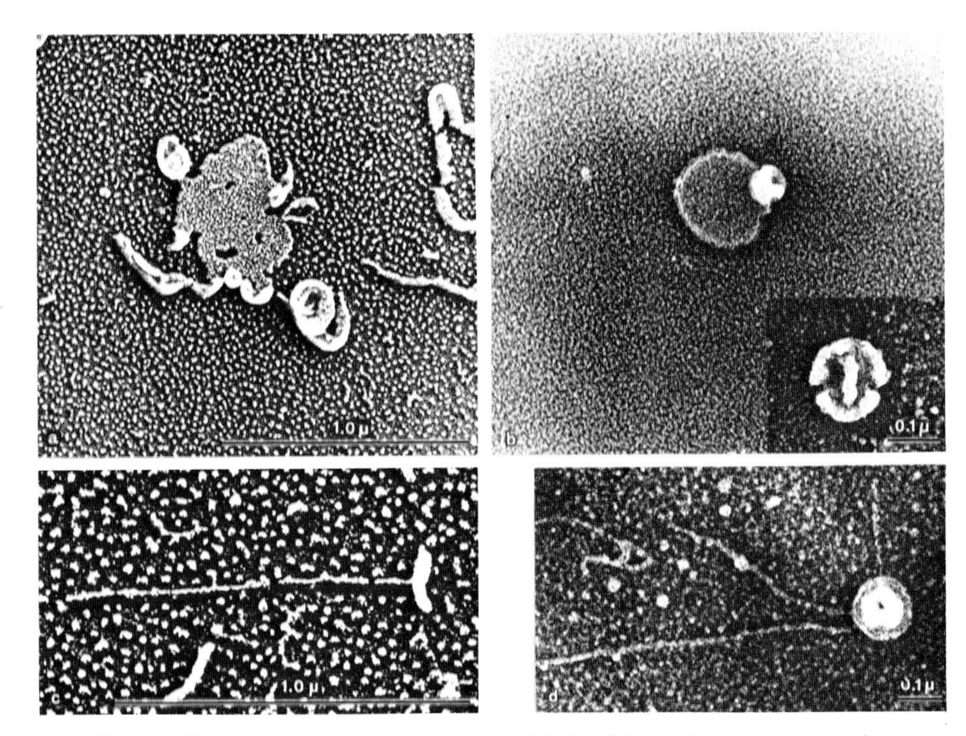

FIG. 5. Visualization of membrane structures. (a) Material from the dense outer membrane band prepared by sucrose density separation of *Escherichia coli* membranes (Osborn *et al.*, 1972) visualized by formamide surface spreading. (b) Material from the intermediate density, associated inner and outer membrane band and (inset) light inner membrane band visualized by tungsten evaporation. (c) M13 phage associated with an outer membrane fragment isolated from rifamycin-treated cells, surface spread. (d) Complex of membrane and extruding M13 isolated from 1-hour infected cells, direct tungsten visualization.

analysis for RNA content provides an estimate of the maximal ribosomal contamination.

Interphase chromatin from many tissues including the following was purified and visualized by tungsten evaporation: rat liver, pea bud, calf thymus, chick erythrocytes and the protozoan *Tetrahymena pyriformis*. In all tissues, with two exceptions, three general classes of chromosomal fibers have been found: (1) large macro fibers 230 Å in diameter; these fibers show a periodicity of condensed substructure; (2) long uniform fibers 30 Å in width; these fibers are often found in a supercoiled configuration, the supercoil having a pitch of 120 Å; (3) fibers roughly 30 Å in diameter characterized by the presence of attached knobs 100–250 Å in diameter. Examples of these types are shown in Fig. 6.

The presence and appearance of these three classes is not visibly affected by prior fixation with formaldehyde or by mounting in the presence of formamide. We believe that the 30 Å fibers consist of a single DNA duplex complexed with the chromosomal histone and nonhistone proteins. The large macro fibers appear indistinguishable from the basic fibers of chromosomes described by DuPraw and others (DuPraw and Bahr, 1969).

Shearing to produce nucleohistone eliminates the 230 Å macro fibers and gives rise to the 30 Å uniform fibers.

Chromatin of different sources differ in their distribution of the three types of fibers. Thus, chromatin from chicken erythrocytes, a fully heterochromatic chromatin, and inactive as a template for *in vitro* RNA transcription by added *E. coli* RNA polymerase, consists predominantly of the 230 Å macro fibers. Chromatin from macronuclei of tetrahymena, a fully euchromatic source and highly active as a template for transcription exhibits no electron microscopic evidence of 230 Å macro fibers.

A third class of chromosomal fibers is easily distinguished by its possession of clustered knobs (Fig. 6a). The knobs may be found bound to single fibers in nuclear lysates as well as in the crude or purified chromatins and nucleohistone. They range in size up to 250 Å in diameter and vary in shape from spherical to oblong or prismatic. The knobs are usually clustered within gene-length distances of each other and are almost always separated by at least 200 Å. These knobs have been correlated with the nonhistone proteins by the following procedure.

Treatment of chromosomal fibers with 0.2 N H_2SO_4 removes 95% or more of the histone proteins leaving the nonhistone protein bound to chromosomal DNA (Bonner *et al.*, 1968). Nucleohistone was treated with acid in this manner followed by centrifugation into 1.2 M sucrose to separate the DNA–nonhistone protein complex from histone, and observed by direct visualization. This treatment does not remove the knobs (Fig. 7a). Other methods of

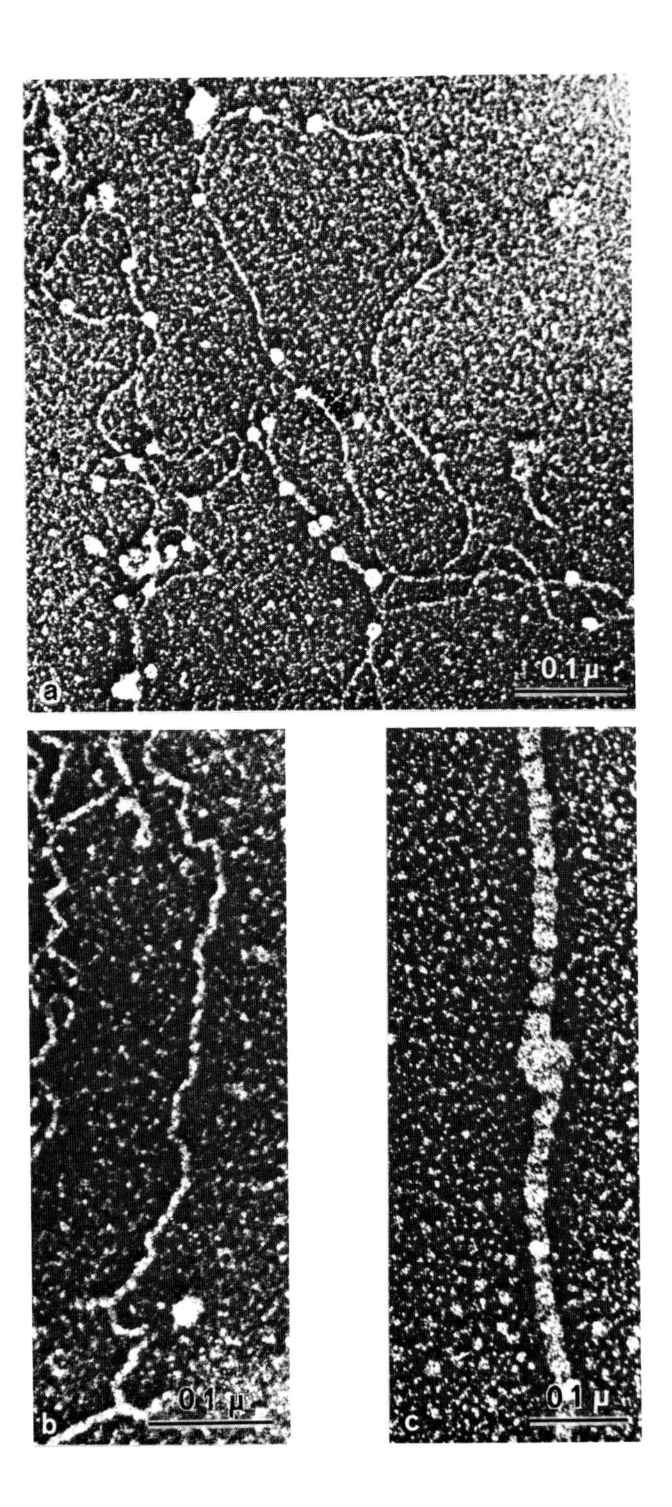
a
0.1 μ
b
0.1 μ
c
0.1 μ

removing total proteins from nucleohistone, centrifugation through 6 *M* CsCl, through 3 *M* NaCl, or treatment of chromatin at 0.1 mg of DNA per milliliter with preincubated pronase, 2 mg/ml for 2 hr at 37°C, does remove the knobs.

The surface spreading techniques also provide a means of visualizing regions of the chromosomal fibers free of associated chromosomal protein. When the sample phase contains salt and cytochrome *c* and the sample is spread on a hypotonic water trough, bare DNA binds a coat of denatured cytochrome. Because the binding is ionic, DNA which is already neutralized by bound protein should not form such a macrocomplex. The DNA–cytochrome is roughly 200 Å in diameter. When nucleohistone fibers are surface spread in this manner, any regions free of histone or nonhistone protein and long enough to undergo cooperative cytochrome binding (we estimate 200 Å or more) should form a complex with cytochrome *c* and be clearly detectable, since this region will be 150–200 Å thicker than the histone-covered DNA.

Nucleohistone and pure DNA of pea bud and rat liver were spread in the above described way, using a layering solution of 0.2 *M* NaCl (in which no histones and few nonhistone proteins dissociate from DNA) and a hypophase of 0.1 *M* ammonium acetate. Although pure DNA under these conditions forms the characteristic Kleinschmidt complex, nucleohistone does not, and exhibits no detectable cytochrome-binding regions 200 Å or more in length. It would appear that essentially all DNA of chromatin is complexed either with histone or with other proteins which prevent cytochrome from binding to it.

Spread from sample phases containing increasing concentrations of NaCl, the nucleohistone fibers bind increasing amounts of cytochrome in a uniform manner. To demonstrate that histone-free DNA would be isolated by these procedures, crude chromatin was prepared from purified nuclei of CV-1 cells (a stable cell line from African green monkey kidney) isolated 36 hours after a lytic SV40 infection. The chromatin was then spread by both Kleinschmidt procedures. At ionic strengths below that of 0.5 *M* NaCl some circular SV40 DNA molecules were visible. A portion of the lysate exposed to 2 *M* NaCl then returned to low ionic strength and spread is shown in Fig. 7b. The chromatin fibers are now free of histone and bind cytochrome in the same manner as the SV40 DNA.

The 230 Å macro fibers that comprise most of the chromatin of chicken

FIG. 6. Direct visualization of isolated chromatin fibers. Examples of the three classes of chromatin fibers observed in eukaryote chromatins, isolated from rat liver (a), (b), and from chicken erythrocytes (c). (a) Thin 20 Å to 30 Å fibers showing large protein knobs. (b) Uniform 30 Å fibers and (c) 250 Å macro fibers.

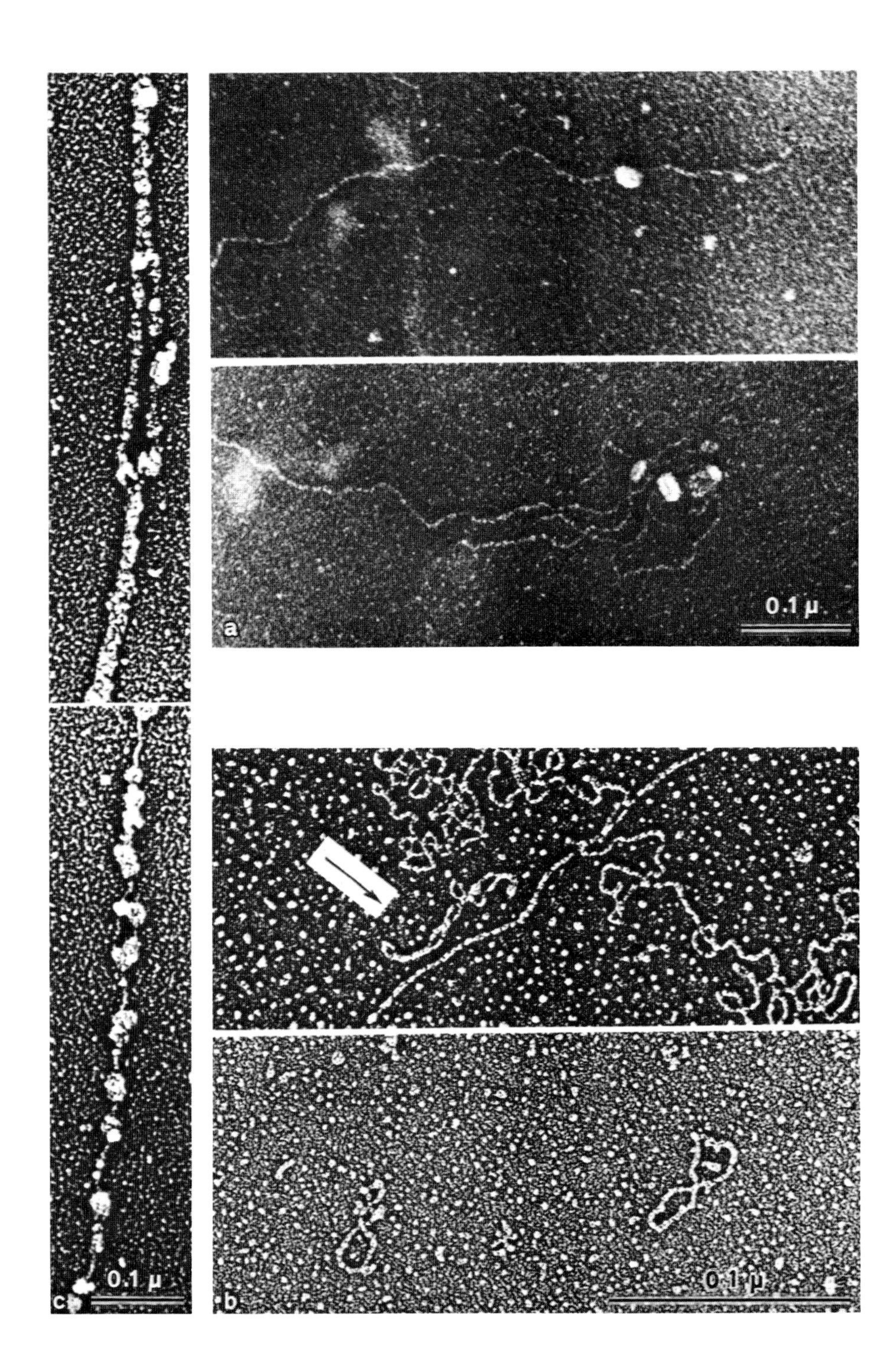
0.1 µ
a
0.1 µ
c
0.1 µ
b

erythrocyte nuclei are characteristic of the heterochromatic state. These fibers prove to be very labile to shearing. Their basic structure may be investigated by gently lysing the nuclei in a disrupting medium and visualization by tungsten evaporation. The examples shown in Fig. 7c shows that the macro fibers exhibit basic doubleness.

ACKNOWLEDGMENTS

This work was supported in part by awards from the National Institutes of Health and the National Science Foundation.

The author wishes to thank Dr. James Bonner, Dr. Benjamin Siegel, and Dr. Arthur Kornberg for providing the facilities and help which have made this work possible.

REFERENCES

Albertsson, P. (1971). "Partition of Cell Particles and Macromolecules." Almqvist & Wiksell, Stockholm.

Bonner, J., Chalkley, G. R., Dahmus, M., Fambrough, D., Fujimura, F., Huang, R. C., Huberman, J., Jensen, R., Marushige, K., Ohlenbusch, H., Olivera, B., and Widholm, J. (1968). *In* "Methods in Enzymology," Vol. 12: Nucleic Acids (L. Grossman and K. Moldave, Part B, p. 3. Academic Press, New York.

Brutlag, D., Schekman, R., and Kornberg, A. (1971). *Proc. Nat. Acad. Sci. U.S.* **69**, 1826.

Davis, R., Simon, M., and Davidson, N. (1971). *In* "Methods in Enzymology," Vol. 21: Nucleic Acids, Part D (L. Grossman and K. Moldave, eds.), p. 413. Academic Press, New York.

DuPraw, E. J., and Bahr, G. F. (1969). *Acta Cytol.* **13** (4), 188.

Englund, P. T., Kelly, R. B., and Kornberg, A. (1969). *J. Biol. Chem.* **244**, 3045.

Fambrough, D., and Bonner, J. (1966). *Biochemistry* **5**, 2563.

Green, M., Miller, H. I., and Hendler, S. (1971). *Proc. Nat. Acad. Sci. U.S.* **68**, 1032.

Griffith, J. D. (1969). Ph.D. Thesis, California Institute of Technology.

Griffith, J. D., and Chamberlin, M. (1972). Unpublished.

Griffith, J. D., and Kornberg, A. (1972). Unpublished.

Griffith, J. D., Huberman, J. A., and Kornberg, A. (1971). *J. Mol. Biol.* **55**, 209.

Griffith, J. D., Reichart, L., and Kaiser, A. D. (1972). Unpublished.

Jovin, T. M., Englund, P. T., and Kornberg, A. (1969). *J. Biol. Chem.* **244**, 3009.

Kleinschmidt, A. K., and Zahn, R. K. (1959). *Z. Naturforsch. B* **14**, 770.

FIG. 7. Selective disruption of chromatin fibers. (a) Isolated pea chromatin treated to remove histone protein. Prominent nonhistone knobs (see Fig. 6) remain. Direct tungsten visualization. (b) Chromatin from cells undergoing a lytic SV40 infection surface spread from a low ionic strength (0.1 *M* Tris) top phase with (top panel) and without (bottom panel) prior treatment with 2 *M* NaCl. The singular appearance of a portion of the SV40 DNA molecules without salt-effected histone removal is taken as evidence for the absence of histone-free regions on the native chromatin fibers. (c) Gentle disruption of 250 Å macro fibers from chicken erythrocyte nuclei. The nuclei are lysed directly on the microscope grid, briefly exposed to 0.1 *M* mercaptoethanol and dehydrated through graded ethanol series and visualized by tungsten evaporation.

Koller, T., Harford, A. G., Lee, Y. K., and Beer, M. (1969). *Micron* **1**, 110.
Marvin, D. A., and Hohn, B. (1969). *Bacteriol. Rev.* **33**, 172.
Murti, K. G., Prescott, D. M., and Pene, J. J. (1972). *J. Mol. Biol.* **68**, 413.
Osborn, K. J., Gander, J. E., Parisi, E., and Carson, J. (1972). *J. Biol. Chem.* **247**, 3962.
Pratt, D., Laws, P., and Griffith, J. (1972). *J. Mol. Biol.* In preparation.
Ptashne, M., and Gilbert, W. (1970). *Sci. Amer.* **222**, 36.
Ptashne, M., and Hopkins, N. (1968). *Proc. Nat. Acad. Sci. U.S.* **60**, 1282.
Schnaitman, C. A. (1970). *J. Bacteriol.* **104**, 882.
Williams, P. G., and Fenwick, M. L. (1967). *Nature* (*London*) **214**, 712.

Chapter 7

Autoradiography of Individual DNA Molecules

D. M. PRESCOTT AND P. L. KUEMPEL

Department of Molecular, Cellular and Developmental Biology, University of Colorado, Boulder, Colorado

I. Introduction

Autoradiography is frequently used in studies of chromosome structure and replication in which the properties of very high molecular weight DNA are being investigated. The technique that has usually been used in such studies is that originally described by Cairns (1962, 1963a,b). With this technique, Cairns was able to label the *Escherchia coli* chromosome with thymidine-^{3}H, gently extract the DNA from the cells with either Duponol (Cairns, 1962, 1963a) or lysozyme (Cairns, 1963b) treatment, affix the DNA to Millipore filters (50 nm pore size), and obtain autoradiographs of the chromosomes by covering the filters with stripping film. Up to 1% of the chromosomes were displayed as more or less tangled circles, and at least one autoradiographic pattern clearly displayed a chromosome in the process of replication as a closed loop (1963b). These experiments demonstrated that the chromosome has a circular structure; the length was approximately 1100–1400 μm. Other investigators have repeated these experiments and obtained essentially the same results (Bleecken *et al.*, 1966).

Cairns (1966) also applied this method of DNA autoradiography to

HeLa cells and estimated the rate of travel of the replication forks. The low rate of fork travel showed that DNA replication must be initiated at multiple points in each chromosome. Subsequently, Huberman and Riggs (1968) used this method with HeLa and Chinese hamster cells to obtain reasonably accurate measurements of the number and size of the replicating units in these chromosomes. They also showed that DNA replication proceeded in a bidirectional fashion from the individual replication origins.

Wake (1972) and Gyurasits and Wake (1972) have recently applied the Cairns procedure to studies of chromosome replication in *Bacillus subtilis*. These studies have shown that the chromosome is replicated bidirectionally over at least half of its total length, and that initiation occurs in a specific region of the chromosome. Rodriguez, Dalbey, and Davern (1972) have also recently used the Cairns procedure to demonstrate that the terminus of replication of the circular *E. coli* chromosome is directly opposite the origin. Consequently, replication must be bidirectional. Rodriguez *et al.* estimated that approximately 0.1% of the chromosomes were displayed in an interpretable fashion.

The basic method of DNA autoradiography has been greatly simplified by Lark, Consigli, and Toliver (1971), who repeated the studies of Huberman and Riggs. Instead of collecting the extracted DNA on Millipore filters, however, Lark *et al.* lysed the labeled mammalian cells with sodium dodecyl sulfate directly on microscope slides. The cell lysate was then smeared out on the slide and allowed to dry, rinsed in trichloroacetic acid, and coated with autoradiographic emulsion. With this procedure the lengthy DNA isolation step and the use of stripping film are eliminated. The method is simple and quick, but more important, much of the DNA is preserved in high molecular weight form and can be readily analyzed.

We (Prescott and Kuempel, 1972; Kuempel *et al.*, 1973) have used the method of Lark, Consigli, and Toliver to study the chromosomes of *E. coli*, T4 bacteriophage, vaccinia virus, and various eukaryotic cells. The method appears to be applicable to any virus or cell that can be treated with detergent to release the DNA. We describe here in detail how we used the method to analyze several kinds of chromosomes for the size of the replicating units, the direction and rate of travel of the replication forks, and the properties of the replication terminus.

II. Labeling of DNA

For autoradiography the DNA must be labeled to a high specific activity in order to obtain results in a reasonable time, e.g., 10 weeks or less. In practice the lower limit of specific activity useful for labeling *E. coli* thymine-requiring mutants is about 5 Ci/mmole of thymine-^{3}H or thymidine-^{3}H. This specific activity produces approximately 0.1 grain per μm of DNA per month with NTB3 emulsion. It is usually desirable to have at least one grain per 3 to 4 μm of DNA in order to have well-defined grain tracks. Much higher specific activities have been used in some experiments (50 Ci/mmole), and the required exposure times were correspondingly shorter. When such high specific activities are used, however, it must be recalled that prolonged growth in the presence of the thymine-^{3}H or thymidine-^{3}H can apparently cause some cell death (Cairns, 1962).

A specific activity of 5 Ci/mmole is also close to the lower practical limit for autoradiography of thymidine-^{3}H in DNA from various kinds of cultured eukaryotic cells (Chinese hamster, chick, duck, trout, *Drosophila*, and *Tetrahymena*). Thymidine auxotrophs are usually unavailable for eukaryotic cells, but the reduction by endogenous synthesis of the specific activity of thymidine-^{3}H utilized from the medium can be minimized by adding 5-fluorodeoxyuridine to the medium to inhibit the activity of thymidylate synthetase.

III. Spreading of DNA

Spreading is done on microscope slides coated with a thin layer of gelatin applied as a subbing solution [see Caro (1964)]. DNA released from cells with *n*-lauroyl sarcosine adheres much better to subbed slides than to clean glass. A drop (approximately 5 μl) of cells is placed at one end of the subbed slide. If bacteria are being used, the cell wall must first be digested with lysozyme (Kuempel, 1972). An equivalent amount of lysing solution is then placed adjacent to the drop of cells. We have achieved the best lysis and spreading using a 2% solution of *n*-lauroyl sarcosine containing 0.01 *M* EDTA and buffered at pH 8.1 with 0.01 *M* Tris. The two drops are allowed to fuse, and the mixture is left undisturbed for 10–15 minutes at room temperature to permit lysis of cells or spheroplasts and release of high molecular weight DNA. The cell lysate is then spread over the microscope slide with a second microscope slide in the same manner that a blood smear is made. This is accomplished by holding the second or "spreading" slide at an angle of approximately 15° to the subbed slide and slowly drawing the "spreading" slide up to the lysate until the lysate uniformly fills the acute angle between the two slides. The "spreading" slide is then slowly pushed in the opposite direction down the length of the slide, spreading the lysate behind it. The slides should then stand upright until dry.

IV. Autoradiography of DNA

After the lysate has dried, the slide is rinsed in 95% ethanol, air-dried again, washed in 5% trichloroacetic acid for 5 minutes at 0°C to remove unincorporated label, rinsed in several changes of 95% ethanol to remove all traces of the acid, and air-dried. Staining racks and staining dishes are very convenient for conducting these washing procedures. The slides are next dipped into autoradiographic emulsion. We routinely use Kodak NTB3 emulsion, but Kodak AR 10 stripping film is satisfactory. The grains produced with Ilford L-4 emulsion are too small to be easily observed by light microscopy at 250–400 × (see Fig. 2). Less sensitive emulsions are less useful for obvious reasons. The dipped slides are stored at room temperature in light-tight slide boxes sealed with black plastic tape. Some black plastic slide boxes are made of such thin plastic that they permit enough light to enter to expose the emulsion. Boxes should be tested with blank, dipped slides before use, or stored in the dark.

Autoradiographs are developed by one of the usual procedures. We use 2-minute development in Kodak D-11 or D-19 at 24°C followed by a rinse in water, fixation in acid fixer for 5–10 minutes, and washing in water for 20 minutes. For microscopic observation the autoradiographs are usually mounted with a coverslip, using one of the usual mounting media. We have used bright-field illumination for examining the autoradiographs.

V. Examples of Autoradiographs

Examples of autoradiographs of *E. coli* chromosomes are shown in Fig. 1. The chromosome in Fig. 1a was labeled with thymine-^{3}H (5 Ci/mmole) for several minutes beginning at the time of initiation of the DNA replication cycle and then for 8 minutes with thymidine-^{3}H (50 Ci/mmole). The autoradiograph was exposed for 6 weeks. The autoradiographic image is light (thymine-^{3}H) in the central region and dense for long regions at both ends (thymidine-^{3}H). The experiment demonstrates that the replication of the *E. coli* chromosome proceeds from the replication origin in bidirectional fashion. The two labeled daughter duplexes are discernible, and the two extremities of the autoradiographic pattern represent the positions of the two replication forks. Unlabeled parental DNA presumably extends from both ends of the grain track. Experiments of this type also demonstrate that the two replication forks travel at approximately the same rate, and the experiments can also be used to estimate the rate of fork travel (Prescott and Kuempel, 1972; Kuempel *et al.*, 1973).

Figure 1b shows the autoradiographic pattern of a chromosome of *E. coli* labeled with thymine-^{3}H (5 Ci/mmole) for 5 minutes and then with thymidine-^{3}H (50 Ci/mmole) for 2 minutes shortly before replication was about to be completed by the meeting of two replication forks in the circular chromosome. The autoradiographs of the daughter duplexes are clearly visible behind each replication fork. The two forks are apparently still separated by 3 μm of unreplicated DNA, which contains the replication terminus. We have observed a number of these patterns, and autoradiography should be very useful in determining whether or not the replication terminus is a defined genetic locus and for studying chromosome segregation in bacteria.

The spreading method has not been very useful for examining the full expanse of the chromosome of *E. coli* because the circular chromosome collapses into a linear configuration due to the spreading forces. We have had some success in visualizing the closed loop form of the *E. coli* chromosome by lysing spheroplasts on a slide as described above, mixing the lysate with liquid autoradiographic emulsion, and spreading the entire mixture on the slide. An autoradiograph of an intact chromosome prepared in this way is shown in Fig. 1c. This technique has the advantage of a 2- to 3-fold faster production of an autoradiographic image because the DNA is embedded in autoradiographic emulsion. The cells must be washed very well if this procedure is to be used, since the preparations cannot be washed on the slide to remove unincorporated thymine-^{3}H or thymidine-^{3}H.

An example of the application of our usual spreading technique to vaccinia virus is shown in Fig. 2. In this experiment chick cells incubated in the presence of 5-fluorodeoxyuridine to block viral DNA replication were infected with vaccinia virus. The inhibition was released 3 hours after infection by the addition of 100 μCi/ml of thymidine-^{3}H (50 Ci/mmole), and labeling was allowed to continue for 60 minutes. Labeling was stopped by replacing the medium with cold balanced solution. The monolayer of cells was scraped from the culture vessel and washed free of exogenous thymidine-^{3}H by 2 centrifugation washes in cold, balanced salt solution (600 *g* for 5 minutes). The cells were finally suspended in balanced salt solution and then lysed with *n*-lauroyl sarcosine and autoradiographed. Autoradiographic images of the type shown in Fig. 2 were scattered individually over the slide. In control experiments, cellular DNA of chick cells formed grain tracks arranged in tandem series as described by Huberman and Riggs (1968) for mammalian cells. The tandem arrangement reflects multiple replicating units in a single piece of cellular DNA.

Figure 3 shows an autoradiograph of DNA from cultured rainbow trout

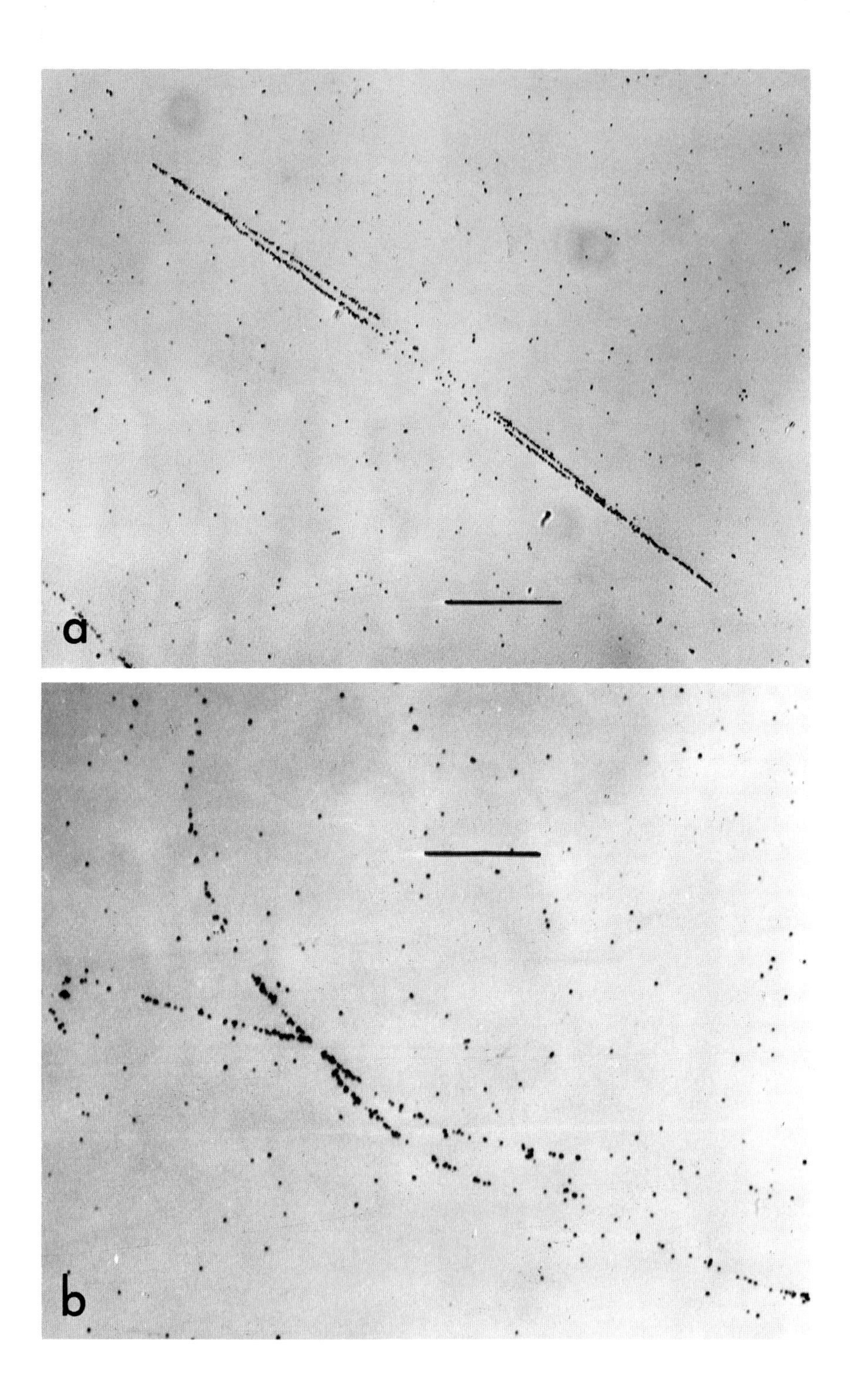
a
b

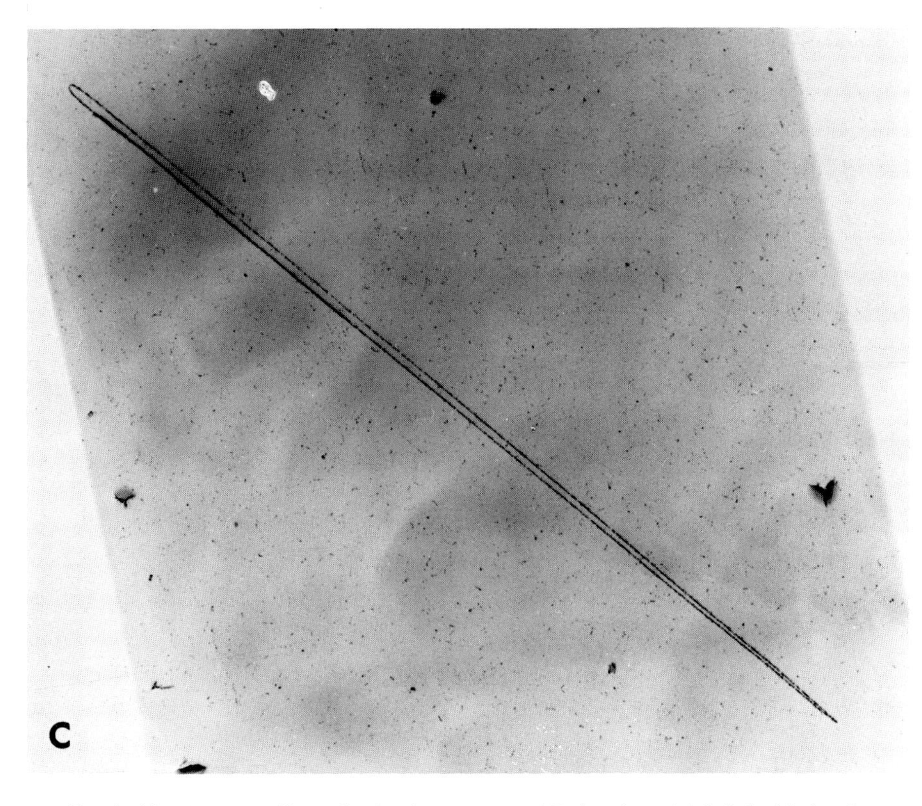

FIG. 1. (a) An autoradiograph of a chromosome of *Escherichia coli* labeled with thymine-^{3}H (5 Ci/mmole) for several minutes, beginning at the time of initiation of the DNA replication cycle, and then for 8 minutes with thymidine-^{3}H (50 Ci/mmole). The origin of chromosome replication is located in the region of the light, central autoradiographic image produced by the thymine-^{3}H. The denser, long images at both ends were produced by the thymidine-^{3}H. Replication forks are located at the two ends of the autoradiograph. The two daughter duplexes are discernible, the unlabeled parental DNA presumably extends from the two replication forks. Exposure time, 6 weeks. The bar is 60 μm.

(b). An autoradiographic pattern produced by a chromosome of *E. coli* labeled with thymine-^{3}H (5 Ci/mmole) for 5 minutes and then with thymidine-^{3}H (50 Ci/mmole) for 2 minutes shortly before replication was about to be completed by the meeting of the two replication forks in the circular chromosome. The autoradiographic images of the daughter duplexes are visible behind the approaching replication forks. The parental DNA between the forks is 3 μm long, corresponding to 6×10^{6} daltons of DNA, and contains the replication terminus. Exposed for 5 weeks. The bar is 24 μm.

(c). An autoradiograph of an intact chromosome of *E. coli* labeled with thymidine-^{3}H (50 Ci/mmole) for about one and a half cell generations. The denser image (arrow) may represent a replicated region in which the two growing daughter duplexes are collapsed together. The autoradiograph was prepared by mixing lysed cells with liquid NTB3 emulsion and smearing the mixture on a microscope slide. By this procedure the chromosome has been buried within the emulsion. The chromosome is 2540 μm long.

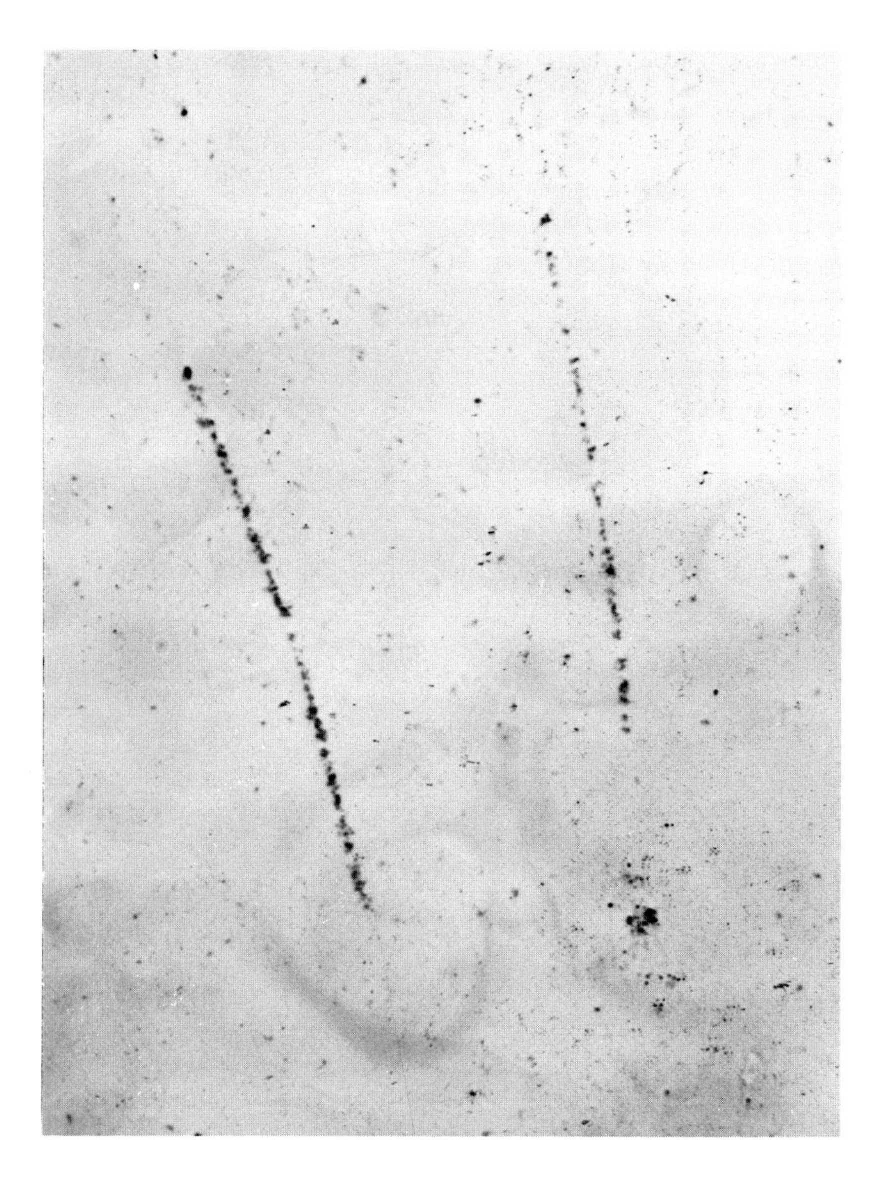

FIG. 2. Autoradiographic images produced by chromosomes of vaccinia virus labeled with thymidine-^{3}H. The tracks measure 73 and 80 μm, corresponding to DNA molecules of 146 and 160 $\times$ 10^{6}. The slide was dipped in Ilford L-4 emulsion, and the grains are so small that the tracks are not sharply defined.

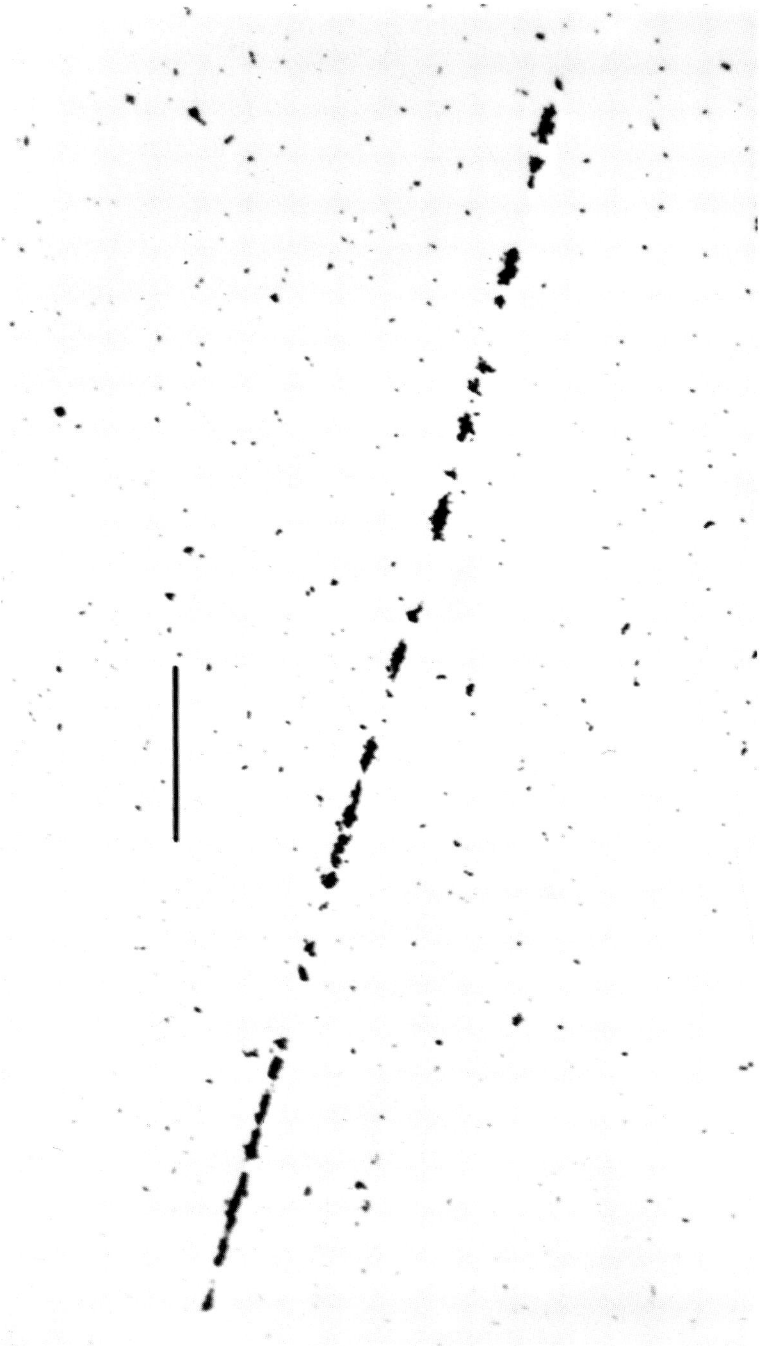

FIG. 3. An autoradiographic image of a portion of a DNA molecule of a replicating chromosome in a cultured cell of the rainbow trout. The cell was labeled with thymidine-^{3}H (50 Ci/mmole; 100 μCi/ml) for 20 minutes in the presence of 0.1 μg of fluorodeoxyuridine per milliliter. Many separate replicating regions are present. The bar is 10 μm.

cells labeled with 100 μCi of thymidine-^{3}H per milliliter (50 Ci/mmole) for 20 minutes and processed as described above for chick cells. The arrangement of multiple replicating units in tandem is apparent.

These experiments will be described in full elsewhere.

Acknowledgments

This work was supported by a National Science Foundation grant No. GB 32232 to D. M. Prescott and National Institutes of Health grant No. GM-15905 to P. L. Kuempel.

References

Bleecken, S., Strokbach, G., and Sarfert, E. (1966). *Z. Allg. Mikrobiol.* **6**, 121.
Cairns, J. (1962). *J. Mol. Biol.* **4**, 407.
Cairns, J. (1963a). *J. Mol. Biol.* **6**, 213.
Cairns, J. (1963b). *Cold Spring Harbor Symp. Quant. Biol.* **28**, 43.
Cairns, J.(1966). *J. Mol. Biol.* **15**, 372.
Caro, L. G. (1964). *In* "Methods in Cell Physiology" (D.M. Prescott, ed.), Vol. I, pp. 327–363. Academic Press, New York.
Gyurasits, E. B., and Wake, R. G. (1973). *J. Mol. Biol.* **73**, 55.
Huberman, J. A., and Riggs, A. D. (1968). *J. Mol. Biol.* **32**, 327.
Kuempel, P. L. (1972). *J. Bacteriol.* **112**, 114.
Kuempel, P. L., Prescott, D. M., and Maglothin, P. (1973). *In* "DNA Synthesis *in vitro*" (R. Wells and P. Inman, eds.). University Park Press, Baltimore, Maryland. In press.
Lark, K. G., Consigli, R., and Toliver, A. (1971). *J. Mol. Biol.* **58**, 873.
Prescott, D. M., and Kuempel, P. L. (1972). *Proc. Nat. Acad. Sci. U.S.* **69**, 2842.
Rodriguez, R. L., Dalbey, M., and Davern, C. (1973). *J. Mol. Biol.* **74**, 599.
Wake, R. G. (1972). *J. Mol. Biol.* **68**, 501.

Chapter 8

HeLa Cell Plasma Membranes

PAUL H. ATKINSON[1]

Department of Pathology and Department of Developmental Biology and Cancer, Albert Einstein College of Medicine, New York, New York

---

[1]Supported in part by United States Public Health Service Research Grants CA 13402 to P. H. Atkinson, AI-07140 to D. F. Summers, and CA 06576 to A. B. Novikoff. The author was recipient of Damon Runyon Memorial Fund Award DRF-538-T during part of the progress of this work.

I. Introduction

The isolation of plasma membranes from animal cells, from *in vivo* and tissue culture sources, has been the subject of several recent and extensive reviews (Wallach, 1967; Steck and Wallach, 1970; Warren and Glick, 1971), which in general cover the cell types used, criteria for identification, and criteria for purity. There is no need in the present article for duplicating these reviews except for a brief recapitulation of principles. I have listed some representative publications, including greater numbers of more recent ones, according to the type of cell/tissue used as starting material. Few, if any, methods are ubiquitously applicable to all cell types, thus, such a listing may be of more general use to those who have a particular cell/tissue other than HeLa cells in mind as their starting material. Work on red cell plasma membranes is not discussed here.

The plasma membrane product of various methods has one of three main morphologies, namely: fragments, ghosts, and vesicles reflecting in a broad sense the innovations of the groups of Neville, Warren, and Wallach, respectively. The first, pioneered by Neville (1960), utilizes gentle disruption of the liver cells in hypotonic solution followed by the isolation of plasma membranes as large fragments possibly "bile fronts" (Steck and Wallach, 1970) as identified in the phase contrast and electron microscope. This approach became the basis for many modifications in the isolation of plasma membranes from liver tissues of various mammals. In general such modifications also produced large fragments rather than intact ghosts and Warren *et al.* (1966) first demonstrated the production of intact ghosts from cultured mammalian cells. These preparations were made from mouse fibroblasts grown in tissue culture and utilized agents which strengthen the plasma membrane for subsequent isolation. Of special note is their use of fluorescein mercuric acetate (FMA), which has been used successfully by numbers of other workers for the preparation of plasma membrane ghosts from tissue culture cells. Criteria for identification and estimation of purity of fragment and ghost preparations have included the ghost morphology itself in light and electron microscopy, enrichment of enzymes deemed specific for the cell surface [5′-nucleotidase, (Na^+, K^+)-activated ATPase], absence of enzymes characteristic of organelles such as mitochondria, endoplasmic reticulum or lysosomes, presence of specific surface antigens such as mouse H2 and human HLA, estimations of the quantities of DNA, RNA, protein, lipid, and carbohydrate.

A second and basically different methodology (Wallach and Kamat, 1964; Kamat and Wallach, 1965) does not preserve the ghost structure, but rather disrupts the cells vigorously in isotonic buffer, converting the entire surface membrane into vesicles. In the process of homogenization, membrane vesicles from other sources also become mixed with the surface membranes. In buffer whose pH, Ca^{2+}, and Mg^{2+}ion content and hypotonicity is carefully defined, vesicles derived from different types of membranes have different densities in Ficoll solution due to the titration of charge groups of the proteins and carbohydrates characteristic of a given type of membrane. It is therefore possible to separate the different types of vesicles by isopycnic centrifugation (Kamat and Wallach, 1965). Plasma membranes are then defined by their content of 5′-nucleotidase, (Na^+, K^+)-ATPase, surface antigens, and virus receptors.

In assessing these two main approaches, the advantages of producing plasma membranes as ghosts include direct identification by morphological means, ease of quantitation on a per cell surface basis and the probability of greater retention of surface configurations which might be lost in a more vigorous disruption. Loss of surface molecules as a result of plasma membrane isolation is a subject reviewed at some length by Kraemer (1971). Weighing against these considerations is the difficulty in obtaining ghosts, even with strengthening agents, in some cell types, low and possibly selective yield from a random population of cells (i.e., does one select for cells in certain phases of the mitotic cycle due to a transitorily increased strength of membrane?), and the possibility of trapping extraneous particles in the convolutions of the ghost. A further consideration is the possible insolubility of plasma membrane polypeptides after the use of FMA. Ghosts were prepared from HeLa cells infected with vesicular stomatitis virus and treated with FMA. Certain species of polypeptide and glycoprotein behaved anomalously after solubilization in 2% SDS, 1% mercaptoethanol and electrophoresis on SDS-polyacrylamide gels (unpublished observations). On the other hand, no solubility problems were encountered with ghosts made from FMA-treated L cells (Hecht and Summers, 1972; Hubbard, 1972).

The advantages of preparing plasma membrane material as vesicles by the procedures of Wallach and co-workers include higher yield, less trapping of other organelles, and in theory it should be possible to obtain plasma membranes from any cells by these methods. (There is no such guarantee when one sets out to prepare ghosts.) Disadvantages of these methods include the greater likelihood of loss of plasma membrane material due to the more intensive disruption of cells, possible trapping of extraneous soluble protein in sealed vesicles, loss of surface configuration in the reorganization involved in vesicle formation and the loss of an important characteristic for separation, identification and quantitation; namely, morphology. The vesicle product is primarily identified by the presence of an enzyme such as 5′-nucleotidase. The reliability of this enzyme as a cell surface marker (Essner *et al.*, 1958; Sabatini *et al.*, 1963) in plasma membrane preparations from liver cells of different species is not without exception (Lauter *et al.*, 1972). Furthermore, others have observed 5′-nucleotidase in rough and smooth endoplasmic reticulum of rat liver cells (Widnell and Unkeless, 1968; Widnell, 1972). Nevertheless, in a number of different cell types, copurification of this enzyme with more readily identifiable surface components, such as the ghost structure or surface antigens, has been demonstrated, and thus its acceptance as a surface marker has gained veracity. A more widespread application of the elegant techniques available in electron microscope histochemistry, especially to cultured cells, should confirm the point and also generalize the evidence concerning the distribution of 5′-nucleotidase.

A criticism of both approaches to the purification of plasma membranes is the possible loss of cell surface components due to the release of degradative enzymes during cell disruption (Kraemer, 1971). As pointed out by this author, techniques which derive surface material without the disruption of cells may be preferred to isolation of the entire surface in many instances. However, the use of proteases, for example, will preclude studies on intact cell surface glycoproteins and polypeptides thus isolation of plasma membranes would be preferred, unless molecules could be removed by some other method without disruption of cells or the molecules.

The *in vivo* sources from which plasma membrane preparations have been made include *rat liver* (Neville, 1960; Emmelot *et al.*, 1964; Takeuchi and Terayama, 1965; Ashworth and Green, 1966; Coleman *et al.*, 1967; Song and Bodansky, 1967; Barclay *et al.*, 1967; Stein *et al.*, 1968; Graham *et al.*, 1968; Weaver and Boyle, 1969; Berman *et al.*, 1969; Ray, 1970; Evans, 1970; Touster *et al.*, 1970; Simon *et al.*, 1970; Pricer and Ashwell, 1971), *rat hepatoma* (Benedetti and Emmelot, 1967) *lumenal cells of rat or rabbit urinary bladder* (Hicks and Ketterer, 1970; Chlapowski *et al.*, 1972) *rat isolated fat cells* (Rodbell, 1967; McKeel and Jarett, 1970; Combret and Laudat, 1972), *Ehrlich ascites carcinoma cells* (Wallach and Kamat, 1964), *rabbit peritoneal leukocytes* (Wieneke and Woodin, 1967), *pig lymphocytes* (Allan and Crumpton 1970; Ferber *et al.*, 1972), *thymocytes* (Allan and Crumpton, 1972), *calf lymphocytes* (Ferber *et al.*, 1972), *lymphoid cells* (Artzt *et al.*, 1968), *calf thyroid tissue* (Turkington, 1962), *rat intestinal microvilli* (Forstner *et al.*, 1968; Leslie and Rowe, 1972; Quigley and Gotterer, 1969), *rat muscle* (McCollester, 1962; Kono *et al.*, 1964; Rosenthal *et al.*, 1965; it should be noted that muscle preparations contain collagen fibrils and basement membrane as well as plasma membrane), *rat lung cells* (Ryan and Smith, 1971), *mouse intestinal mucosa* (Fujita *et al.*, 1972), *guinea pig liver, kidney, and small intestinal mucosa* (Coleman and Finean, 1966), *bovine mammary glands* (Keenan *et al.*, 1970), *guinea pig pancreas* (Meldolesi *et al.*, 1971), *myelin* [reviewed by Mokrasch (1971)], *isolated neurones of rabbit cerebral cortex* (Henn *et al.*, 1972), *bovine thyroid* (Wolff and Jones, 1971; Yamashita and Field, 1970), *isolated rat and guinea pig liver cells* (Solyom *et al.*, 1972), *bovine anterior pituitary gland* (Labrie *et al.*, 1972), *bovine adrenal cortex* (Finn *et al.*, 1972), *mouse liver* (Evans, 1970; Evans and Bruning, 1971; Goodenough and Stoeckenius, 1972), *human liver cells* (Wilson and Amos, 1972), *comparative preparations from the livers of rat, guinea pig, rabbit, and calf* (Lauter *et al.*, 1972), and *pig kidneys* (Campbell *et al.*, 1972).

Preparations have also been made from cells cultures *in vitro* including *L cells* (Warren *et al.*, 1966; Heine and Schnaitman, 1971; Brunette and Till, 1971; Wagner *et al.*, 1972; Hubbard, 1972), *KB cells* (Gerner *et al.*, 1970), *HeLa cells* (Bosmann *et al.*, 1968; McLaren *et al.*, 1968; Philipson *et al.*, 1968; Boone *et al.*, 1969; Atkinson and Summers, 1971), *3T6 cells* (Barland and Schroeder, 1970), *3T3 cells and virus transformed 3T3 cells* (Wu *et al.*, 1969; Sheinin and Onodera, 1972), *chick embryo fibroblasts* (Perdue and Sneider, 1970; Quingley *et al.*, 1971; Lazarowitz *et al.*, 1971; Bose and Brundige, 1972; Bingham and Burke, 1972), *chick embryo fibroblasts transformed with viruses* (Warren *et al.*, 1972a; Perdue *et al.*, 1972), *monkey kidney cells* (Klenk and Choppin, 1969), *hamster kidney fibroblasts* (Klenk and Choppin, 1969; Buck *et al.*, 1970; Renkonen *et al.*, 1971; Gahmberg, 1971; Warren *et al.*, 1972b), *human epidermoid carcinoma (Hep-2) cells* (Heine *et al.*, 1972), *cells derived from a human liver malignancy and peripheral lymphocytes* (Wilson and Amos, 1972).

The following article describes a method for isolation of HeLa cell plasma membranes. The method has the advantage of being rapid (with practice, a product can be obtained in about 45 minutes) and does not require the use of uncommon equipment. Some observations on the disposition of fucose-containing glycoproteins in HeLa cells are also included, together with preliminary data on the kinetics of their arrival in the plasma membrane.

II. Preparation of HeLa Cell Plasma Membranes

A. Design of Method

The general design of the method described below for the preparation of HeLa cell plasma membranes is based on a series of rate zonal sedimentations which serves to differentially separate cellular components of disrupted cells. The choice of buffers, homogenization conditions, and some centrifugation steps are based on the experiences of Penman (1966) in the isolation of HeLa cell nuclei. Thus, gentle disruption of cells in hypotonic buffer produces a homogenate which contains plasma membrane ghosts, nuclei, undisrupted cells, mitochondria, lysosomes, membrane vesicles, and fragments as the principal large components. A very short centrifugation serves to pellet the nuclei and whole cells, leaving most of the plasma membrane ghosts in the supernatant. A subsequent short centrifugation through

30% sucrose substantially separates the ghosts from smaller, though more dense, particles, such as mitochondria and lysosomes. Since such separations depend on only small differences in sedimentation rate, the values of various parameters given below must be fairly closely adhered to. In illustration, centrifugation of a homogenate for twice the stated period of time would result in most of the ghosts being found in the pellet along with the nuclei and whole cells.

B. Growth of Cells

HeLa-S_3 cells are grown in suspension culture in Eagle's minimal essential medium (Eagle, 1959) in the presence of penicillin and streptomycin (Joklik Modified). All media used for maintaining cells or growing cells under experimental conditions are supplemented with 3.5% calf serum and 3.5% fetal calf serum. For the isolation of membranes and for radioactive labeling of cell constituents, cells are harvested from cultures when the cells reach a density of 3.0 to 4.0 $\times$ 10^5 cells per milliliter. Membranes can be satisfactorily derived from cells harvested at greater densities, but the yield is somewhat less and the ghosts more fragmented. Cells should be free of PPLO, as determined by the standard plate test, and be otherwise healthy as best indicated by daily doubling in population during the days prior to the experiment.

C. Radioactive Labeling and Assay

1. Labeling

For labeling times of one cell cycle or greater (16 to 24 hours) cells are suspended at a density of 1.5 $\times$ 10^5 cells per milliliter in minimal essential medium. For short labeling times of up to 4 hours, cells are concentrated by centrifugation and resuspended in minimal essential medium at a density of 15 to 20 $\times$ 10^5 cells per milliliter. Under these conditions, maximum incorporation and linear uptake of the radioactively labeled precursors is achieved. ^{14}C-labeled amino acid mixture (New England Nuclear, Boston, Massachusetts, 15 amino acids, 80–400 mCi per mmole) is added to a final concentration of 0.5–2.0 μCi/ml, and ^{3}H-labeled amino acid mixture (New England Nuclear, 15 amino acids, 0.4–60 Ci/mmole) to a concentration of 5–10 μCi/ml. Fucose-L-^{3}H (New England Nuclear, 4.3 Ci per mmole) is used at concentrations of 2 to 5 μCi/ml. Uridine-^{14}C (New England Nuclear, 0.50 mCi/mmole) is used at a concentration of 0.1 μCi/ml, and the medium is made 0.02 m*M* with respect to unlabeled uridine when the labeling period exceeds one cell cycle (about 20 hours). Thymidine-methyl-^{3}H (New England Nuclear, 6.7 Ci/mmole) is used at a concentration of 1 μCi/ml and fucose-L-^{14}C (New England Nuclear, 50.8 mCi/mmole) at 0.1 μCi/ml.

2. ASSAYS

a. Whole Cells. Duplicate 0.1-ml aliquots of whole cells in minimal essential medium are washed with 2 ml of cold (4°C) Earle's solution. The cells are collected by centrifugation and lysed by the addition of 1.0 ml of distilled water. Approximately 0.1 ml of bovine serum albumin (0.25 mg/ml) is added as carrier followed immediately by 1.0 ml of 10% trichloroacetic acid (TCA). The precipitate is collected on Whatman (W. R. Balston Ltd., England) glass filters (GFA), washed several times with approximately 5 ml of 5% TCA and once with approximately 5 ml of 95% ethanol. The procedure thus far is carried out at 4°C. The filters are placed without drying in plastic scintillation vials (Nuclear Associates, Westbury, New York) with 6 ml of scintillation fluid consisting of 1 volume of Triton X100 (industrial grade, Rohm and Haas, Philadelphia, Pennsylvania), 2 volumes of toluene with 4 g of 2,5-diphenyloxazole per liter. The samples are counted in a liquid scintillation spectrometer (Beckman Instruments Inc., Fullerton, California).

b. Cell Fractions. With the exception of sucrose gradient and acrylamide gel fractions, the same assay procedure as described for whole cells is utilized for all cell fractions including the plasma membrane fractions, except that the initial washing step is omitted in assays requiring direct comparison of cell fractions with whole cells.

c. Aliquots from Sucrose Gradients. Aliquots (5 or 10 μl) of sucrose gradient fractions are pipetted into 1 ml of water in plastic scintillation vials. Scintillation fluid (6 ml) is added, and the radioactivity is counted as described above.

D. Buffers and Solutions

1. EARLE'S WASH

Cells are washed in Earle's balanced salts (Earle, 1943):

Component	gm/liter
NaCl	6.8
KCl	0.4
Glucose	1.0
$NaHCO_3$	2.2
$NaH_2PO_4 \cdot H_2O$	0.14
CaCl	0.2
$MgCl_2 \cdot 6H_2O$	0.18

Separate stock concentrations of Ca^{2+} and Mg^{2+} solutions (20×) and the other salts (10×) are stored and mixed appropriately for use. The 10× salts must be tightly stoppered otherwise a more alkaline pH results and causes precipitates on mixture and dilution.

2. Hypotonic Buffer

Stock 1.0 *M* Tris-HCl buffer is made by adjusting the pH of a 1.0 *M* solution of Trizma Base (Sigma Chemical Co., St. Louis, Missouri) to pH 8.4 with concentrated HCl. This stock is used at a 1:100 dilution (10 m*M*) in which the pH falls to about 8. This buffer may be adjusted to 15 m*M* sodium iodoacetate. The resulting buffer should not be stored since the iodoacetate is slowly photooxidized to free iodine. Iodoacetate is added as a degradative enzyme inhibitor.

3. RSB

Reticulocyte standard buffer (RSB) is kept as a 10-fold concentrated stock solution of 100 m*M*, NaCl, 100 m*M*, Tris-HCl pH 8.0, 30 m*M* $MgCl_2$ (Penman, 1966; except his final [Mg^{2+}] is 1.5m*M*, not 3.0m*M*, and pH is 7.4, not 8.0).

4. Sucrose Solutions

Sucrose solution, 30% w/w, consists of 33.8 gm sucrose in 100 ml of solution buffered with 10 m*M* tris-HCl, pH 8 (Tris buffer) 45% w/w sucrose is 54.1 gm of sucrose in 100 ml of solution. These are made up fresh on the day of isolation since even storage at 4°C and the presence of azide is no guarantee against microorganism contamination in the 30% sucrose.

E. Collecting and Washing Cells

This and all subsequent steps are performed at 0–4°C. Cells are collected by centrifugation in the International Equipment Company (IEC) PR2 or PRJ centrifuge. A moderate-sized preparation is started with two 175-ml heavy-walled bottles with screw caps, designed to fit in the IEC No. 259 rotor head. Cells in this volume (approximately 1.2×10^8 cells) are sedimented to a pellet in about 5 minutes at 1500 rpm, resuspended in ice cold Earle's balanced salts and recentrifuged in one bottle only. Cells are then suspended in only 25–30 ml Earle's salts and transferred to a 40-ml Pyrex conical heavy-walled graduated glass centrifuge tube (Corning Cat. No. 8140). (There is nothing obligatory in this choice of glassware; it is chosen to conveniently fit the centrifuges.) The cells are centrifuged to a pellet again by use of the IEC No. 253 rotor head with the appropriate rubber cushions (IEC Cat. No. 676) in the cups. With these cushions, Pyrex heavy-walled conical tubes may be spun up to 2000 rpm in the No. 253 rotor; without them the tubes collapse at a much lower speed. A speed of 1500 rpm for about 5 minutes produces a pellet of cells of approximately 0.5 ml. The supernatant Earle's wash is discarded, and the tube is inverted on a paper towel for 30 seconds. The insides of the tube are then wiped dry with a paper towel, serving to remove excess salts which might otherwise raise the salt concentration of the hypotonic buffer (below). It is not known whether more washing steps are necessary to rid the cells of adventitious material. In one respect it is counterproductive in that better yields are obtained when isolations are performed as soon as possible after the cells are removed from culture.

F. Hypotonic Swelling

The cell pellet is resuspended in 20–25 times its volume (10 ml, for the quantity of cells described here) of 10 m*M* Tris-HCl, pH 8.0, and if desired, 15 m*M* sodium iodoacetate; it is then allowed to swell, in the ice bucket, for 5 minutes. The use of Tris-HCl is not obligatory; 7.5 m*M* sodium phosphate buffer pH 7.2 was equally satisfactory in the production of ghosts. I have not tried other buffers, but I would assume that any buffer in the pH range 7–8 at about 10 m*M* concentration, in the absence of divalent cations, would suffice. Cells do swell in RSB [3 m*M* Mg^{2+} or 1.5 m*M* as originally described by Penman (1966)] but the subsequent production of ghosts is poor, owing to a need for longer swelling time, and to the comparatively vigorous homogenization needed to break most of the cells. Too great a dilution of the cell pellet and prolonged swelling in hypotonic buffer in the absence of Mg^{2+} results in cells and nuclei bursting, and successful preparations are not possible.

G. Homogenization

Homogenization is acheived by very slow strokes in a stainless steel Dounce homogenizer (clearance of 0.002 inch). A tight glass Dounce homogenizer has also been found satisfactory. If from 80 to 100% of the cells are not broken by 5–20 strokes (1 stroke = up and down) with formation of ghosts (Fig. 1A), more vigorous homogenization usually does not help. In this case, greater dilution of the cell pellet obtained after washing is required, or waiting a small interval before continuing homogenization.

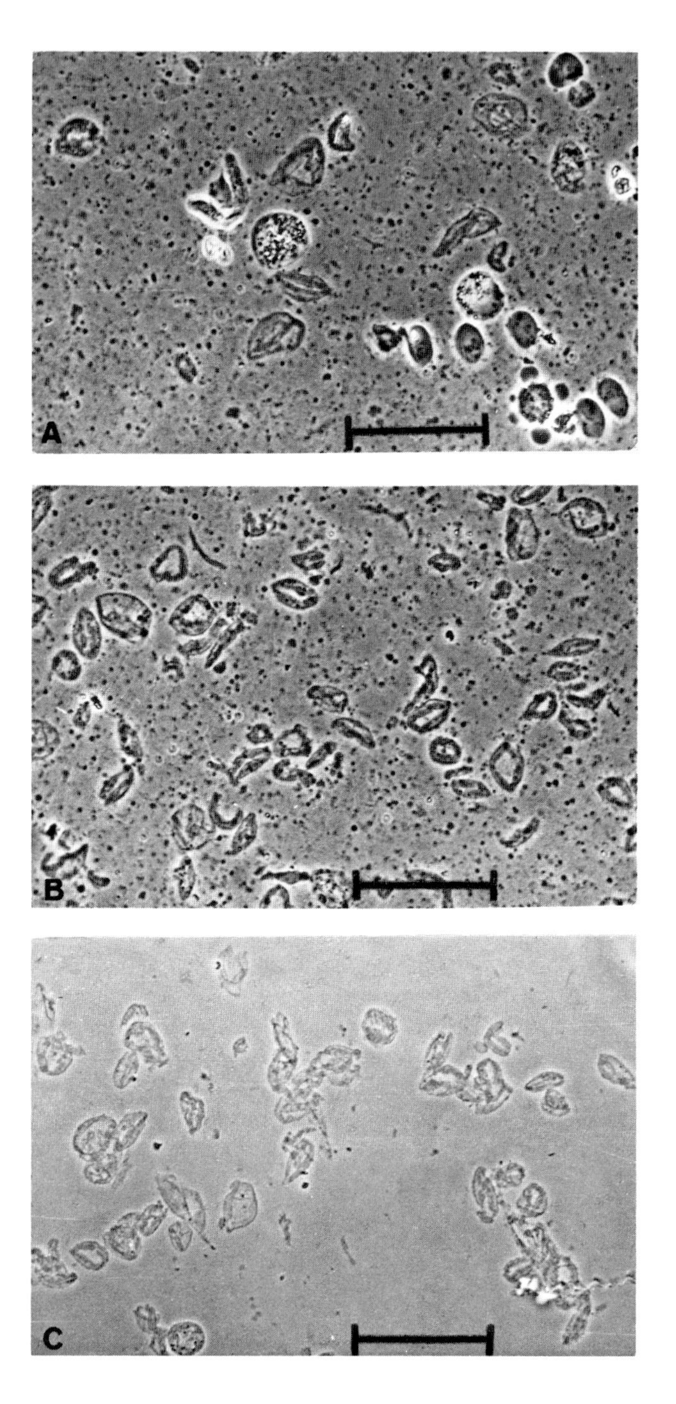
A
B
C

H. Observation of Material with Phase Microscopy (Figs. 1–3)

1. Optics

Good phase optics are required to observe ghosts in light microscopy since the ghosts are not visible in bright field. I have found the Zeiss Planapochromat 25X Ph2 lens with 12.5 X eyepieces and a Phako IVZ7 condenser (Carl Zeiss, W. Germany) very convenient and useful for observing ghosts. However, almost any good quality phase microscope will do. Poor quality lenses, even on good instruments, will make the identification of ghosts very difficult.

2. Preparation of Slide

Approximately 2 drops of the homogenate or material to be observed is placed on a glass slide before the coverslip is lowered, one edge in contact with the solution and slide, onto the sample. Too little liquid or random dropping of coverslips results in squashed components, making it impossible to judge the state of homogenization.

3. Representative Fields

The extent of homogenization is best estimated from stationary though still floating cells and components, since it is here that sample composition approaches what is in the Dounce homogenizer. I have found the Zeiss 16 X Planachromat lens useful for estimating percent breakage of cells, because a very large flat field is observed.

4. When to Stop Homogenization

When further Dounce homogenization fails to improve on the ratio of free nuclei to remaining whole cells (ideally there should be only free nuclei), homogenization should be discontinued. Further homogenization only serves to disrupt nuclei and fragment already free ghosts. Percent breakage varies from 80 to 100%.

FIG. 1. Phase contrast micrographs of cell fractions. (A) Cells were Dounce-homogenized as detailed in the text. A whole cell and nuclei are visible. Compare the size of ghosts visible in the three plates. The size is consistent throughout the purification. (B) Nuclei removed by 1000 *g*, 30-second centrifugation. (C) Ghosts purified by two cycles of zonal centrifugation. All photographs were taken of material suspended in Tris buffer. Bar = 25 μm. Reprinted from Atkinson and Summers (1971) by permission of the American Society of Biological Chemists Inc., Baltimore, Maryland.

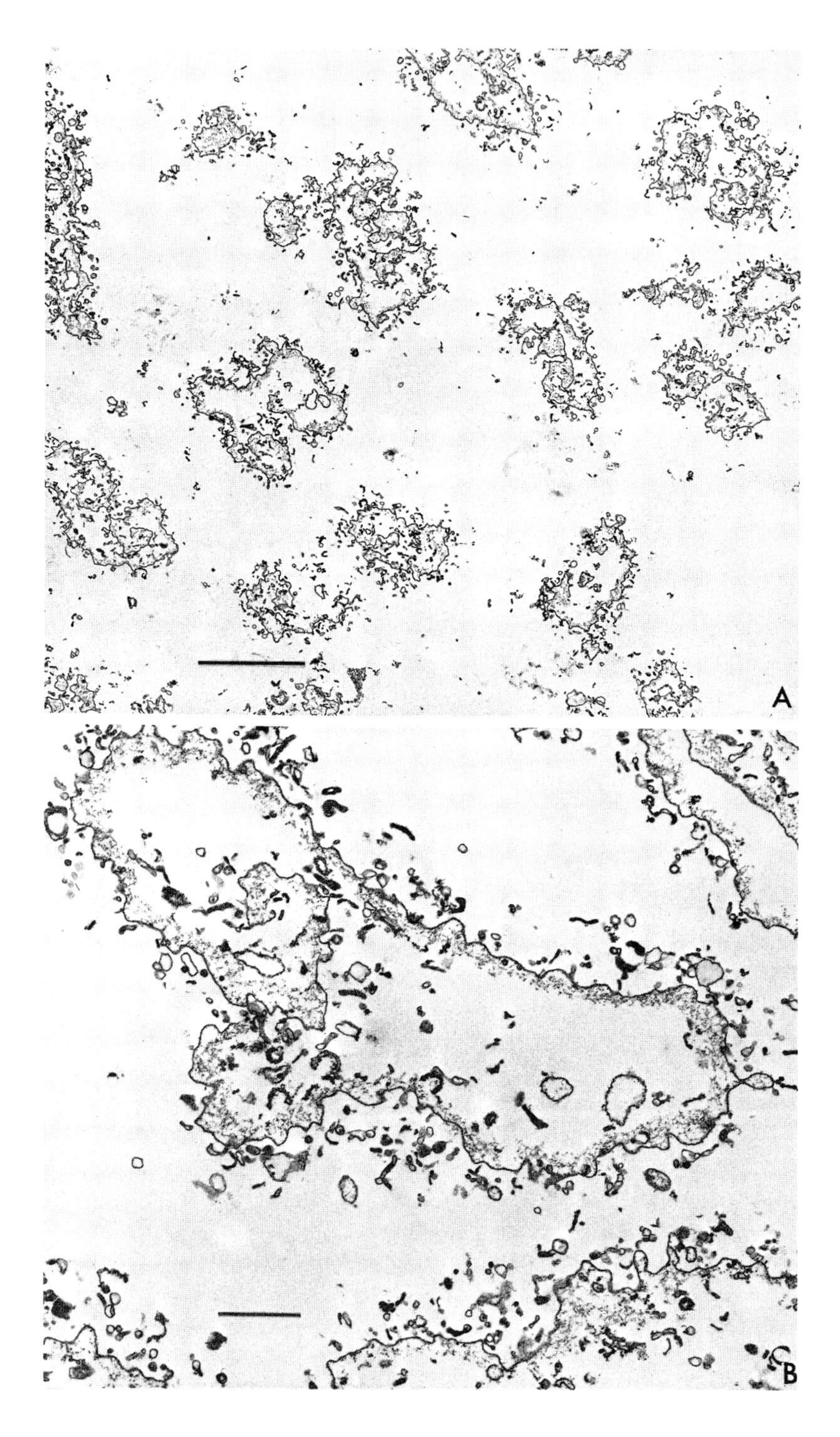
A
B

I. Removal of Nuclei

1. Stabilization

Nuclei are unstable in the hypotonic buffers lacking cations, and prolonged swelling eventually results in their lysis. Thus, the whole homogenization procedure, after the addition of hypotonic buffer, should not take more than 10 to 15 minutes, preferably less. Once maximum cell breakage is achieved, the homogenate is adjusted to 3 m*M* Mg^{2+} 10 m*M* NaCl in order to stabilize the nuclei. Observation in the phase microscope will confirm that the nuclei have compacted from amorphous swollen spheres, increased in contrast and nucleoli and have become visible. Nuclei in this condition are relatively stable. Unbroken cells remain swollen in this buffer and their nuclei though not compact display nucleoli.

2. Centrifugation

A very brief centrifugation separates ghosts from nuclei and whole cells [cf. Penman (1966)]. The conditions for the centrifugation are as follows: 8–10 ml homogenate in a 40-ml conical centrifuge tube is centrifuged at 2000 rpm (ca. 1000 *g*) in an IEC No. 253 rotor for 30 seconds from the time speed is attained at maximum acceleration. This step pellets more than 95% of the nuclei and whole cells and leaves most of the ghosts in the supernatant (Fig. 1B, cf. Fig. 1A). The correct sedimentation time gives a sharply defined interface between the pellet and the cloudy supernatant. The supernatant is then carefully drawn off with a 10-ml disposable plastic syringe equipped with a 13- to 14-gauge blunted needle. Care must be taken to avoid disturbing the very loose pellet in this aspiration, and, since the next step involves washing the pellet once, perhaps twice, to increase recovery of ghosts, not all supernatant over the pellet need be removed. The supernatant should be examined at this point for contamination by nuclei. Should large numbers remain, the centrifugation was not long enough. Nuclei are not stable in 30% sucrose in the absence of Mg^{2+}, thus their presence in too great a quantity after this step makes the subsequent purification of ghosts minimal. It should be noted that in preparations made in 15 m*M* iodoacetate, components sediment much faster and about half the stated times are sufficient to partition ghosts and nuclei at this step.

FIG. 2. Electron micrograph of partially purified HeLa cell ghosts. Plasma membranes were purified by one cycle of zonal centrifugation and Tris buffer wash as detailed in the text. Electron microscopy was performed as detailed in Fig. 3, except that sections were photographed in a Philips 300 electron microscope. Note the absence of large organelles, the presence of microvilli and blebs on the outer surface, the presence of some rough endoplasmic reticulum and the presence of a fuzzy coat on the inner surface of the membrane. (A) Bar = 5 μm; (B) bar = 1 μm.

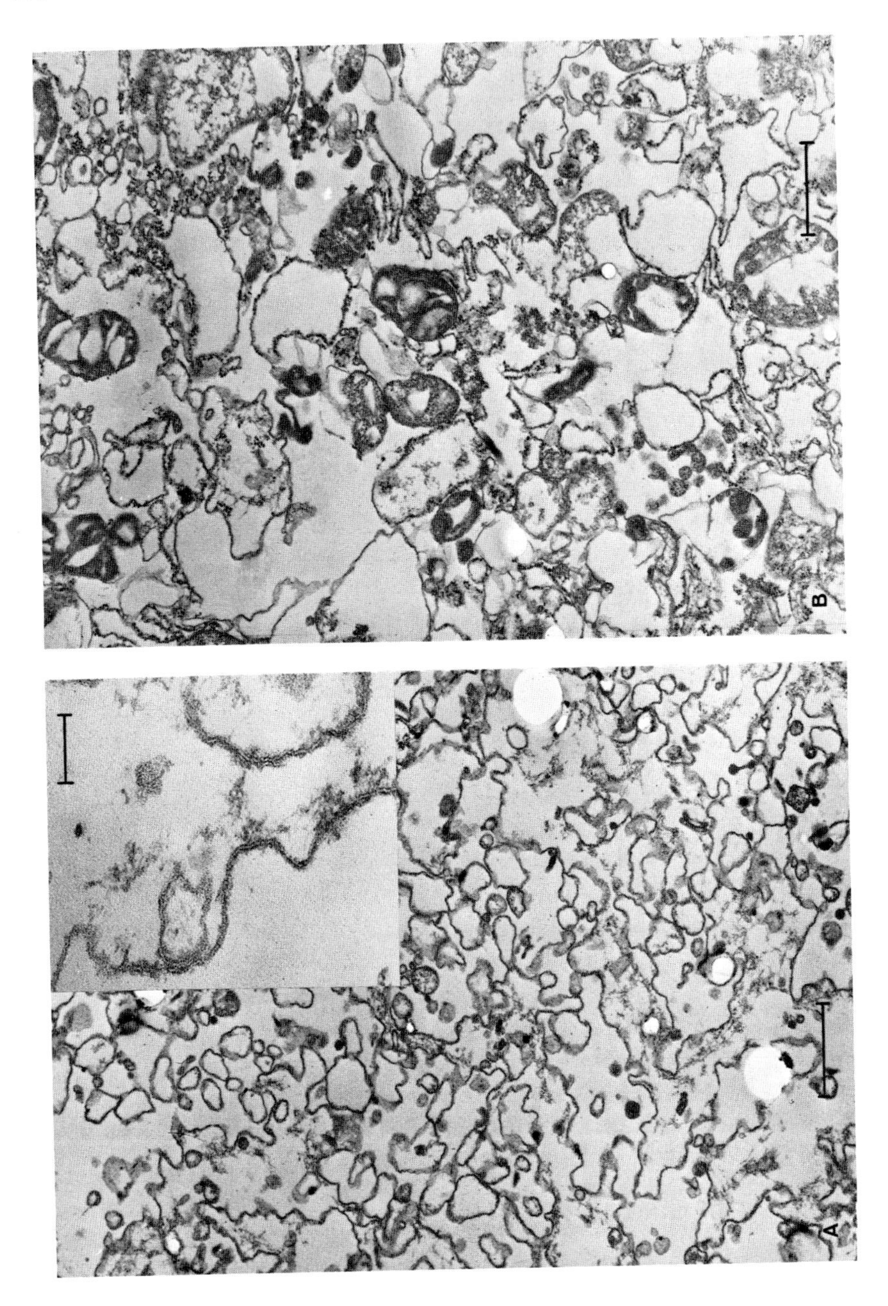
B
A

3. Washing the Pellet

The pellet is resuspended in four times its volume of RSB and recentrifuged as before at ca. 1000 *g*, this time for 10 seconds only. Ghosts are once again observed in the supernatant. The nuclei and whole cells are observed in the pellet. This process may be repeated again should there seem to be large numbers of ghosts still in the pellet.

4. Problem of Aggregation of Cell Components after Hypotonic Disruption

If this initial centrifugation is too extensive, all the ghosts, nuclei, and whole cells and quite a large amount of the small particulate material will be found in the pellet. Subsequent attempts to separate the components often fail due to an extensive aggregation of the components, and it becomes difficult if not impossible to separate the components purely on the basis of sedimentation rate. Similarly, should some nuclei lyse due to prolonged exposure to hypotonic medium lacking Mg^{2+}, separation is again impaired if not prohibited, owing to aggregation of the various organelles in the homogenate.

J. Isolation of Ghosts

1. Formation of the Step Sucrose Gradient for Rate Zonal Centrifugation

For the number of cells started with (in this case, ca. 10^8) a combined supernatant of approximately 10 ml results from the previous steps. To purify ghosts from this, it is appropriate to use four gradients in 30-ml round-bottomed Corex tubes (Corning Cat. No. 8445). Fewer gradients (or more material) lead to aggregation of cellular components as a result of overloading the cross-sectional area of the gradient, and little separation is subsequently achieved. The gradients are most easily formed by pipetting into the tubes 15 ml of 30% sucrose first; 5 ml of 45% sucrose is then underlayered by the use of a plastic disposable syringe, and the same 14-gauge needle used for other operations in the procedure.

FIG 3. Electron micrographs of cell fractions. Cell fractions were embedded and sectioned for electron microscopy as detailed in the text. The sections were stained with ethanolic uranyl acetate, double stained with Reynolds lead citrate (Reynolds, 1963), and examined in a Siemens electron microscope. (A) Plasma membrane ghosts purified by two zonal centrifugations and a 0.5 *M* NaCl wash. Bar = 1.0 μm. Inset: Same ghosts at high magnification. Bar = 0.1 μm. (B) Organelles found in the 30% zonal fraction (Fig. 4 fraction III) and collected by centrifugation at 12,000 *g* for 20 minutes. Bar = 1.0 μm. Reprinted from Atkinson and Summers (1971) by permission of the American Society of Biological Chemists Inc., Baltimore, Maryland.

2. Centrifugation

The combined supernatant from the previous steps is divided among the four gradients and centrifuged at 7500 rpm in the HB4 rotor of a Sorvall RC2 centrifuge, 15–20 mintues at 4°C. (The same centrifugation could be achieved at 6000 rpm for ca. 10 minutes is a Spinco SW 27 rotor, with a sucrose gradient of similar proportions in a SW 27 cellulose tube). Centrifugation in the RC2 should not be extended beyond 20 minutes. Under the right conditions of centrifugation, the plasma membrane ghosts are observed at the 30–45% interface as a white band (Fig. 4, IV). Shorter times

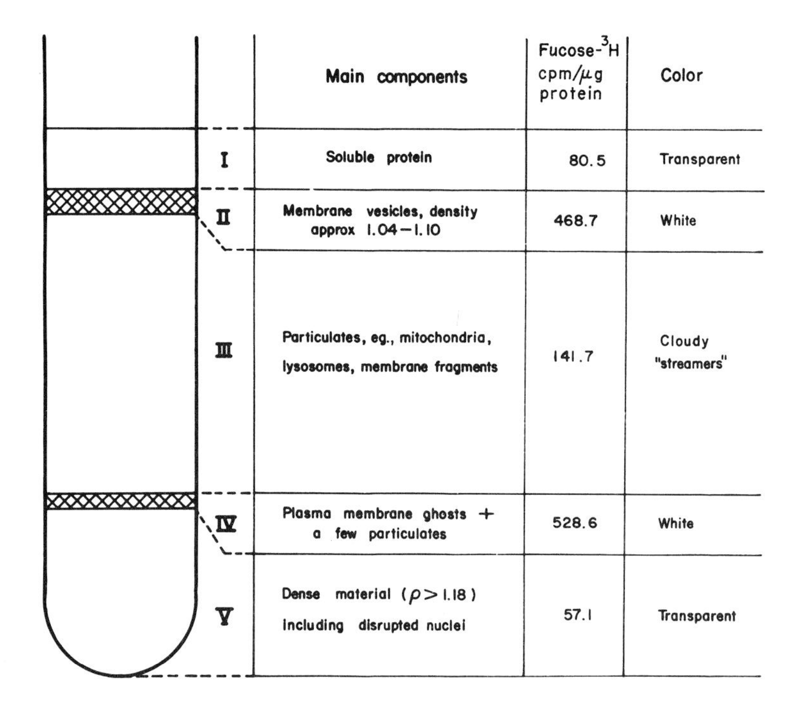

Fig. 4. Characteristics of the sucrose gradient after the first zonal centrifugation. At a density of 2×10^5 cells/ml, 500 ml of cells were labeled for 18.5 hours with fucose-^{3}H at a concentration of 2 μCi/ml. Cells were harvested when they reached a density of 3.5×10^5 cells/ml. Cells were homogenized and the components subfractionated as detailed in the text, using one cycle of zonal centrifugation. The figure describes the appearance of one such gradient after centrifugation in a Sorvall HB-4 rotor at 7500 rpm for 20 minutes. The gradients were subfractionated (by use of a syringe and 14-gauge blunted needle) into fractions I–V. Radioactivity and protein content were determined in duplicate 0.1-ml aliquots. The figure expresses the specific radioactivity (^{3}H-labeled fucose, counts per minute per microgram of protein) of the various fractions.

(12–15 minutes) result in fewer ghosts reaching the 30–45% interface and also much fewer contaminating, smaller, more dense particles. However, longer times result in most of the particles (Fig. 3B) of the overlaying supernatant arriving at the interface with the ghosts, and the purification is minimal. Another white band (Fig. 4, II) can be seen at the overlay 30% sucrose interface and cloudy "streamers" of material reaching almost to the 30–45% interface. In general, the uppermost white band (mostly membrane vesicles) is broader and more prominent than the plasma membrane band at the 30–45% interface (Fig. 4). The latter band is removed by use of a syringe and the 14-gauge blunted needle by aspirating in the 45% sucrose just in and below the plane of the interface. Slight scraping and continued aspiration removes the small amount of material adhering to the glass at the 30–45% interface. Gelatinous aggregates adhering to the glass at the interface are due to lysed nuclei, insufficiently removed in previous steps. Visibility for these operations is improved by working in a cold room or by repeated removal of condensation with a paper towel. With practice, it is possible to aspirate the plasma membrane band from all four gradients in about 10–15 ml mainly of the 45% sucrose. In this way, excessive contamination from the 30% layer of sucrose is avoided and thus the smaller particles, such as membrane vesicles and fragments, mitochondria, and lysosomes are largely separated. As noted above, however, overloading the gradients with too much material results in a large amount of the contaminating small particles (Figs 1B, and 3B) to be found with the ghosts at the interface.

K. Washing and Concentration of Ghosts

The approximately 45% sucrose containing the ghosts obtained in the previous step is diluted 3- to 4-fold with Tris buffer in order to lower the density below that of the ghosts (i.e., below 1.16 gm/ml, see Fig. 6). The ghosts are then pelleted by centrifugation in the HB4 rotor for 5 minutes at 7500 rpm. It is not necessary to centrifuge for a longer time, in fact, this is undesirable since small particles will also be found in the pellet. Thus, this step serves to concentrate, wash, and purify ghosts. This washing procedure may be multiple, but the yield of ghosts falls each time owing to fragmentation and to ghosts sticking to the walls of the glass tube. From a starting number of 10^8 cells, a small pellet 1–2 mm deep covering approximately 25% of the bottom of the 30-ml Corex tube is observed. This is suspended in 1–2 ml of Tris buffer and observed in the phase microscope. A good preparation reveals a homogeneous population of whole ghosts with a few contaminating small particles (cf. Fig. 1C). When viewed in the electron microscope the ghosts contain few contaminant organelles besides rough endoplasmic reticulum (Fig. 2). Should the ghost population be heterogeneous in size (due to fragmentation) or grossly contaminated with particles, another cycle of rate zonal sedimentation on the discontinous sucrose gradient is applied. The material from the previous step is overlaid on 7 ml of 30% sucrose, 3 ml of 45% sucrose in a 15-ml Corex tube (Corning Cat. No. 8442). Centrifugation, collection, and washing of the ghosts is as above yielding a more homogeneous product size (Fig. 1C).

L. Estimation of Yield

Ghosts, concentrated as above are too numerous to count in the hemacytometer so a 1:10 dilution is made with Tris buffer. A drop of this is applied to a hemacytometer (Spencer Bright Line, A. H. Thomas Cat. No. 2936M25, 3 mm thickness) and the ghosts observed in the phase microscope. Unless the objective and condenser have sufficiently long working distrances (as the above-described instrument has), it may be difficult to adjust the instrument for best phase. However, a nearly focused condenser does almost as well or, failing this, one may remove the condenser top lens and rack the condenser all the way down to achieve an approximation to phase. If the ghosts are observed immediately after the drop is applied, they appear in many planes of focus, using a 25 X objective, making them difficult to count. However, if they are observed 10 minutes later, they have settled in the chamber to one plane of focus, thus simplifying the counting. After one sucrose gradient, a 20% recovery of cell surfaces is standard. With speed, care, and good cells, this figure may reach 40%. There are usually fragments of ghosts present, and either an estimate of their contribution should be made or another sucrose gradient cycle added to achieve a more homogeneous population of whole ghosts.

M. Electron Microscopy

Ghosts may be fixed and embedded by standard procedures, (Atkinson and Summers, 1971) except that centrifugation must be used to collect the ghosts after each step. Without this precaution, ghosts will be thrown away with, for example, the glutaraldehyde, owing to resuspension of ghosts in the fluid, since fixing agents do not aggregate ghosts into a permanently cohesive pellet. Ghosts purified by only one cycle of centrifugation and one Tris wash are contaminated with small amounts of rough endoplasmic reticulum and appear to have underlying fuzzy material attached to them (Fig. 2). Note the association of vesicles and sections of microvilli with the outer surface of the plasma membrane. A further centrifugation, Tris wash, and wash in 0.5 *M* salt causes some convolution, vesiculation, and loss of some of the underlying fuzzy material (Fig. 3A) and loss of most of the ribosomes. The loss of the ribosomes is reflected by a further 3- to 4-fold purification of fucose label compared to uridine label (see Table I).

N. Incorporation of Radioactive Precursors to Various Cellular Components

Fucose-^{3}H and ^{14}C-labeled amino acids are incorporated into TCA-insoluble material of concentrated HeLa cells in a linear fashion (Fig. 5). Subsequent hydrolysis and chromatography performed as described by Kaufman and Ginsburg (1968) of the fucose-^{3}H confirms their observation that radioactive fucose is a precursor only for fucose in HeLa cell macromolecules (probably glycoproteins, Trujillo and Gan, 1971; Atkinson and Summers, 1971). In a preceding paper (Atkinson and Summers, 1971) on the distribution of incorporated (TCA insoluble) radioactive fucose in cells after labeling periods of 4 hours or more, it was shown that much of the fucose can be accounted for in the plasma membrane. Here it is shown that the specific radioactivity of fucose in the various particulate fractions separated in the first rate zonal gradient is maximal in the plasma membrane fraction (Fig. 4). Since the bulk of cell protein is probably not fucosylated, since DNA as labeled by radioactive thymidine is not expected to be a major component of surface membranes, and since the bulk of components labeled with radioactive uridine will be RNA, it should be possible to demonstrate an enrichment of radioactive fucose compared to these labels and to the amount of protein present as purification proceeds. Experiments along these lines show that in fact this is the case (Table I). The specific radioactivity of incorporated fucose (counts per minute per microgram of protein) increases 5-fold in the plasma membrane after one sucrose gradient and wash and increases 12-fold after the second gradient and wash (Table I). At this step the plasma membrane preparation, when corrected for yield, contains 4–7% of the cell protein, less than 1.3% of the cell DNA, less that 1% of the cell RNA and most of the fucosylated glycoprotein. The subsequent salt wash results in further enrichment of the fucose label (Table I), but yield estimates were not possible due to ghost fragmentation.

O. Isopycnic Centrifugation

One or two rate-zonal centrifugation steps are usually sufficient to obtain a purified product in that other organelles are eliminated and a distinct cell fraction as defined by presence of radioactive precursors and virus molecules (Table I, Fig. 8) is produced. However, a product of homogeneous density may be obtained by isopycnic centrifugation on a 10–45% w/w continuous sucrose gradient; 0.3–0.5 ml of the concentrated product labeled with

TABLE I

COMPARISON OF THE INCORPORATION OF RADIOACTIVE FUCOSE INTO VARIOUS SUBCELLULAR FRACTIONS WITH OTHER LABELED PRECURSORS[a]

Cell fractions	Ratio[b] amino acids-^{14}C to L-fucose-^{3}H	Ratio thymidine-^{3}H to L-fucose-^{14}C	Ratio uridine-^{14}C to L-fucose-^{3}H	Cpm fucose-^{14}C per μg protein
Homogenized cells	26.4	16.9	8.3	9.7
500 *g* min. supernatant from homogenized cells (mainly ghosts, small particles and soluble material)	10.0	1.4	3.6	12.7
500 *g* min. pellet from homogenized cells (mainly nuclei)	ND	123.6	ND	7.5
Nuclei washed with detergent, final concentration approx. 1.0% Tween, 0.5% DOC (Penman, 1966)	72.3	435.5	43.1	3.2
Material washed from nuclei with detergent, final concentration approx. 1.0% Tween, 0.5% DOC	20.0	0.9	3.1	7.2
Plasma membrane ghosts after one discontinuous gradient	3.0	ND	0.8	19.4
Plasma membrane ghosts after one discontinuous gradient and a wash in 10 m*M* Tris, pH 8	2.0	0.6[c]	0.7	52.5

Plasma membrane ghosts after two discontinuous gradients and wash in 10m*M* Tris, pH 8	1.3	0.2[d]	0.4	123.9[f]
Plasma membrane ghosts as above and after a wash in 0.5 *M* NaCl, 10 m*M* Tris, pH 8	1.3	0.1[e]	0.2	ND

[a]These data are reprinted in part from Atkinson and Summers (1971) by permission of the American Society for Biological Chemists, Inc., Baltimore, Maryland.

[b]Ratio of trichloracetic acid precipitable radioactive amino acids to precipitable radioactive fucose in each of the isolated cell fractions.

[c]The amount of incorporated thymidine-^{3}H corrected for yield of plasma membranes, expressed as a percentage of that in whole cells = 2.8% fucose-^{14}C expressed similarly = 72%

[d]As for footnote *c*, but thymidine-^{3}H = 1.3% and fucose-^{14}C = 90–100%.

[e]A further 2-fold purification of thymidine-^{3}H to fucose-^{14}C was obtained at this step, thus, though ghosts were not counted, the amount of thymidine-^{3}H present per membrane must be less than 1% that per cell.

[f]The amount of protein in the membranes as percentage of the cell protein, corrected for yield, is 4–7% at this step. ND = not determined. The time of incorporation of all the labeled precursors (amino acids-^{14}C, L-fucose-^{3}H, thymidine-^{3}H, L-fucose-^{14}C, and uridine-^{14}C) was 20–24 hours.

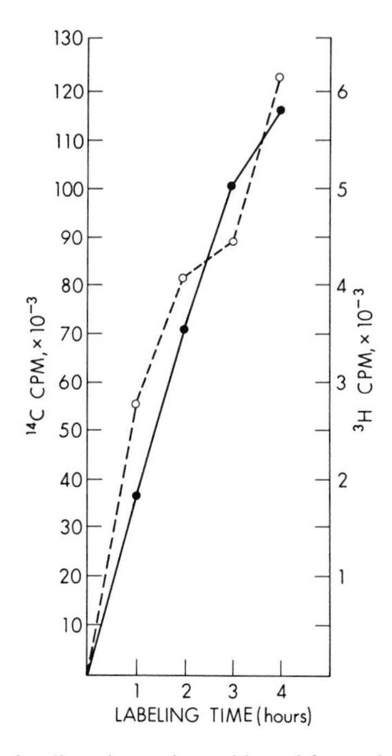

FIG. 5. Incorporation of radioactive amino acids and fucose into HeLa cells. HeLa cells, 500 ml, were grown to a density of 3×10^5 cells/ml, collected by centrifugation, and resuspended in 100 ml of MEM ($0.2 \times$ normal concentration amino acids), 3.5% calf and 3.5% fetal calf serum in a 125-ml conical flask. The pH level was maintained near 7 by placing or removing a stopper; 300 μCi of ^{14}C-labeled amino acids mixture and 600 μCi of fucose-^{3}H were added. At hourly intervals, duplicate 0.1-ml samples were removed and mixed with 1.0 ml of distilled water. TCA was immediately added to a final concentration of 5%; the precipitate was collected on Whatman filters (GFA) and washed with 5% TCA, 95% ethanol. The radioactivity was counted as given in the text. ●——●, ^{14}C-labeled amino acids; ○-----○, fucose-^{3}H.

fucose-^{3}H and amino acids-^{14}C from previous steps is overlayered on a 4.5-ml 10 to 45% sucrose gradient in a 5-ml Spinco SW 65 plastic tube. The gradient is centrifuged at 45,000 rpm for 4 hours at 4°C in a Spinco SW 65 rotor, fractionated into 32 fractions (approximately 11 drops through a 19-gauge needle per fraction), and an aliquot is analyzed for radioactivity. The ghosts band sharply around a density of 1.16 gm/ml (Fig. 6). Ghosts banding at various densities between 1.155 gm/ml and 1.165 gm/ml had very similar profiles (Atkinson and Summers, 1971) when assayed by SDS-acrylamide gel electrophoresis. The ghosts did not appear morphologically different from those obtained after rate-zonal sedimentation on discontinuous gradients.

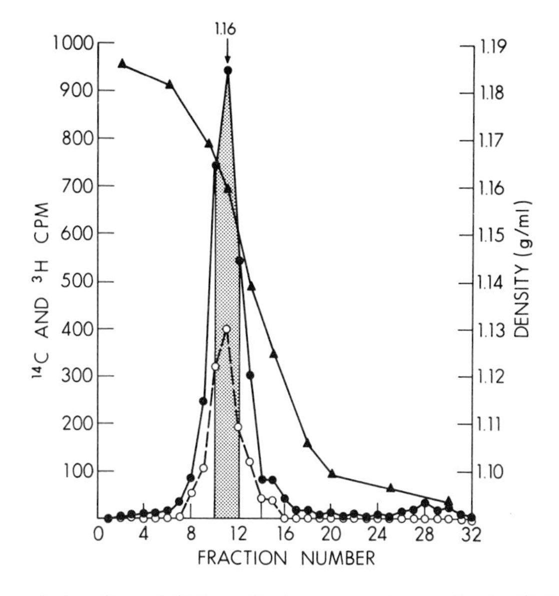

FIG. 6. Isopycnic banding of HeLa cell plasma membrane ghosts. HeLa cells, 250 ml, at a density of 3.5×10^5 cells/ml were collected by centrifugation and resuspended in 50 ml of medium containing ($0.2 \times$ MEM concentration of amino acids), 3.5% calf and 3.5% fetal calf serum, 4 μCi/ml of fucose-^{3}H and 2 μCi/ml of ^{14}C-labeled amino acids. The cells were incubated 4 hours at 37°C, and plasma membranes were purified by 2 cycles of zonal centrifugation as outlined in the text. After the final Tris wash, ghosts were resuspended in 1.0 ml of Tris buffer and 0.5 ml of the suspension was overlaid on a 4.5 ml 10–45% w/w sucrose gradient buffered with Tris buffer. The gradient was centrifuged at 45,000 rpm for 4 hours at 4°C in a Spinco SW 65 rotor and fractionated into 11-drop fractions by use of a 19-gauge needle. Densities of fractions were determined by weighing in 100-μl disposable pipettes. Radioactivity of 5-μl aliquots was determined as outlined in the text. ▲——▲, Density; ●——●, ^{14}C-labeled amino acids; ○-----○, fucose-^{3}H amino acids. Reprinted from Atkinson and Summers (1971) by permission of the American Society for Biological Chemists, Inc, Baltimore, Maryland.

III. Uses of the Method

A. Plasma Membrane Precursors

When the percentage of the total cell incorporated radioactive fucose found in the plasma membrane is determined as a function of labeling time starting 5 to 10 minutes after labeling has begun, there is a lag of about 15 minutes in the incorporation into the membrane macromolecules compared to those in the whole cells (Fig. 7). The implication is that fucose is incorporated into glycoproteins at some site other than plasma membranes inside

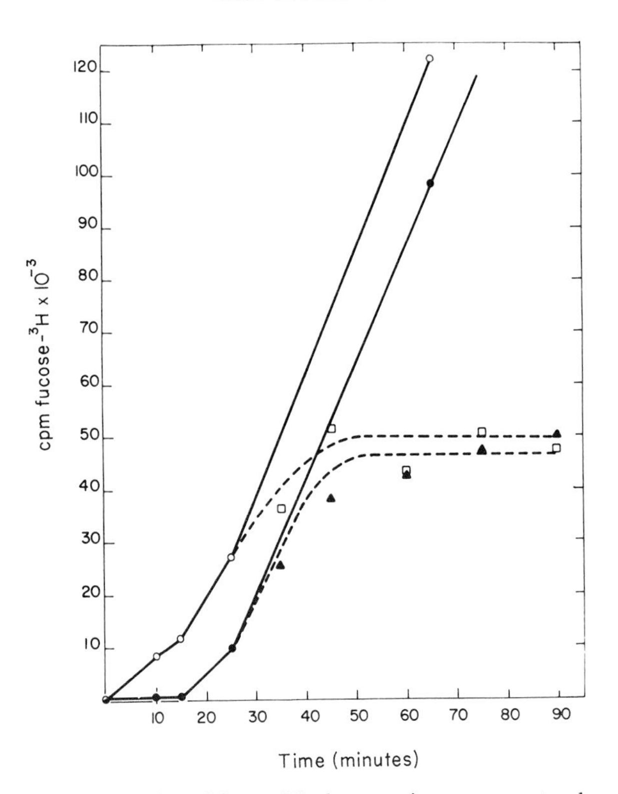

FIG. 7. Pulse-chase kinetics of fucose-^{3}H glycoprotein precursors to plasma membrane glycoproteins. HeLa cells, 1000 ml, at a density of 2.0×10^5 cells/ml were grown for 16 hours in MEM supplemented with 3.5% calf, 3.5% fetal calf serum in the presence of 0.1 μCi/ml fucose-^{14}C. Cells were then collected by centrifugation and resuspended in 200 ml of medium of the same composition. The cells were allowed to equilibrate 1 hour, 0.5 ml of 1.0 *M* HEPES (Calbiochem) was added to help maintain the pH; 2.0 mCi of fucose-^{3}H was then added (zero time in the figure). The culture was immediately divided, using appropriately sized rubber-stoppered flasks, into 2 cultures of 175 ml (A) and 25 ml (B). At 25 minutes after the addition of the fucose-^{3}H the (A) culture was adjusted to 40 m*M** with respect to nonradioactive fucose (Sigma Chem. Co. St. Louis, Missouri). This results in approximately a 40,000-fold dilution of the label in the medium. At 10, 15, 25, 35, 45, 60, 75, and 90 minutes after the addition of the fucose-^{3}H, 20-ml aliquots of the (A) culture were removed and stored briefly in an ice bucket subsequent to the isolation of plasma membranes. At 65 minutes after the addition of fucose-^{3}H, a 20-ml aliquot was removed from culture (B) for the same purpose. Plasma membranes were prepared by one cycle of zonal centrifugation as outlined in the text. ^{14}C and ^{3}H radioactivity was determined in aliquots of the Dounce homo-

*In a growth study over 3 days in which cell number was determined in a Coulter Counter control cells grew from 15×10^4 cells/ml to 29×10^4 cells/ml, 57×10^4 cells/ml, 87×10^4 cells/ml in 1, 2, 3 days of growth, respectively. Cells grown in the presence of 40 m*M* L-fucose grew from 15×10^4 cells/ml, to 29×10^4 cells/ml, 48×10^4 cells/ml, 75×10^4 cells/ml in the same period. Thus 40 m*M* L-fucose is only slightly inhibitory to the growth of HeLa cells as grown here.

the cell. A pulse-chase experiment in which cells are labeled for a short period (25 minutes) and then further incorporation of radioactive fucose into glycoproteins is stopped by the addition of excess nonradioactive fucose shows that fucose-labeled macromolecules inside the cell are transported to the plasma membrane (Fig. 7). Thus, it appears radioactive fucose can be used as a relatively specific label for HeLa cell plasma membrane precursors. It is not known whether these precursors are free glycoproteins, bound glycoproteins, or even perhaps complete membrane subcomponents containing lipids, proteins, and carbohydrates. This type of experiment would be very difficult to perform were it not for the rapidity of the isolation procedure (the experiment had nine preparations, each taking 45 to 60 minutes) and for the fact that the distribution of radioactive fucose in cells and cell surfaces can be quantitated by counting ghosts.

B. Kinetics of Synthesis of Membrane Maturing Viruses

Vesicular stomatitis virus (VSV) matures by budding from the host cell surface membrane. To more fully understand this process, Cohen *et al.* (1971) followed the incorporation of virus polypeptides into HeLa cell plasma membranes as a function of short-term labeling times. In this, as in the previous experiment, events were occurring on a time scale of minutes, necessitating the preparation of plasma membranes at a number of close-spaced time points. The arrival of the nucleocapsid protein (III, Fig. 8B) appears to follow that of the spike glycoprotein (II, Fig. 8B) and the membrane protein (V, Fig. 8B). The possible adventitious association of V during the isolation of the membranes could not be ruled out, however, because

---

genate and was a measure of radioactivity in the whole cells. Radioactivity was also determined in aliquots of the plasma membrane ghost preparations. Ghosts were counted to determine the yield as detailed in the text. When the percentage of whole cell fucose-^{14}C radioactivity in the ghosts was determined, it was found to be a similar figure to the yield estimate. In illustration, in several preparations the figures for ghost count percent of whole cells versus ^{14}C percent of whole cells were: 18.6 vs 17.8%, 17.9 vs 20.7%, 24.2 vs 23.6%, 19.8 vs 20.7%, respectively. Thus the yield of plasma membranes was more simply determined by the percent whole cell fucose-^{14}C obtained in each preparation.* The figure therefore expresses fucose-^{3}H TCA-insoluble radioactivity in whole cells and in the purified plasma membranes corrected for yield of plasma membranes. ○——○, Whole cells, no chase; □——□, whole cells, after adjustment of medium to 40 m*M* cold fucose; ●——●, plasma membranes, no chase; ▲-----▲, plasma membranes, after adjustment of medium to 40 m*M* cold fucose.

*This is an assumption, largely borne out by the ghost counts, based on the observation that much of the incorporated radioactive fucose after a long labeling period (16 hours in this case) is found in HeLa cell plasma membranes [Atkinson and Summers (1971); see also Table I].

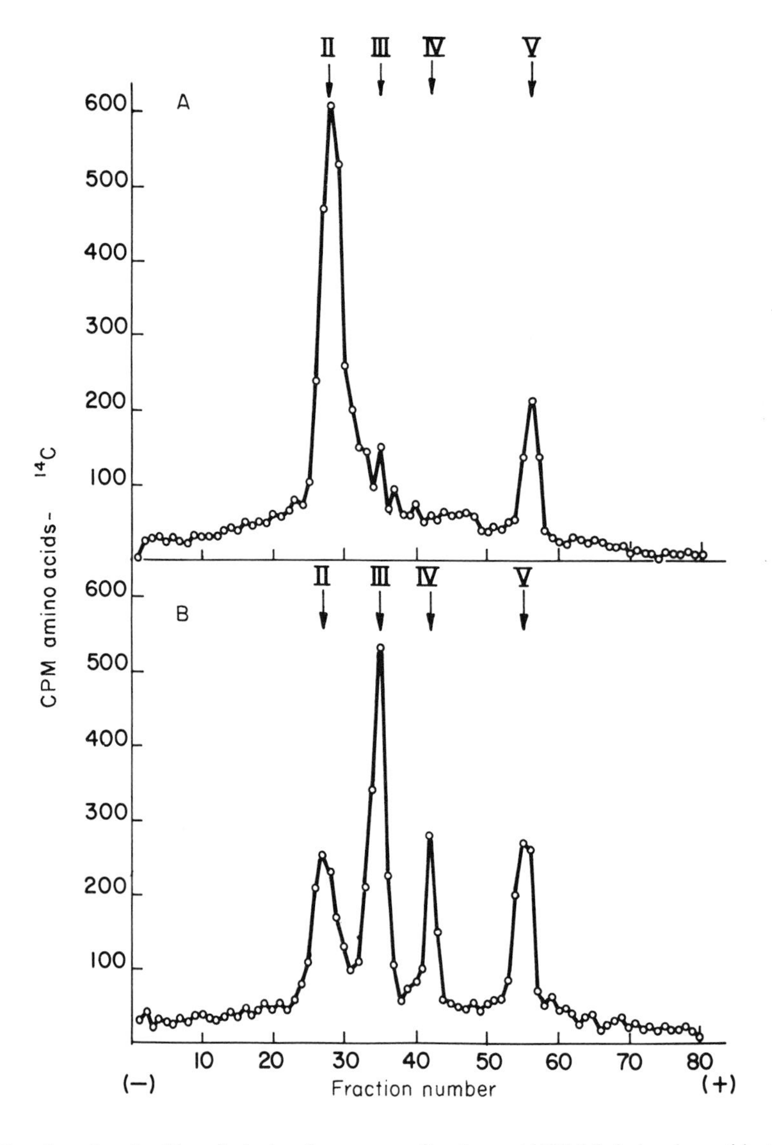

FIG. 8. Acrylamide gel electropherograms of amino acid-^{14}C-labeled polypeptides from VSV infected HeLa cells and plasma membranes. Conditions for infection, labeling and preparation of plasma membranes are described elsewhere (Cohen *et al.*, 1971). (A) Infected cells plasma membranes pulse-labeled for 5 minutes. (B) Cytoplasm from the same cells. II, III, IV, V are the positions that amino acid-^{3}H-labeled polypeptides from purified virus migrated in the gel. Reprinted from Cohen *et al.* (1971), by permission of Macmillan, New York.

mixing radioactive infected cell supernatant and unlabeled plasma membranes demonstrated an *in vitro* association of V. An interesting point brought out by the data in this paper is that at 5 minutes of labeling time the pattern of virus polypeptides in the plasma membrane differed distinctly from the pattern in the cell as a whole (Fig. 8A, cf. Fig. 8B). The whole cell pattern was similar to that of purified virus (Cohen *et al.*, 1971; Mudd and Summers, 1970). We can conclude therefore that plasma membranes isolated as detailed here are relatively "pure" and distinct as regards the distribution of VSV polypeptides.

C. Cell Cycle Studies

A study aimed at determining the time of maximum incorporation of radioactive fucose during the cell cycle (Nowakowski *et al.*, 1972) also required the preparation of a number of plasma membrane fractions (8) at close-spaced intervals. In this way, we were able to determine that maximum incorporation occurred in the late S, G2 phase of growth. Whether this reflects an increased rate of glycoprotein synthesis is being investigated by measurement of the specific radioactivity of precursor pools.

IV. Discussion

A. Other Cell Types

Success in preparing plasma membrane ghosts similar to those shown in Fig. IC from other cell types has been somewhat limited and the method detailed here should be regarded as useful mainly for HeLa spinner cells. Successful preparations have been made from mouse myeloma cells (strain 45–6 kindly donated by Dr. M. D. Scharff, Department of Cell Biology), which is also grown in suspension culture. I have not tried other suspension cultures, but Hubbard (1972) used this method of preparation of L-cell plasma membranes. A mouse cell line grown in monolayer (JLSV9 and JLSV9 chronically infected with Rauscher leukemia virus) also yielded satisfactory ghosts on a regular basis in experiments aimed at studying virus polypeptides and virus-specific enzymes in the plasma membrane fraction. However, in all these preparations ghosts are not yet characterized, as for HeLa cells, so their distinctness as a cell fraction is not known. Human embryo diploid lung fibroblasts (KL2, kindly donated by Dr. E. Levine, Department of Cell Biology) also yielded satisfactory ghosts in many preparations, but at times some cultures of KL2 were totally refractory, for unknown reasons. Another human embryo diploid fibroblast (kindly donated by Dr. H. Klinger, Department of Genetics) on several occasions did not yield whole ghosts on Dounce homogenization. Balb/c 3T3 cells yielded ghosts on homogenization, but further purification was not attempted. I have not attempted to define the critical factors for preparation of ghosts from monolayer cultures, and neither have successful products been much characterized. However, some points seem worthy of note. First, ghosts can be derived from monolayer cells after their removal by the usual procedures of trypsinization. Cells that were not regularly spherical after trypsinization gave fragmented ghosts at low yield. Spherical monolayer cells can be achieved by a combination of thorough trypsinization and increased swelling time in hypotonic buffer. Second, only limited success was achieved in separating whole monolayer cells and their nuclei from ghosts by brief centrifugation. The ghosts apparently had a very similar sedimentation rate to these components. All mouse cells displayed similar characteristics. This step has not been explored, and it may be possible to partition ghosts and nuclei as for HeLa cells by paying greater attention to dilution of the homogenate, presence of cations, sucrose, and so on. Others have had more success (Quigley *et al.*, 1971), and the ghosts derived from chick embryo fibroblasts did partition substantially from nuclei and whole cells in a brief centrifugation (200–300 *g* 3 minutes).

B. Plasma Membranes and Fucose

Other authors have reported that fucose is enriched in plasma membrane fractions. Gahmberg (1971) observed a 9- to 12-fold concentration of fucose-^{3}H in the plasma membranes of hamster-kidney fibroblasts as prepared by a modification of the method of Kamat and Wallach (1965). He concluded that, in these cells radioactive fucose is as good a marker for the plasma membranes as (Na^+, K^+)-activated ATPase, gangliosides, and surface antigens. Heine *et al.* (1972) used the Kamat and Wallach (1965) procedure to purify plasma membranes from human epidermoid carcinoma cells (Hep-2) infected with Herpes virus. Compared to the microsomal fraction, radioactive fucose became purified 14-fold in the plasma membrane fraction, and 5′-nucleotidase was purified 26-fold. Using the Zn^{2+} and FMA method of Warren *et al.* (1966) for the purification of plasma membranes from L cells, Glick *et al.* (1970) observed $2.3 \pm 2.2 \times 10^{-10}$ μmole (36% of the cell total) and $0.7 \pm 0.6 \times 10^{-10}$ μmole (11%) of fucose per surface membrane by each of these methods, respectively. Shen and Ginsburg (1968) were able to release a large amount of the total cell fucose (40%) by treatment of intact HeLa spinner cells with trypsin, and 31% from HeLa cells grown in monolayer culture. The enzyme treatment was believed not to affect cell viability, and thus a large amount of HeLa cell fucose was located on the cell surface. Bosmann *et al.* (1969) studied the uptake of fucose-^{14}C and glucosamine-^{14}C into HeLa cells and the plasma membrane fraction. In a 60-minute labeling period, the specific fucose radioactivity of the homogenate was 5.7×10^{-4} cpm/mg protein and in the plasma membrane fraction (S7) the figure was 7.5×10^{-4} cpm/mg protein. In an experiment in which I pulse-labeled HeLa cells for 40 minutes (under similar conditions to Fig. 7) with fucose-^{3}H the specific radioactivity of the homogenate was 6.9 cpm/μg protein while the plasma membrane fraction was 45.3 cpm/μg protein. This indicates a specific enrichment of radioactive fucose in plasma membranes, even in short labeling times, as prepared here but not as prepared by Bosmann *et al.* (1969). The reason for this discrepancy is not clear. However, when Bosmann *et al.* pulsed HeLa cells for 30 minutes with glucosamine-^{14}C and chased with an excess of glucosamine for 3.5 hours, the specific radioactivity of the smooth membrane fractions (S1 and S2) decreased 2- to 3-fold while that of the plasma membrane fraction increased about 5- to 6-fold. This is analogous to the situation I have observed when pulse-chasing radioactive fucose (Fig. 7). After a 25-minute label, the specific radioactivity of fucose-^{3}H incorporated in the plasma membrane was 16.9 cpm/μg of protein, and after a 45-minute chase with excess fucose the specific radioactivity was 96.8 cpm/μg protein. From Fig. 7, I surmise that it takes 15 to 20 minutes for newly completed fucose containing macromolecules to arrive in the plasma membrane and that it takes 20 to 30 minutes to dilute the internal radioactive pool of precursors (fucose, fucose 1-phosphate, fucose-GDP) sufficiently to prevent further incorporation of radioactive fucose into macromolecules. Bennett (1971) determined the distribution of silver grains over columnar cells of rat duodenal villi at various time intervals after injection of fucose-^{3}H. When the percentage of total grains was determined over various organelles, it was found that radioactive molecules moved from the Golgi apparatus to the surface membranes in about the same time as is indicated by the lag between labeling whole cells and plasma membranes (15 to 20 minutes) in HeLa cells (Fig. 7). Rambourg (1971) reported a similar time course for the distribution of fucose-^{3}H in epithelial cells. These studies and those of Bennett and Leblond (1970) are thus in agreement with those reported here; that much of the radioactive fucose incorporated in cells eventually becomes localized in the cell surface. I assume injection of radioactive fucose into an animal results in a "pulse" of this precursor, in the order of minutes, followed by a chase caused by a subsequent rapid dilution of the isotope. Thus these *in vivo* studies should approximate to the situation described here for HeLa cells. The plasma membrane can be regarded therefore as a sink for fucose-containing macromolecules which are made inside the cell possibly in the Golgi apparatus as is indicated by the data of Bennett (1971) and Bennett and Leblond (1970). The subsequent fate of these molecules may include release, excretion, interiorization, or accumulation in the plasma membrane.

Acknowledgments

I wish to thank Ellen Sikora for technical assistance, Phyllis M. Novikoff for electron micrographs, and Cleveland Davis for preparation of material for electron microscopy.

References

Allan, D., and Crumpton, M. J. (1970). *Biochem J.* **120**, 133.
Allan, D., and Crumpton, M. J. (1972). *Biochim. Biophys. Acta* **274**, 22.
Ashworth, L. A. E., and Green, C. (1966). *Science* **151**, 210.
Atkinson, P. H., and Summers, D. F. (1971). *J. Biol. Chem.* **246**, 5162.
Artzt, K. J., Sanford, B. H., and Marfey, P. S. (1968). *J. Lab. Clin. Med.* **72**, 350.
Barclay, M., Barclay, R. K., Essner, E. S., Shipski, V. P., and Terebus-Kekish, O. (1967). *Science* **156**, 665.
Barland, P., and Schroeder, E. A. (1970). *J. Cell Biol.* **45**, 662.
Benedetti, E. L., and Emmelot, P. (1967). *J. Cell Sci.* **2**, 499.
Bennett, G., and Leblond, C. P. (1970). *J. Cell Biol.* **46**, 409.
Bennett, G. C. (1971). Ph. D. Thesis, McGill University.
Berman, H. M., Gram, W., and Spirtes, M. A. (1969). *Biochim. Biophys. Acta* **183**, 10.
Bingham, R. W., and Burke, D. C. (1972). *Biochim. Biophys. Acta* **274**, 348.
Boone, C. W., Ford, L. E., Bond, H. E., Stuart, D. C., and Lorenz, D. (1969). *J. Cell Biol.* **41**, 378.
Bose, H. R., and Brundige, M. A. (1972). *J. Virol.* **9**, 785.
Bosmann, H. B., Hagopian, A., and Eylar, E. H. (1968). *Arch. Biochem. Biophys.* **128**, 51.
Bosmann, H. B., Hagopian, A., and Eylar, E. H. (1969). *Arch. Biochem. Biophys.* **130**, 573.
Brunette, D. M., and Till, J. E. (1971). *J. Membrane Biol.* **5**, 215.
Buck, C. A., Glick, M. C., and Warren, L. (1970). *Biochemistry* **9**, 4567.
Campbell, B. J., Woodward, G., and Borberg, V. (1972). *J. Biol. Chem.* **247**, 6167.
Chlapowski, F. J., Bonneville, M. A., and Staehelin, L. A. (1972). *J. Cell Biol.* **53**, 92.
Cohen, G. H., Atkinson, P. H., and Summers, D. F. (1971). *Nature (London) New Biol.* **231**, 121.
Coleman, R., and Finean, J. B. (1966). *Biochem. Biophys. Acta* **125**, 197.
Coleman, R., Michell, R. H., Finean, J. B., and Hawthorne, J. N. (1967). *Biochim. Biophys. Acta* **135**, 573.
Combret, Y., and Laudat, P. (1972). *FEBS (Fed. Eur. Biochem. Soc.) Lett.* **21**, 45.
DePierre, J. W., and Karnovsky, M. L. (1973). *J. Cell Biol.* **56**, 275.
Eagle, H. (1959). *Science* **130**, 432.
Earle, W. R. (1943). *J. Nat. Cancer Inst.* **4**, 165.
Emmelot, P., Bos, C. J., Benedetti, E. L., and Rumke, PH. (1964). *Biochim. Biophys. Acta* **90**, 126.
Essner, E., Novikoff, A. B., and Masek, B. (1958). *J. Biophys. Biochem. Cytol.* **4**, 711.
Evans, W. H. (1970). *Biochem. J.* **116**, 833.
Evans, W. H., and Bruning, J. W. (1971). *Immunology* **19**, 735.
Ferber, E., Resch, K., Wallach, D. F. H., and Imm, W. (1972). *Biochim. Biophys. Acta* **266**, 494.
Finn, F. M., Widnell, C. C., and Hofman, K. (1972). *J. Biol. Chem.* **247**, 5695.
Forstner, G. G., Sabesin, S. M., and Isselbacher, K. J. (1968). *Biochem. J.* **106**, 381.

Fujita, M., Ohta, H., Kawai, K., Matsui, H., and Nakao, M. (1972). *Biochim. Biophys. Acta* **274**, 336.
Gahmberg, C. G. (1971). *Biochim. Biophys. Acta* **249**, 81.
Gerner, E. W., Glick, M. C., and Warren, L. (1970). *J. Cell. Physiol.* **75**, 257.
Glick, M. C., Comstock, C., and Warren, L. (1970). *Biochim. Biophys. Acta* **219**, 290.
Goodenough, D. A., and Stoechenius, W. (1972). *J. Cell Biol.* **54**, 646.
Graham, J. M., Higgins, J. A., and Green, C. (1968). *Biochim. Biophys. Acta* **150**, 303.
Hecht, T., and Summers, D. F. (1972). *J. Virol.* **10**, 578.
Heine, J. W., and Schnaitman, C. A. (1971). *J. Cell Biol.* **48**, 703.
Heine, J. W., Spear, P. J., and Roizman, B. (1972). *J. Virol.* **9**, 431.
Henn, F. A., Hansson, H.-A., and Hamberger, A. (1972). *J. Cell Biol.* **53**, 654.
Hicks, R. M., and Ketterer, B. (1970). *J. Cell Biol.* **45, 542.**
Hubbard, A. (1972). Personal communication.
Kamat, V. B., and Wallach, D. F. H. (1965). *Science* **148**, 1343.
Kaufman, R. L., and Ginsberg, V. (1968). *Exp. Cell Res.* **50**, 127.
Keenan, T. W., Morre, D. J., Olson, D. E., Yunghans, W. N., and Patton, S. (1970). *J. Cell Biol.* **44**, 80.
Klenk, H.-D., and Choppin, P. W. (1969). *Virology* **38**, 255.
Kono, T., Kakuma, F., Homma, M., and Fukuda, S. (1964). *Biochim. Biophys. Acta* **88**, 155.
Kraemer, P. M. (1971). *In* "Biomembranes" (L. A. Manson, ed.), Vol. I, pp. 67–190. Plenum, New York.
Labrie, F., Barden, N., Poirier, G., and DeLean, A. (1972). *Proc. Nat. Acad. Sci. U.S.* **69**, 283.
Lauter, P., Solyom, A., and Trams, E. C. (1972). *Biochim. Biophys. Acta* **266**, 511.
Lazarowitz, S. G., Compans, R. W., and Choppin, P. W. (1971). *Virology* **46**, 830.
Leslie, G. I., and Rowe, P. B. (1972). *Biochemistry* **11**, 1696.
McCollester, D. L. (1962). *Biochim. Biophys. Acta* **57**, 427.
McKeel, D. W., and Jarett, L. (1970). *J. Cell Biol.* **44**, 417.
McLaren, L. C., Scaletti, J. V., and James C. G. (1968). *In* "Biological Properties of the Mammalian Surface Membrane" (L. A. Manson, ed.), Monograph 8, pp. 123–135. Wistar Inst. Press, Philadelphia, Pennsylvania.
Meldolesi, J., Jamieson, J. D., and Palade, G. E. (1971). *J. Cell Biol.* **49**, 109.
Mokrasch, L. C. (1971). *In* "Methods of Neurochemistry" (R. Fried, ed.), Vol. I, pp. 2–23. Dekker, New York.
Mudd, J. A., and Summers, D. F. (1970). *Virology* **42**, 328.
Neville, D. M., Jr. (1960). *J. Biophys. Biochem. Cytol.* **8**, 413.
Nowakowski, M., Atkinson, P. H., Summers, D. F. (1972). *Biochim. Biophys. Acta* **266**, 154.
Penman, S. (1966). *J. Mol. Biol.* **17**, 117.
Perdue, J. F., and Sneider, J. (1970). *Biochim. Biophys. Acta* **196**, 125.
Perdue, J. F., Kletzein, R., and Wray, V. L. (1972). *Biochim. Biophys. Acta.* **266**, 505.
Philipson, L., Lonberg-Holm, K., and Petterson, U. (1968). *J. Virol.* **2**, 1064.
Pricer, W. E., Jr., and Ashwell, G. (1971). *J. Biol. Chem.* **246**, 4825.
Quigley, J. P., and Gotterer, G. S. (1969). *Biochim. Biophys. Acta* **173**, 456.
Quigley, J. P., Rifkin, D. B., and Reich, E. (1971). *Virology* **46**, 106.
Rambourg, A. (1971). *Int. Rev. Cytol.* **31**, 57.
Ray, T. K. (1970). *Biochim. Biophys. Acta* **196**, 1.
Renkonen, O., Kääräinen, L., Simons, K., and Gahmberg, C. G. (1971). *Virology* **46**, 318.
Reynolds, E. S. (1963). *J. Cell Biol.* **17**, 208.
Rodbell, M. (1967). *J. Biol. Chem.* **242**, 5744.
Rosenthal, S. L., Edelman, P. M., and Schwarz, I. L., (1965). *Biochim. Biophys. Acta* **109**, 512.
Ryan, J. W., and Smith, U. (1971). *Biochim. Biophys. Acta* **249**, 177.
Sabatini, D. D., Bensch, K., and Barrnett, R. J. (1963). J. Cell Biol. **17**, 19.

Sheinin, R., and Onodera, K. (1972). *Biochim. Biophys. Acta* **274**, 49.
Shen, L., and Ginsburg, V. (1968). *In* "Biological Properties of the Mammalian Surface Membrane" (L. A. Manson, ed.), Monograph No. 8, pp. 67–71. Wistar Inst. Press, Philadelphia, Pennsylvania.
Simon, T. R., Blumenfeld, O. O., and Arias, I. M. (1970). *Biochim. Biophys. Acta* **219**, 349.
Solyom, A., Lanter, C. J., and Trams, E. C. (1972). *Biochim. Biophys. Acta* **274**, 631.
Song, C.`S., and Bodanksy, O. (1967). *J. Biol. Chem.* **242**, 694.
Steck, T. L., and Wallach, D. F. H. (1970). *In* "Methods in Cancer Research" (H. Busch, ed.), Vol. 5, pp. 93–153. Academic Press, New York.
Stein, Y., Widnell, C., and Stein, O. (1968). *J. Cell Biol.* **39**, 185.
Takeuchi, M., and Terayama, H. (1965). *Exp. Cell Res.* **40**, 32.
Touster, O., Aronson, N. N., Jr. Dulaney, J. T., Hendrickson, H. (1970). *J. Cell Biol.* **47**, 604.
Trujillo, J. L., and Gan, J. C. (1971). *Biochim. Biophys. Acta* **230**, 610.
Turkington, R. W. (1962). *Biochim. Biophys. Acta* **65**, 386.
Wagner, R. R. Kiley, M. P., Snyder, R. M., and Schnaitman, C. A. (1972). *J. Virol.* **9**, 672.
Wallach, D. F. H. (1967). *In* "The Specificity of Cell Surfaces" (B. D. Davis and L. Warren, eds.), pp. 129–163. Prentice-Hall, Englewood Cliffs, New Jersey.
Wallach, D. F. H., and Kamat, V. B. (1964). *Proc. Nat. Acad. Sci. U.S.* **52**, 721.
Warren, L., and Glick, M. C. (1971). *In* "Biomembranes" (L. A. Manson, ed.), Vol. I, pp. 257–288. Plenum, New York.
Warren, L., Glick, M. C., and Nass, M. K. (1966). *J. Cell. Physiol.* **68**, 269.
Warren, L., Critchley, D., and Macpherson, I. (1972a). *Nature* (*London*) **235**, 275.
Warren, L., Fuhrer, J. P., and Buck, C. A. (1972b). *Proc. Nat. Acad. Sci. U.S.* **69**, 1838.
Weaver, R. A., and Boyle, W. (1969). *Biochim. Biophys. Acta* **173**, 377.
Weineke, A. A., and Woodin, A. M. (1967). *Biochem. J.* **105**, 1039.
Widnell, C. C. (1972). *J. Cell. Biol.* **52**, 542.
Widnell, C. C., and Unkeless, J. C. (1968). *Proc. Nat. Acad. Sci. U.S.* **61**, 1050.
Wilson, L. A., and Amos, D. B. (1972). *Tissue Antigens* **2**, 105.
Wolff, J., and Jones, A. B. (1971). *J. Biol. Chem.* **246**, 3939.
Wu, H. C., Meezan, E., Black, P. H., and Robbins, P. W. (1969). *Biochemistry* **8**, 2509.
Yamashita, K., and Field, J. B. (1970). *Biochem. Biophys. Res. Commun.* **40**, 171.

ADDENDUM

A very extensive and constructively critical review of methods for the isolation and characterization of plasma membranes from mammalian cells has been published (De Pierre and Karnovsky, 1973) since this article went to press.

Chapter 9

Mass Enucleation of Cultured Animal Cells

D. M. PRESCOTT AND J. B. KIRKPATRICK

Department of Molecular, Cellular and Developmental Biology, University of Colorado, Boulder, Colorado

I. Introduction

The nuclei of animal cells growing as monolayers can be removed from the cells by centrifugation in nutrient medium containing the drug cytochalasin B. A preliminary account of the method has been published (Prescott *et al.*, 1972). The procedure to be described here contains modifications of the original method that permit more rapid and more efficient enucleation. Three other procedures for enucleation with cytochalasin B have been described (Wright and Hayflick, 1972; Wright, this volume; Poste and Reeve, 1972; Poste, 1972; Poste, this volume).

The availability of enucleated cells permits a number of otherwise difficult experiments on the individual roles of the nucleus and cytoplasm in a variety of cell activities. Enucleated cells, which survive for 1–3 days, can be useful for many types of investigations. The technique could facilitate turnover studies of cytoplasmic constituents, e.g., messenger RNA, the analysis of cell-virus interactions, and various kinds of cell fusion studies. We (Prescott *et al.*, 1971) have used enucleated cells to show the absence of a nuclear role in the replication of DNA of vaccinia virus.

The nuclei that are removed from cells remain metabolically active for some hours and may be useful in some types of experiments—for example, for the construction by fusion of cells composed of cytoplasm from one kind of cell and a nuclei from another.

II. Cells

We have enucleated Chinese hamster ovary cells, mouse L cells, and chick cells with about equal efficiency. The method is probably applicable to any type of cell that will grow as a monolayer. Cells are grown in the usual way on glass or plastic coverslips. Plastic coverslips are preferable because the cells attach more firmly to plastic than to glass and because the problem of breakage of coverslips during centrifugation is eliminated. We have also grown cells on octagonal shaped pieces of glass cut with a diamond pencil from ordinary glass slides. These thicker pieces of glass do not break during centrifugation.

III. Manufacture of Plastic Coverslips

Plastic coverslips are cut from plastic tissue culture dishes. We use Falcon plastic tissue culture dishes, but undoubtedly any source of plastic vessels for tissue culture would be suitable. The coverslips are punched out with the instrument shown in Fig. 1. The instrument consists of a cylindrical steel sleeve, resembling a cork borer, sharpened at the one end for cutting. The sleeve is heated in a gas flame and then pressed against the bottom of a plastic petri dish to cut a coverslip out of the dish. The cutting should be done against a surface that will not dull the punch, e.g., against wood. The coverslip is in reality melted out rather than cut out. With practice one learns how much to heat the sleeve to accomplish the cutting. If the sleeve is not sufficiently hot, cutting will be incomplete and the plastic usually cracks. With an overheated sleeve, the coverslip is usually distorted by partial melting.

A newly formed coverslip is removed from the steel sleeve using a rod inserted through a hole in the wooden handle as shown in Fig. 1. The end of this rod is covered with a piece of felt to prevent scratching of the plastic.

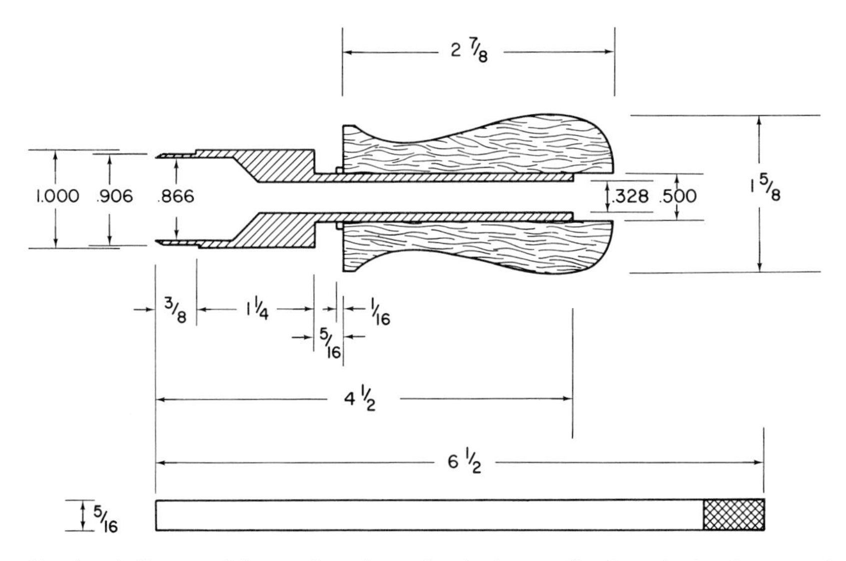

FIG. 1. A diagram of the punch used to make plastic coverslips from plastic culture vessels. The dimensions are given in inches. The central rod (below) is tipped with felt and used to push newly cut coverslips from the punch.

Best results are obtained if the newly formed coverslip is removed from the sleeve as rapidly as possible. This minimizes the buildup of a lip of plastic around the edge of the coverslip due to continued melting by the heat of the sleeve.

In the case of Falcon plastic culture dishes, only the inner surface of the bottom portion of petri dishes has been treated by the manufacturer so that cells will adhere to the surface. Coverslips must be cut with this in mind. The surface that is not suitable for cell attachment can be marked with a scratch before or after cutting to avoid later confusion.

The plastic coverslips are sterilized by a rinse in 95% alcohol and exposure to a germicidal UV lamp for 15 minutes. We have not found a commercial supplier willing to provide plastic coverslips suitable for cell culture and thick enough for centrifugation.

IV. Enucleation

Cells are enucleated in nutrient medium containing 10 μg/ml of cytochalasin B. Cytochalasin is poorly soluble in aqueous media, and therefore the drug is prepared as a stock solution in dimethyl sulfoxide (DMSO). One milligram of cytochalasin B is dissolved in 1 ml of DMSO, and 1 ml of this solution is added to 99 ml of nutrient medium to yield a concentration of 10 μg of drug per milliliter.

Cells growing as monolayers on plastic coverslips are placed in 8.0 ml of nutrient medium with cytochalasin in a polycarbonate centrifuge tube (IEC No. 1650) (29 × 104 mm) at 37°C. The coverslip is positioned with the cells facing the bottom of the centrifuge tube. The positioning of coverslips into centrifuge tubes is easily done using a triceps planchet holder (Fisher Scientific No. 10–317B). The removal of coverslips from centrifuge tubes can be done using ordinary forceps sufficiently long to reach the bottom of the centrifuge tube.

The cells are centrifuged at 12,000 rpm (17,369 *g*) in the SS-34 rotor or at 10,000 rpm (16,319 *g*) in the HD-4 rotor for 15 minutes in the Sorvall RC2B centrifuge. Centrifugation should be done close to 37°C to achieve the maximum effect (up to 99% of the cells enucleated). The lower the temperature the lower the percentage of cells that becomes enucleated. At 4°C no more than a few percent of the cells become enucleated. The temperature of the RC2B centrifuge can be raised to 37°C by running the machine with an empty, prewarmed rotor at 1700 rpm for 20 minutes. The temperature control should be set for 37°C ±2°C. The gauge will rarely read 37°C, but the rotor remains close to 37°C because of its mass.

After centrifugation the coverslips are removed from the centrifuge tubes, washed in nutrient medium to remove cytochalasin, and incubated in fresh medium at 37°C to permit recovery from the morphological changes induced by cytochalasin. Return of the cells to normal morphology is completed in 15–30 minutes. Although most cells become enucleated during centrifugation, some cells lose their nuclei during the return to normal morphology. This occurs in cells in which the nucleus has remained attached to the main body of the cell by a thin stalk of cytoplasm. The cytoplasmic stalk usually breaks during recovery from cytochalasin. Enucleated Chinese hamster cells obtained by the above procedure are shown in Fig. 2. When the centrifugation causes a Chinese hamster cell to detach from the coverslip, the cell leaves behind a number of small pieces of cytoplasm. An example of these fragments is shown in Fig. 3.

The enucleated cells contain all the usual cytoplasmic components in approximately the usual amounts. Figure 4 shows an electron micrograph of an enucleate Chinese hamster cell fixed immediately after recovery from the cytochalasin.

The nuclei removed from cells can be recovered from the bottom of the centrifuge tube. The pellet also contains the whole cells stripped from the coverslip by the centrifugation. Figure 5 contains an electron micrograph of a typical removed nucleus. A thin layer of cytoplasm with ribosomes and an

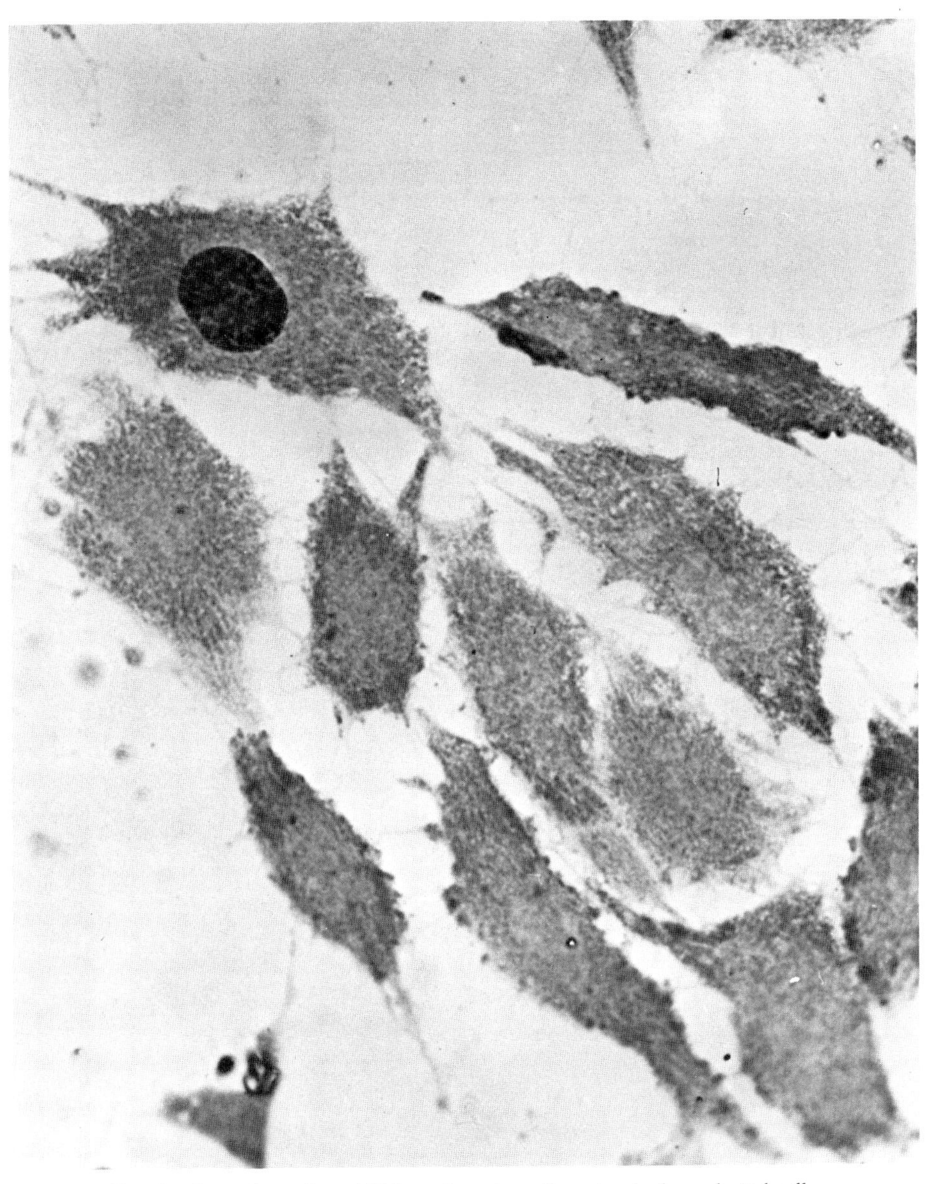

FIG. 2. Several enucleated Chinese hamster cells and a single nucleated cell.

occasional mitochondrion accompany the nucleus during its removal from the cell.

The centrifugation conditions described above were determined to yield the highest rate of enucleation with a minimum of cell loss from the coverslip.

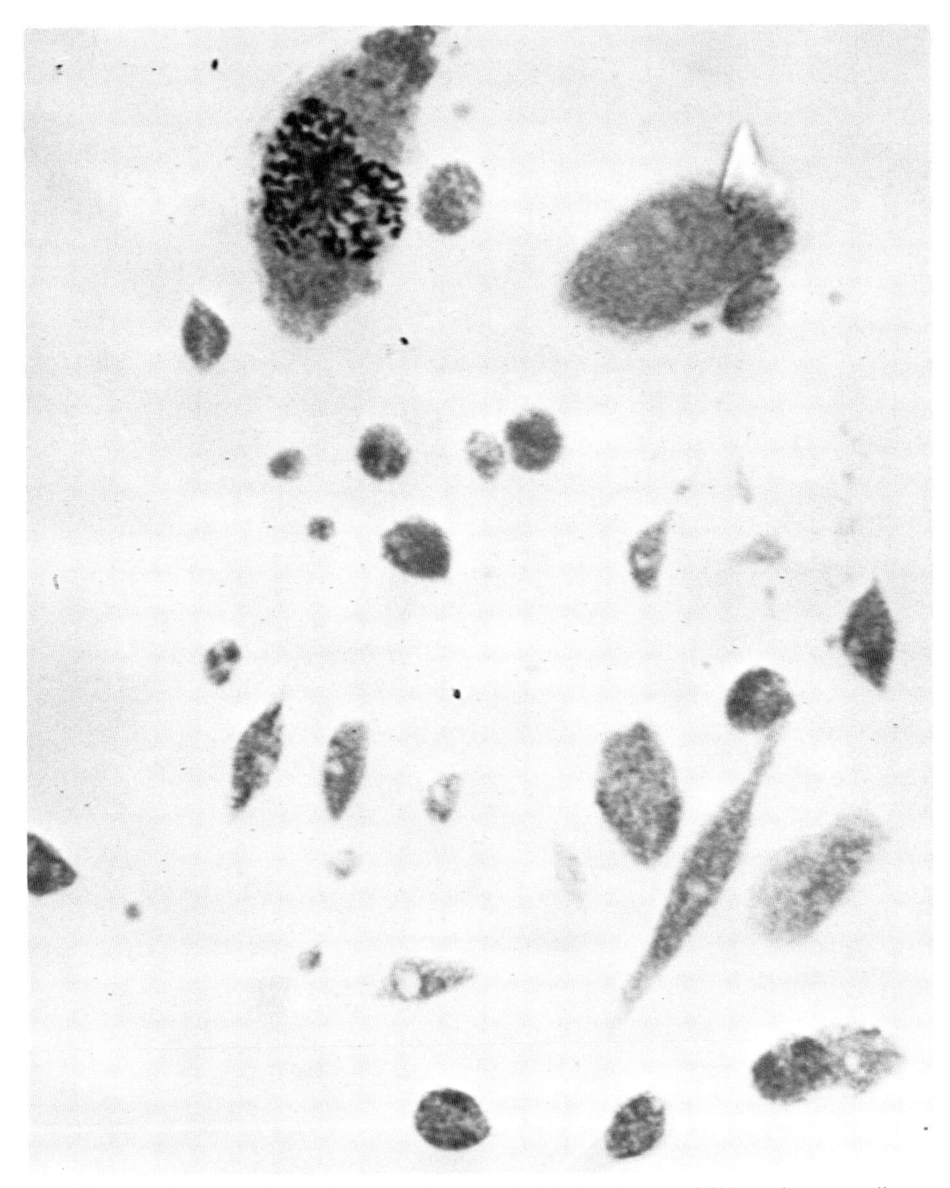

FIG. 3. Small cytoplasmic fragments left behind with detachment of Chinese hamster cells during centrifugation. The remaining nucleated cell is in prophase.

Consistently, more than 90%, and frequently 99%, of the cells that remain attached to the coverslip are enucleated. Occasionally, the enucleation of chick cells grown on plastic coverslips is 100%.

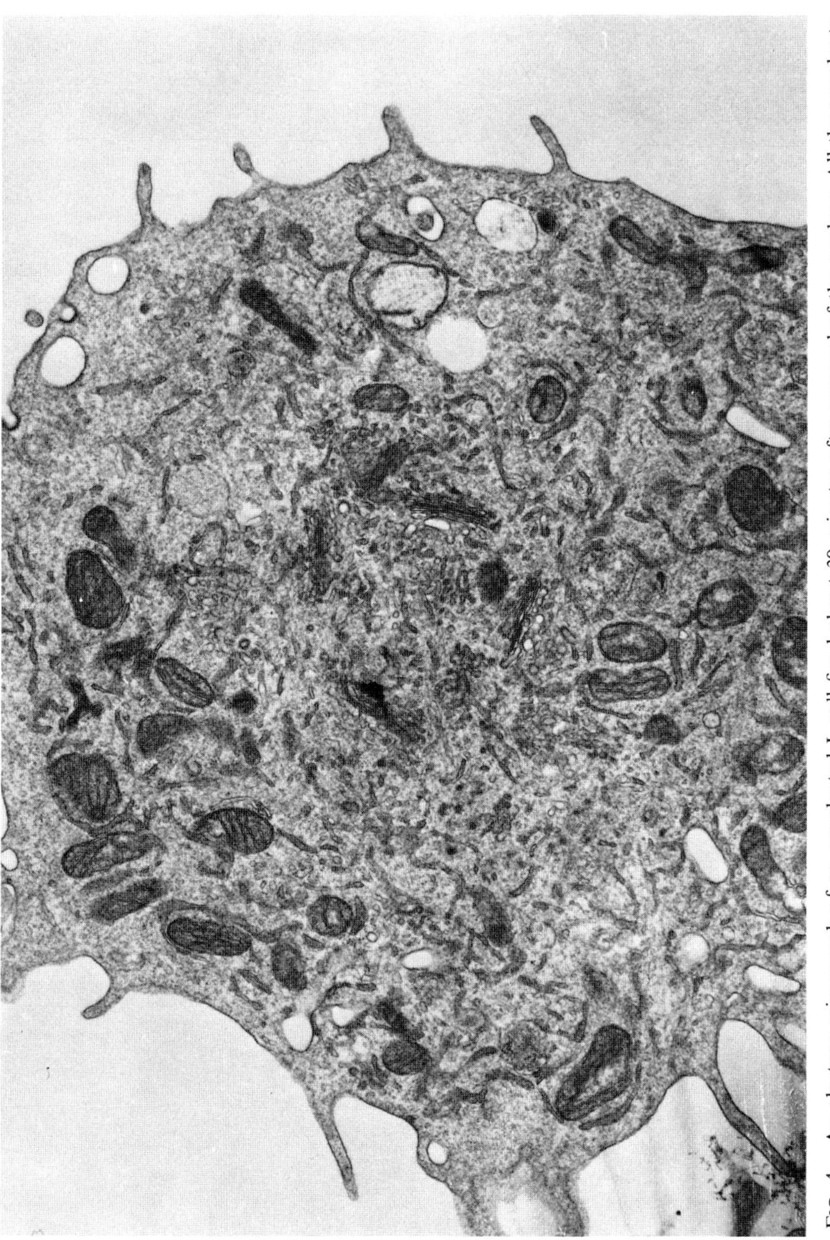

FIG. 4. An electron micrograph of an enucleated L cell fixed about 30 minutes after removal of the nucleus. All the usual cytoplasmic components are present. × 18,000 (Photograph by G. Wise.)

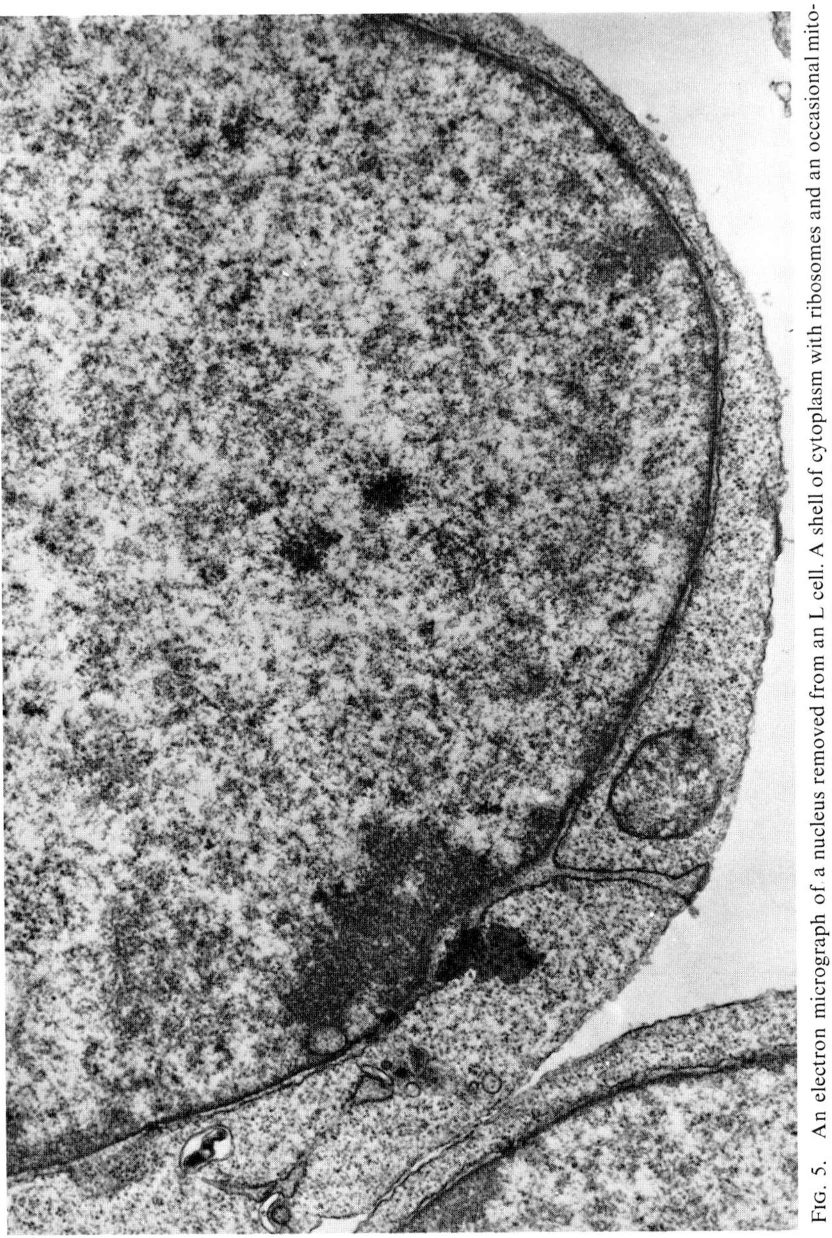

FIG. 5. An electron micrograph of a nucleus removed from an L cell. A shell of cytoplasm with ribosomes and an occasional mitochondrion accompanies the nucleus during enucleation. × 25,000 (Photograph by G. Wise.)

Centrifugation at speed greater than those given above or for times longer than 15 minutes usually results in an unacceptable amount of cell loss from the coverslip and still does not consistently accomplish 100% enucleation of the cells that remain attached to the coverslip. The optimal conditions for enucleation may be somewhat different for different kinds of cells, but the conditions are easily worked out by trial and error. We have also noted that cells in *confluent* monolayers tend to come off in sheets. The best enucleation results have been obtained with slightly subconfluent cultures.

V. Viability of Enucleated Cells

Enucleated Chinese hamster cells, L cells, and chick cells remain attached to coverslips for up to at least 3 days, during which time they become much smaller. As demonstrated by autoradiography, enucleated cells incorporate ^{3}H-labeled amino acids at a high rate during the first several hours after removal of nuclei (Fig. 6). By 12 hours after enucleation, the rate of incorporation of ^{3}H-labeled amino acids is severely reduced, and by 18 hours many enucleates no longer show detectable incorporation. Enucleated cells do not incorporate detectable amounts of either uridine-^{3}H (Fig. 7) or thymidine-^{3}H at any time after enucleation.

VI. Virus Activities in Enucleated Cells

Freshly enucleated L cells and chick cells can be infected with vaccinia virus with the subsequent development of about the same number of virus DNA factories per enucleate as per nucleated cell. L cells exposed to virus 12 hours after enucleation develop very few virus DNA factories. Exposure to vaccinia virus at 18 hours after enucleation does not lead to virus DNA synthesis. Enucleated chick cells can support vaccinia virus DNA synthesis somewhat longer; a few enucleated cells still develop virus DNA factories 24 hours after enucleation.

Enucleated chick cells infected with vaccinia virus immediately after infection also incorporate uridine-^{3}H into RNA, presumably RNA synthesized on virus DNA templates. Uninfected cells do not incorporate uridine-^{3}H. Figure 8 shows an autoradiograph of an enucleated chick cell labeled for 60 minutes beginning 30 minutes after removal of the nucleus. The incorporated uridine-^{3}H can be removed with RNase.

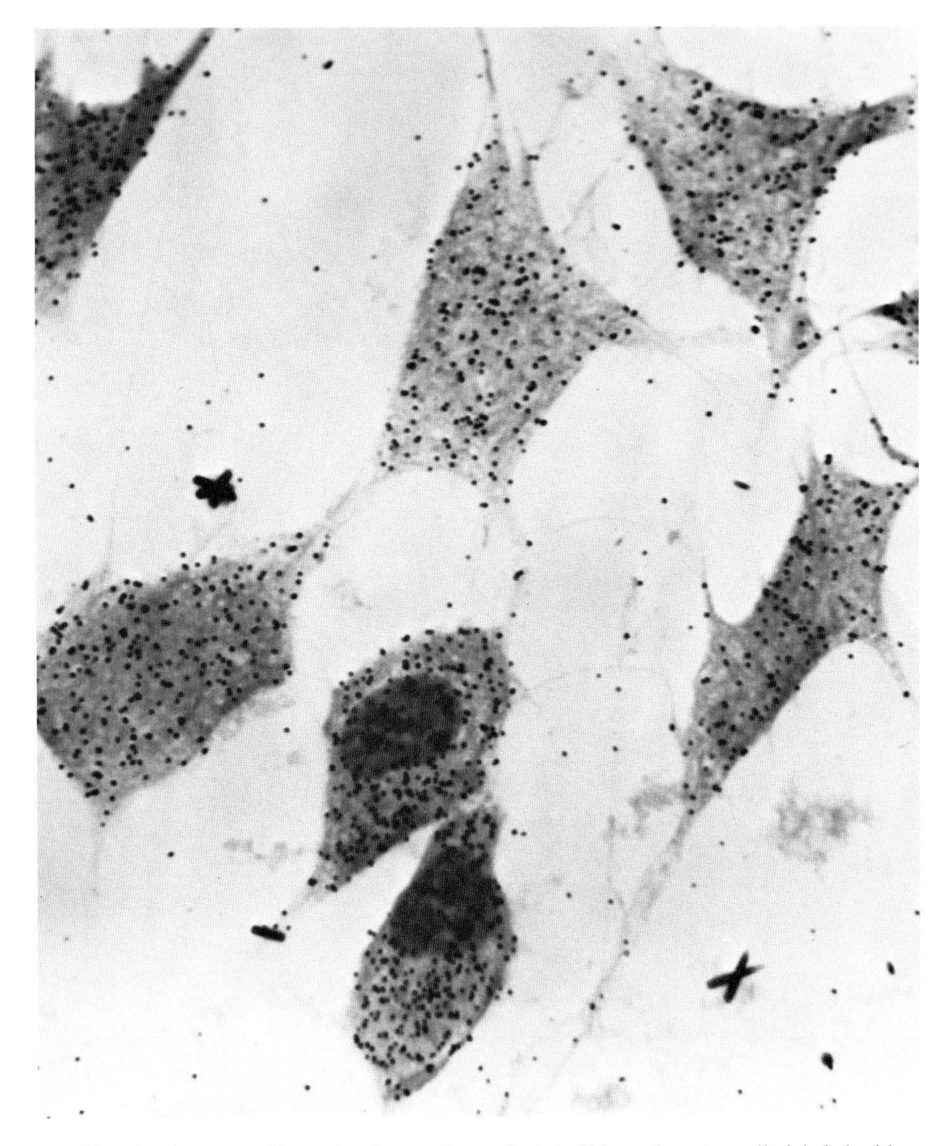

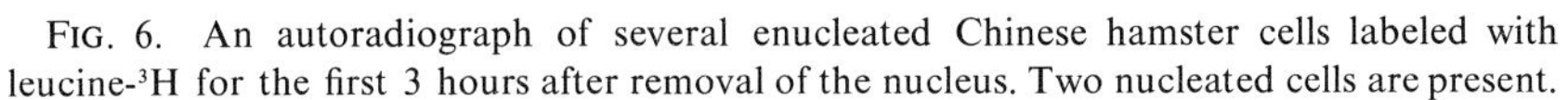

FIG. 6. An autoradiograph of several enucleated Chinese hamster cells labeled with leucine-^{3}H for the first 3 hours after removal of the nucleus. Two nucleated cells are present.

Uridine-^{3}H is also incorporated by enucleated chick cells infected with Sindbis virus (an RNA virus). The enucleated cell shown in Fig. 9 was exposed to Sindbis virus for 45 minutes, 30 minutes after enucleation, and labeled with uridine-^{3}H for 4 hours after infection. This observation paral-

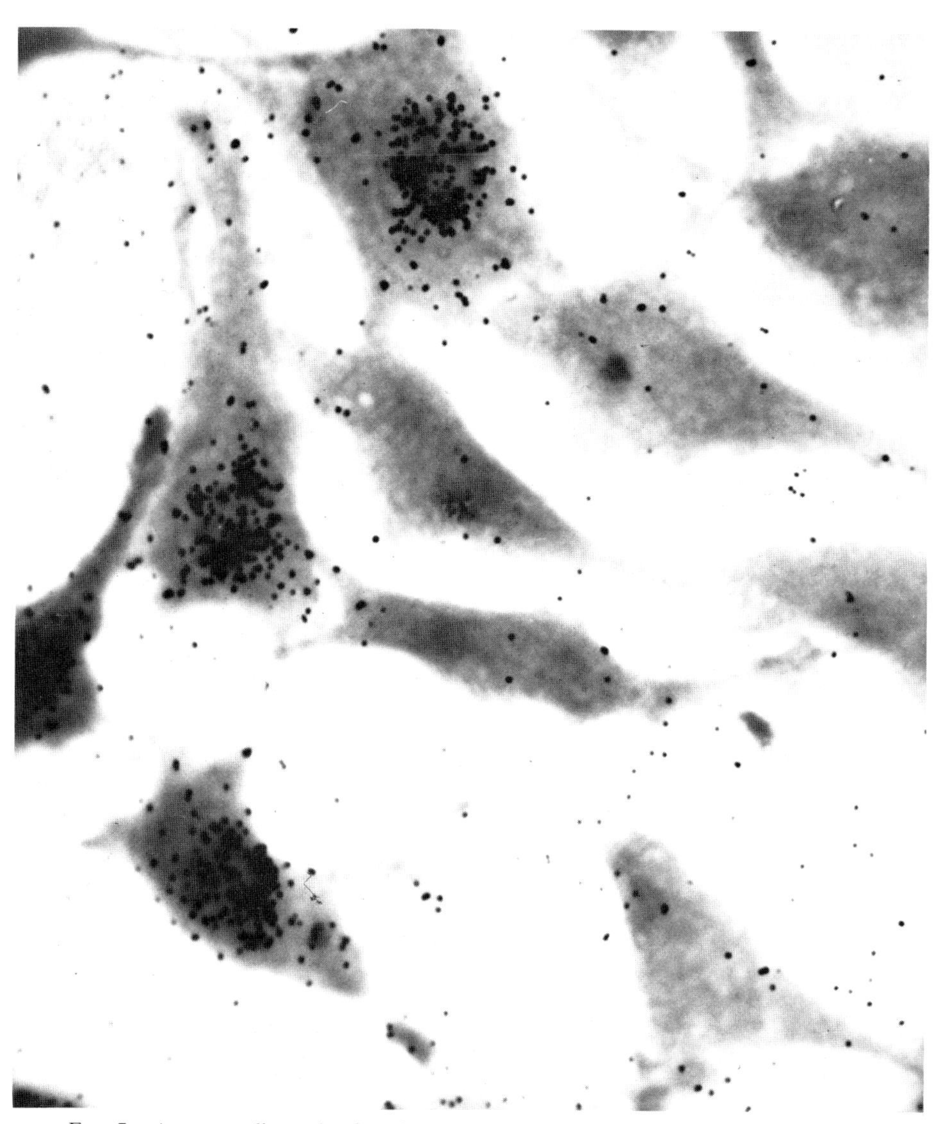

FIG. 7. An autoradiograph of nucleated and enucleated Chinese hamster cells labeled with uridine-^{3}H for 30 minutes after centrifugation. Enucleated cells do not incorporate label into an acid-insoluble form.

lels the findings of Marcus and Freiman (1966) on the synthesis of RNA in enucleated HeLa cells infected with poliovirus. In the latter case, the enucleated cells were produced by cutting of HeLa cells with a glass-cutting wheel.

VII. Viability of Nuclei Removed by Centrifugation

As shown in the electron micrograph in Fig. 5, nuclei separated from cells are surrounded by a thin layer of cytoplasm. These nuclei continue to incorporate uridine-^{3}H for 24–32 hours after enucleation. Thymidine-^{3}H is incorporated into DNA by a fraction of the nuclei for up to 24–32 hours after removal from cells. Since it seems unlikely that nuclei initiate DNA synthesis after removal from cells, the continuation of incorporation of thymidine-^{3}H for 24–32 hours may reflect a greatly slowed rate of DNA synthesis in those nuclei engaged in synthesis at the time of removal from cells. Finally, the nuclei and accompanying shell of cytoplasm continue to incorporate ^{3}H-labeled amino acids for 24–32 hours after the enucleation procedure.

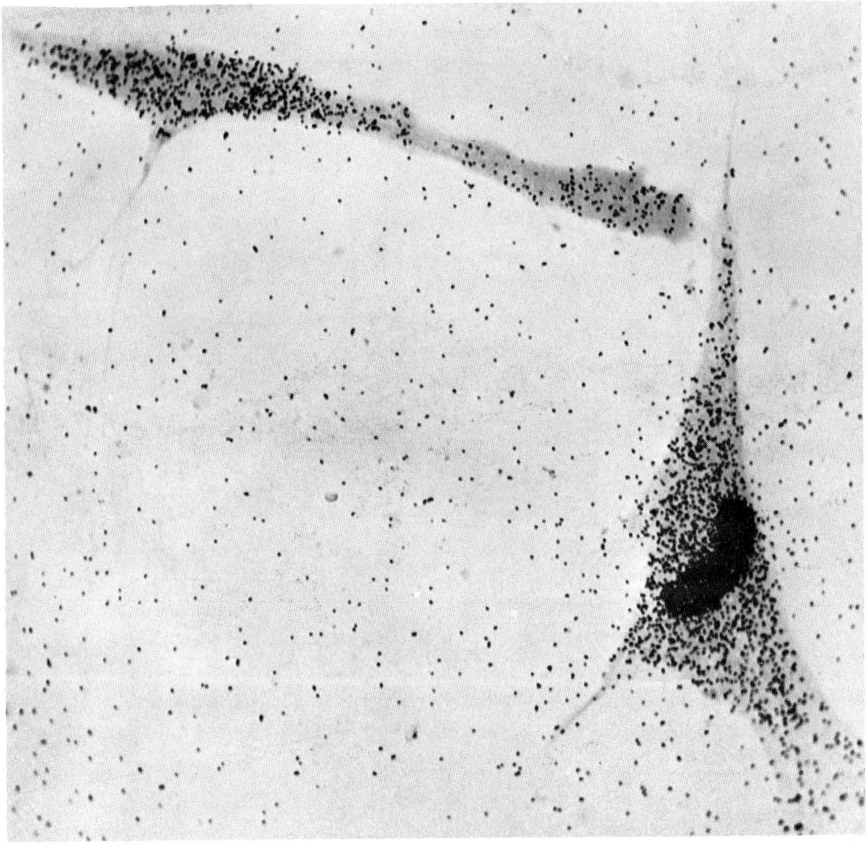

FIG. 8. An autoradiograph of a nucleated and an enucleated chick cell infected with vaccinia virus 30 minutes after centrifugation and labeled with uridine-^{3}H for 60 minutes, beginning 30 minutes after infection.

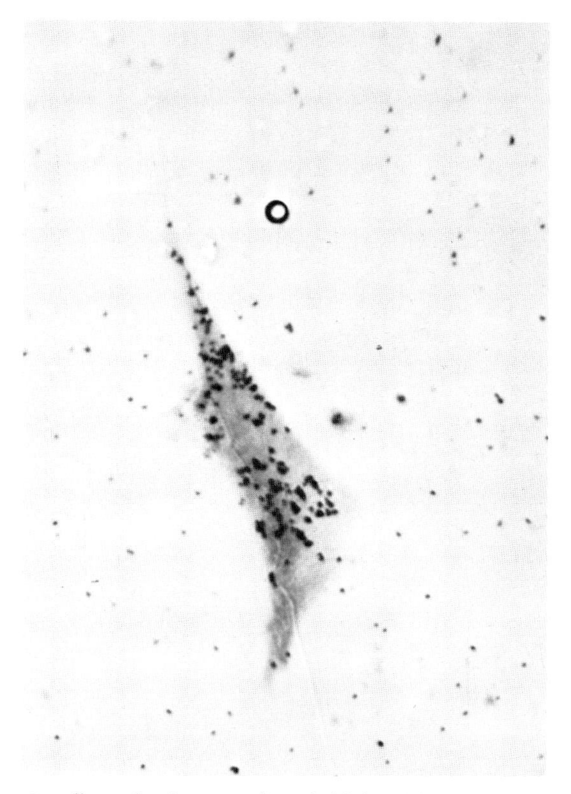

FIG. 9. An autoradiograph of an enucleated chick cell infected with Sindbis virus and labeled with uridine-^{3}H.

Presumably, this protein synthesis occurs in the thin shell of cytoplasm surrounding the nucleus. The demonstrated viability of removed nuclei suggests the possibility of using such nuclei in cell fusion experiments, i.e., recombining various kinds of cytoplasm with various kinds of nuclei.

An electron microscope study of enucleated Chinese hamster cells and separated nuclei will be published elsewhere (Wise and Prescott, 1973).

ACKNOWLEDGMENTS

This work was supported by a National Cancer Institute grant no. 5 RO1 CA 12302-02 to Dr. David M. Prescott.

REFERENCES

Marcus, P. I., and Freiman, M. E. (1966). *In* "Methods in Cell Physiology" (D. M. Prescott, ed.), Vol. II, pp. 93–111. Academic Press, New York.

Poste, G. (1972). *Exp. Cell Res.* **73**, 273.

Poste, G., and Reeve, P. (1972). *Exp. Cell Res.* **73**, 287.
Prescott, D. M., Kates, J., and Kirkpatrick, J. B. (1971). *J. Mol. Biol.* **59**, 505.
Prescott, D. M., Myerson, D., and Wallace, J. (1972). *Exp. Cell Res.* **71**, 480.
Wise, G. E., and Prescott, D. M. (1973). *Proc. Nat. Acad. Sci. U.S.* **70**, 714.
Wright, W. E., and Hayflick, L. (1972). *Exp. Cell Res.* **74**, 187.

Chapter 10

The Production of Mass Populations of Anucleate Cytoplasms[1]

WOODRING E. WRIGHT

Department of Medical Microbiology, Stanford University School of Medicine, Stanford, California

---

I. Introduction

Until recently, techniques for obtaining anucleate cytoplasms were limited to microsurgical manipulations. This restricted the use of anucleate cytoplasms to experiments in which only a very small number of cells could be used. In 1967, Carter reported the discovery of cytochalasin B (CB), a metabolite of the mold *Helminthosporium dematiodium.* This metabolite had a number of novel biological properties. In relatively low concentrations, 1–5 μg/ml (2.1 to 10.5 $\times$ 10^{-6} *M*), it inhibited cytokinesis without arresting nuclear division, thus causing cells to become multinucleate. At higher concentrations (10 to 20 μg/ml), the plasma membrane appeared to collapse around the nucleus until only a thin cytoplasmic bridge connected the nuc-

[1]Supported by Research Grant HD 04004 from the National Institute of Child Health and Human Development and by Medical Scientist Training Program grant no. GM 1922 from the National Institute of General Medical Sciences.

leus to the rest of the cytoplasm. Occasionally this connection ruptured, leaving behind a totally anucleate cytoplasm. Poste and Reeve (1971) were able to obtain 40% to 80% enucleation using mouse peritoneal macrophages or L–929 cells. Many other actions of cytochalasin have since been described (Wessells *et al.*, 1971; Carter, 1972).

We are using CB for preparing anucleate cytoplasms of the human fetal diploid cell strain WI–38 for cell fusion studies. In our initial experiments, WI–38 proved highly resistant to enucleation by CB. Even at concentrations of 500 μg/ml anucleate cells were only very rarely observed (Wright and Hayflick, 1972). In the following sections we will describe a method, combining the actions of CB and high *g* forces, by which we were able to consistently obtain 99+% enucleation.

II. Preparation of Lucite Plates

Special Lucite plates that withstand high *g* forces can be used as a substrate for the cells. One end of a $\frac{1}{4} \times 1 \times 2\frac{1}{2}$-inch Lucite plate is rounded so that the plate will stand upright in a centrifuge tube (Fig. 1a). These dimensions were chosen to fit tubes for Beckman SW 25.1 or SW 27 rotors, but can be modified to fit any desired tube size. Because WI–38 cells do not attach well to Lucite surfaces, glass cover slips were glued with clear epoxy cement to one surface of the Lucite plates. It is important to mix the epoxy resin and hardener very thoroughly, otherwise the bond will contain many refracting interfaces that make examination of the cells difficult. Although Lucite will not withstand autoclaving, these Lucite–cover slip plates can be conveniently sterilized by flaming in 95% ethanol.

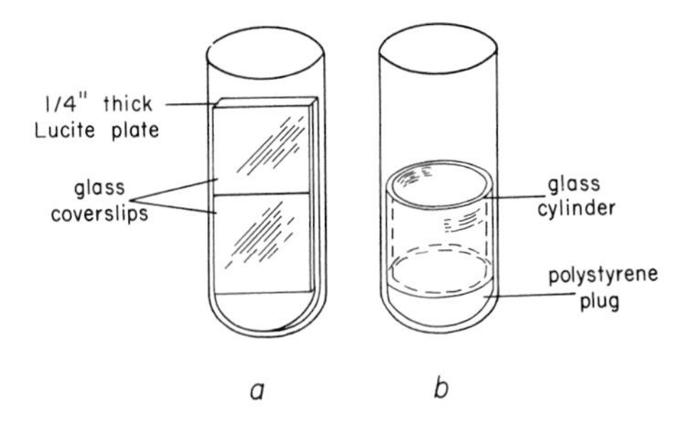

FIG. 1. Diagram of Lucite–coverslip plate (a) and glass cylinders (b) used for enucleating WI-38 cells at high *g* forces in the presence of cytochalasin B.

III. Plating of Cells

After flaming, the Lucite plates are placed coverslip side up in any vessel deep enough so that the plates can be covered by at least 1 cm of medium. If less medium is used, the bulk of the medium will rest between the Lucite plates. Thus, when the cell suspension is added, most of the cells will settle between the plates rather than upon them. If sterility is important, sterile 1000-ml beakers fitted with a petri dish lid can be used. Alternatively, rectangular plastic boxes disinfected with 70% alcohol are perfectly adequate, and more convenient since a greater number of plates can be fitted into an equivalent surface area, saving time, cells, and medium. Cells can be seeded at any desired density, since no obvious difference was observed in enucleating confluent or partially confluent monolayers. However, the cell suspension should be thoroughly aspirated, since clumps of cells appear to form multilayers that reinforce each other and reduce the efficiency of enucleation. The cells are allowed to attach overnight at 37° and are ready for enucleation the following morning.

IV. Enucleation

Cytochalasin B (Imperial Chemical Industries Ltd., Macclesfield, Cheshire, England) is poorly soluble in water, and is thus first dissolved in dimethyl sulfoxide (DMSO). Stock solutions of 1 mg of CB per milliliter of DMSO (2.1×10^{-3} M) maintain their activity for long periods of time when stored at -70°C. Portions of this stock solution are added to BME to obtain final concentrations of 0 to 10 μg of CB per milliliter of medium. CB is stable in medium at 4°C for at least one month.

Approximately 25 ml of CB medium is added to each centrifuge tube containing one Lucite–cover slip plate. The tubes are balanced to within 0.1 gm of each other with additional CB medium. The Lucite–cover slip plates are then centrifuged in a Beckman SW 25.1 or SW 27 rotor for 30 minutes at various speeds. Incubating the Lucite–cover slip plates in cytochalasin medium for 1 hour before centrifugation did not improve the efficiency of enucleation. Immediately after centrifugation the cells have a strung-out appearance, with a tail of cytoplasm extending toward the bottom of the tube. If the Lucite–cover slip plates are incubated in CB-free medium for 0.5 hour at 37°C after centrifugation, the cells assume a more normal appearance. The cells are then fixed in 95% ethanol, stained with Giemsa, and examined for the efficiency of enucleation (Fig. 2).

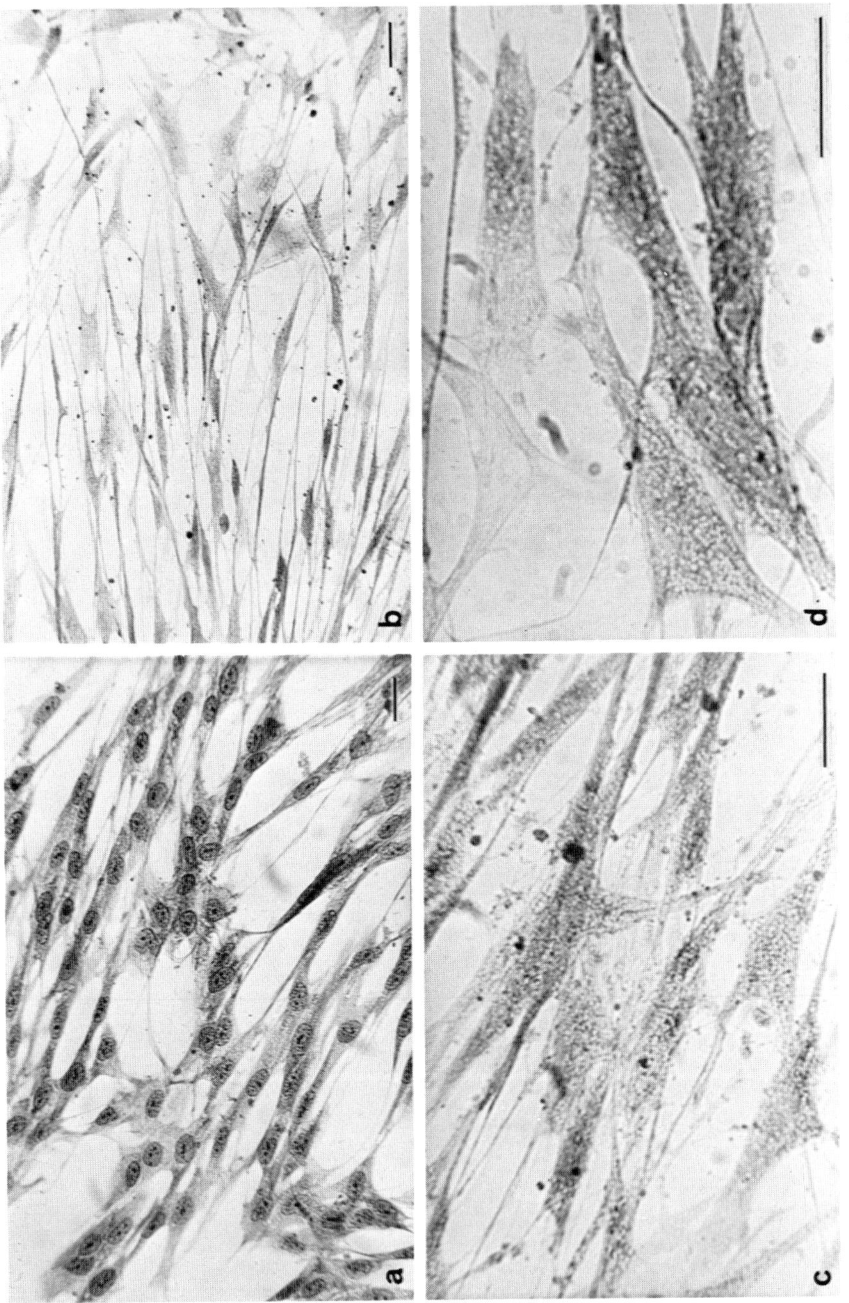

FIG. 2. WI-38 cells centrifuged at 25,000 g for 30 minutes at 25°C. (a) no cytochalasin B: (b, c, and d) 4 μg of cytochalasin B/ml. Bars indicate 20 μ.

Figure 3 shows the percent enucleation for cells centrifuged for 30 minutes at different *g* forces and CB concentrations. Figure 4 shows the relationship beteen *g* forces and rpm at various radii of spin. Since the Lucite–cover slip plates rest upright in the centrifuge tube, there can be a large difference between the force at the top and bottom of a single plate. The curves in Fig. 3 were obtained by determining the percent enucleations at several points on each Lucite–cover slip plate, and computing the *g* force at each point.

Surprisingly, very few cells are stripped off the slide at forces below 30,000 *g*. Above that point, the factors determining whether a cell will strip off are apparently independent of whether it will enucleate, since the curves in Fig. 3 for 0 to 2 μg/ml (> 30,000 *g*) are comparable to those from 2 to 10 μg (< 30,000 *g*). At forces above 80,000 *g*, most of the cells have stripped off the Lucite–cover slip plate.

The enucleation of cells is highly temperature dependent between 0°C and 25°C with much greater efficiency at the higher temperatures. Since the efficiency of enucleation is virtually the same at 25°C as at 35°C, ambient temperature is recommended as the most convenient temperature compatible with a high efficiency of enucleation.

Figure 5 shows the dependence of enucleation on time of centrifugation.

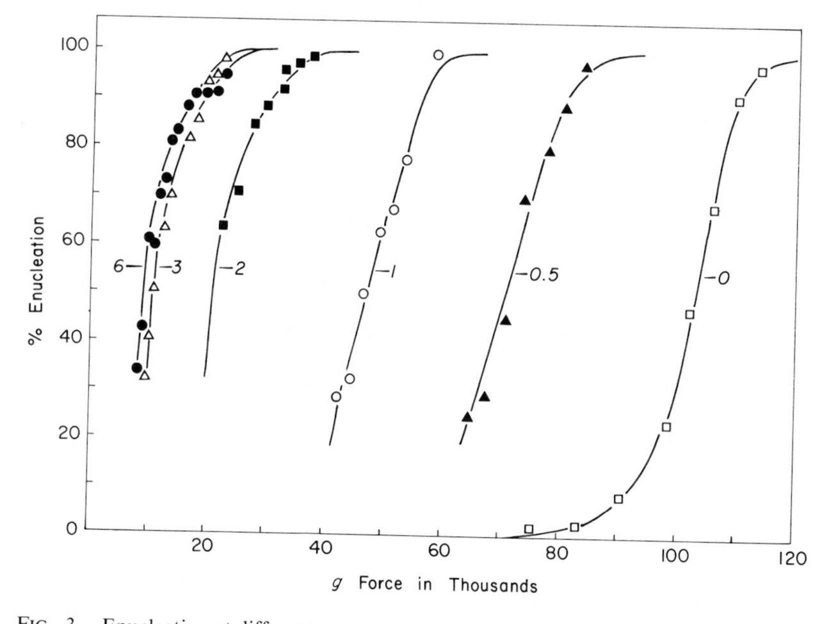

FIG. 3. Enucleation at different concentrations of cytochalasin B. Cells growing on Lucite–cover slip plates were centrifuged for 30 minutes at 25°C. The number beside each curve indicates the concentration of CB (μg/ml).

Since the distance moved by an object under a centrifugal force is $\frac{1}{2}$ atm^2, it is not surprising that enucleation is proportional to the log of the time of centrifugation.

V. Preparation of Large Numbers of Anucleate Cytoplasms

Because of their good optical qualities, Lucite–cover slip plates are very useful for determining the efficiency of enucleation under various conditions. However, the use of glass cylinders as a cell substrate is more satis-

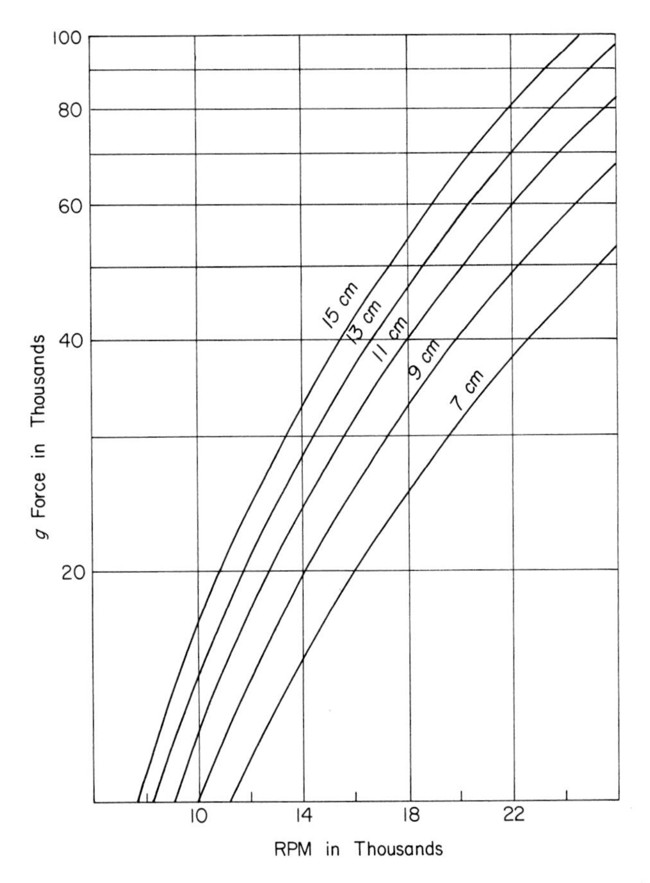

FIG. 4. Relationship of centrifugal speed to centrifugal force. The number beside each curve indicates the radius of spin.

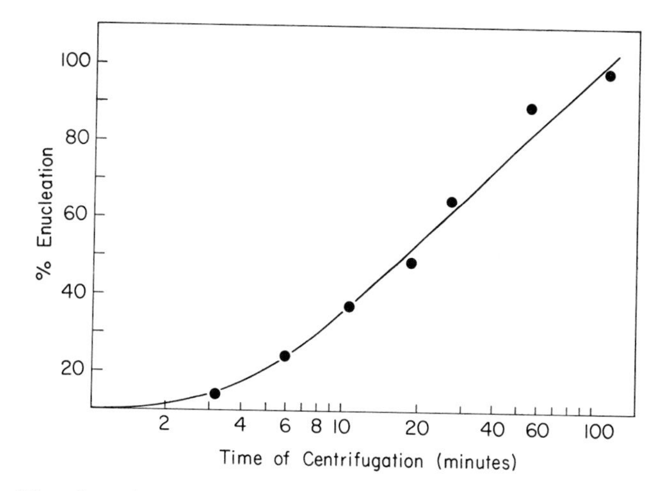

FIG. 5. Time dependence of enucleation. Cells growing on Lucite–cover slip plates were centrifuged at 22,500 *g* at 25°C in the presence of 2 μg CB/ml medium. Centrifugation times are corrected for rotor acceleration time.

factory for the preparation of large numbers of anucleate cytoplasms, since they provide a greater surface area per centrifuge tube.

Glass tubing, 25 mm, is cut into 1- or 2-inch lengths. Care must be taken to have the cut surfaces flat and perpendicular to the long axis of the tubing, otherwise stresses developing during centrifugation will cause the glass cylinders to break. Either Pyrex or flint glass is satisfactory. Since commercial glass tubing varies slightly in diameter and most pieces of "25 mm tubing" are about 0.2 mm too large to fit in the centrifuge tubes for the SW 25.1 or SW 27 rotor, the smallest of an assortment of tubing should be selected. Since most stockrooms carry 25 mm test tubes, these are a convenient source from which to select the smallest diameter tubing. Alternatively, Pyrex tubing can be shrunk on a mandrel in order to reduce its diameter so that it will fit into the centrifuge tubes. The glass cylinders should slide easily in and out of the centrifuge tubes, since tight-fitting cylinders have a tendency to break. The centrifuge tubes can be stretched slightly by filling them with hot water and centrifuging for 30 minutes at approximately 24,000 rpm. Often this is sufficient to get the desired close fit between the glass tubing and the centrifuge tube. The glass cylinders may be reused many times.

The sterile glass cylinders are placed in any convenient receptacle (capped centrifuge tubes, 30 mm glass tubing closed at each end with #6 silicon stoppers, 100 ml graduated cylinders closed with sterile Saran wrap held around the end of the cylinder with rubber bands, etc.), cells and medium are added, and the receptacle is rotated overnight at 1 rpm at 37°C to allow the cells to attach to the inner surface of the glass cylinders. Rotation is effected by placing the receptacle between the bottles on a roller bottle apparatus (Bellco Glass Inc., Vineland, New Jersey).

Lucite or polystyrene plugs are shaped into hemispheres that fit snugly into the bottom of the centrifuge tubes, providing a flat surface to support the glass cylinders (Fig. 1b). For enucleation, one cylinder is placed in each tube fitted with a plastic plug, and appropriate CB medium is added. Conditions established with the Lucite–cover slip plates are then used to obtain the desired enucleation with the glass cylinders. Using six 2-inch glass cylinders and 4 μg of CB per milliliter of medium for 30 minutes at 25°C and 15,000 rpm in the SW 27 rotor, 2–6 million cytoplasms can be obtained.

VI. Conclusion

As is evident from Figs. 3 and 5 and the above discussion, the efficiency of enucleation is dependent on CB concentration, *g* force, time of centrifugation, and temperature. Any or all of these parameters can be manipulated to meet the requirements of the particular experiment at hand. With WI-38 cells, we have settled on 4 μg CB per milliliter at $\geq$ 23,000 *g* for 30 minutes at $\geq$25°C as a satisfactory combination of conditions, although other permutations would have served as well.

References

Carter, S. B. (1967). *Nature* (*London*) **213**, 261–264.
Carter, S. B. (1972). *Endeavour* **31**, 77–82.
Poste, G., and Reeve, P. (1971). *Nature* (*London*), *New Biol.* **229**, 123–125.
Wessells, N. K., Spooner, B. S., Ash, F. F., Bradley, M. O., Luduena, M. A., Taylor, E. L., Wrenn, J. T., and Yamada, K. M. (1971). *Science* **171**, 135–143.
Wright, W. E., and Hayflick, L. (1972). *Exp. Cell Res.* **74**, 187–194.

Chapter 11

Anucleate Mammalian Cells: Applications in Cell Biology and Virology

GEORGE POSTE

Department of Experimental Pathology, Roswell Park Memorial Institute, Buffalo, New York

---

I. Introduction

The value of anucleate cells and nuclear transplantation in providing information on the interaction between the cell nucleus and the cytoplasm has been emphasized on many occasions (Hämmerling, 1963; King, 1966;

Keck, 1969; Gurdon, 1970; Gurdon and Woodland, 1971; Jeon and Danielli, 1971; Brachet, 1972). Until recently, most experimental approaches to the production of anucleate cells have relied upon removal of the nuclei by microsurgery. Successful enucleation by this method has been confined, however, to large cells, such as protozoa (Jeon and Danielli, 1971; Tartar, 1972), algae (Hämmerling, 1963; Brachet and Bonotto, 1970), and oocytes from amphibians (Gurdon and Woodland, 1971; Ecker, 1972) and sea urchins (enucleation by high speed centrifugation) (Harvey, 1936; Tyler, 1966; Craig, 1972).

These methods have been of limited value in the enucleation of much smaller mammalian somatic cells cultured *in vitro*. The difficulty of exact microdissection and the removal of nuclei from mammalian cells *in vitro* has dictated that efforts to enucleate these cells by microsurgery have concentrated instead on the production of anucleate cell fragments by the surgical separation of a portion of the peripheral cytoplasm from the remaining cytoplasm containing the nucleus (Goldstein *et al.*, 1960a; Marcus, 1962; Crocker *et al.*, 1964; Marcus and Freiman, 1966; Cheyne and White, 1969). Apart from the considerable technical difficulties and the tedious nature of the microsurgical approach to the enucleation of mammalian cells, the production of anucleate cell fragments rather than whole cells introduces the problem that there may be a preferential or atypical distribution of organelles and metabolic products in the fragment compared with the cytoplasm as a whole. A more important drawback to microsurgical techniques, however, is that they cannot provide the large numbers of anucleate cells required for detailed biochemical studies on the properties of the anucleate cell state and the interaction between the nucleus and the cytoplasm.

In 1967, Carter reported that cytochalasin B (CB), a metabolite obtained from cultures of the fungus *Helminthosporium dematioideum*, induced enucleation of mouse L cells cultured *in vitro*. This observation has since been confirmed in other laboratories using both L cells and a variety of other mammalian cell types (Ladda and Estensen, 1970; Poste, 1972a; Poste and Reeve, 1971, 1972a; Goldman, 1972; Goldman *et al.*, 1972; Prescott *et al.*, 1972; Wright and Hayflick, 1972). Importantly, methods are now available in which CB can be used to enucleate most cells in populations of mammalian somatic cells cultured *in vitro* (Section II,B). The anucleate cells obtained by these methods provide cell biologists with a potent tool for the investigation of the stability of cytoplasmic macromolecules in the absence of a nucleus (Section V,A), the exchange of material between nucleus and cytoplasm (Section V,B), and the respective importance of the nucleus and the cytoplasm in virus infections of the cell (Section V,C). Also, the recent development of methods for the formation of hybrid cells and heterokaryons by fusion of anucleate cells with nucleated cells of a different type (Section

III) provides a further area in which the use of anucleate cells should extend the already significant contribution of cell fusion techniques to the study of nucleocytoplasmic interactions and the control and expression of specific cell functions (Barski, 1970; Harris, 1970; Poste, 1972b).

In this review, methods for the isolation and characterization of anucleate mammalian cells and the techniques for the fusion of anucleate and nucleated cells will be described in detail and an attempt made to outline certain of the potential experimental applications of anucleate mammalian cell systems.

II. Enucleation of Mammalian Cells by Cytochalasin B

A. The Cytochalasins

The cytochalasins are a group of fungal metabolites with a related macrolide structure (Fig. 1) and a number of common biolgical properties (Carter, 1972). Cytochalasins A and B were the first compounds of the series to be isolated and were detected in filtrates of the mold *Helminthosporium dematioideum* (Aldridge *et al.*, 1967). Cytochalasin B has since been found to be identical with phomin, a compound isolated from *Phoma* species (Rothweiler and Tamm, 1970). Cytochalasins C and D were isolated later from *Metarrhizium anisopliae* (Aldridge and Turner, 1969), and cytochalasins

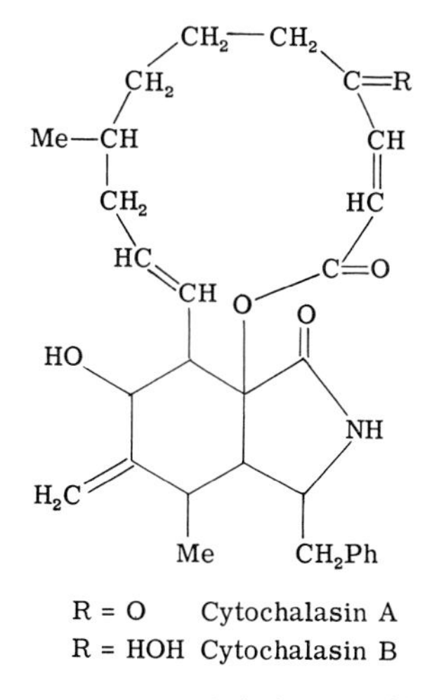

FIG. 1. Cytochalasins A and B.

E and F have been identified even more recently in filtrates of *Rosellina necatrix* and *H. dematioidium*, respectively (Aldridge *et al*., 1972).

All the cytochalasins share certain common properties in relation to their effects on cells cultured *in vitro* (Carter, 1972), but most experimental observations on their action on cells, and all the studies on cell enucleation have been done with cytochalasin B (CB). Unless stated otherwise in the following sections the use of CB may be assumed.

Cytochalasin B [Imperial Chemical Industries (I.C.I.), Pharmaceutical Division, Macclesfield, Cheshire, England] is poorly soluble in water and is normally dissolved in dimethyl sulfoxide (DMSO) in which it has a solubility at 24°C of 371 mg/ml of saturated solution. Active preparations of a known concentration can be prepared in DMSO and then diluted in aqueous cell culture media at the required lower concentrations for experimental use. Solutions of CB in DMSO are highly stable under normal conditions and may be stored for up to 3 years at 4°C without significant loss of activity (I.C.I.Technical Data Sheet—supplied with preparations of cytochalasin B).

B. Spontaneous and Induced Cellular Enucleation by Cytochalasin B

For descriptive convenience a distinction will be made between the "spontaneous" and "induced" enucleation of cells *in vitro* by CB. The former represents the type of enucleation response observed when cells cultured on glass or plastic surfaces extrude their nuclei following the addition of high doses of CB (20–50 μg/ml) to the culture medium. However, many cell types do not enucleate spontaneously following exposure to CB but can be "induced" to enucleate by exposure to both CB and high *g* forces. The term "induced" enucleation used in this article refers to this latter type of enucleation in which cells are exposed to a dose of CB insufficient to produce significant spontaneous enucleation ($\leq$ 10 μg/ml) and then centrifuged at high *g* forces to "induce" expulsion of their nuclei.

"Spontaneous" cellular enucleation by CB was first described by Carter (1967) in cultures of mouse L929 cells and has since been documented in other murine cell cultures and in cell cultures from a wide variety of other species (Ladda and Estensen, 1970; Goldman, 1972; Poste, 1972a). The kinetics of spontaneous enucleation by CB in a range of diploid, heteroploid, and viral and chemically transformed cell cultures and the various morphological changes accompanying enucleation have been described in detail by Poste (1972a).

Apart from certain exceptions such as mouse L929 cells and mouse peritoneal macrophages (MPM) the frequency of spontaneous cellular enucleation following exposure to CB is low and many cell types do not enucleate spontaneously (Table I). The need to treat cell cultures with

TABLE I
SUSCEPTIBILITY OF MAMMALIAN CELLS TO SPONTANEOUS ENUCLEATION BY CYTOCHALASIN B[a,b]

Designation	Description	Mean % anucleate Cells ± SE[c]
Primary and secondary cell strains		
MPM	Mouse Peritoneal Macrophages	32.6 ± 11.5
MEF	Mouse embryo fibroblasts	12.7 ± 5.5
DKC	Dog kidney cells	7.9 ± 3.3
HEL	Human embryonic lung cells	6.1 ± 3.7
CE	Chick embryo fibroblasts	5.3 ± 3.6
BKC	Bovine kidney cells	0
MKC	Monkey (Patas) kidney cells	0
R. Ly.	Rabbit lymphocytes	0
Established cell lines		
L929	Clone of mouse L line	28.4 ± 9.8
HeLa	Human epitheloid carcinoma line	12.7 ± 4.6
HEp-2	Human epidermal carcinoma line	9.2 ± 3.7
BHK-21	Spontaneous hamster cell line	6.8 ± 3.2
NIL-2E	Spontaneous hamster cell line	6.3 ± 2.4
RK-13	Rabbit kidney line	5.2 ± 3.5
MDCK	Madin: Darby canine kidney line	0
3T3	Spontaneous mouse cell line	0
Transformed cell lines		
TT2	Polyoma virus-transformed mouse embryo fibroblasts	15.8 ± 5.3
TT3	Polyoma virus-transformed mouse embryo fibroblasts	11.1 ± 3.7
RSV-NIL	Rous sarcoma virus (Schmitt-Ruppin)-transformed NIL 2E cells	8.1 ± 3.4
HSV-NIL	Hamster sarcoma virus-transformed NII 2E cells	7.2 ± 3.2
PY-BHK	Polyoma virus-transformed BKH cells	7.1 ± 4.7
PY-NIL	Polyoma virus transformed NIL 2E cells	4.4 ± 2.6
Klein T	Methylcholanthrene-transformed mouse embryo fibroblasts	4.1 ± 2.6
SV-3T3	SV 40 virus-transformed 3T3 cells	0

[a] Modified from Poste (1972a).
[b] Cells were incubated in cytochalasin B, 40 μg/ml, for 8 hours at 37°C.
[c] Mean value derived from six separate experiments and counts of a minimum of 6000 cells in each experiment.

high doses of CB (30–40 μg/ml) for up to 8 hours to obtain the maximum yield of anucleate cells by spontaneous enucleation also dictates that this type of enucleation does not lend itself to many of the potential experimental applications of anucleate cells described in Section V in which it is necessary to remove the nuclei from cells within less than 1 hour. Also, the failure to obtain spontaneous enucleation of *all* cells within a susceptible population introduces the need for a further time-consuming density-gradient centrifugation procedure to separate the anucleate and nucleated cells (Poste and Reeve, 1971; Poste, 1972a).

These difficulties have prompted the search for methods to increase the frequency and rapidity of cellular enucleation by CB and to achieve enucleation of entire cell populations. This goal has been realized within the last year by two groups of independent investigators using very similar methods (Prescott *et al.*, 1972; Wright and Hayflick, 1972). These methods for "induced" enucleation involve the treatment of cells with CB followed by high speed centrifugation. This type of "induced" enucleation has the significant advantage over "spontaneous" enucleation that a uniformly high proportion (>90%) of most cell populations can be enucleated within 30–45 minutes. The details of these methods of "induced" enucleation using high speed centrifugation are described elsewhere in this volume by Prescott and Kirkpatrick and by Wright, further details are therefore unnecessary. The present author has developed a modification of these techniques which is now used routinely to induce enucleation in more than 80% of cells in most populations (Fig. 2). A further advantage of these "induced" enucleation techniques is that large numbers of anucleate cells can be obtained from cultures of cell types such as mouse 3T3 cells which are completely resistant to spontaneous enucleation by CB (G. Poste, unpublished observations, 1972).

In the present author's method for "induced" enucleation, cells are grown on glass plates, which can be inserted into a special holder constructed from polypropylene and stainless steel that fits tightly into an ultracentrifuge tube (Fig. 3). It has been found that a holder with the cross-sectional profile shown in Fig. 3 provides excellent mechanical support for the glass plate and enables *g* forces as high as 80,000 to be used without breakage of the plate. The holder may also be reused after each preparation and can be conveniently sterilized with alcohol. Holders have been constructed so far for use with the MSE 3 × 23 and Beckman SW 25.1 swingout ultracentrifuge rotors. Initial experiments in which the cells were grown on conventional glass flying coverslips resulted in breakage of the coverslip in the holder during centrifugation. This problem has been overcome by using glass plates prepared from microscope slides designed for use in fluorescence microscopy which are thinner (0.9 mm) than conventional slides. These do not break during centrifugation and provide excellent optical resolution for microscopic examination of the cells.

Cells to be enucleated are inoculated onto the glass plates in disposable plastic petri dishes and allowed to grow to the appropriate density (see below). The plates with the attached cells are then transferred to the holders,

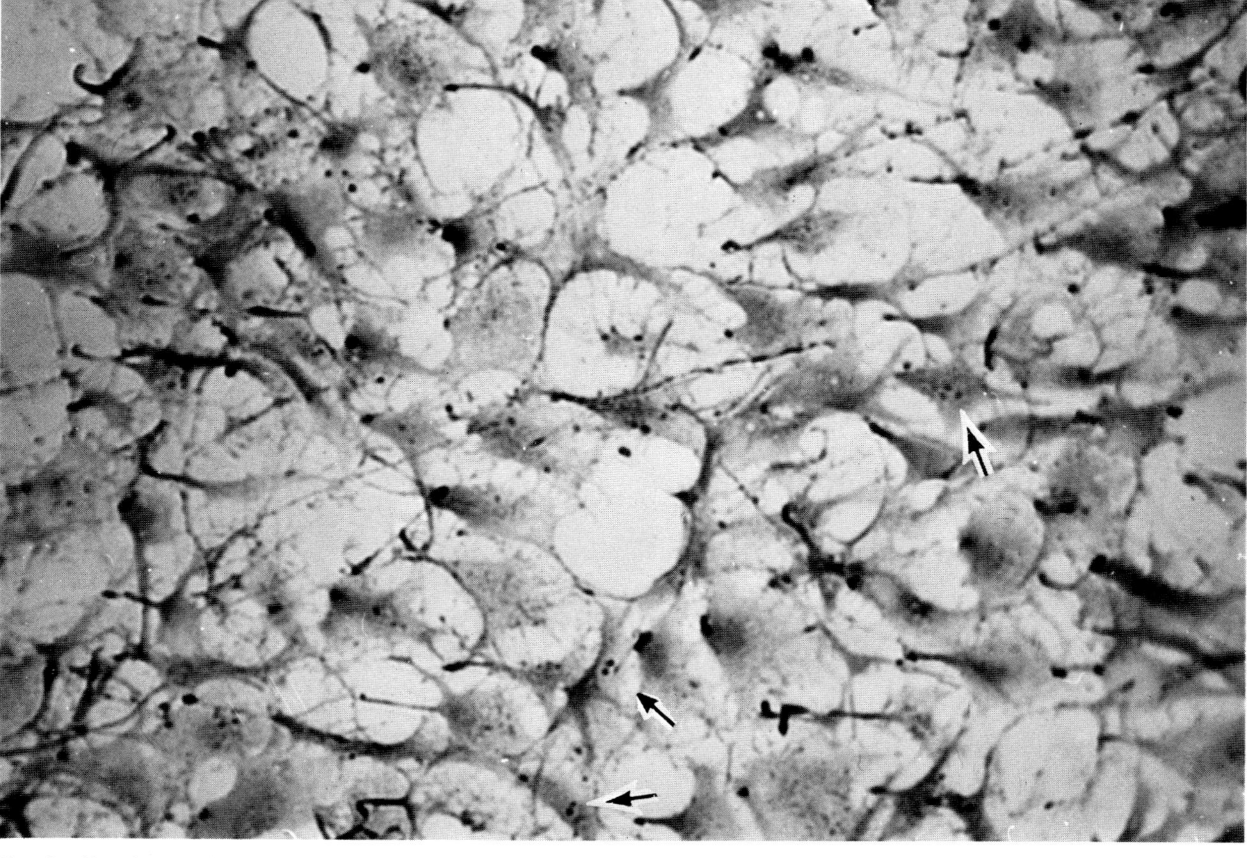

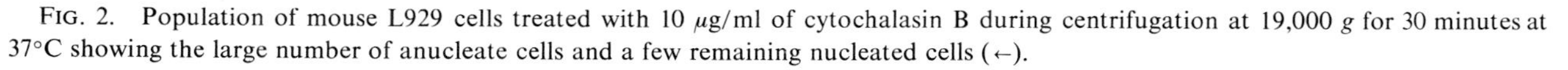

FIG. 2. Population of mouse L929 cells treated with 10 μg/ml of cytochalasin B during centrifugation at 19,000 *g* for 30 minutes at 37°C showing the large number of anucleate cells and a few remaining nucleated cells (←).

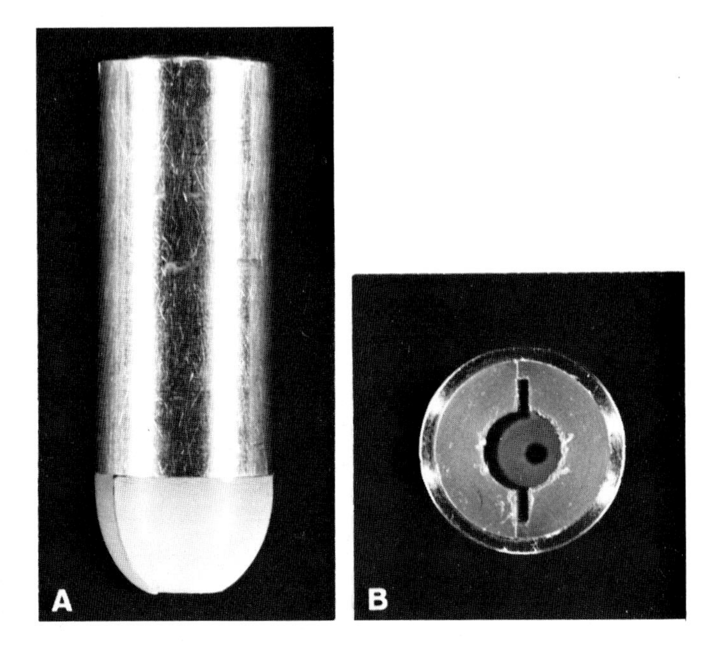

FIG. 3. Stainless steel-polypropylene holder designed for use with a MSE 3 × 23 swing-out ultracentrifuge rotor in the enucleation of cells by cytochalasin B at high *g* forces. (A) Side view of holder showing outside stainless steel case around a central milled plug of polypropylene. The bottom portion of the holder is polypropylene alone and is shaped to fit the contour of the ultracentrifuge tube. Length = 60 mm. (B) Top view of the holder showing the central slit for insertion of the glass plate plus the attached cells. The central cylindrical core is filled with culture medium containing cytochalasin B. A small drainage hole is present in the bottom of the tube (slightly off-center in the photograph). Outside diameter = 20 mm; slit = 1 × 14 mm; and hollow central core diameter = 8 mm.

and serum-free culture medium containing 10 μg of CB per milliliter is added to fill the central cylinder well of the holder (Fig. 3). The centrifuge tubes with the holders are balanced and centrifuged at 15,000 *g* for 30 minutes in a prewarmed rotor (> 30°C). After centrifugation, the glass plates are removed from the holders and either fixed and stained for examination of the cells by light microscopy or returned to fresh medium without CB to allow recovery of the cells for use in experiments.

When examined immediately after centrifugation the anucleate cells have a highly altered elongated "spindle" morphology due to marked deformation of the cell along the line of the centrifugal force. However, when returned to fresh culture medium without CB, the anucleate cells assume their normal morphology within 15–30 minutes. Centrifugation for longer than 40 minutes increases significantly the loss of cells from the plate. Few cells (< 10%) are stripped from the plate at forces below 25,000 *g*, but above this point cell loss is much more marked and all cells are stripped at forces above 60,000 *g*. There is a more marked cell loss during centrifugation at 25,000 *g* from cell monolayers that have been confluent for two or more days and the use of nearly confluent populations is more successful in reducing cell loss. However, problems due to cell loss have been encountered more frequently in the centrifugation of certain virus-transformed cell lines which grow as heaped-up colonies, and the above technique of induced enucleation has so far been only of limited value in the production of anucleate cells from cultures which form these colonies.

Since the glass plates to which the cells are attached are in a vertical position in the holder there is a significant difference in the centrifugal force exerted on cells at the top and bottom of the plate. This gradient is reflected in the significantly higher number of anucleate cells found at the bottom of the plate where the *g* forces are highest during centrifugation. The effect of centrifugal force and the dose of CB on the yield of anucleate cells in mouse 3T3 cell cultures obtained with this technique is shown in Fig. 4. The curves shown in Fig. 4 for the efficiency of cellular enucleation at different concentrations of CB were calculated by measuring the percentage of cells enucleated at several points on the glass plate and computing the *g* forces at the same points. As shown in Fig. 4 it is possible to obtain large numbers of anucleate cells within as little as 30 minutes using doses of CB of 10 μg/ml or less. A similar efficiency of enucleation by CB in relation to dose and

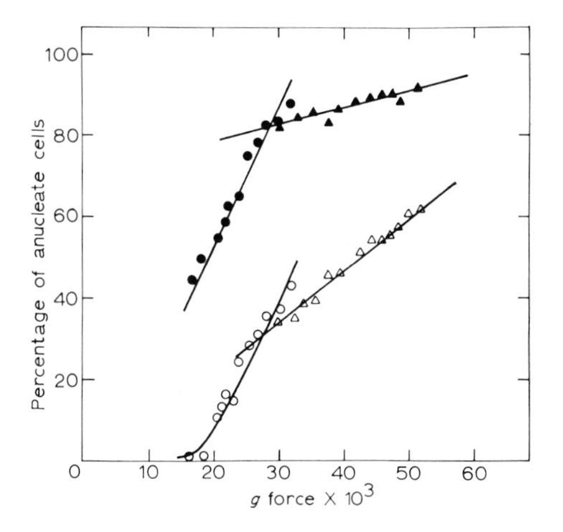

FIG. 4. The efficiency of enucleation of mouse 3T3 cells by different doses of cytochalasin at different *g* forces. ●——● and ▲——▲ = 10 μg/ml of cytochalasin B; ○——○ and △——△ = 1 μg/ml of cytochalasin B.

centrifugal force has been found in cultures of mouse L cells and human HEp-2 cells. The results obtained by this method are in general agreement with the findings of Prescott, Myerson, and Wallace (1972) and Wright and Hayflick (1972).

The composition of the culture medium, variations in pH, differences in the method of growing the cells before treatment with CB and differences in cell density do not appear to influence significantly the susceptibility of cells to either spontaneous (Poste, 1972a) or induced (G. Poste, unpublished observations, 1972) enucleation by CB. However, temperature is important in both spontaneous and induced enucleation. Enucleation is optimal in the temperature range 30–39°C, but is reduced significantly at room temperature (18°C) and is inhibited completely at 4°C (Poste, 1972a; Prescott *et al.*, 1972; Wright and Hayflick, 1972).

In view of the success of these methods of "induced" enucleation in obtaining yields of more than 80% anucleate cells from mammalian cell populations, it is surprising that the period of centrifugation used by Poste (1972a) to separate anucleate and nucleated cells following spontaneous enucleation by CB did not produce a similar increase in the proportion of anucleate cells. Two possible explanations can be put forward to account for this apparent discrepancy. First, in the density gradient separation method, the cells were centrifuged in a 10–30% (w/v) Ficoll solution and CB was not present, whereas in the "induced" enucleation methods cells were exposed to CB for the entire period of centrifugation. This difference may be important in determining the yield of anucleate cells, since many of the biological effects of CB are known to be rapidly reversed when the cells are transferred to fresh medium without CB (see Section II,D). Second, in the Ficoll gradient separation method the cells were centrifuged in suspension while in the methods for rapid enucleation by high speed centrifugation the cells are attached to solid substratum, and this may lead to the deformation of the cell in a different way which favors expulsion of the nucleus.

In cells attached to a substratum a reduction in the amount of cytoplasm overlying the nucleus would facilitate contact and possible fusion of the nuclear and plasma membrane leading to nuclear expulsion (Figs. 5A, B). The finding that certain cell types treated with CB have a broader and more flattened cytoplasm (Carter, 1967; Gail and Boone, 1971; Krishan, 1971; Poste, 1972a) could be interpreted as being likely to produce this effect, although definitive evidence on this possibility is still awaited. If the assumption is made that CB does facilitate a general spreading of the cell and that this results in a reduction in the layer of cytoplasm overlying the nucleus, then it might be expected that the deformation and compression of the cell onto the substratum during centrifugation would further enhance this process and increase the opportunities for contact between the nucleus

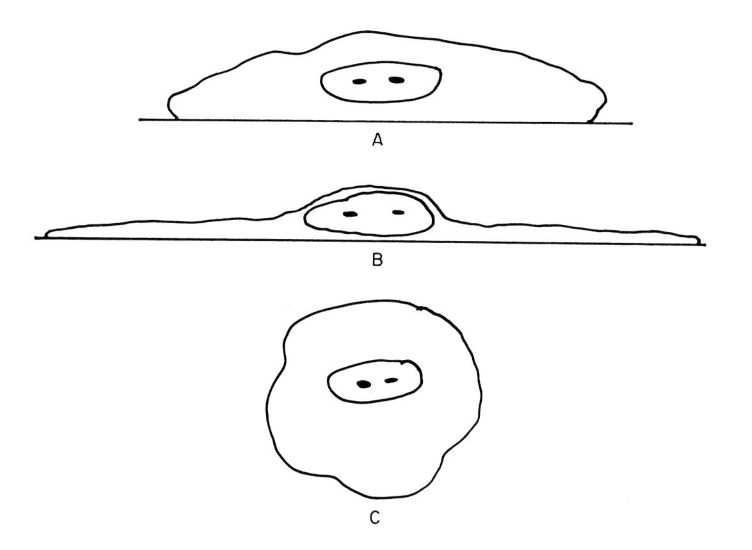

FIG. 5. Diagrammatic representation of the possible effects of cytochalasin B on the profile of mammalian cells grown in monolayer and suspension cultures. (A) Normal cell attached to a solid surface showing the position of the nucleus relative to the plasma membrane on the upper surface of the cell. (B) Attached cell following treatment with cytochalasin B showing increased spreading and "flattening" of the cell and a reduction in the thickness of the layer of cytoplasm between the nucleus and the plasma membrane. (C) Cell from a suspension culture treated with cytochalasin B showing the central position of the nucleus and the thick layer of cytoplasm between the nucleus and the plasma membrane.

and the plasma membrane necessary for nuclear extrusion (Fig. 5B). In contrast, the spherical form adopted by cells in suspension dictates that the nucleus would be surrounded in all directions by a relatively uniform thickness of cytoplasm (Fig. 5C), which might hinder contact between the nucleus and the plasma membrane and limit the opportunities for nuclear extrusion. If these proposals are correct it should be possible to enucleate attached cells by merely centrifuging them at very high speeds without CB. It is of interest to note therefore that Wright (this volume) has observed enucleation of Wl-38 cells at 100,000 *g* in the absence of CB.

C. Properties of Anucleate Cells

The absence of nuclear material and the true anucleate status of cells enucleated by CB have been confirmed by light and electron microscopy, Feulgen microspectrophotometry, and the inability of the anucleate cells to incorporate thymidine-^{3}H (Goldman *et al.*, 1972; Poste, 1972a; Prescott *et al.*, 1972). Cells enucleated by CB attach to both glass and plastic surfaces and are capable of active movement and endocytosis (Carter, 1967; Goldman *et al.*, 1972; Poste and Reeve, 1972a). Anucleate cells obtained by both the spontaneous and the induced enucleation methods continue to synthesize protein for up to 12 hours after enucleation as judged by the incorporation of leucine-^{3}H, but do not synthesize DNA or RNA (Poste, 1972a; Poste and Reeve, 1972a; Prescott *et al.*, 1972). These properties parallel the earlier observations of Goldstein *et al.* (1960a,b) on anucleate mammalian cells obtained by microsurgical enucleation.

Mean survival times for anucleate cells of between 16 and 30 hours were reported by Poste (1972a) for several cell types produced by spontaneous enucleation with CB, and similar survival times of 8–36 hours were found by Wright and Hayflick (1972) in cultures of anucleate WI-38 cells obtained by the induced enucleation technique following centrifugation at 23,000 *g* for 30 minutes. Longer survival times of 3 days have been reported by Prescott and his colleagues for anucleate L929 and Chinese hamster cells enucleated by CB at 10 μg/ml during centrifugation at 3000 *g* for 40 minutes. The better survival times found in the latter study may reflect the lower centrifugal force used to induce enucleation. Attempts to increase the survival time of CB-induced anucleate cells *in vitro* by the use of conditioned medium, cultivation in vessels containing only a limited volume of medium or plating onto feeder layers of irradiated cells have been unsuccessful (Poste, 1972a).

Successful subcultivation and transfer of anucleate cells has been reported using a 0.2% trypsin 0.2% EDTA solution to detach the cells (Poste and Reeve, 1972a; Goldman *et al.*, 1972), but Wright and Hayflick (1972) found that anucleate WI-38 cells subcultured with trypsin were unable to reattach to the substratum though successful subculture of the same cells was obtained using an EDTA solution alone.

D. Cytotoxic Effects of Cytochalasin B

Cytochalasin B affects a wide variety of cellular activities at doses (≤ 2 μg/ml) below those normally used to achieve spontaneous or induced enucleation. These include: inhibition of cell movement (Carter, 1967; Allison *et al.*, 1971; Spooner *et al.*, 1971; Armstrong and Parenti, 1972); alterations in the attachment of cells to surfaces (Gail and Boone, 1971; Weiss, 1972); inhibition of cell separation after division leading to multinucleation (Krishan, 1972); and alterations in membrane permeability as indicated by inhibition of the uptake of glucose (Cohn *et al.*, 1972; Kletzien *et al.*, 1972; Zigmond and Hirsch, 1972a) and polynucleotides (Plagemann and Estensen, 1972). The inhibition of certain of these cellular functions by CB can take place extremely quickly. Kletzien *et al.* (1972) recorded a 95% inhibition of 2-deoxy-D-glucose uptake in chick embryo cells within 2 minutes of treatment with 10 μg/ml CB. Despite its inhibitory effect on a wide variety of cellular activities, CB does not inhibit cellular DNA, RNA or protein synthesis (Estensen, 1971; Shepro *et al.*, 1970; Yamada *et al.*, 1970, 1971; Cohn *et al.*, 1972; Poste, 1972a; Prescott *et al.*, 1972; Sanger and Holtzer, 1972a; Raff, 1972) and has no significant effect on intracellular ATP levels (Warner and Perdue, 1972). Importantly, most of the effects of CB on the cell appear to be reversible and cells rapidly return to normal following removal of CB from the medium (Carter, 1967; Estensen, 1971; Davis *et al.*, 1971; Gail and Boone, 1971; Malawista, 1971; Wagner *et al.*, 1971; Estensen and Plagemann, 1972; Goldman, 1972; Kletzien *et al.*, 1972; Poste, 1972a; Sanger and Holtzer, 1972a,b; Zigmond and Hirsch, 1972a,b).

The rapidity with which CB is effective and the swift recovery of cells after removal of CB suggests that it is either metabolized rapidly or is only loosely bound within the cell so that when exogenous CB is removed the cells revert immediately to the normal condition. That the latter possibility is more likely is suggested by the observation that the culture medium recovered from cells treated with very low doses of CB (0.1 and 1 μg/ml) retains a constant inhibitory potential over several transfers when added to fresh cell cultures (Mayhew and Maslow, 1973).

The literature on the overall cytotoxicity of CB is somewhat contradictory. Part of the confusion on this subject is no doubt due to differences in the type of cells used by different investigators, the culture conditions and, most importantly, the time for which the cells were exposed to CB. Estensen *et al.* (1971) found that CB at 10 μg/ml was toxic for both "epithelial" and "fibroblastic" cells (identity not specified) after 10–12 hours at 37°C. In contrast, Wessells *et al.* (1971a) failed to observe cytotoxic effects in glial cells exposed to 50 μg/ml for 18 hours. Similarly, exposure of mouse peritoneal macrophages (Wills *et al.*, 1972), human, horse, and rabbit polymorphonuclear leukocytes (Zigmond and Hirsch, 1972b), and BHK-21 cells (Wang, 1972) to CB at 40–50 μg/ml for up to 4 hours did not impair cell viability, and specialized cellular activities such as endocytosis were observed. Poste (1972a) did not find any significant reduction in the viability of a wide range of diploid, heteroploid, and virus-transformed cell cultures exposed to CB at 40 μg/ml for 8 hours, but exposure to 50 μg/ml for more than 14 hours resulted in a rapid decline in cell viability and death of all cells in the culture within 24 hours. However, cells can survive for significantly longer periods in the presence of lower doses of CB.

Sanger and Holtzer (1972a) reported that muscle cell cultures were still viable and capable of active contraction after exposure to CB at 5 μg/ml for 3 days. Similarly, unpublished observations in the present author's laboratory have shown that mouse L929 and 3T3 cells remain viable for 2 days in medium containing CB at 10 μg/ml while NIL2E hamster cells and a mouse (L929)–human (WI-38) hybrid cell line can survive for up to 60 hours under the same conditions. Carter (1967) stated that "cells remain viable for many days in the continuous presence of cytochalasin B" (1 μg/ml). Krishan (1971) found that mouse L929 cells were viable for 15 days in the presence of CB at 1 μg/ml but by this time many of the cells contained eight or nine nuclei, and other factors such as alterations in the nuclear/cytoplasmic ratio might be more important as a cause of cell death [cf. Poste (1972b)] rather than any cumulative toxic effect of CB per se.

It would appear therefore that most mammalian cell cultures can tolerate exposure to doses of CB as high as 50 μg/ml for several hours and exposure to lower doses (≤ 10 μg/ml) for several days.

E. Mechanism of Cell Enucleation by Cytochalasin B

The mechanism of cellular enucleation by CB is still unknown. Wessells and his colleagues (1971b) have put forward a unitary theory to explain the many and varied effects of CB on the cell. This theory proposes that CB acts on intracellular contractile microfilaments and that "sensitivity to the drug implies the presence of some type of contractile system." As Carter (1972) has pointed out, this proposal is premature and is not supported fully by the available experimental evidence. The results of studies concerning the action of CB on microfilament systems are contradictory [for discussion, see Holtzer and Sanger (1972)] and it is now clear that certain effects of CB are not mediated by action on cellular microfilaments (Cohn *et al.*, 1972; Estensen and Plagemann, 1972; Forer *et al.*, 1972; Goldman, 1972; Kletzien *et al.*, 1972; Plagemann and Estensen, 1972; Sanger and Holtzer, 1972a,b; Zigmond and Hirsch 1972a,b). Much of the confusion on this aspect of CB activity appears to be due to differences in the criteria used by different investigators to identify microfilament systems. In addition, before the mechanism(s) of CB action on different cellular functions can be solved it will be necessary to establish whether CB enters the cell and, if it does, where it acts. The investigation of this problem is frustrated presently by the lack of a suitable radioactively labeled preparation of CB.

In an attempt to explain the enucleation of cells by CB, Carter (1967) suggested that it might modify the adhesive properties of cell membranes. Carter suggested that the adsorption of CB onto the nuclear and plasma membranes might lower the boundary tension between them, thereby increasing the chance of stable areas of adhesion being formed between the apposed membranes followed by membrane fusion and expulsion of the nucleus by exocytosis. As discussed in Section II,B, CB may also increase the opportunity for contact and subsequent fusion between the nuclear and plasma membranes by facilitating the spreading of cells, thereby reducing the amount of cytoplasm between the nucleus and the plasma membrane.

Cytochalasin B has been found recently to reduce the electrophoretic mobility of cultured Ehrlich ascites tumor cells (E. Mayhew, unpublished observations, 1972) and *Rana pipiens* embryo cells cultured *in vitro* (Schaeffer and Brick, 1972). A similar lowering of the surface charge of the nuclear and plasma membranes might therefore facilitate contact between them by reducing the repulsion betwen them. In this respect, it is also of interest to note that Skutelsky and Danon (1969) in a study of the natural expulsion of nuclei from mammalian erythroblasts found that the negative surface charge of the membrane of the expelled nuclei was reduced significantly. However, further comments on these possibilities must await more detailed information on the interaction of CB with biological membranes.

In direct contrast to Carter's (1967) proposal for the enhancement of membrane fusion in CB-treated cells, several other authors (Davis *et al.*, 1971; Estensen, 1971; Rifkin *et al.*, 1972) have proposed that CB produces its highly varied effects on the cell by inhibition of membrane fusion behavior. While the inhibitory effect of CB on endocytosis (Allison *et al.*, 1971; Davies *et al.*, 1971; Malawista *et al.*, 1971; Wagner *et al.*, 1971; Zigmond and Hirsch, 1972b) and exocytosis (Williams and Wolff, 1971; Néve *et al.*, 1972; Orr *et al.*, 1972; Rifkin *et al.*, 1972; Thoa *et al.*, 1972) could possibly be interpreted on the basis of inhibition of the fusion between vesicle membranes and the plasma membrane, there is no direct evidence that CB can specifically inhibit the membrane fusion stage in these processes. Indeed, it is equally plausible that these processes are inhibited by the action of CB on microfilament systems associated with the membranes of the endo- and exocytotic vesicles which might be involved in the intracellular movement of the vesicles toward or away from the plasma membrane in exo- and endocytosis, respectively, and before and after fusion per se (Freed and Lebowitz, 1970; Poste and Allison, 1971, 1973).

It is difficult to see how the inhibition of membrane fusion could account for cell enucleation by CB. The extrusion of the nucleus from the cell, intact in its nuclear membrane (Ladda and Estensen, 1970; Poste, 1972a; Prescott *et al.*, 1972), would require initial fusion between the nuclear and plasma membranes during nuclear expulsion followed by fusion of two segments of the plasma membrane to restore its integrity after expulsion of the nucleus. That inhibition of membrane fusion behaviour represents the principal action of CB on the cell is also considered unlikely since CB does not inhibit virus-induced cell fusion (Poste and Reeve, 1972a), the fusion of cytoplasmic organelles (Hoffstein *et al.*, 1972; Wang, 1972 Wills *et al.*, 1972), and certain endo- and exocytotic processes which require membrane fusion can still take place in the presence of CB (Diegelmann and Peterkofsky, 1972; Wang, 1972; Wills *et al.*, 1972). Similarly, the inhibition of cytokinesis by low doses of CB, which was interpreted initially as being due to an inhibition of membrane fusion causing failure of cleavage furrow formation (Estensen, 1971), has been shown recently to be caused not by the failure of furrow formation, but to a failure of the new daughter cells to separate which then *fuse* to form a multinucleate cell (Krishan, 1972).

Even if in some cases inhibition of cleavage furrow formation does occur following exposure to CB, it is clear that other types of membrane fusion can still occur in the same cell since karyokinesis is completed normally which would require the formation of a new nuclear envelope by the fusion of separate vesicular membrane segments (Longo and Anderson, 1969; Ikeuchi *et al.*, 1971). These criticisms concerning the proposal for inhibition of membrane fusion by CB do rely, however, on the assumption that fusion occurring in different membrane systems involves a common mechanism (Poste 1972b; Poste and Allison, 1971, 1973; Poste and Reeve, 1972b).

Nuclear expulsion takes place naturally in the extrusion of polar bodies from the oocyte during meiosis (Austin, 1961) and in the maturation of mammalian erythroblasts (Skutelsky and Danon, 1970, 1972). Extrusion or "budding" of large lobes or segments of the nucleus into the cytoplasm also occurs in heteroploid mammalian somatic cell lines cultured *in vitro* (Hsu and Lou, 1959; Elston, 1963; Longwell and Yerganian, 1965). Unfortunately, our understanding of these processes offers little insight into the possible mechanism of nuclear extrusion in cells treated with CB.

In these naturally occurring examples the process of nuclear extrusion is inhibited by microtubular disruptive drugs such as colchicine (Skutelsky and Danon, 1970). One possible interpretation of the inhibitory effects of colchicine on enucleation in these situations is that microtubules may play a role in "anchoring" the nucleus (Norberg, 1971; Woodcock, 1971) and that drugs such as colchicine might interfere with these structures and prevent intracellular movement of the nucleus, thereby limiting the opportunities for contact and fusion between the nuclear and plasma membranes prior to nuclear extrusion. However, treatment of mammalian cell cultures with colchicine does not produce any significant change in the incidence of enucleation over that found in cells treated with CB alone (Poste, 1972a).

Whatever the precise mechanism of cellular enucleation by CB, it is clear that the anucleate cells produced by the use of this drug provide a potentially valuable tool for the investigation of problems in several areas of cell biology and virology (see Section V).

III. Fusion of Anucleate and Nucleated Cells

The use of viruses to induce fusion of different types of cell to form heterokaryons and hybrid cells has created a powerful experimental tool for the study of the interaction between nucleus and cytoplasm in mammalian cells (Harris, 1968; Barski, 1970; Watkins, 1971; Poste, 1972b). The study of metabolic regulation and phenotypic expression in cell hybrids and heterokaryons produced by fusion of mammalian cells with widely differing properties has established the role of cytoplasmic factors in regulating aspects of nuclear activity and the importance of macromolecular exchange between the nucleus and the cytoplasm. The interpretation of experimental data on metabolic and genetic regulation in cell hybrids and heterokaryons is complicated, however, by the fact that the expression of any property in these cells results from an interplay between two types of nuclei and two types of cytoplasm. The recent development of methods for the production of hybrid cells and heterokaryons by fusion of anucleate cells with nucleated cells of a different type (Poste and Reeve, 1971, 1972a) enables this problem to be overcome and offers a new experimental system to complement existing methods for cell fusion using nucleated cells. Ultimately, it is to be hoped that methods will be available for the fusion of isolated nuclei into anucleate cells (see Section IV).

A. Terminology

Cells formed by the fusion of anucleate and nucleated cells are not hybrid cells or heterokaryons in the conventionally accepted sense since two different genomes are not represented. These terms have been used, however, with the acknowledged reservation that fusion between anucleate and nucleated cells only produces hybridization or heterokaryosis of the cytoplasmic components (Poste and Reeve, 1972a).

The term *hybrid cell* will be used in the remainder of this section to describe any cell having a single nucleus in which the properties of both the anucleate and nucleated parent cell types are expressed. The term *heterokaryon* will be applied to any multinucleate cell showing a similar expression of the properties of both parent cell types. The term *homokaryon* will be used in its conventional sense to describe any multinucleate cell produced by the fusion of two or more cells of the same type.

B. Cell Fusion Techniques

The range of cell types susceptible to fusion by inactivated Sendai virus is extremely wide. Susceptibility extends over a considerable range of the vertebrate phylum and the type of cell used in cell fusion studies is virtually limited only by the requirements and the ingenuity of the experimental scientist. This, together with the already extensive list of highly different cell types fused successfully, dictates that it is difficult to difficult to describe a single method of cell fusion. For this reason, only a few general comments will be made here using the published examples of fusion between anucleate and nucleated cells (Poste and Reeve, 1971, 1972a) to illustrate general principles and methods of the cell fusion technique.

1. Virus

Sendai virus for use in cell fusion experiments is grown in 10- to 11-day-old embryonated hens eggs since virus propagated in mammalian cell cultures has negligible cell fusion activity (Okada, 1969; Poste, 1972b). The chorioallantoic membrane (CAM) of the egg is inoculated with 0.1 ml of Sendai virus, incubated at 36°C with the pointed end of the egg down for 3 days, and then maintained at 4°C overnight. Storage at 4°C serves to kill the chick embryo and also limits contamination of the virus-containing allantoic fluid by blood from ruptured blood vessels. Heavy contamination of the allantoic fluids by blood is a serious problem since the virus particles will adsorb to the erythrocytes and will be lost during the initial centrifugation (see below). After storage at 4°C, the infected allantoic fluids are harvested aseptically. The fluids from infected eggs inoculated with the same batch of virus may be pooled. The pooled fluids are first centrifuged at 500 *g* for 10 minutes, and the Sendai virus hemagglutination titer of the suspension is determined by the standard hemagglutination assay using sheep red blood cells in Perspex hemagglutination trays [for details of method, see Rosen (1969)], and the result is expressed as the number of viral hemagglutinating units (HAU) per milliliter of infected material. A known volume of the infected fluids is then centrifuged at 30,000 *g* for 30 minutes, the suspension is discarded, and the pellet is resuspended in a balanced saline solution to give a known titer of virus.

A convenient concentration used by the author for a stock Sendai virus preparation is 40,000 HAU/ml, which can be diluted easily to appropriate concentrations for experimental use. Stock virus can be stored at −70°C after rapid freezing. The addition of 0.5% bovine serum albumin to the stock virus before freezing avoids the rapid decay of its cell fusion capacity (Neff and Enders, 1968; Pedreira and Taurosa, 1969), and virus can be stored routinely for 6–9 months under these conditions with no significant loss of its fusion capacity.

Sendai virus preparations used for fusing cells are normally inactivated to avoid the problem of replication of infective virus within the fused cells and the accompanying virus-induced effects on cellular morphology and metabolism. The virus can be inactivated either by irradiation with ultraviolet (UV) light or by treatment with β-propiolactone (BPL). The former method is more simple: a 1-ml aliquot of the infective concentrated virus preparation is placed in a watch glass or 25-mm petri dish and irradiated for 10–15 minutes with UV light from a 15-W germicidal lamp. The virus-containing fluids should be agitated using a sterile Pasteur pipette at 5-minute intervals during the period of irradiation. Accurate studies on the inactivation of Sendai virus indicate that a UV dose of 1500 ergs/cm^2/second is adequate to eliminate more than 90% of the viral infectivity in 10 minutes (Harris *et al.*, 1966). Empirically, this dose of irradiation can be achieved by placing the dish or watch glass 25 cm from the UV tube (the lamp should be changed at regular intervals to ensure significant emission at 2600 Å).

Alternatively, Sendai virus preparations can be inactivated with BPL (Neff and Enders, 1968), A 10% solution of BPL (Betaprone, Fellows Testagar Co., Detroit, Michigan) is freshly prepared and diluted in saline bicarbonate (1.68 gm of NaHCO3 and 0.5 ml of phenol red in 100 ml of isotonic saline) at a final concentration of 0.1%. The virus suspension (precooled to 4°C) is then treated with the dilute BPL solution (also precooled to 4°C) using 9 parts of virus to one part of BPL. The mixture is shaken for 10 minutes in an ice bath, incubated for 2 hours at 37°C with shaking at 10-minute intervals followed by incubation at 4°C overnight to allow hydrolysis of the BPL. The preparation is then ready for use.

2. Cell Fusion

The degree of cell fusion that takes place after treatment with inactivated Sendai virus is related to the concentration of virus used, the number of cells and their inherent susceptibility to fusion, and, to a lesser extent, the conditions of incubation during the interaction between the cells and the virus. The effect of these factors on the overall efficiency of cell fusion have been reviewed recently by Poste (1970, 1972b) and need not be discussed further.

Cell fusion by inactivated Sendai virus can be achieved with methods suitable for cells maintained in suspension (Okada, 1962; Harris *et al.*, 1966), monolayers (Kohn, 1965; Poste *et al.*, 1972), or as a combination of both methods in which a suspension of one parent cell type is added to a monolayer of another cell type at the same time as the virus (Davidson, 1969; Klebe *et al.*, 1970). As a further modification of the last method, the virus can be added to the monolayer first followed by the suspension of the second cell type (Poste and Reeve, 1972a).

Cellular susceptibility to fusion by inactivated Sendai virus is probably the most important factor in determining both the efficiency of cell fusion and the size of any polykaryocytes produced by fusion. As a general rule, high concentrations of cells are necessary for the fusion of cells in suspension, and at least 1×10^6 to 1×10^7 cells per milliliter of each cell type have been used in most successful experiments. The fusion of cells in suspension is achieved more easily than when the same cells are grown as a monolayer; the latter requires a larger dose of virus to induce fusion and the yield of heterokaryons is also less than that in suspension cell cultures treated with the same dose of virus. It is also possible to control the extent of fusion between the different parent cell types in both monolayers and suspension cultures by varying the concentration of virus and the proportions of each parent cell type.

Although trial-and-error technical manipulations of this type may alter the efficiency of cell fusion and the yield of heterokaryons, it must be recognized that there are marked variations in the facility with which different cell types will fuse. Consequently, it may be stated with reasonable confidence that methods that are successful in producing fusion between certain cell types may not work with other cell combinations, and attempts to fuse each new mixture of cells may require a reappraisal of the experimental conditions for fusion.

The dose of virus, cell numbers, and the ratio of anucleate to nucleated cells used successfully by Poste and Reeve (1972a) to obtain optimum yields of hybrid cells and heterokaryons by fusion of a range of anucleate and nucleated cells is shown in Table II. The methods used to achieve fusion between these cell combinations are referred to in Table II (bottom line) as methods 1–3. In method 1, anucleate and nucleated cells are mixed together, inoculated onto glass coverslips and allowed to attach to glass and then treated with inactivated Sendai virus. In method 2 a suspension of anucleate cells is added to a virtually confluent monolayer of nucleated cells, allowed to attach and then treated with inactivated Sendai virus.

TABLE II

CELL NUMBERS AND VIRUS CONCENTRATION USED TO PRODUCE FUSION OF ANUCLEATE AND NUCLEATED MAMMALIAN CELLS[a]

Cell type, virus concentration, and fusion method[b]	Number of cells/ml					
Nucleated macrophages	—	—	2×10^6	—	—	—
Anucleate macrophages	3×10^6	1×10^7	—	—	—	4×10^6
Nucleated L929 cells	1.5×10^6	—	—	—	—	—
Anucleate L929 cells	—	—	1×10^7	5×10^6	3×10^6	—
Nucleated HEp-2 cells	—	4×10^6	—	2×10^6	—	—
Anucleate HEp-2 cells	—	—	—	—	—	—
Chick embryo erythrocyte ghosts	—	—	—	—	3×10^7	3×10^7
Virus dose (HAU/ml)	8000	12,000	12,000	10,000	8000	8000
Fusion method	1, 2, and 3	2	1	1, 2	3	3

[a] Reproduced, with permission, from Poste and Reeve (1972a).

[b] Refers to different cell fusion methods (see text).

In method 3 the anucleate cells are first centrifuged onto a glass cover slip using a cytocentrifuge (Shandon Scientific Corporation, London, England) either together with the nucleated cells or alone followed by the addition of the nucleated cells and inactivated Sendai virus. In all three methods, cell mixtures are treated with virus for 1 hour at 37°C, after which the cultures are washed twice with prewarmed phosphate-buffered saline (PBS) to remove excess virus and then returned to fresh culture medium at 37°C for further incubation and use in experiments.

C. Identification of Anucleate Cell Properties in Hybrid Cells and Heterokaryons

To demonstrate that fusion has occurred between two different cell types, it is necessary to identify characteristics from each parent cell in the fused cell. By the appropriate choice of species, and cell lines within a species, it is possible to exploit an infinite number of marker characteristics to identify the presence of components from both parent cells in the cell hybrid or heterokaryon produced by fusion. The range of cellular characteristics used in the identification of fusion between nucleated cells of different types is shown in Table III. Similarly, most of these markers could

TABLE III

CHARACTERISTICS USED IN THE IDENTIFICATION OF HYBRID CELLS AND HETEROKARYONS[a]

1. Production of enzymes and other proteins derived from each parent cell type (species-specific isoenzymes and various cell-specific gene products)
2. Presence of chromosomes from both parent cell types
3. Detection of species-specific cell surface antigens from both parent cell types in inter-specific hybrid cells and heterokaryons
4. Detection of cell-specific histocompatibility antigens from both parent cells in intra-specific hybrid cells and heterokaryons
5. Ability of hybrid cells and heterokaryons to grow in selective media due to intergenic complementation between deficient parent cells (drug-resistant mutants; nutritional auxotrophs)
6. Presence of morphologically distinct properties from each parent cell in hybrid cells and heterokaryons (differences in nuclear and nucleolar morphology between the parent cell types)
7. Use of artificial cell markers to label one parent cell type and subsequent demonstration of labeled and unlabeled cells in the heterokaryons (radioisotopes)

[a]Detailed examples of the use of these methods to identify hybrid cells and heterokaryons are given in several recent reviews (Harris, 1970; Migeon and Childs, 1970; Ruddle, 1972; Poste, 1972b).

be used to detect fusion occurring between anucleate and nucleated cells of different types.

1. Fusion of Anucleate Mouse Macrophages with Nucleated Mouse L Cells or Human HEp-2 Cells

Mouse peritoneal macrophages (MPM) are not detached from a solid surface on which they are growing by treatment with trypsin (Gordon and Cohn, 1970), and they possess on their surfaces specific receptor sites that enable them to ingest sheep red blood cells coated with antibody (Lay and Nussenzweig, 1969; Mauel and Defendi, 1971). These properties were used by Poste and Reeve (1972a) to identify the presence of macrophage surface components in hybrid cells and heterokaryons produced by fusion of MPM enucleated by CB with nucleated mouse L cells and human HEp-2 cells. Detection of these "macrophage" characteristics in nucleated cells within the mixed cell culture after treatment with inactivated Sendai virus confirmed that fusion had occurred. Furthermore, the trypsin resistance of the macrophages provided a simple and effective selection procedure for separating the fused cells from the unfused single nucleated cells and L or HEp-2 homokaryons present in the same culture, since the latter were removed from the glass by the trypsin treatment. Additional proof that fusion had taken place between the anucleate macrophages and nucleated L cells was established using L cells whose nuclei had been labeled with thymidine-^{3}H before treatment with Sendai virus so that any hybrid cells and heterokaryons could be recognized both by their trypsin-resistance and phagocytic properties and by the presence of a radioactive nuclear label (detected by radioautography).

2. Fusion of Anucleate Mouse L Cells and Nucleated Mouse Peritoneal Macrophages or Human HEp-2 Cells

The detection of species-specific antigens on the surface of hybrid cells and heterokaryons can be used to provide direct proof of interspecific cell fusion. Watkins and Grace (1967) and Kano *et al.* (1969) used the mixed hemadsorption (HAD) technique of Espmark and Fagraeus (1965) to demonstrate species-specific antigens on heterokaryons between nucleated cells from species as diverse as mouse, chick, and man. This method offers a relatively crude but convenient technique for determining whether antigens of a particular species are present on a cell. Mixed HAD may be observed microscopically at the level of the single cell, and this enables studies to be made on the number, distribution, and persistence of certain antigens on the surface of hybrid cells and heterokaryons (Harris, 1967; Harris *et al.*,

1969). The mixed HAD technique may also be used without further modification to detect species-specific antigens on cell hybrids and heterokaryons produced by interspecific fusion of anucleate and nucleated cells. Thus, Poste and Reeve (1972a) used this method to identify murine antigens on the surface of single and multinucleate cells in mixed cultures of anucleate mouse L cells and nucleated human HEp-2 cells treated with inactivated Sendai virus.

3. Fusion of Chick Erythrocytes and Anucleate Mouse Macrophages

Fusion between cells with morphologically distinct nuclei can be used to demonstrate cell fusion. Unfortunately, among mammalian cell cultures there are very few cell strains or lines that possess sufficiently distinct chromatinic and nucleolar characteristics to allow these properties to be used with certainty to identify the contribution of each cell type to a heterokaryon. However, the nucleus of the chick erythroblast is significantly smaller than the nucleus of most mammalian cell lines and this difference can be used to identify fusion occurring between chick erythrocytes and various mammalian cells. Post and Reeve (1972a) used the distinctive size and morphology of the chick erythrocyte to demonstrate fusion between erythrocyte ghosts and anucleate mouse cells. The appearance in mixed cell cultures of mouse and chick cells after treatment with inactivated Sendai virus of single and multinucleate cells containing erythrocyte nuclei and which were attached to the glass provided suggestive evidence for fusion between the different cell types. Unequivocal proof of interspecific fusion was provided however, by the demonstration by mixed HAD of chick-specific antigens on the surface of these cells.

These methods are still at an early stage of development, and further improvements will almost certainly be introduced. Also, the range of anucleate and nucleated cells fused successfully is still very limited. However, the methods for induced cellular enucleation by CB described in Section II,b give every indication that they can be used to enucleate most types of mammalian cells, and it is likely that a wider range of anucleate cell types will soon be available for similar cell fusion experiments.

Fusion of anucleate and nucleated mammalian cells *in vitro* has the significant advantage over methods for the fusion of nucleated cells that mononuclear hybrid cells are formed immediately after fusion. This is of considerable practical importance since it enables experiments to be carried out on hybrid cells immediately and avoids the need for the nuclear fusion sequence that must take place to form a hybrid cell after fusion of nucleated cells (Harris, 1970).

IV. The Transfer of Isolated Nuclei to Anucleate Cells

The cell fusion experiments described in the previous section were prompted by the need for an experimental system to investigate the behavior of a single nucleus in a cytoplasm derived from two different cell types. Although the full potential of these methods has yet to be realized, it is recognized that even greater experimental opportunities would be available if methods could be devised for the transfer of isolated mammalian cell nuclei to anucleate mammalian cells. The fascinating and valuable results obtained by similar nuclear transfer experiments in protozoa (Goldstein and Prescott, 1967a,b; Jeon and Danielli, 1971) and oocytes (Gurdon and Woodland, 1971) are adequate testimony to the value of this approach in the study of nucleocytoplasmic interactions and the regulation of gene activity.

A series of experiments have been done recently in the author's laboratory to assess the feasibility of nuclear transfer in mammalian somatic cells cultured *in vitro*. The results and problems encountered in these experiments will be discussed briefly.

Many methods are available for the preparation of isolated nuclei from mammalian cells and tissues. Most of these are designed, however, for use in biochemical investigations on purified cell fractions and are unsuitable for the purpose of isolating nuclei for potential transplantation to a new cell. In the preliminary experiments conducted in the author's laboratory, nuclei were obtained from confluent monolayer cultures of HeLa and L929 cells by the method of Penman (1969) which involves rapid breaking of the cells in reticulocyte standard buffer followed by low speed centrifugation to clarify the homogenate and separate the nuclei from most of the cytoplasm. The nuclei obtained by this method are well preserved morphologically, possess a high level of NAD-pyrophosphorylase activity and are able to incorporate thymidine-^{3}H-triphosphate in the presence of adenosine triphosphate, deoxyadenosine triphosphate, deoxyguanosine triphosphate, and deoxycytidine triphosphate. This method of nuclear isolation suffers from the major limitation, however, that the nuclei are exposed to aqueous solutions (see below).

Nucleated mammalian cells attached to a solid substrate *in vitro* are capable of ingesting isolated nuclei by endocytosis (Sekiya *et al.*, 1969). This finding has been confirmed in the author's laboratory using L929 and HeLa cell nuclei isolated by the method described above. However, the frequency of the uptake of isolated nuclei by endocytosis is very low, occurring in less than 5% of the cells in populations of nucleated cells (L, HEp-2, BHK-21) and in less than 1% of cells in anucleate cell populations (peritoneal macrophages). In an attempt to increase the frequency of incorporation of isolated nuclei into cells, isolated nuclei (10^8/ml) have also been added to monolayer cultures and then treated with inactivated Sendai virus (8000 HAU/ml). This increased significantly the number of isolated nuclei incorporated into the cells, but even under these conditions only 10–12% of the cell population incorporated nuclei. More importantly, it has been found consistently that the uptake of nuclei by the cells is followed by rapid degeneration and fragmentation of the nuclei within the cytoplasm.

Microscopic examination of incorporated nuclei has revealed evidence of degeneration from as little as 2 hours after their uptake by the cell, and by 24 hours most nuclei are completely fragmented. This degenerative change has been observed consistently in nuclei incorporated both by endocytosis and by fusion, inactivated Sendai virus being used. Degeneration does not appear to be due simply to the introduction of the isolated nuclei into a heterologous cytoplasm, since it occurs in nuclei incorporated into both homotypic and heterotypic cells. Preliminary electron microscopic studies indicate that an additional membrane forms around the nuclei soon after they are introduced into the cytoplasm, suggesting that a component of the cell vacuolar apparatus (lysosomes?) may have contributed to the degeneration of the nuclei (lysosomal nucleases?). In this respect, it is of interest to note that similar rapid fragmentation has been observed in isolated mammalian chromosomes introduced into the cytoplasm of mammalian cells *in vitro* (Burkholder and Mukherjee, 1970; Ebina *et al.*, 1970; Kato *et al.*, 1971).

The factors responsible for the degeneration of mammalian cell nuclei introduced into the cytoplasm of the same or different cell types remain to be identified. However, the difficulties encountered in these initial experiments serve to focus attention on the many technical problems that will have to be overcome before successful nuclear transfer in mammalian cells can be achieved. In view of the enormous experimental possibilities offered to cell biologists if successful methods were developed, it is considered pertinent to outline briefly some of the more important factors that might influence the success of nuclear transfer in mammalian cell cultures.

The difficulties encountered in the transfer of isolated nuclei in mammalian cell systems contrasts with the successful transplantation of nuclei in oocytes and various species of protozoa (see above) and the even more spectacular success of the reassembly of viable amebae from separated nuclei, cytoplasm and membranes derived from three different donor cells (Jeon *et al.*, 1970). The success of nuclear transfer in these cells probably reflects the fact that experiments are done on individual cells, rather than large populations, and the transfer of the nuclei takes place quickly with microsurgical insertion of the nucleus into its new host immediately after its removal from the donor cell so that the nucleus is not exposed to the external environment. Clearly, these microsurgical methods do not lend themselves to experiments on much smaller mammalian cells in culture, and any methodology for the large-scale transfer of nuclei in mammalian cell populations *in vitro* will require the development of a suitable medium in which nuclei can be stored without loss of viability for periods of up to 1 hour.

One lesson that can be learned from the descriptions of nuclear transplantation in protozoa and oocytes is the extreme sensitivity of the nucleus to exposure to external aqueous media. Indeed, in all successful transplantation experiments, nuclei have been transferred surrounded by an appreciable amount of cytoplasm and, even then, transfer had to be completed quickly to ensure viability of the nuclei. The importance of the cytoplasm in maintaining the viability of nuclei has been stressed in microsurgical nuclear transplantation experiments on protozoa (Jeon and Danielli, 1971), amphibian oocytes (King, 1966) and *Drosophila* (Zalokar, 1971). Even under these conditions, the transplanted nuclei may show abnormalities in activity and artifactual macromolecular exchange between the nucleus and the cytoplasm (Legname and Goldstein, 1972).

In the few reports where nuclei obtained from cells cultured *in vitro* have been transferred by microsurgery to large cells such as amphibian oocytes (King, 1966; Gurdon and Woodland, 1971) and mammalian muscle cells (Gräbmann, 1970) substantial amounts of cytoplasm were also transferred at the same time. Indeed, in certain experiments of this type claiming transfer of "isolated nuclei" by microsurgery, examination of the methods shows that whole cells were in fact used (Gurdon and Laskey, 1970a,b).

The transfer of large amounts of cytoplasm along with the nucleus may be unimportant when transfer is made to cells with a large volume of cytoplasm such as oocytes or amebae, but it is clear that more stringent requirements must be fulfilled in nuclear transfer experiments with much smaller mammalian cells. The transfer of a substantial quantity of cytoplasm to a mammalian cell at the same time as a nucleus may well negate the purpose of the experiment. For example, a nucleated mammalian cell receiving a transferred nucleus plus considerable contaminating cytoplasm may differ very little from the product of fusion between two nucleated cells. Similarly, the presence of significant amounts of cytoplasm around a nucleus transferred to an anucleate cell may hinder the study of macromolecular exchange between the cytoplasm of the anucleate cell and its new nucleus.

In view of the injurious effects of aqueous solutions on nuclei and the apparent protective effect of cytoplasm it seems likely that any method for the transfer of nuclei in mammalian cell cultures will have to reach a degree of compromise concerning the amount of cytoplasmic contamination and "carry over" that would be acceptable during nuclear transfer. Alternatively, in order to isolate viable nuclei for transfer experiments that have minimal cytoplasmic contamination it may be necessary to explore the value of nuclear isolation and transfer in nonaqueous solutions, such as the nontoxic fluorocarbon oils used by Kopac (1955) in his experiments on microsurgical transfer of nuclei and nucleoli in amebae.

To the best of the present author's knowledge there are no reports in the literature of the successful large-scale transfer of isolated nuclei into mammalian cells *in vitro*. Harris *et al.* (1966) reported successful fusion of nuclei from chick erythrocyte ghosts with mouse macrophages, and Poste and Reeve (1972a) reported similar success in the fusion of chick erythrocyte ghosts with anucleate mouse macrophages and L929 cells. Since the plasma membrane and a limited amount of cytoplasm are still present in erythrocyte ghosts [see Harris and Brown (1971)] the fusion of these cells with other cells does not represent the transfer of an isolated nucleus in its strictest sense. However, since the majority of the cytoplasm is undoubtedly lost from most ghosts, and little cytoplasm remains attached to the outer nuclear membrane (Harris and Brown, 1971), the use of these cells in fusion experiments goes some way toward achieving the transfer of an isolated nucleus.

In view of the problem of the degeneration of transferred nuclei discussed earlier it is of interest to note that a high proportion of nuclei from erythrocyte ghosts also degenerate soon after fusion (Harris *et al.*, 1966). However, a number of the ghost nuclei do survive, show no morphological damage, and are capable of synthesizing both DNA and RNA (Harris, 1970). Why the nuclei in certain erythrocyte ghosts degenerate after fusion while others do not has not been identified, but it is tempting to speculate that differences in survival may be related to the time a particular nucleus is exposed to the aqueous culture medium or to the amount of cytoplasm that remained fortuitously in the ghost.

In addition to the obvious importance of the viability of the nucleus in determining the success of nuclear transfer to anucleate cells, the properties of the cytoplasm in the anucleate cell may prove to be equally important. The role of cytoplasmic factors in the initiation and control of nucleic acid synthesis in the nucleus has been demonstrated in many types of cell, and there is now considerable evidence to indicate that the absence or loss of these factors from the cytoplasm may limit nuclear activity (DeTerra, 1969; Harris, 1970; Gurdon and Woodland, 1971). Thus, one possible limiting factor in the success of nuclear transfer to anucleate mammalian cells might be that the various cytoplasmic RNA and protein species involved in regulating nuclear activity might decay rapidly in the absence of a nucleus, so that even if a viable nucleus could be introduced into the cytoplasm, it might fail to function normally owing to the state of the cytoplasm.

This proposal gains some support from the experimental results of Lorch and Jeon (1969), who transplanted nuclei from amebae treated with actinomycin D (AMD) to normal untreated anucleate amebae and from untreated amebae to enucleate amebae pretreated with AMD. Normal anucleate cells that received the AMD-treated nuclei recovered their normal nuclear function and reproduced. In contrast, transfer of nuclei from normal cells to anucleate amebae pretreated with AMD resulted in loss of nuclear activity, loss of reproductive capacity and eventual cell death. Similarly, Sawicki and Godman (1971, 1972) have proposed that differences in the rate of the decay of the cytoplasmic factors involved in the regulation of nuclear activity are responsible for the differential capacity of mammalian cell cultures to recover from AMD treatment *in vitro*. Sawicki and Godman suggested that there are several species of RNA present in the cytoplasm essential for normal nuclear activity and that these have a significantly shorter half-life in cell types (HeLa) that fail to recover from AMD treatment than in cells (Vero; Wl-38; L292) that can resume normal nuclear activity after removal of AMD.

The enucleation of a cell leading to the suppression of the transcription, synthesis and transfer to the cytoplasm of the species of RNA and protein required for normal nuclear activity might therefore create a metabolic state in the anucleate cell that could seriously impair its capacity to initiate and regulate nuclear activity. Consequently, if the decay of these essential factors in the cytoplasm fell below a critical concentration, it might mean that, even if a viable nucleus could be introduced into the cytoplasm, expression of its normal activity would be frustrated. This explanation could well account for the failure to observe reactivation of DNA and RNA synthesis in the nuclei of chick erythrocytes fused with anucleate mouse L929 cells (Poste and Reeve, 1972a). In contrast, reactivation of both nuclear DNA and RNA synthesis occurred when the same cells were fused with nucleated cells that were actively synthesizing both DNA and RNA (Johnson and Harris, 1969).

Despite the various problems discussed above which presently hinder successful nuclear transfer in mammalian cells *in vitro*, it is to be hoped that these problems will be solved in the not too distant future. It is important that previous microsurgical studies have demonstrated the feasibility of nuclear transfer techniques in cells from a wide range of species, and it would appear therefore that most of the problems associated with nuclear transfer are of a technical nature and do not reflect any inherent inability of cells to survive with a new type of nucleus. The unique experimental opportunities that would be available to cell biologists if successful methods for nuclear transfer in mammalian cells were developed does not require emphasis. Undoubtedly, this problem will receive the detailed future attention it deserves.

V. Experimental Applications of Anucleate Mammalian Cells

A. The Study of Messenger RNA Activity

Anucleate cells provide an excellent system for the study of the post-transcriptional control of cellular metabolism, and of protein synthesis in particular. The ability of anucleate cells to synthesize proteins is a consequence of the persistence in the cytoplasm of messenger RNA (mRNA) molecules, ribosomes, and transfer RNAs (tRNA) that were synthesized in the nucleus before it was removed. Measurements of the rate of decay of the synthesis of individual protein species in anucleate cells can therefore provide information on the longevity of protein-specific mRNA molecules within the cytoplasm. Anucleate cells have a further advantage as experimental systems over intact cells for this purpose since the mechanisms controlling translation in the cytoplasm can be investigated without the superimposed complexity of the transcription process in the nucleus. This use of anucleate cells to study the stability of mRNA activity is based, however, on the restrictive assumption that mRNA is the only limiting factor in determining the rate and duration of protein synthesis in the absence of the nucleus and that the concentration of tRNAs, amino acids, enzyme donors and other components of the protein synthetic machinery do not act as limiting factors. The validity of these assumptions and the factors that may contribute to deviations from the ideal relationship between mRNA activity and protein synthesis in anucleate cells have been reviewed by Keck (1969).

Studies of the anucleate state in eukaryotic cells has provided strong evidence that the various protein-specific mRNAs in the cell differ greatly in their functional lifetimes (Keck, 1969; Brachet, 1972). However, apart from studies on the behavior of mRNA species in mammalian reticulocytes that lose their nuclei naturally during differentiation (Lane *et al.*, 1972; Williamson, 1972), the lack of suitable enucleation techniques has dictated that estimates of the lifetime of mRNA species in mammalian somatic cells have been based on the duration of protein synthesis in intact cells following so-called "physiological enucleation" (Davidson, 1968) by actinomycin D (AMD) which inhibits DNA-dependent RNA synthesis.

There are a number of serious limitations to the use of AMD and similar metabolic inhibitors to study mRNA longevity, which may result in erroneous values for the lifetimes of these molecules in the cytoplasm (see below). First, the susceptibility of cellular RNA synthesis to inhibition by AMD varies significantly in different cell types and the suppression of RNA synthesis by this drug may not be complete when AMD is used at the low doses (≤ 1 μg/ml) necessary to avoid cytotoxic effects [for references, see Sawicki and Godman (1972)]. Second, even at these low doses, inhibition of

RNA synthesis by AMD may be accompanied by a secondary inhibition of cellular DNA (Magee and Miller, 1968; Bacchetti and Whitmore, 1969) and protein synthesis (Earl and Korner, 1966; Soeiro and Amos, 1966; Sawicki and Godman, 1971). Finally, the use of intact cells in studies with AMD introduces the possibility that transitory mRNA in the nucleus might still enter the cytoplasm after inhibition of RNA synthesis by AMD thereby postponing the time of effective "enucleation."

Estimates of mRNA lifetimes in intact mammalian cells based on the duration of protein synthesis or polysome degradation after treatment with AMD give values for the half-life of mRNA species of between 2 and 4 hours (Penman *et al.*, 1963; Staehelin *et al.*, 1963; Trakatellis *et al.*, 1965; Cheevers and Sheinin, 1970; Craig *et al.*, 1971). However, the recent discovery that mRNA molecules in eukaryotic cells (Burr and Lingrel, 1971; Darnell *et al.*, 1971; Sheldon *et al.*, 1972) and viruses (Phillipson *et al.*, 1971; Bachenheimer and Roizman, 1972; Gillespie *et al.*, 1972) contain a length of 3′ terminal polyadenylic acid, poly (A), provides a powerful new tool for the accurate measurement of mRNA metabolism. Thus, the results of two very recent studies in which mRNA was measured in mammalian cell systems by hybridization with polyuridylic acid, poly (U), has shown that despite rapid decay of cellular protein synthesis following treatment with AMD, mRNA activity remained stable and had a half-life of at least two to three times that calculated from the duration of protein synthesis (Greenberg, 1972; Singer and Penman, 1972). The results of these two studies indicate instead that the decay of protein synthesis in AMD-treated cells is due not to the decay of functional mRNA, but to a failure in the initiation of the translation process.

In view of these serious limitations on the value of AMD in studies of mRNA activity, the use of large-scale populations of anucleate mammalian cells enucleated by the techniques described in Section II,B and in other chapters of this book may well emerge as an important technique in future studies of the stability of mRNA species in mammalian cells under different conditions.

B. Nucleocytoplasmic Interactions

The importance of the exchange of macromolecules between the nucleus and the cytoplasm in eukaryotic cells in regulating gene expression in the nucleus and protein synthetic activity in the cytoplasm has been demonstrated by a variety of experimental methods (Goldstein and Prescott, 1967a,b; DeTerra, 1969; Harris, 1970; Wyngaarden, 1970; Johnson and Rao, 1971). As mentioned in the introduction to this chapter, anucleate cells and nuclear transplantation methods have contributed significantly to our understanding of this subject in nonmammalian cells, but the impact of these two methods in the experimental study of the interaction of nucleus and cytoplasm in mammalian somatic cells has been restricted by the lack of methods for cell enucleation. It is anticipated therefore that the methods described here for the enucleation of large numbers of mammalian cells will remedy this deficiency and open the way for a wide range of ambitious experiments on anucleate cells that were only possible previously with very large cells amenable to individual enucleation by microsurgery.

C. The Role of the Nucleus and the Cytoplasm in Virus Replication

Viruses can replicate only in the complex environment of the living cell and cannot synthesize proteins independently. Host cell enzymes usually synthesize the simple precursors and energy-rich molecules needed for the synthesis and assembly of viral components, and to reproduce successfully the viral genome must divert the metabolism of the host cell towards the synthesis of viral macromolecules at the expense of host cell constituents. A fundamental question arises in the study of virus replication: What is the source of the genetic information needed to specify those macromolecules in an infected cell which lead to the production of new infective progeny virus? The search for the answer to this question and an understanding of the "strategy of the viral genome" (Subak-Sharpe, 1971) is not merely of academic interest and has important implications for the mechanisms by which viruses are able to damage cells and for the development of effective antiviral agents that will be able to block virus replication by acting selectively on the metabolism of the virus rather than that of the cell.

The replication of certain RNA-containing viruses takes place entirely within the cytoplasm with no involvement of the transcription mechanisms within the nucleus. In contrast, certain other RNA-containing viruses and all the DNA-containing viruses replicate both in the nucleus and in the cytoplasm, with a number of early stages in the replication cycle occurring either within the nucleus or requiring functions directed by the nucleus. Although the absolute requirements of most viruses from the different classification groups for replication in the nucleus and the cytoplasm have now been identified, there are many aspects of the role of the nucleus and the interaction between nucleus and cytoplasm during the replication of nuclear-dependent viruses that are still unclear. The comparative study of the replication of viruses in both nucleated and anucleate cells offers an excellent experimental system for the identification of virus-specific components and new patterns of host-specific macromolecules within the nucleus at different stages in the replication cycle and whether these remain in the nucleus or are transported to and from the cytoplasm.

The value of anucleate mammalian cells in the study of animal virus replication has long been recognized, but the use of this approach experimentally has been frustrated by the difficulties of obtaining sufficient numbers of anucleate cells. Several attempts have been made to define the role of the nucleus in virus replication using anucleate fragments of mammalian cells obtained by microsurgery.

Marcus and Freiman (1966) showed that two RNA-containing viruses, Newcastle disease virus (NDV) and poliovirus, were able to replicate normally in anucleate fragments of X-ray-induced giant HeLa cells *in vitro*. A similar demonstration of the ability of poliovirus to replicate in anucleate cells was made by Crocker *et al.* (1964), and Cheyne and White (1969) confirmed that NDV could multiply successfully in HeLa cells in the absence of a nucleus. Cheyne and White also demonstrated that influenza virus, another RNA virus, could only replicate in intact cells confirming an earlier proposal for a nuclear-dependent stage in the replication of this virus based on the inhibition of virus growth in intact cells treated with actinomycin D (Barry *et al.*, 1962).

Anucleate mammalian cells obtained by microsurgical separation of large fragments of cytoplasm have little value in the study of virus replication other than to demonstrate whether a particular virus is capable of replicating in the absence of a nucleus. The small number of anucleate cells obtained by microsurgery dictates that virus replication has to be assessed in individual cells using radioautography, hemadsorption, and light or electron microscopy. Sophisticated biochemical studies to identify the various species of virus-coded and host-specific macromolecules present in the infected cell are impossible. However, the large-scale methods for the rapid enucleation of mammalian cells with CB overcome this problem, and it is anticipated that detailed biochemical studies on the replication of viruses in CB-enucleated cells will become increasingly common.

Prescott and his colleagues at the University of Colorado have already made a start in this direction, and have clarified the long-standing confusion on the role of host-coded functions in the early stages of vaccinia virus infection by showing that both uncoating of the virus and the initiation of virus-specific DNA synthesis in the cytoplasm could occur in L929 cells enucleated by CB (Prescott *et al.*, 1971).

The use of anucleate cells in the study of the role of the nucleus and the cytoplasm throughout the entire virus replication cycle extending over a number of hours means that methods must be available for the rapid removal of nuclei from cells at intervals after virus infection. The techniques for induced cellular enucleation by CB described in Section II,B enable a high proportion of the cell population to be enucleated within 30 minutes; this is comparable to the time scale for the "physiological enucleation" of cells by low doses of actinomycin D used widely in virology to eliminate host-cell transcription. The feasibility of removing the nuclei from mammalian cells at intervals after virus infection has been demonstrated recently by C. Colby (personal communication), who has used this approach to demonstrate the essential role of the nucleus in both the initiation and the maintenance of the antiviral state induced in cells by treatment with interferon.

Apart from the investigation of the role of the nucleus in the replication of different viruses, there are a number of other questions of current interest to virologists that could be investigated using anucleate cells. As Prescott *et al.* (1971) pointed out, it would be of considerable interest to examine the growth of nuclear-dependent viruses in anucleate cells coinfected with cytoplasmic viruses that could provide the enzymes capable of transcribing or replicating the nucleic acid of the nuclear virus. Similarly, the replication of cytoplasmic RNA-containing viruses in anucleate cells could be used to examine the changes that might occur in progeny virus with depletion and decay of various host-specific functions at different rates. Finally, the formation of hybrid cells and heterokaryons by fusion of anucleate and nucleated cells (Section III) could be used in the study of the factors that determine the varying susceptibility of different cell types to virus infection employing a similar protocol to the recent studies on this problem made with hybrid cells and heterokaryons formed by fusion of nucleated cells (Green *et al.*, 1971; Poste, 1972b). This use of heterokaryons formed by fusion of anucleate and nucleated cells to study the mechanism of cellular permisiveness to specific virus infections should enlarge our understanding of virus-induced cytopathogenicity. For example, recent experiments in the author's laboratory have established that infective SV40 virus can be "rescued" from SV40-transformed 3T3 cells following fusion with anucleate permissive monkey cells (G. Poste, unpublished observations, 1972). These results indicate that SV 40 does not need to replicate in the permissive cell nucleus in order to complete its replicative cycle. Further experiments of this kind will hopefully clarify the role of the permissive cell component in the rescue of oncogenic viruses from heterokaryons of permissive and nonpermissive transformed cells.

D. The Autonomy of Cytoplasmic Organelles

The identification of DNA in mitochondria, chloroplasts, and centrioles (Sager, 1972) poses a number of important questions concerning the contribution of this DNA to the autonomous replication of these organelles and the relationship between nuclear DNA and genetic information in cytoplasmic organelles. Anucleate cells provide unique opportunities for the study of the replication and synthetic abilities of these organelles. Mitochondria, chloroplasts, and centrioles all appear to retain activities such as transcription and translation, and they even undergo limited replication in anucleate cells (Brachet, 1972; Bresch, 1972; Craig, 1972). Although these activities are lower than in normal nucleated cells, further comparison of activity in both anucleate and nucleated cell systems should enlarge our understanding of the interaction between the nucleus and DNA-containing cytoplasmic organelles.

VI. Conclusions

The importance of the continuous exchange of macromolecules between the nucleus and the cytoplasm in regulating the metabolic activity of eukaryotic cells is now well accepted. Most forms of cytoplasmically utilized RNA (ribosomal, messenger, and transfer) are synthesized within the nucleus, and there is now strong evidence to indicate that cells possess mechanisms for selecting those species of RNA that are transported to the cytoplasm and those that remain in the nucleus. The transfer of proteins from the nucleus to the cytoplasm, often complexed with RNA in the form of ribonucleoprotein, has been demonstrated by a variety of techniques in a wide range of cell types. Protein transport in the reverse direction, from cytoplasm to nucleus, is equally well documented. Cytoplasmic proteins may enter the nucleus and act either as structural components or as regulators of nuclear activity. The majority of histone and nonhistone proteins are synthesized in the cytoplasm and are transported subsequently to the nucleus where at least some of them probably act to regulate gene expression in the nucleus. A relationship between changes in gene activity and the action of cytoplasmic proteins on the nucleus has also been demonstrated by elegant nuclear transplantation and cell fusion experiments.

The complex interaction between the nucleus and the cytoplasm can be studied in a number of ways: biochemical analysis of purified cell fractions; electron-microscopic cytochemistry and radioautography; and the comparison of the properties of nucleated and anucleate cells of the same type. The value of anucleate cells as an experimental tool in studying the interaction between nucleus and cytoplasm has been demonstrated in several studies with very large cells, such as protozoan species, algae, and amphibian eggs enucleated by microsurgery or ultracentrifugation. The impact of this approach on experimental studies with mammalian cells has been frustrated, however, by the technical problem of cellular enucleation. Even with very large protozoan cells and oocytes the use of microsurgical methods to produce cell enucleation has meant that the amount of anucleate material is usually so small that detailed biochemical analyses are possible only with delicate and time-consuming micromethods. These problems have virtually precluded the use of similar microsurgical techniques to enucleate mammalian somatic cells cultured *in vitro*.

The search for a more suitable method for the large-scale enucleation of cultured mammalian cells has been rewarded in the last few years by the finding that these cells can be enucleated by the fungal metabolite cytochalasin B. Following recognition of this unique property of cytochalasin B in 1967, a number of methods have been developed for the use of this compound to enucleate large numbers of mammalian cells cultured *in vitro*. These methods, as described here and in other chapters of this book, are still at the early stages of their development and further sophistication and technical refinements to improve the efficiency of cellular enucleation will almost certainly be introduced in the next few years.

Further studies are also needed to define the mechanism of cell enucleation by cytochalasin B, and its other manifold effects on the cell. A more detailed characterization of the anucleate cells produced by cytochalasin treatment is required to establish the quality of anucleate cells used in experiments, and to identify the stability and longevity of various important cytoplasmic macromolecules in the absence of a nucleus. The fusion of anucleate and nucleated cells to form hybrid cells and heterokaryons appears to offer promise as an experimental tool for the study of the effect of cytoplasmic macromolecules on nuclear function, the genomic control of synthetic activities within the cytoplasm and the maintenance of the structural organization of the cell and its various organelle systems. It is clear, however, that, if we wish to understand better these aspects of nucleocytoplasmic interactions in mammalian cells, all available methods should be used, and work done on anucleate cell systems should be designed to integrate within a broad framework of experimental methodology.

Acknowledgments

Most of the personal work cited in this chapter was done in the Department of Virology, Royal Postgraduate Medical School, Du Cane Road, London W. 12., and was supported by grants from the Medical Research Council, the Agricultural Research Council, and the Cancer Research Campaign. I am most grateful to Professor A. P. Waterson for advice and encouragement and to Mr. P. Lister for excellent assistance and the construction of the holders for the ultracentrifuge rotors. I also wish to thank Drs. Peter Reeve and Dennis Alexander for their collaboration in certain experiments and for their helpful advice throughout this work.

References

Aldridge, D. C., and Turner, W. B. (1969). *J. Chem. Soc., C* 923.
Aldridge, D. C., Armstrong, J. J., Speake, R. N., and Turner, W. B. (1967). *Chem. Commun.* 26.
Aldridge, D. C., Burrows, B. F., and Turner, W. B. (1972). *Chem. Commun.* 148.
Allison, A. C., Davies, P., and Petris, S. de. (1971). *Nature (London), New Biol.* **232**, 153.
Armstrong, P. B., and Parenti, D. (1972). *J. Cell Biol.* **55**, 542.
Austin, C. R. (1961). "The Mammalian Egg," pp. 52–86. Blackwell, Oxford.
Bacchetti, S., and Whitmore, G. F. (1969). *Biophys. J.* **9**, 1427.
Bachenheimer, S. L., and Roizman, B. (1972). *J. Virol.* **10**, 875.
Barry, R. D., Ives, D. R., and Cruickshank, F. G. (1962). *Nature (London)* **20**, 663.
Barski, G. (1970). *Int. Rev. Exp. Pathol.* **9**, 151.
Brachet, J. (1972). *In* "Biology and Radiobiology of Anucleate Systems" (S. Bonotto, R. Goutier, R. Kirchmann, and J.-R. Maisin, eds.), Vol. 1, pp. 1–26. Academic Press, New York.
Brachet, J., and Bonotto, S., eds. (1970). "Biology of Acetabularia." Academic Press, New York.
Bresch, H. (1972). *In* "Biology and Radiobiology of Anucleate Systems" (S. Bonotto, R. Goutier, R. Kirchmann, and J.-R. Maisin, eds.), Vol. 1, pp. 207–227. Academic Press, New York.
Burkholder, G. D., and Mukherjee, B. B. (1970). *Exp. Cell Res.* **61**, 413.
Burr, H., and Lingrel, J. B. (1971). *Nature (London), New Biol.* **233**, 41.
Carter, S. B. (1967). *Nature (London)* **213**, 261.
Carter, S. B. (1972). *Endeavour* **31**, 77.
Cheevers, W. P., and Sheinin, R. (1970). *Biochim. Biophys. Acta* **204**, 449.
Cheyne, I. M., and White, D. O. (1969). *Aust. J. Exp. Biol. Med. Sci.* **47**, 145.
Cohn, R. H., Banerjee, S. D., Shelton, E. R., and Bernfield, M. R. (1972). *Proc. Nat. Acad. Sci. U.S.* **10**, 2865.
Craig, N., Perry, R. P., and Kelley, D. E. (1971) *Biochim. Biophys. Acta* **246**, 493.
Craig, S. P. (1972). *In* "Biology and Radiobiology of Anucleate Systems" (S. Bonotto, R. Goutier, R. Kirchmann, and J.-R. Maisin, eds.), Vol. 1, pp. 181–205. Academic Press, New York.
Crocker, T. T., Pfendt, E., and Spendlove, R. (1964). *Science* **130**, 432.
Darnell, J. E., Phillipson, L., Wall, R., and Adesnik, M. (1971). *Science* **174**, 507.
Davidson, E. H. (1968). "Gene Activity in Early Development." Academic Press, New York.
Davidson, R. (1969). *Exp. Cell Res.* **55**, 424.
Davis, A. T., Estensen, R. D., and Quie, P. G. (1971). *Proc. Soc. Exp. Biol. Med.* **137**, 161.
DeTerra, N. (1969). *Int. Rev. Cytol.* **25**, 1.
Diegelmann, R. F., and Peterkofsky, B. (1972). *Proc. Nat. Acad. Sci. U.S.* **69**, 829.
Earl, D. C., and Korner, A. (1966). *Arch. Biochem.* **115**, 437.
Ebina, T., Kano, I., Takahashi, K., Homma, M., and Ishida, N. (1970). *Exp. Cell Res.* **62**, 384.
Ecker, R. E. (1972). *In* "Biology and Radiobiology of Anucleate Systems" (S. Bonotto, R. Goutier, R. Kirchman, and J.-R. Maisin, eds.), Vol. 1, pp. 165–179. Academic Press, New York.
Elston, R. N. (1963). *Acta Pathol. Microbiol. Scand.* **59**, 195.
Espmark, J. A., and Fagraeus, A. (1965). *J. Immunol.* **94**, 530.
Estensen, R. D. (1971). *Proc. Soc. Exp. Biol. Med.* **136**, 1256.
Estensen, R. D., and Plagemann, P. G. W. (1972). *Proc. Nat. Acad. Sci. U.S.* **69**, 1430.
Estensen, R. D., Rosenberg, M., and Sheridan, J. D. (1971). *Science* **173**, 356.
Forer, A., Emmerson, J., and Behnke, O. (1972). *Science* **175**, 774.
Freed, J. J., and Lebowitz, M. M. (1970). *J. Cell Biol.* **45**, 334.
Gail, M. H., and Boone, C. W. (1971). *Exp. Cell Res.* **68**, 226.

Gillespie, D., Marshall, S., and Gallo, R. C. (1972). *Nature (London), New Biol.* **236**, 227.
Goldman, R. D. (1972). *J. Cell Biol.* **52**, 246.
Goldman, R. D., Hopkins, N., and Pollack, R. (1972). *J. Cell Biol.* **55**, 87A.
Goldstein, L., and Prescott, D. M. (1967a). *In* "The Control of Nuclear Activity" (L. Goldstein, ed.), pp. 3–17. Prentice-Hall, Englewood Cliffs, New Jersey.
Goldstein, L., and Prescott, D. M. (1967b). *J. Cell Biol.* **33**, 637.
Goldstein, L., Cailleau, R., and Crocker, T. T. (1960a). *Exp. Cell Res.* **19**, 332.
Goldstein, L., Micou, J., and Crocker, T. T. (1960b). *Biochim. Biophys. Acta.* **45**, 82.
Gordon, S., and Cohn, Z. A. (1970). *J. Exp. Med.* **131**, 981.
Gräbmann, A. (1970). *Exp. Cell Res.* **60**, 373.
Green, H., Wang, R., Basilico, C., Pollack, R., Kusano, T., and Salas, J. (1971). *Fed. Proc., Fed. Amer. Soc. Exp. Biol.* **30**, 930.
Greenberg, J. R. (1972). *Nature (London)* **240**, 102.
Gurdon, J. B. (1970). *Proc. Roy. Soc., Ser. B* **176**, 303.
Gurdon, J. B., and Laskey, R. A. (1970a). *J. Embryol. Exp. Morphol.* **24**, 227.
Gurdon, J. B., and Laskey, R. A. (1970b). *J. Embryol. Exp. Morphol.* **24**, 249.
Gurdon, J. B., and Woodland, H. R. (1971). *Curr. Top. Develop. Biol.* **5**, 39.
Hämmerling, J. (1963). *Annu. Rev. Plant Physiol.* **14**, 65.
Harris, H. (1967). *J. Cell Sci.* **2**, 23.
Harris, H. (1968). "Nucleus and Cytoplasm." Oxford Univ. Press (Clarendon), London and New York.
Harris, H. (1970). "Cell Fusion." Oxford Univ. Press (Clarendon), London and New York.
Harris, H., Watkins, J. F., Ford, C. E., and Schoefl, G. I. (1966). *J. Cell Sci.* **1**, 1.
Harris, H., Sidebottom, E., Grace, D. M., and Bramwell, M. E. (1969). *J. Cell Sci.* **4**, 499.
Harris, J. R., and Brown, J. N. (1971). *J. Ultrastruct. Res.* **36**, 8.
Harvey, E. B. (1936). *Biol. Bull. Mar. Biol. Lab., Woods Hole, Mass.* **71**, 101.
Hoffstein, S., Zurier, R. B., and Weissmann, G. (1972). *J. Cell Biol.* **55**, 115A.
Holtzer, H., and Sanger, J. W. (1972). *Develop. Biol.* **27**, 444.
Hsu, T.-C., and Lou, T.-Y. (1959). *In* "Pigment Cell Biology" (J. Whittaker, ed.), pp. 315–325. Academic Press, New York.
Ikeuchi, T., Sanbe, M., Weinfeld, H., and Sandberg, A. A. (1971). *J. Cell Biol.* **51**, 104.
Jeon, K. W., and Danielli, J. F. (1971). *Int. Rev. Cytol.* **30**, 49.
Jeon, K. W., Lorch, I. J., and Danielli, J. F. (1970). *Science* **167**, 1626.
Johnson, R. T., and Harris, H. (1969). *J. Cell Sci.* **5**, 625.
Johnson, R. T., and Rao, P. N. (1971). *Biol. Rev.* **46**, 97.
Kano, K., Baranska, W., Knowles, B. B., Koprowski, H., and Milgrom, F. (1969). *J. Immunol.* **103**, 1050.
Kato, H., Sekiya, K., and Yoshida, T. H. (1971). *Exp. Cell Res.* **65**, 454.
Keck, K. (1969). *Int. Rev. Cytol.* **26**, 191.
King, T. J. (1966). *In* "Methods in Cell Physiology" (D. M. Prescott, ed.), Vol. II, pp. 1–36. Academic Press, New York.
Klebe, R. J., Chen, T. R., and Ruddle, F. H. (1970). *J. Cell Biol.* **45**, 74.
Kletzien, R. F., Perdue, J. F., and Springer, A. (1972). *J. Biol. Chem.* **247**, 2964.
Kohn, A. (1965). *Virology* **26**, 228.
Kopac, M. J. (1955). *Trans. N.Y. Acad. Sci.* **17**, 257.
Krishan, A. (1971). *J. Ultrastruct. Res.* **36**, 191.
Krishan, A. (1972). *J. Cell Biol.* **54**, 657.
Ladda, R. L., and Estensen, R. D. (1970). *Proc. Nat. Acad. Sci. U.S.* **67**, 1528.
Lane, C. D., Marbaix, G., and Gurdon, J. B. (1972). *In* "Biology and Radiobiology of Anucleate Systems" (S. Bonotto, R. Goutier, R. Kirchmann, and J.-R. Maisin, eds.), Vol. 1, pp. 101–113. Academic Press, New York.

Lay, W. H., and Nussenzweig, V. (1969). *J. Immunol.* **102**, 1172.
Legname, C., and Goldstein, L. (1972). *Exp. Cell Res.* **75**, 111.
Longo, F. J., and Anderson, E. (1969). *J. Exp. Zool.* **172**, 69.
Longwell, A. C., and Yergánian, G. (1965). *J. Nat. Cancer Inst.* **34**, 53.
Lorch, I. J., and Jeon, K. W. (1969). *Nature (London)* **221**, 1073.
Magee, W. E., and Miller, O. V. (1968). *J. Virol.* **2**, 678.
Malawista, S. E. (1971). *In* "Progress in Immunology" (B. Amos, ed.), pp. 187–196. Academic Press, New York.
Malawista, S. E., Gee, J. B. L., and Bensch, K. G. (1971). *Yale J. Biol. Med.* **44**, 286.
Marcus, P. I. (1962). *Cold Spring Harbor Symp. Quant. Biol.* **27**, 351.
Marcus, P. I., and Freiman, M. E. (1966). *In* "Methods in Cell Physiology" (D. M. Prescott, ed), Vol. II, pp. 93–111.
Mauel, J., and Defendi, V. (1971). *J. Exp. Med.* **134**, 3335.
Mayhew, E., and Maslow, D. (1973). *Develop. Biol.* In press.
Migeon, B. R., and Childs, B. (1970). *Progr. Med. Genet.* **7**, 1.
Neff, J. M., and Enders, J. F. (1968). *Proc. Soc. Exp. Biol. Med.* **127**, 260.
Nève, P., Ketelbant-Balasse, P., Willems, C., and Dumont, J. E. (1972). *Exp. Cell Res.* **74**, 227.
Norberg, B. (1971). *Scand. J. Haematol.*, Suppl. 14.
Okada, Y. (1962). *Exp. Cell Res.* **26**, 98.
Okada, Y. (1969). *Curr. Top. Microbiol. Immunol.* **48**, 102.
Orr, T. S. C., Hall, D. G., and Allison, A. C. (1972). *Nature (London)* **236**, 350–351.
Pedreira, F. A., and Taurosa, N. M. (1969). *Arch. Gesamte Virusforsch.* **28**, 361.
Penman, S. (1969). *In* "Fundamental Techniques in Virology" (K. Habel and N. P. Salzman, eds.), pp. 35–48. Academic Press, New York.
Penman, S., Scherrer, K., Becker, Y., and Darnell, J. E. (1963). *Proc. Nat. Acad. Sci. U.S.* **49**, 654.
Phillipson, L., Wall, R., Glickman, G., and Darnell, J. E. (1971). *Proc. Nat. Acad. Sci. U.S.* **68**, 2806.
Plagemann, P. G. W., and Estensen, R. D. (1972). *J. Cell Biol.* **55**, 179.
Poste, G. (1970). *Advan. Virus Res.* **16**, 303.
Poste, G. (1972a). *Exp. Cell Res.* **73**, 273.
Poste, G. (1972b). *Int. Rev. Cytol.* **33**, 157.
Poste, G., and Allison, A. C. (1971). *J. Theoret. Biol.* **32**, 165.
Poste, G., and Allison, A. C. (1973). *Biochim. Biophys. Acta.* In press.
Poste, G., and Reeve, P. (1971). *Nature (London), New Biol.* **229**, 123.
Poste, G., and Reeve, P. (1972a). *Exp. Cell Res.* **73**, 287.
Poste, G., and Reeve, P. (1972b). *Exp. Cell Res.* **72**, 556.
Poste, G., Waterson, A. P., Terry, G., Alexander, D. J., and Reeve, P. (1972). *J. Gen. Virol.* **16**, 95.
Prescott, D. M., Kates, J., and Kirkpatrick, J. B. (1971). *J. Mol. Biol.* **59**, 505.
Prescott, D. M., Myerson, D., and Wallace, A. C. (1972). *Exp. Cell Res.* **71**, 480.
Raff, R. A. (1972). *Exp. Cell Res.* **71**, 455.
Rifkin, B. R., Sands, J. A., Jr., and Panner, B. J. (1972). *J. Cell Biol.* **55**, 215A.
Rosen, L. (1969). *In* "Fundamental Techniques in Virology" (K. Habel and N. P. Salzman, eds.), pp. 276–287. Academic Press, New York.
Rothweiler, W., and Tamm, Ch. (1970). *Helv. Chim. Acta.* **53**, 696.
Ruddle, F. H. (1972). *Advan. Human Genet.* **3**, 173.
Sager, R. (1972). "Cytoplasmic Genes and Organelles." Academic Press, New York.
Sanger, J. W., and Holtzer, H. (1972a). *Proc. Nat. Acad. Sci. U.S.* **69**, 253.
Sanger, J. W., and Holtzer, H. (1972b). *Amer. J. Anat.* **135**, 293.
Sawicki, S. G., and Godman, G. C. (1971). *J. Cell Biol.* **50**, 746.
Sawicki, S. G., and Godman, G. C. (1972). *J. Cell Biol.* **55**, 299.

Schaeffer, H., and Brick, I. (1972). *Amer. Zool.* **12**, 700.
Sekiya, Y., Kato, H., and Yoshida, T. H. (1969). *Annu. Rep. Nat. Inst. Genet. Jap.* **20**, 18.
Sheldon, R., Jurale, C., and Kates, J. (1972). *Proc. Nat. Acad. Sci. U.S.* **69**, 417.
Shepro, D., Belamarich, F. A., Robblee, L., and Chao, F. C. (1970). *J. Cell Biol.* **47**, 544.
Singer, R. H., and Penman, S. (1972). *Nature (London)* **240**, 100.
Skutelsky, E., and Danon, D. (1969). *J. Cell Biol.* **43**, 8.
Skutelsky, E., and Danon, D. (1970). *Exp. Cell Res.* **60**, 427.
Skutelsky, E., and Danon, D. (1972). *Anat. Rec.* **173**, 123.
Soeiro, R., and Amos, H. (1966). *Biochim. Biophys. Acta* **129**, 406.
Spooner, B. S., Yamada, K. M., and Wessels, N. K. (1971). *J. Cell Biol.* **49**, 595.
Staehelin, T., Wettstein, F. O., and Noll, H. (1963). *Science* **140**, 180.
Subak-Sharpe, J. H. (1971). *In* "The Strategy of the Viral Genome" (G. E. W. Wolstenholme and M. O'Connor, eds.), Churchill, London.
Tartar, V. (1972). *In* "Biology and Radiobiology of Anucleate Systems" (S. Bonotto, R. Goutier, R. Kirchmann, and J.-R. Maisin, eds.), Vol. 1, pp. 125–144. Academic Press, New York.
Thoa, N. B., Wooten, G. F., Axelrod, J., and Koysin, I. J. (1972). *Fed. Proc., Fed. Amer. Soc. Exp. Biol.* **31**, 586.
Trakatellis, A. C., Montjar, M., and Axelrod, A. E. (1965). *Biochemistry* **4**, 1678.
Tyler, A. (1966). *Biol. Bull.* **130**, 450.
Wagner, R., Rosenberg, M., and Estensen, R. (1971). *J. Cell Biol.* **50**, 804.
Wang, E. (1972). *J. Cell Biol.* **55**, 273A.
Warner, D. A., and Perdue, J. F. (1972). *J. Cell Biol.* **55**, 242.
Watkins, J. F. (1971). *Int. Rev. Exp. Pathol.* **10**, 115.
Watkins, J. F., and Grace, D. M. (1967). *J. Cell Sci.* **2**, 193.
Weiss, L. (1972). *Exp. Cell Res.* **74**, 21.
Wessells, N. K., Spooner, B. S., Ash, J. F., Luduena, M. A., and Wrenn, J. T. (1971a). *Science* **173**, 358.
Wessells, N. K., Spooner, B. S., Ash, J. F., Bradley, M. O., Luduena, M. A., Taylor, E. L., Wrenn, J. T., and Yamada, K. M. (1971b). *Science* **171**, 135.
Williams, T. A., and Wolff, J. (1971). *Biochem. Biophys. Res. Commun.* **44**, 422.
Williamson, R. (1972). *J. Med. Genet.* **9**, 348.
Wills, E. J., Davies, P., Allison, A. C., and Haswell, A. D. (1972). *Nature (London), New Biol.* **240**, 58.
Woodcock, C. L. F. (1971). *J. Cell Sci.* **8**, 611.
Wright, W. E., and Hayflick, L. (1972). *Exp. Cell Res.* **74**, 187.
Wyngaarden, J. B. (1970). *Biochem. Genet.* **4**, 105.
Yamada, K., Spooner, B., and Wessells, N. K. (1970). *Proc. Nat. Acad. Sci. U.S.* **66**, 1206.
Yamada, K. M., Spooner, B. S., and Wessells, N. K. (1971). *J. Cell Biol.* **49**, 614.
Zalokar, M. (1971). *Proc. Nat. Acad. Sci. U.S.* **68**, 1539.
Zigmond, S. H., and Hirsch, J. G. (1972a). *Science* **176**, 1432.
Zigmond, S. H., and Hirsch, J. G. (1972b). *Exp. Cell Res.* **73**, 383.

Chapter 12

Fusion of Somatic and Gametic Cells with Lysolecithin[1]

HILARY KOPROWSKI AND CARLO M. CROCE

The Wistar Institute of Anatomy and Biology, Philadelphia, Pennsylvania

I. Introduction

Sendai virus-induced fusion (Okada, 1962a,b; Harris and Watkins, 1965) between somatic cells and between somatic cells and gametic cells (Johnson, 1970; Sawicki and Koprowski, 1971) has become an established procedure in biological studies of mammalian cells. Problems occur, however, because the "fusion factors" in the Sendai virus preparation are unknown and, therefore, cannot be quantitively determined. Since stocks of Sendai virus vary

[1]This work was supported in part by the National Institutes of Health grant CA 04534 and CORE grant CA 10815.

considerably in their ability to fuse cells, it has been impossible to devise a standard method for the production of such stocks. This applies also to the preservation of Sendai virus stocks, since it has not been possible to relate the occasional loss of ability to fuse cells directly to the condition under which the virus was preserved.

Furthermore, the possible introduction of virus particles in a somatic cell during the fusion process may cause alterations in the host cell which are, at present, unknown. Related to this is the fact that Sendai virus may interfere (see Section VI) with the rescue of a virus which may be present in cells participating in the fusion experiment. For these reasons, it became interesting to follow the observations of Poole *et al.* (1970) and Lucy (1970) that lysolecithin (LL), a phospholipid, is capable of fusing somatic cells, and that viable heterokaryocytes and hybrid cells are produced through fusion in the presence of LL (Croce *et al.*, 1971). Furthermore, Barbanti-Brodano *et al.* (1971) found that lysolecithinase (phospholipase B), which acts specifically on LL, inhibits the fusion and hemolytic activity of Sendai virus.

II. Preparation of Lysolecithin Solution

Egg LL is obtained from commercial sources, but it is essential to specify a purity of not less than 99% of LL.

LL in lipid emulsion is prepared by dissolving LL in chloroform, and removing the chloroform in an atmosphere of nitrogen (Ahkong *et al.*, 1972). The dried LL is then swelled by exposure for 15–30 minutes in a 3:5 mixture of 0.9% NaCl and 0.15 *M* sodium acetate buffer at pH 5.6, or directly in Dulbecco's medium at pH 7.2. The sample is then sonicated under a stream of nitrogen in order to disperse the phospholipid. This preparation must be used within 10 minutes after sonication.

There are two methods of preparing aqueous LL solution for cell fusion purposes: (1) The aqueous solution described by Ahkong *et al.* (1972) is prepared by dissolving 70–560 μg of LL in 1.7 ml of 106 m*M* NaCl and 44 m*M* sodium acetate at final pH of 5.7 or, alternatively by dissolving the desired quantity of LL in Hank's balanced salt solution (Poole *et al.*, 1970). (2) The aqueous solution described by Croce *et al.* (1971) is prepared by dissolving LL in absolute ethanol to make a stock solution of 1 mg/ml. An aliquot of the stock solution is placed in closed tubes and warmed at 70° to 80°C for 5 to 10 minutes. The tubes are then opened, and a flow of nitrogen is directed into the tubes until all the ethanol is evaporated and the LL remains as a sterile powder at the bottom of the tubes. Absolute ethanol (0.5 ml) is added to dissolve the LL. Eagle's minimal essential medium (MEM) at the desired pH (Croce *et al.*, 1972) or phosphate-buffered saline (pH 7.2) containing 5 mg/ml of bovine serum albumin (BSA), (Miles Laboratories, Inc., Kankakee, Illinois, fraction V) is then added. The concentration of ethanol in the final LL solution must be less than 1%. This preparation will be referred to in the text as LL.

III. Cytotoxicity of Lysolecithin Solution

LL in aqueous solution without BSA is highly toxic for cells. Hen erythrocytes exposed to such a preparation, of LL at a concentration of 560 μg/1.7 ml, are lysed within a few minutes; lower concentrations (70 μg/1.7 ml) seemed less toxic, but the preparation was without fusing properties (Ahkong *et al.*, 1972). Toxicity of this preparation for mouse L cells was not established but, following exposure of these cells to 330 μg/ml they had to be fixed within 1 minute or they lysed (Ahkong *et al.*, 1972).

LL in lipid emulsions is less damaging to the cells. Hen erythrocytes exposed to such a preparation gradually lyse within 45 to 60 minutes. Treatment of mouse and hamster cells with LL in lipid emulsion, resulted in the production of hybrid cells. Since the number of hybrid colonies did not exceed that observed in spontaneous fusion of the parental cells, LL may have damaged or destroyed some fraction of the cells (Ahkong *et al.*, 1972).

Toxicity of LL with BSA varies with the cell type used. Hamster cells (F5–1) transformed by SV40 were more resistant to the toxic effect of LL than human fibroblasts (WI-38) and African green monkey (AGM) kidney cells (CV-1) (Fig. 1). The toxic effect of various concentrations of LL with

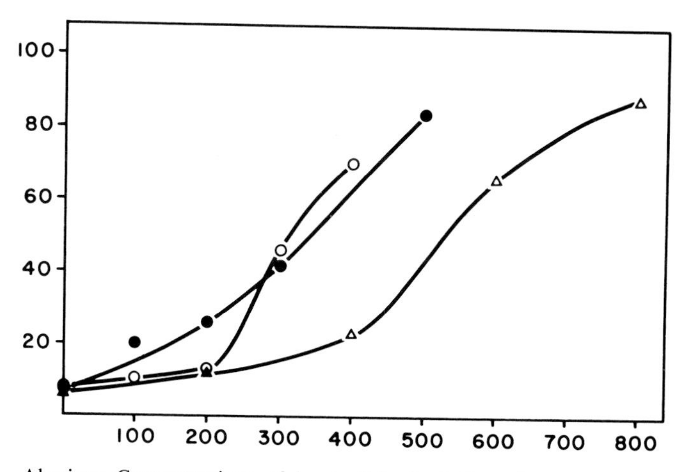

FIG. 1. Abscissa: Concentrations of lysolecithin (μg/ml); ordinate: percentage of cells stained with trypan blue. ○, CV-1 cells; ▲, F5-1 cells; ●, WI-38 cells. Trypan blue exclusion tested after fusion.

TABLE I
TRYPAN BLUE EXCLUSION TEST AFTER TREATMENT OF 10^7 CELLS WITH LYSOLECITHIN

LL treatment (μg/ml)	Percent of cells	
	Stained	Not stained
Control, no LL	22.4	77.6
200	55.6	44.4
400	88.0	12.0
600	93.3	6.7

BSA on a mixture of AGM kidney cells and human lymphoid cells is shown Table I. Approximately 50% of the cells remain viable.

With regard to the cytotoxicity of LL, it should be emphasized that the pH of the solution used in fusion experiments may be important. For example, when fusion is performed at a pH of 5.5 to 6.4 in the presence of either Sendai virus or LL with BSA, cell viability decreases 2- or 3-fold. This may partially explain the high toxicity of LL aqueous solution prepared according to the method of Poole *et al.* (1970) since the final pH of the preparation ranged from 5.0 to 5.6.

IV. Lysolecithin-Induced Fusion

A. Somatic Cells

When LL is prepared in lipid emulsion (Ahkong *et al.*, 1972), monolayers 10^3 cells are treated for 5 minutes at room temperature with 0.8 ml of LL in lipid emulsion in serum-free Dulbecco's medium (Ahkong *et al.*, 1972). The petri dishes are gently rocked during this procedure. The emulsion is then removed, and the cells are washed twice with serum-free Dulbecco's medium. The cells that are to be fused are then added, in suspension, in serum-free Dulbecco's medium, to the monolayer of the treated cells and kept in contact for 7 minutes, while the cultures are gently rocked. The added nonfused cells are then removed from the monolayer and the monolayer is washed twice with 2.5 ml of fetal calf serum (FCS) and selective medium (Littlefield, 1966) is added to the monolayer.

The technique of cell fusion in the presence of aqueous solution of LL with BSA (Croce *et al.*, 1971) is as follows:

1. The cells of the parental cultures are mixed, and the mixture is centrifuged at 180 *g* for 5 to 10 minutes at room temperature.
2. The pellets, containing 0.5 to 1×10^7 cells are treated for 1 minute with 0.1 ml of LL with BSA solution at the desired concentration and pH.

3. During treatment by LL the tubes are shaken in order to assure maximum exposure to LL.

4. Then 1 ml of MEM, containing 30% of FCS, inactivated at 56°C for 30 minutes is added to stop the action of the LL.

5. The tubes are centrifuged again, the supernatants are discarded and 1 ml of MEM is added over the pellets, then the tubes are placed in a water bath at 37°C for 15 to 20 minutes.

6. Finally, the cells are resuspened in MEM containing 10% FCS and seeded in culture in petri dishes or flasks. A number of these cells are exposed to 0.1% trypan blue (Sawicki *et al.*, 1967) in order to determine their viability.

The concentration of LL may vary depending on its toxicity for a given type of cell, and it is advisable to determine this toxicity before fusion is undertaken. In general, between 100 and 600 μg/ml of LL were satisfactory for 0.5 to 1×10^7 cells.

It is important that fusion take place at an alkaline pH (pH 8.0) in order to obtain the maximum number of heterokaryocytes (Croce *et al.*, 1972).

B. Gametic Cells

Ejaculate and cauda epididymis spermatozoa are obtained from rabbits, washed three times in Hank's balanced salt solution and centrifuged at < 180 *g* for 5 minutes. Pellets containing 5×10^7 motile spermatozoa are treated with 0.1 ml of LL at a concentration of 1.2 mg/ml for 1 minute. Under these conditions, more than 95% of the rabbit spermatozoa lose their motility. After adding 4–5 ml of MEM, the spermatozoa are recentrifuged.

In the fusion procedure, 5×10^7 LL-pretreated rabbit spermatozoa are mixed with 5×10^6 somatic cells of any origin, and the mixture is centrifuged at 180 *g* for 5 to 10 minutes at room temperature. The pellets are treated with LL solution using the procedure described in Section IV,A, steps 3–6. It should be noted that, in fusing LL-treated spermatozoa with somatic cells, it is possible to use either LL or a preparation of inactivated Sendai virus.

It was not possible under the conditions of LL with BSA fusion described above to obtain satisfactory fusion between mammalian eggs and either spermatozoa or somatic cells.

V. Quantification Procedures

A. Multinucleated Cell Formation

To determine the percentage of heterokaryocytes, thymidine-^{3}H prelabeled cells are fused with unlabeled cells. At 12 to 18 hours post fusion, the cultures grown on coverslips are fixed in an ice-cold mixture of acetic acid:ethanol (1:3) for 15 minutes and washed with ethanol 70% for 10 minutes. The coverslips are mounted on gelatinized glass slides and covered with Kodak AR 10 stripping film for autoradiography. After 3 to 4 days, the slides are developed and fixed and stained with Giemsa buffered at pH 5.75. The percentage of heterokaryocytes is determined by examing at least 1000 cells.

B. Hybrid Cell Colony Formation

Bromodeoxyuridine-resistant cells (5×10^6) and 8-azaguanine-resistant cells (5×10^6) are fused with either LL or Sendai virus and inoculated at the desired densities in petri dishes or flasks containing HAT selective medium (Littlefield, 1966). The medium is changed after 1, 3, 6, 9, and 12 days. When using medium adjusted to the desired pH for the selection of hybrid cells (Croce *et al.*, 1972), it is necessary to change the medium every two days. After 14 to 15 days, the colonies of hybrid cells are fixed with methanol, stained with Giemsa, and counted.

VI. Discussion

Table II shows the results of fusion between somatic cells of two different origins, at a final concentration of 1×10^7 cells, in the presence of various concentrations of an aqueous solution of LL with BSA. The optimal con-

TABLE II

HOMO- AND HETEROKARYOCYTE FORMATION AFTER FUSION OF CV-1 AND F5-1 CELLS WITH VARIOUS CONCENTRATIONS OF LYSOLECITHIN[a]

Concentration[b] of lysolecithin (μg/ml)	Percent multinucleated cells[c]		Percent homokaryocytes CV-1[d]		Percent homokaryocytes F5-1[d]		Percent heterokaryocytes CV-1 + F5-1[d]	
	Exp. 1	Exp. 2	Exp. 1	Exp. 2	Exp. 1	Exp. 2	Exp. 1	Exp. 2
0	4.5	6.3	33.3	40.0	62.0	52.0	4.7	8.0
400	15.0	18.5	19.8	23.2	43.6	48.0	36.6	28.8
600	18.7	22.4	13.7	21.4	60.5	46.6	25.8	32.0
1000	17.7	25.4	20.4	23.5	50.0	45.1	29.6	31.4

[a]From Croce *et al.* (1971), by permission of Academic Press, New York.
[b]0.1 ml/5×10^6 CV-1 and 5×10^6 F5-1 cells.
[c]Percent of total number of cells.
[d]Percent of total number of multinucleated cells.

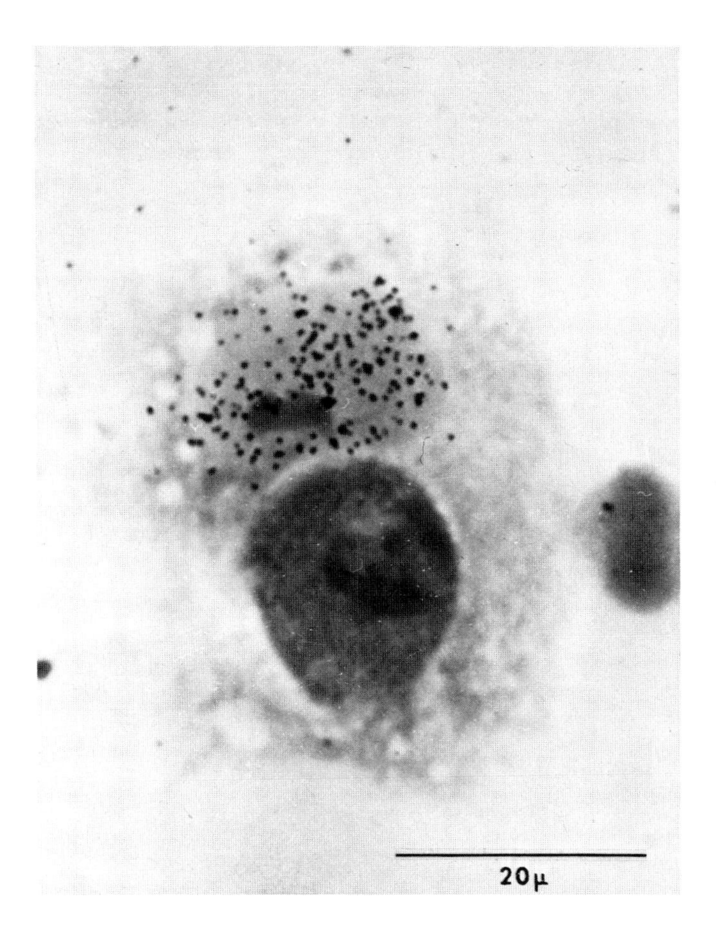

FIG. 2. Autoradiograph of heterokaryocyte of CV-1 cell labeled with tritiated thymidine and unlabeled F5-1 cell.

centration of LL for the production of heterokaryocytes (Fig. 2) was 400 to 600 μg/ml (Table II).

In fusing CV-1 cells with human lymphoid cells (Table III), the best results were obtained with 600 μg/ml with a yield of multinucleated cells of 33.5% and of heterokaryocytes of 15.5%.

More recently, Keay *et al.* (1972) were able to successfully fuse two fibroblast cells of mouse origin with two genetic markers in monolayers using LL solution at a concentration of 50 to 500 μg/ml.

The pH affects not only the fusion process itself, but also the production of cell hybrids, as shown on Table IV. In order to obtain high numbers of hybrid

TABLE III

FUSION OF CV-1 CELLS WITH HUMAN LYMPHOID CELLS (WIL-2) INDUCED BY LYSOLECITHIN

Concentration of LL (μg/ml)	Percent multinucleate cells[a]	Percent homokaryocytes CV-1[b]	Percent homokaryocytes WIL-2[c]	Percent heterokaryocytes[b]
0	9.0	96.7	0	3.3
400	31.6	59.8	0	40.2
600	33.5	58.6	0	41.4

[a] Percentage of multinucleate cells = (number multinucleate cells)/(number mononucleate + multinucleate cells) × 100.
[b] Percentage of homo- or heterokaryocyte = (number homo- or heterokaryocytes)/(number multinucleate cells) × 100.
[c] The WIL-2 cells grow only in suspension.

cell colonies in HAT selective medium, an alkaline pH medium is necessary not only during the fusion process but also for 3 to 5 days after fusion during the period of the selection of hybrid cells.

Whereas fusion of untreated rabbit spermatozoa with somatic cells by Sendai virus did not produce any activation of the spermatozoan nuclei (Sawicki and Koprowski, 1971), the fusion of LL-treated rabbit spermatozoa with hamster cells in the presence of LL, resulted in the activation of DNA

TABLE IV

EFFECT OF ENVIRONMENTAL pH ON FORMATION OF WI-18VA2 (HUMAN) AND C1-1D (MOUSE) HYBRID CELL COLONIES[a]

pH[b] of: Medium	pH[b] of: Culture fluid after fusion, 1 Day	pH[b] of: Culture fluid after fusion, 3 Days	Average number of hybrid colonies per petri dish (10^6 cells) after 14 days; fusion induced by: Cocultivation	Lysolecithin	Sendai virus
6.4	6.5	6.4	0	0	0
7.2	7.0	6.9	0	0	0
7.6	7.4	7.3	2.3	6.4	12.1
8.0	7.8	7.7	14.6	43.5	50.5

[a] From Croce *et al.* (1972), by permission of the National Academy of Sciences, Washington, D.C.
[b] Cells were exposed to medium at indicated pH for 12 to 24 hours before fusion, and fused and inoculated at that pH. Medium changed 1, 3, 6, 9, and 12 days after fusion.

synthesis of the nucleus of the spermatozoa. Figure 3 shows the presence of silver grains over the spermatozoal head inside F5-1 cytoplasm following autoradiography of F5-1 cells fused with LL-treated spermatozoa and exposed to thymidine-^{3}H (Gledhill *et al.*, 1972).

Although the technique of cell fusion induced by Sendai virus has been used for a number of years to rescue latent virus from somatic cells (Koprowski *et al.*, 1967), recently it became apparent that the presence of inactivated Sendai virus may actually inhibit the rescue of a related virus.

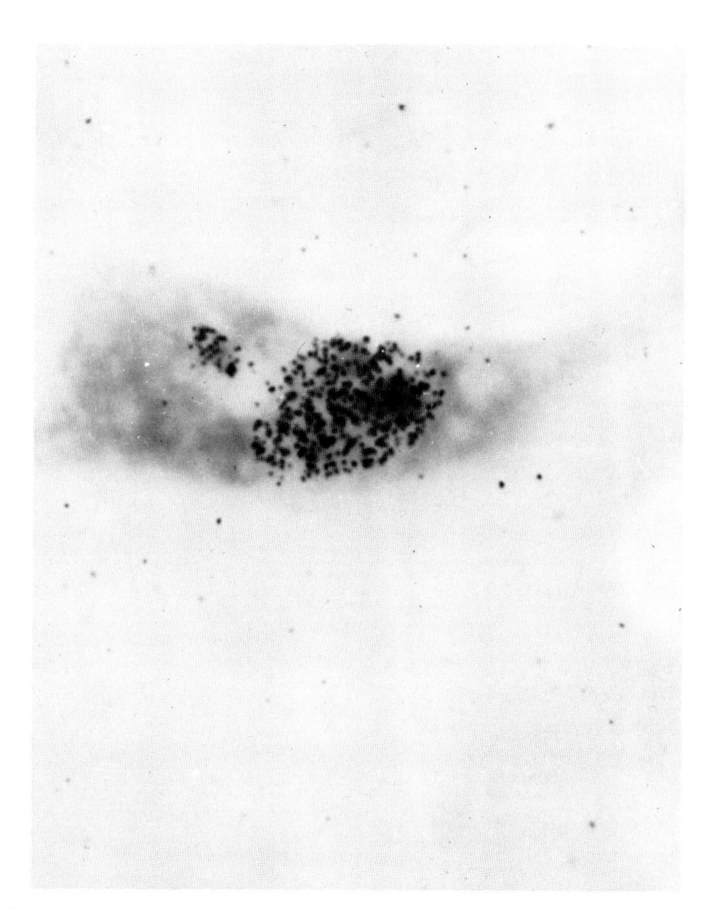

FIG. 3. Autoradiograph (about ×750) of a F5-1: cauda epididymis spermatozoan heterokaryocyte cultured for 20 hours in the continuous presence of thymidine-^{3}H. This autoradiograph was developed after 42 days' exposure and was stained with Light Green SF. Note the significant development of silver grains over the F5-1 nucleus and the smaller sperm body indicating incorporation of the DNA precursor by both structures. From Gledhill *et al.* (1972), by permission of Academic Press, New York.

Rescue of parainfluenza virus type 1 from cultured brain cells from multiple sclerosis patients which had been fused with CV-1 cells (ter Meulen *et al.*, 1972) did not take place in the presence of inactivated Sendai virus, but did occur when LL was used instead of Sendai virus.

References

Ahkong, Q. F., Cramp, F. C., Fisher, D., Howell, J. I., and Lucy, J. A. (1972). *J. Cell Sci.* **10**, 769.

Barbanti-Brodano, G. J., Possati, L., and La Placa, M. (1971). *J. Virol.* **8**, 796.

Croce, C. M., Sawicki, W., Kritchevsky, D., and Koprowski, H. (1971). *Exp. Cell Res.* **67**, 427.

Croce, C. M., Koprowski, H., and Eagle, H. (1972). *Proc. Nat. Acad. Sci. U.S.* **69**, 1953.

Gledhill, B. L., Sawicki, W., Croce, C. M., and Koprowski, H. (1972). *Exp. Cell Res.* **73**, 33.

Harris, H., and Watkins, J. F. (1965). *Nature* (*London*) **205**, 640.

Johnson, R. T., Rao, P. N., and Hughes, S. D. (1970). *J. Cell. Physiol.* **76**, 151.

Keay, L., Weiss, S. A., Cirulis, N. and Wildi, B. S. (1972), *In Vitro* **8**, 19.

Koprowski, H., Jensen, F., and Steplewski, Z. (1967). *Proc. Nat. Acad. Sci. U.S.* **58**, 127.

Littlefield, J. W. (1966). *Exp. Cell Res.* **41**, 190.

Lucy, J. A. (1970). *Nature* (*London*) **227**, 815–817.

Okada, Y. (1962a). *Exp. Cell Res.* **26**, 98.

Okada, Y. (1962b). *Exp. Cell Res.* **26**, 119.

Poole, A. R., Howell, J. I., and Lucy, J. A. (1970). *Nature* (*London*) **227**, 810.

Sawicki, W., and Koprowski, H. (1971). *Exp. Cell Res.* **66**, 145.

Sawicki, W., Kieler, J., and Briand, P. (1967). *Stain Technol.* **42**, 143.

ter Meulen, W., Koprowski, H., Iwasaki, Y., Käckell, Y. M., and Müller, D. (1972). *Lancet* **2**, 1.

Chapter 13

The Isolation and Replica Plating of Cell Clones

RICHARD A. GOLDSBY AND NATHAN MANDELL

Department of Chemistry, University of Maryland, College Park, Maryland, and Yale University, New Haven, Connecticut

---

I. Introduction

The vigorous prosecution of genetic studies in somatic cell cultures requires the isolation and study of clonal populations. A general solution (Goldsby and Zipser, 1969) and modifications (Robb, 1970; Suzuki *et al.*, 1971), all employing systems of complementary matrices, have been applied to the closely related problems of mass clone isolation and the replica plating of clonal populations. This chapter will describe a system that provides for the mechanical performance of such necessary operations in clone isolation and replica plating as seeding, feeding, pipetting, and the precise transfer of measured aliquots.

II. The Incubation Matrix

The basic culture vessel employed in clone isolation and replica plating contains a matrix of 96 optically clear, flat-bottom wells and is shown in

Figs. 1 and 2. (These dishes and their covers may be obtained as sterile combinations from either the Linbro Chemical Company, Cat. No. ISFB 96 TC, 960 Dixwell Ave, New Haven, Connecticut or an equivalent dish, the Microtest II tissue culture plate and cover from Falcon Plastics Company, Los Angeles, California). For some applications it is desirable to compose a matrix of individual removable vessels using a plastic frame and 0.25-dram vials.

The additional flexibility provided by a matrix of movable elements is clear. With such an arrangement, an isolated population (one vessel) may be removed and used as an element in the composition of a new matrix of clonal population. The 0.25-dram vials used to compose such matrices are readily sterilized by immersion in absolute alcohol or by autoclaving prior to their insertion into the plastic frame.

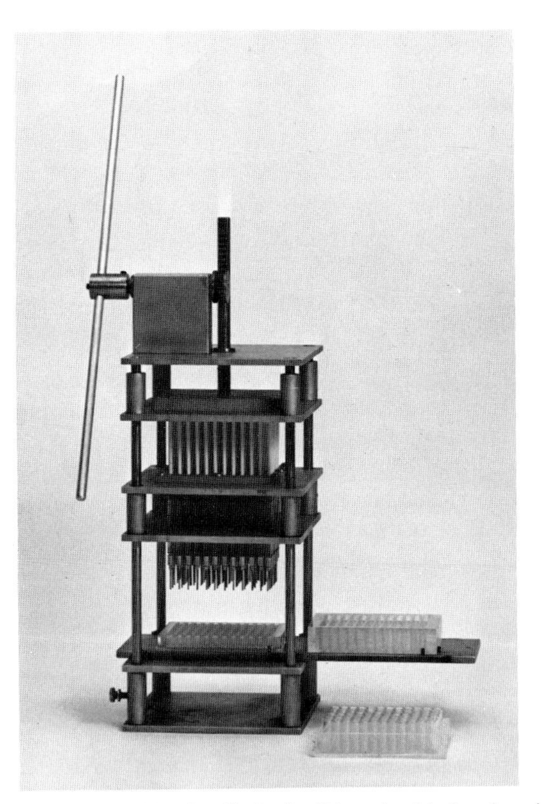

FIG. 1. Syringe replicator with 96-well plastic dish and a 24-chambered trough on ways (to the right). Below and to the right of the replicator is a collapsible matrix of 96 quarter-dram vials.

III. Seeding for the Isolation of Single Clones

Generally it is desirable to seed the matrix at an average density of 1 cell/well by adding 0.1 ml of a cell suspension which contains 10 cells/ml. At such low cell densities some wells receive no cells, many receive one cell, and a few receive more than one cell. While this seeding operation can be conducted with Pasteur pipettes outfitted with dropping bulbs, a considerable saving in time and effort, as well as a more uniform pattern of addition, is obtained when one uses the syringe replicator shown in Fig. 1. (Syringe replicators may be obtained by contacting Mr. Nathan Mandell, Director of the Biology Department Machine Shop, Yale University, New Haven, Connecticut.) The syringe replicator is a matrix of 96 disposable tuberculin syringes outfitted with dropping tips and concentric with the 96 wells of the plastic dish shown in Fig. 2.

Mechanical seeding is accomplished by immersing the syringe replicator's matrix of transfer elements into a cell suspension of the desired concen-

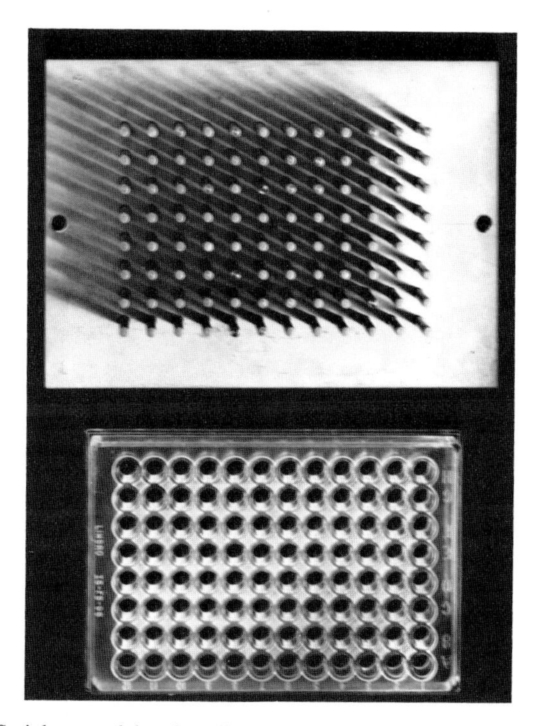

FIG. 2. Stainless steel hand replicator and 96 fixed-place incubation matrix.

tration and drawing an appropriate volume of the seeding suspension into the syringes by rotation of the handle which drives the rack-and-pinion assembly. The matrix of vessles to be loaded is then placed on the ways and moved into position under the transfer elements of the syringe replicator in such a fashion as to place its 96 wells directly beneath the 96 transfer elements, as illustrated in Fig. 1. The adjustable platform of the replicator is then raised so as to place the transfer elements of the replicator in the wells of the matrix. The syringe replicator is then made to deliver the desired volume of suspension ($\pm 7\%$) to the wells of the incubation matrix by appropriate rotation of the handle driving the rack-and-pinion assembly.

Following the seeding operation the matrices are incubated under desired conditions of atmosphere and temperature until, after several days, microscopic examination reveals that one has clones growing in isolation. As an alternative to the use of the syringe replicator, one can employ the hand replicator shown in Fig. 2 to seed previously prepared matrices. Because the volume of suspension transferred by the hand replicator is small (ca. 20–40 μl) it is necessary first to add the appropriate growth medium, usually in the neighborhood of 0.2 ml, to the elements of the matrix prior to use of the hand replicator are then immersed in a cell suspension containing 20 cells/ml and subsequently transferred to the incubation matrix. While this manual procedure can be applied to clone isolation on a small scale when only two or three matrices are involved, its use on a large scale involving many incubation matrices is too tedious and time consuming.

Whichever technique of cell seeding is used, mechanical or manual, Poisson considerations make it inevitable that some wells will contain more than one clone. Where it is essential for purposes of the experiment, those wells containing multiple clones may be located by rapidly scaning the matrix at low power ($20\times$) and, once located, can be eliminated by the addition of one drop of 1 N NaOH to the well containing the multiple clone. In those cases where one intends to clone cell types which have inherently low cloning efficiencies, such as primary human, diploid fibroblasts, it is necessary to seed the wells of the matrices with a larger number of cells in order to compensate for the poor cloning efficiency of such cell lines. In practice one usually does not have advance, precise knowledge of the cloning efficiency of the cell line under consideration prior to seeding matrices. This difficulty is obviated by seeding several matrices at each of a number of different cell concentrations. Thus, in a typical experiment involving a given cell type suspected of a low cloning efficiency, such as human, diploid fibroblasts, one would seed 6 matrices at each of the following concentrations; 1 cell/well, 5 cells/well, 20 cells/well and 50 cells/well. After a suitable period of incubation for clone maturation, it is possible to reject those sets of matrices that have been seeded at densities that are too low and hence give very few or no clones per matrix and those densities that are too high and hence yield a frequency of multiple clones.

IV. Replica Plating

The replica plating process consists of three stages, detachment, dispersion, and transfer. Detachment is accomplished by removing the serum-containing growth medium from the plate to be replicated by quickly inverting the plate and shaking it sharply. A matrix of glass vials may be emptied by inverting it onto sanitary napkins which have been previously sterilized by autoclaving. After the removal of growth medium, 0.05 ml of 25°C, 0.25% trypsin is added to each well of the plate and left for 15 minutes. The matrix is then examined microscopically to determine if detachment has taken place. If further incubation with trypsin is needed, the plates are placed at 37°C and again examined after 10 minutes. Following detachment, the trypsin is quenched by the addition of 0.1 ml of cold serum-containing growth medium. Detached cells can be dispersed manually by stirring the contents of the wells with the matrix of 96 rods shown in Fig. 2. However this process of dispersion is more efficiently carried out by simultaneously pipetting the contents of each of the 96 wells rapidly back and forth with the syringe replicator.

Replica plating is accomplished by first immersing the pattern of transfer elements into the suspensions of the master plate and then immersing them in a second plate, the replica, which contains 0.2 ml of growth medium in each well. Although once again the replication may be accomplished manually with the simple matrix of rods, the operation, including the refilling of the recipient replicas with growth medium, can be accomplished with considerably more convenience and consistency by employing the syringe replicator. With this device it is possible to deliver simultaneously in a single operation, a measured amount of medium (which may contain a selective agent if desired) to the plates which are to serve as replicas.

When the syringe replicator is used in the production of replica plates, the master plate is placed on the ways, and the tubes of the syringe replicator are inserted into the wells. By rotating the handle shown in Fig. 1 rapidly back and forth, it is possible to prepare a smooth suspension of the detached cells in the wells of the matrix. After dispersion, a sample of each cell suspension, usually on the order of 0.025 to 0.05 ml of each cell suspension on the master plate, is pulled into the syringe replicator and expelled into a second plate containing the desired incubation medium.

Three or four replicas of a single master plate may be obtained by using this procedure within the space of 5 to 10 minutes. The syringe replicator is freed of contaminating cells between series of replications from different master plates by first filling the syringes with sterile 0.85% NaCl and subsequently pipetting 70% alcohol to kill any residual cells that may have been trapped in the apparatus and finally by pipetting sterile saline to remove all traces of alcohol. This same procedure is used to sterilize the replicator prior to use in a series of replications or plate fillings.

V. Capacities of the Technique

Using an earlier version of the mechanical replicator, cloning efficiencies ranging from 40 to 50% were obtained for Chinese hamster cells. Table I demonstrates that this approach, employing complementary matrices and a mechanical system for transfering aliquots, is capable of replicating clonal patterns of Chinese hamster or BHK-21 cells with a fidelity of replication in excess of 90%. The vitality of the replicated cultures is demonstrated by the observation that the use of such a replica as a secondary master for subsequent replications gave excellent transfer efficiency. As many as six secondary replicas can be taken at one time from a primary replica with fidelity of replication exceeding 95% (see Table II).

It is possible to preserve the clonal patterns in replica plates by decanting the growth medium and adding 0.05 ml of the same type of growth medium containing 10% glycerol and subsequently storing them at $\times 120°$ F in the deep freeze. Growth may be reinitiated by warming the dish to room temperature and adding approximately 0.2 ml of growth medium without glycerol. In our experience, approximately 70% of the clonal populations on a replica which had been subjected to freezing and thawing remained viable and were successfully replicated to other plates.

As a final procedural note the successful application of the technique of clone isolation and replication described herein requires that certain critical considerations be met. First of all, it is essential that the cell suspension employed for the statistical isolation of cell clones have no clumps of aggregated cells. Such clumps markedly increase the frequency of multiple clones. Second, it is important not to overtrypsinize the cells prior to transfer since this causes a loss in viability and can lower the fidelity of replication. We have found that it is essential to add a 2-fold excess of cold, serum-containing medium to the plate soon after the trypsin has detached the cells. Finally, since the addition of trypsin detaches but does not disperse the cells,

TABLE I

FIDELITY OF REPLICA PLATING[a]

Master plate	No. of clones on master	No. transferred to identical positions on replica 1	Percent fidelity of replication	Transferred to identical positions on replica 2	Percent fidelity of replication	Average percent fidelity replication
a. Replica plating of Don clones[b]						
1	42	42	100.0	40	95.2	97.6
2	50	48	96.0	46	92.0	94.0
3	51	48	94.1	48	94.1	94.1
4	43	37	86.0	42	97.7	91.1
5	59	56	94.9	52	88.1	91.5
6	43	40	93.0	40	93.0	93.0
7	39	38	97.4	32	82.1	89.8
8	31	25	80.6	31	100.0	90.3
9	51	45	88.2	41	90.2	89.2
10	46	40	87.0	41	89.1	88.0
b. Replica plating of BHK-21 clones[c]						
A	19	17	89.5	18	94.7	92.1
B	16	15	93.8	16	100.0	96.9
C	30	26	86.7	26	100.0	86.7
D	15	15	100.0	15	100.0	100.0
E	32	29	90.6	30	93.8	92.2

[a]Suspensions of Don Chinese hamster cells (10 cells/ml) and BHK-21 cells (10 cells/ml) were plated into the wells of Linbro dishes (0.1 ml suspension/well) and allowed to grow into mature clones. The clonal patterns were replicated to two other Linbro dishes using the replicator as described in methods. After 5 days the patterns in the replicas were compared to the original pattern in the master.

[b]Total Don clones replicated 455; overall fidelity of replication, 91.7%

[c]Total BHK-21 clones replicated, 112; overall % fidelity of replication, 93.5%.

TABLE II

FIDELITY OF MULTIPLE REPLICA PLATING[a]

Primary replica (master plate)	No. clones on master	No. transferred to identical positions on replica						Percent fidelity of replication
		R_1	R_2	R_3	R_4	R_5	R_6	
No. 3, replica 1	48	47	46	47	46	46	46	96.5
No. 9, replica 2	41	39	40	40	41	40	40	97.6

[a]Replica plates of Don cell clones were selected at random from the primary replication described in Table Ia. Primary replica plates were then used as master plates, from which 6 secondary replicas were made in succession.

Total Don clones replicated, 89; overall fidelity of replication, 97.1%.

it is necessary to disperse partially and resuspend the trypsinized cells either by vigorously stirring them with the hand replicator or, much more satisfactorily, by rapidly pipetting back and forth the cell suspension with the aid of the syringe replicator prior to replication.

The mechanical procedure described herein for rapid clone isolation and replica plating should find wide application in the genetic analysis of cultured cells. Replica plating shoudl facilitate the isolation of useful mutants and the investigation of mutagenesis in somatic cell systems, just as it has in bacterial and fungal systems. Such a procedure should be useful in studies of temperature-sensitive somatic cell lines, in the isolation and analysis of clonal lines of drug resistant and nutritionally auxotrophic cells, and in studies of cell hybridization. Through the use of the mechanical technique described here it is possible to expose a large number of clonal populations (on the order of 1000) to a variety of selective conditions without sacrificing the parent clone. The use of a mechanical device for pipetting and filling eliminates much of the drudgery that would normally be involved in manipulating and tending the dozens of discrete clonal cultures growing in the wells of the incubation matrices. Also, the introduction of chambered troughs which can accommodate simultaneously a number of different cell cultures or a number of different selective agents increases the flexibility of the technique.

REFERENCES

Goldsby, R. A., and Zipser, E. (1969). *Exp. Cell Res.* **54**, 271–275.

Robb, J. A. (1970). *Science* **170**, 857–858.

Suzuki, F., Kashimoto, M., and Horikawa, M. (1971). *Exp. Cell Res.* **68**, 476–479.

Chapter 14

Selection Synchronization by Velocity Sedimentation Separation of Mouse Fibroblast Cells Grown in Suspension Culture

SYDNEY SHALL

Biochemistry Laboratory, University of Sussex, Brighton, Sussex, England

I. Introduction

A relatively simple method of establishing synchronous cultures of animal cells would facilitate the analysis of a variety of cell phenomena. To obtain synchronously growing cells, we have available two sets of methods: (1) induction or coercion, and (2) selection (Stubblefield, 1968). Induction or coercion methods usually rely on inhibition of DNA synthesis or of mitosis, but deprivation of selected amino acids has also been used (Tobey and Ley, 1971). All available coercive methods induce distortions of cell metabolism. In particular, those that inhibit DNA synthesis obviously interfere with and distort DNA synthesis and related processes. Consequently, methods which coerce the cells into synchrony by the use of metobolic inhibition of one sort or another are best avoided unless specific indications suggest their use. On the other hand, investigations of the mode of synchronization by inhibitors may be very informative.

Mitotic cells are an excellent source of a synchronous culture when the cells are grown in monolayers (Terasima and Tolmach, 1963; Petersen, Anderson, and Tobey, 1968). The selection of mitotic cells from a monolayer culture by the method of Terasima and Tolmach is certainly the method of choice whenever it can be applied. However, for suspension cultures or to meet a requirement for a very large number of synchronized cells, separation by velocity sedimentation can be used. This method, originally developed by Mitchison and Vincent (1965) with yeast, was extended to L cells by Sinclair and Bishop (1965). Schindler *et al.* (1970) improved this technique by using a gradient of 2 to 10% sucrose (w/v) in complete medium from which sufficient sodium chloride was removed to maintain an isotonic solution throughout the gradient. We have modified the technique further (Shall and McClelland, 1971) by adopting the improvement of Schindler *et al.* (1970), but using normal gravity to sediment the cells.

This simple procedure has several advantages. It works by selection, and therefore environmental perturbations are minimal. There is no gross change in the medium, and the temperature is easy to control. The method yields sufficient material for biochemical analysis. There is no apparent obstacle to scaling up the procedure. The method would seem to be of general use for all cells grown in suspension culture.

Macdonald and Miller (1970) have described an elegant method for obtaining synchronous cultures of mouse L cells by a velocity sedimentation technique at unit gravity in a gradient of fetal calf serum.

II. Principle of Velocity Sedimentation Separation

The principle of the method is to sediment cells through a column of liquid so that the cells become distributed throughout the column. The column of liquid may then be fractionated and the cells recovered. This procedure will separate cells on the basis of their volume or their density or a combination of these properties. As described below, the present method separates cells by volume. This assumes that there are changes in volume through the cell cycle, but it does not necessarily rely on a linear change in volume. A nonlinear change in cell volume during the cell cycle is acceptable, and will be reflected in the distribution of cells in the separating column.

The rate at which spherical cells sediment through a column of liquid is:

$$S = \frac{2}{9} \times g \times \frac{(\rho c - \rho s)}{\eta} \times \mathrm{r}^2 \times 3.6 \times 10^{-4} \tag{1}$$

where S is the rate of sedimentation (mm/hour), ρc is the density of the cells (gm/ml), ρs is the density of the medium (gm/ml), η is the viscosity of the medium in poises, r is the radius of the cell (μm), and g is the gravitational acceleration (981 cm S^{-2}). Assimilating this value of g, Eq. (1) reduces to

$$S = 0.0785 \times \frac{(\rho c - \rho s)}{\eta} \times r^2 \tag{2}$$

If all the cells in the population have very similar densities (ρc), then the cells will separate because of the differences in their radii (r). Cells of different types often show quite marked differences in density, and this has been used to separate one cell type from another quite successfully [reviewed by Shortman (1972)]. Cells of the same type at different stages of the cell cycle, however, do not show large differences in density, and recorded densities are usually in the range 1.05 to 1.10 gm/ml (Macdonald and Miller, 1970). If, in addition, the density of the gradient (ρs) does not vary much, then $(\rho c - \rho s)/\eta$ will effectively be a constant, with a value of about 1 to 5 gm/ml per poise. The rate at which the cells sediment will then depend almost entirely on their radii. If the radius becomes doubled, it will sediment four times faster. We assume that the volume of a cell increases as the cell progresses through a cell cycle. A 2-fold increase in volume is equivalent to about a 1.6-fold increase in r^2 and therefore a 1.6-fold increase in sedimentation rate. Thus, the technique can be used to separate cells according to their volumes. The larger, older cells will sediment more rapidly and the younger, smaller cells will be found nearer the top of the gradient.

III. Preparation of the Gradient

It is necessary to include a gradient of some kind in the column of liquid to stabilize it against convection. The gradient is made by preparing two solutions, one of 10.0% (w/v) sucrose and the other of 2.72% (w/v) sucrose in complete medium. We use sucrose because it is cheap and pure. A constant osmotic pressure must be maintained throughout the gradient. Therefore, the concentration of sodium chloride is decreased by 146 m*M* (8.54 gm/liter) in the 10% sucrose solution and by 40 m*M* (2.32 gm/liter) in the 2.72% sucrose solution. This requires the preparation of special media containing decreased sodium chloride concentrations. The gradient is made with the simple glass apparatus shown in Fig. 1.

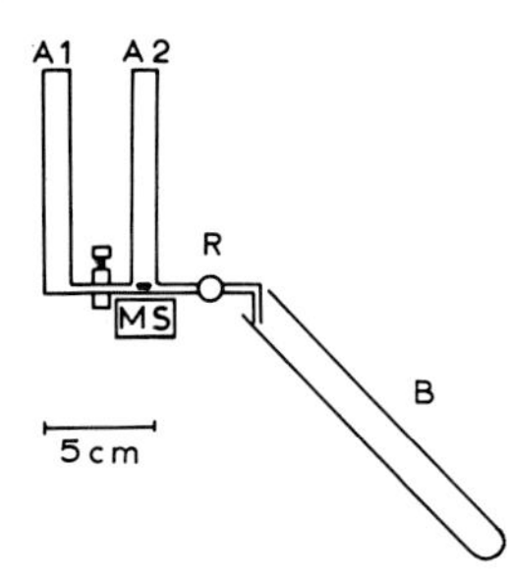

FIG. 1. Apparatus for generation of a gradient on a small scale. The cell separation is carried out in a glass test tube (B) 150 × 16 mm. The gradient is made in cylinders A1 and A2 connected by a short length of silicone tubing with a clip to open and close it. The outflow is controlled by a Rotaflo tap (R.). Rotaflo taps (manufactured by Quickfit & Quartz Ltd.) are sterilizable and permit fine control of flow rate. The dimensions of the cylinders A1 and A2 are 12 mm i.d. and 100 mm in height. Cylinder A2 contains a small bar magnet controlled by a magnetic stirrer (MS) on which the apparatus is placed.

The gradient is formed by transferring heavy solution (10% sucrose) from the right-hand chamber (A2, Fig. 1) to the gradient tube (B, Fig. 1). As solution leaves the right-hand chamber, hydrostatic pressure will force lighter solution (2.72% sucrose) from the left to the right chamber. The right-hand chamber is vigorously stirred. The concentration of sucrose in the right chamber decreases progressively as heavy liquid leaves and is replaced by lighter solution. If the two chambers are of the same dimensions, then the gradient of sucrose will be linear from 10% to 2.72%. If growth experiments are to be done, the entire apparatus is autoclaved before use.

We have used mouse LS cells selected by Dr. J. Paul of Glasgow. Scotland, for their ability to grow in suspension culture. The medium in which the cells are grown does not seem to be critical; similar synchronization results have been obtained with different growth media.

IV. Separation of Cells by Velocity Sedimentation

Complete medium (7.5 ml) containing 10% (w/v) sucrose (but minus 146 m*M* NaCl) is placed in the gradient vessel A2 (Fig. 1) containing the bar magnet; 7.5 ml of the complete medium containing 2.72% (w/v) sucrose (but minus 40 m*M* NaCl) is placed in the other gradient vessel A1 (Fig. 1). The two vessels are kept level and the magnet is set to stir vigorously. The clip between the two reservoirs is released, and then the Rotaflo tap is opened gradually so that the solution drips out quite slowly into a 150 × 16 mm test tube (B, Fig. 1). The solution is allowed to run down the side of the test tube to the bottom. The gradient will form and nearly fill the tube.

Cells in exponential growth are collected by centrifugation at about 100 *g* for 5 minutes. The medium is pipetted off and kept. The cells are resuspended in 1.0 ml of this used medium and layered on the gradient in the test tube with a wide-mouthed pipette held against the wall of the tube. The cells should run slowly down the side of the tube and across the surface of the gradient. There may be some turbulence at the top of the gradient, but no streaming of cells into the gradient should occur. Usually about 30×10^6 cells are applied to the gradient. The tube is then left standing upright in the incubator at 37° for 50 minutes while the cells sediment down the gradient. The topmost 1 ml of the gradient may then be removed with a pipette, or the whole column may be fractionated using the device shown in Fig. 2. The topmost 1 ml contains the youngest and smallest cells. The sucrose should be removed by centrifugation and the cells returned to normal growth medium. These cells will then show synchronous growth for one or two generations (Fig. 5–7).

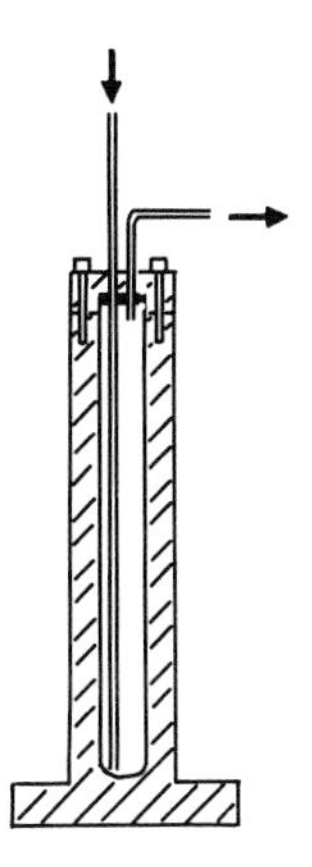

FIG. 2. Apparatus to fractionate the gradient used in Fig. 1. The unit is made from Perspex and is hollowed out so that the test tube used (150 × 16 mm) fits in exactly. The top is screwed tightly to the holder with two screws. The rim of the test tube inserts into a depression in the top, where a rubber gasket is used to form a watertight seal. Two thin metal tubes pass through the top; one goes to the base of the test tube, the other is flush with the inside of the top. The gradient is displaced from the test tube by pumping 25% (w/v) sucrose through the long tube to the bottom of the test tube. The gradient will be displaced upward through the short tube. Fractions of 1.0 ml may be collected.

If the cells are quite large, for example, with an average volume of about 1500 μm^3 and have a long doubling time, e.g., 24 hours, then the sedimentation may be done conveniently at unit gravity in an incubator at 37°C. If the cells are smaller or have shorter doubling times the separation may be done in the cold overnight (10 to 15 hours) or in a centrifuge at 37°C for 3 to 5 minutes at 100 *g* to 200 *g*. An approximate estimate of the sedimentation velocity can be obtained from Eq. (2). At lower temperatures the cell volume may decrease, and in such cases the cells will sediment more slowly. The cell density may be assumed to be 1.05 to 1.10 gm/ml unless known to be otherwise. The sucrose gradient described has a density gradient from 1.01 gm/ml at the top to 1.03 gm/ml at the bottom.

If the sedimentation period is more than 5 to 10% of the doubling time, then the sedimentation should be done at 25°C or 4°C to slow or stop cell growth during the separation period.

V. Size of Separated Cells

The volumes of the cells through the gradient may be measured microscopically with phase contrast microscopy and a graticule or with a Coulter Counter and a cell volume plotter.

The distribution of cell volumes in exponential and selected cultures of LS cells, as determined with a Coulter Counter and a cell volume plotter, are shown in Fig. 3. The velocity sedimentation separates the small cells (lower continuous line) from the bigger cells (upper continuous line). The volume distributions in the three populations are shown in Table I. The mean volume of the selected cells is 1330 $\pm$ 404 μm^3, that is, a mean diameter of 13.6 μm.

The sucrose is not essential but is desirable. This is shown by repeating the experiment without the sucrose. The cell separation is carried out as before but in a column of complete medium with no sucrose. With care the same result may be obtained (Fig. 4). Again, the smaller cells (dashed line) are selected, the mean volume of the top cells is 1281 $\pm$ 486 μm^3, that is, a mean diameter of 13.5 μm^3. In this experiment the cells were measured by phase contrast microscopy, so the full range of sizes is revealed (Table II).

Experience has shown that the sucrose is desirable in order to achieve reproducible results. In the absence of the sucrose there is more mixing, and it is more difficult to maintain the separation once achieved. Also, some "streaming" always occurs when the sucrose is omitted.

VI. Synchronous Growth of Separated Cells

The topmost layer of cells (small cells) are centrifuged down, and the medium is pipetted off and discarded. The cells are suspended in 40% (v/v) conditioned medium. They now display a degree of synchronous growth (Fig. 5). In some, but not all, experiments, a burst of cell division occurred immediately after the selected cells were transferred to the new medium. One generation later, a second burst of mitosis was observed. The cultures show moderate to good synchronous growth during one generation. DNA synthesis at each time was estimated by incubating the cells with tritiated thymidine for 30 minutes at 37°C and then measuring the acid-precipitable radioactivity. A peak in the rate of DNA synthesis is observed midway between two divisions. (Fig. 6). The position and duration of this S phase corresponds to reported data.

The initial burst of mitosis (Fig. 5) was unexpected and not predicted

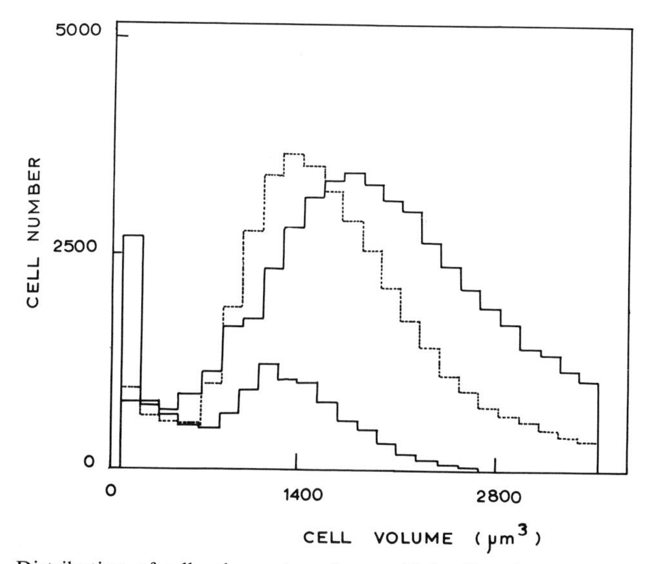

FIG. 3. Distribution of cell volumes in cultures of LS cells. The synchronizing gradient was prepared from complete medium by the omission of sodium chloride and by its replacement by a 2-fold molar equivalent of sucrose. A 15.0 ml linear gradient of 2.5 to 10% (w/v) sucrose in complete medium was prepared in sterile 150 × 16 mm test tubes. Cells were layered on top of this gradient and allowed to sediment at 37°C for 50 minutes, after which the topmost 1 ml was pipetted off. The rest of the gradient served as a control. The peak at the left is machine noise. The cell size counter cut off at 3500 μm^3, but there were larger cells. The dashed line is the original asynchronous culture; the lower solid line is the selected, topmost 1 ml; the upper solid line is the remaining 15 ml.

TABLE I
VOLUME DISTRIBUTION IN LS CELLS[a]

Parameter	Asynchronous culture	Top 1.0 ml of column	Remaining cells in column
Modal volume	1260 μm^3	1120 μm^3	1820 μm^3
Mean volume	1769 μm^3	1331 μm^3	2037 μm^3
Standard error of mean volume	621 μm^3	404 μm^3	702 μm^3
Coefficient of variance of mean volume	0.35	0.30	0.34
Modal diameter	13.4 μm	12.9 μm	15.1 μm
Mean diameter	15.0 μm	13.6 μm	15.7 μm

[a]Analysis of the distribution of volumes of cells in an asynchronous culture, in the top 1.0 ml of the synchronizing column, and in the remainder of the column. Cell volumes and numbers were determined with a Model B Coulter Counter and "J" Plotter.

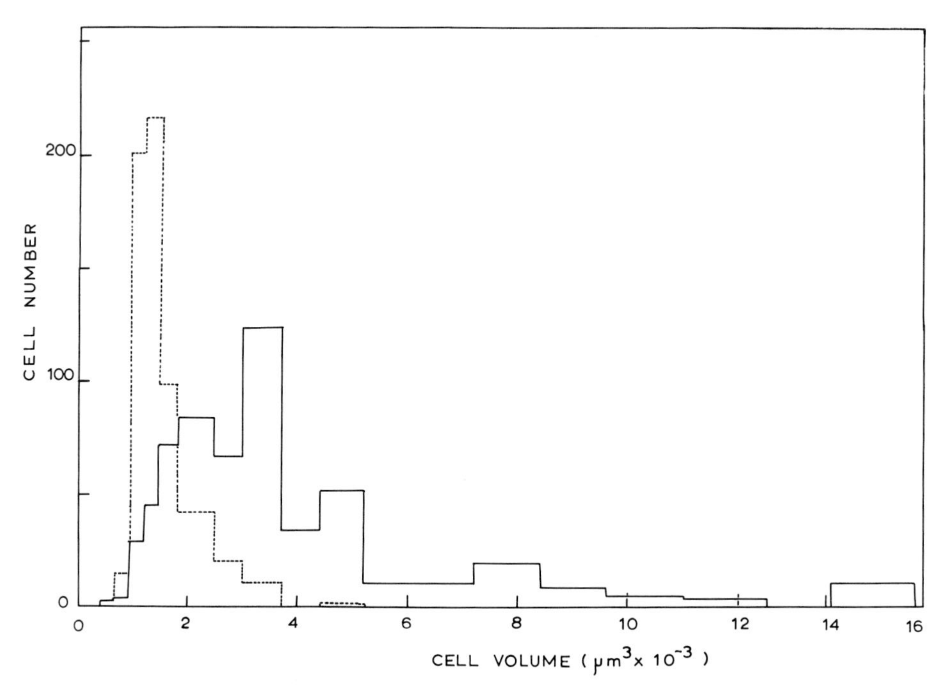

FIG. 4. Distribution of cell volumes in cultures of LS cells separated on a column of complete medium, without any sucrose. The dotted line is the topmost 1.0 ml; the full line is the remaining 12 ml. The cell numbers and diameters were measured by direct microscopic examination with phase contrast microscopy on unstained unfixed cells.

TABLE II
VOLUME DISTRIBUTION IN LS CELLS[a]

Parameter	Asynchronous culture	Top fraction
Mean volume	3077 μm^3	1281 μm^3
Standard error of mean volume	2316 μm^3	486 μm^3
Coefficient of variance of mean volume	0.753	0.379
Mean diameter	18.1 μm	13.5 μm

[a] Analysis of the distribution of volumes in an asynchronous culture and in the top of the synchronizing column. This synchronizing column contained only medium with no sucrose. The top fraction contained 7% of the total cells. The cell numbers and cell diameters were measured by direct microscopic examination under phase contrast on unstained, unfixed cells.

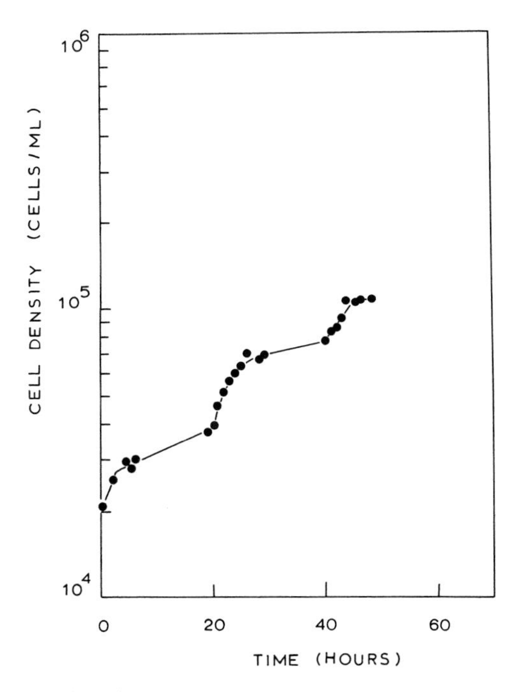

FIG. 5. Synchronization of LS cells on a sucrose gradient. The cells were separated as described in the text, and the topmost 1.0 ml of cells was resuspended in medium and allowed to grow. The culture was sampled periodically. Three bursts of cell division are apparent.

by the simple theory outlined in Section II. The explanation for the selection of these premitotic and mitotic cells is not clear. At this time, I can only draw attention to the heterogeneity of the LS cell cultures; Figs. 3 and 4 show that the observed cell volumes extend well beyond a simple 2-fold range. The selection of premitotic and mitotic cells, as well as early postmitotic cells, raises the possibility that the former cells experience a transient but marked fluctuation in either cell volume or cell density.

Cells separated on a column of medium without sucrose also grow synchronously. Exponential cells were resuspended in 0.75 ml of serum-free medium. Instead of a gradient, a sterile tube was filled with 12 ml of complete medium. The cell suspension was layered on to the column of medium with a wide-mouthed pipette, slowly traversing around the wall of the tube so that the suspension spread over the surface toward the

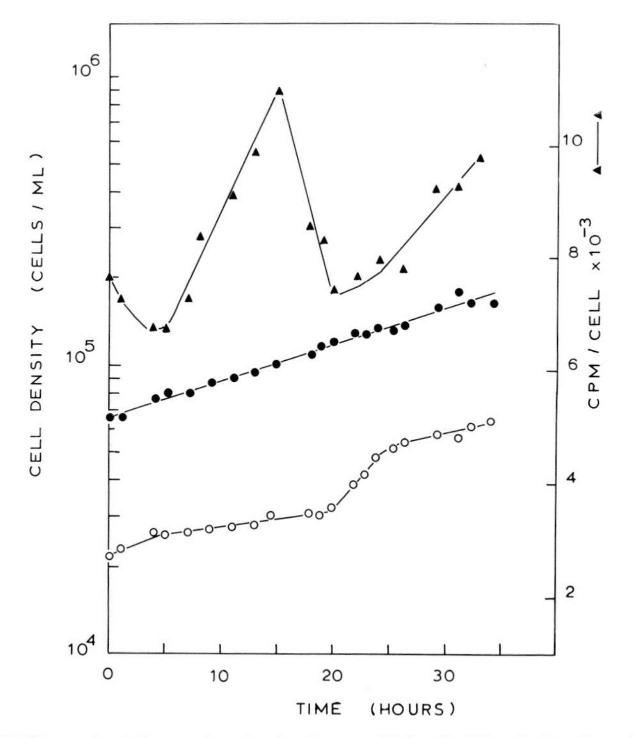

FIG. 6. DNA synthesis in synchronized cultures of LS cells. The circles show the increase in cell number in the unsynchronized (●) and the synchronized (○) cultures. The triangles show the incorporation of thymidine into acid-insoluble material in the synchronized culture during a 30 minute incubation. (Unpublished work of A. J. McClelland and S. Shall.)

center. Some turbulence was observed, but a layer of cells was formed which was gently stirred with the tip of the pipette. The layer produced had the same thickness as when a sucrose gradient was used. Beneath the layer of cells there was a "streamer" of cells. The cells were allowed to sediment in a 37°C incubator for 50 minutes. The top 1 ml of the column, which contained 2.25% of the total load of cells, was then collected. The remaining 12 ml was mixed and used as a control. Both sets of cells were collected by centrifugation and resuspended in conditioned medium. The growth patterns of these cells (Fig. 7) show that the topmost cells (small cells) (Fig. 4) do grow synchronously.

VII. Estimation of the Degree of Synchrony

The degree of synchrony may be estimated in several ways. The most direct is to count the number of cells periodically and to plot the results as in Figs. 5 to 7. This immediately gives a qualitative picture of the degree of synchrony by comparison with exponential growth. The quantitation of the data is described below.

To measure DNA synthesis the asynchronous culture is incubated with

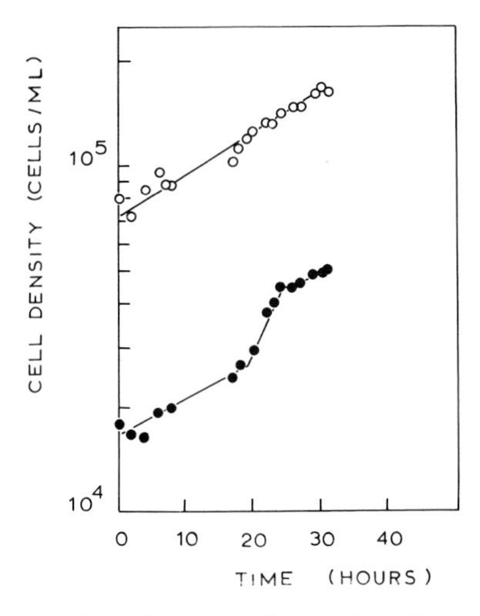

FIG. 7. Synchronization of LS cells on a column of medium without sucrose. The top 1.0 ml after 50 minutes at 37°C (●) shows synchrononus growth. The remaining 12 ml (○) show exponential growth.

tritiated thymidine (5.0 μCi/ml) for 30 minutes before sedimentation. S phase cells may then be identified after sedimentation by measuring the radioactivity insoluble in 5% (w/v) trichloroacetic acid in each fraction.

The mitotic cells may be identified and the mitotic index of each fraction estimated. The procedure we use is as follows: spin down the cells at about 100 *g* for 5 minutes *in the cold.* Remove the supernatant with a Pasteur pipette and discard. Add several milliliters of precooled 0.6% (w/v) sodium citrate. Resuspend the cells by gentle shaking and leave for 20 minutes at 0°C. Centrifuge the cells *in the cold* as before and discard the supernatant. Add several milliliters of ice cold fixative (acetic acid : methanol, 3:1 v/v) dropwise down the side of the tube onto the cell pellet. Allow to stand for 5 minutes at 0°C and centrifuge. Remove and discard the supernatant. Add 1.0 ml of fixative and allow to stand for 2 minutes at 0°C. Centrifuge and discard the supernatant. Add 0.2 ml of ice cold fixative and resuspend the cells by gentle shaking. Place several drops of this cell suspension on to a clean cover-slip and allow to dry completely in air. Stain for 10 minutes in natural aceto-orcein, wash twice in methoxyethanol and twice in Euparol essence. Leave to dry in air. Mount in Euparol on clean slides. This preparation gives permanent slides. Count several fields, scoring the number of mitotic cells. It is advisable to count at least 400 cells and preferably 1000.

It is very desirable that any description of synchronous growth should be accompanied by a quantitative estimate of the degree of synchrony. The most satisfactory, currently available method of estimating the degree of synchrony was devised by Engelberg (1961, 1964). The principle of the method is to compare the growth rate of the synchronous culture with that of the asynchronous culture and then to estimate the integral over the time during which the synchronous culture grows faster than the asynchronous culture. In order to standardize the procedure for different cell densities the normalized rate of cell division R, is defined as

$$R = (dn/dt)/n \tag{3}$$

R is then the fraction of cells dividing per unit time. R is derived from a table of cell density against time by calculating the increment in cell number, Δn for a given *small* interval in time, Δt, and then $R = (\Delta n/\Delta t)/n$. Plot R, calculated in this way, against time, and evolve a graph such as Fig. 8. With a low *rate* of increase in cell number, R is small; a burst of cell division will give a peak of R. A constant rate of cell division, as in exponential growth, yields a horizontal line for the normalized growth rate. The area enclosed by the line R and two time points during which the cell number doubles is equal to $\ln 2 = 0.693$, for all possible growth curves. Consequently, for exponential growth the value of $R = \ln 2$/doubling time. Engelberg has named the area under the R curve above the exponential level, the overlap area. He then defined the *percentage synchronization* of a culture to be:

$$\text{Percentage synchronization} = \frac{\text{overlap area}}{\text{total area under } R \text{ between two time points during which the cell number doubles}} \times 100 \quad (4)$$

The data shown in Fig. 5 are used to calculate an Engelberg index in Fig. 8. Exponential growth occurs at a constant growth rate and hence gives a constant value of R. The synchronized culture has three steps or bursts of mitosis during which the growth rate is greater than the exponential culture. These are separated by periods of slower increase in cell number, and during this time the growth rate sinks below that of the exponential culture. The Engelberg index for the central part of this growth curve is 62%.

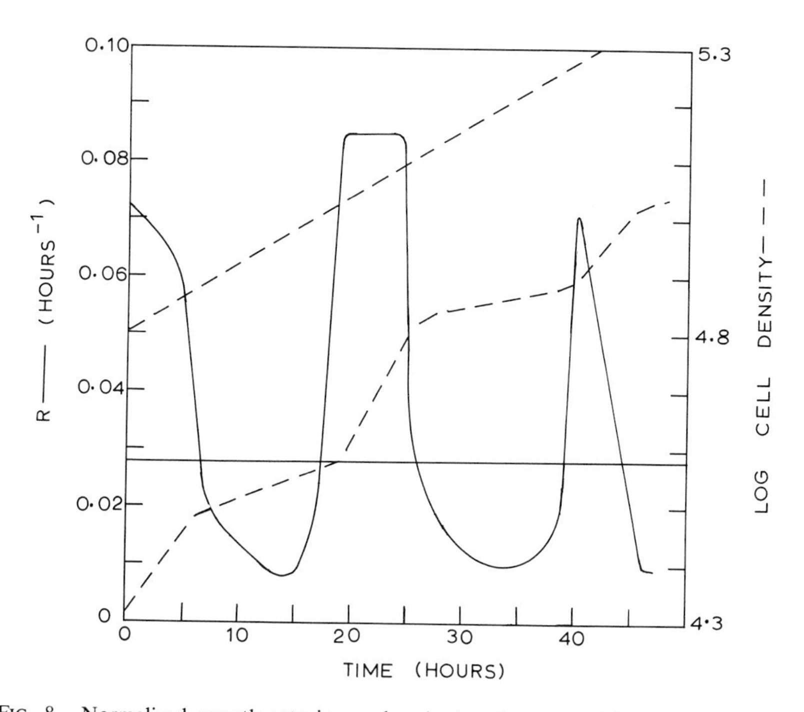

FIG. 8. Normalized growth rate in synchronized and exponential cultures of LS cells. The data are from Fig. 5. The dashed lines show the logarithms of the cell density; the solid lines show the respective normalized growth rates (R). The linear exponential growth of the control culture generates a constant normalized growth rate, the horizontal solid line. The synchronous growth curve, the lower discontinuous line, generates a normalized growth curve with three peaks corresponding to the three bursts of mitosis. The "overlap area" is the area enclosed by the normalized growth rates of the two cultures, that is, above the horizontal solid line and below the fluctuating solid line.

The overlap area may be measured by counting squares on graph paper, by cutting up the graph and weighing the respective pieces or by using a planimeter. Remember that the denominator of Eq. (4) always equals 0.693 if the correct dimensions are used.

Engelberg (1964) has drawn attention to the method by which one can make a similar calculation of percentage synchronization from estimates of the mitotic index during synchronous growth. For this calculation one needs to know the length of mitosis (T_m) which may be derived from the exponential culture.

$$T_m = [T_d \ln(1 + I_m)]/\ln 2 \quad (5)$$

where T_d is the doubling time and I_m is the mitotic index. The normalized rate of cell division can then be calculated,

$$R = I_m/T_m \quad (6)$$

and a plot of R against time can be constructed. Then proceed as before.

There are other methods of estimating the degree of synchrony, but they are all inferior in some respect to the Engelberg method.

Zeuthen (1958) suggested the following formula:

$$\text{Percentage synchrony} = \frac{T_d/2 - D_{0.50}}{T_d/2} \times 100 \quad (7)$$

where T_d is the doubling time for the culture and $D_{0.50}$ is the shortest time during which 50% of the population doubles.

Scherbaum (1959) defined a *synchronization index* as

$$\left(\frac{N_2 - N_1}{N_1}\right) \times \left(1 - \frac{\Delta t}{T_d}\right) \quad (8)$$

where N_1 and N_2 are the cell densities just before and after synchronous division, and Δt is the duration of the synchronous division.

Blumenthal and Zahler (1962) defined the fraction F, which measures the fraction of the population which divides during a time interval t in excess of that expected to divide during logarithmic growth in the same time interval.

$$F = (N_2/N_1) - (t/2T_d) \quad (9)$$

where N_1 and N_2 are cell densities at the beginning and end of time interval t; T_d is the doubling time of the culture. Note that there is a similarity in the principle of this index and that of Engelberg.

The data in Figs. 5–7 are used to calculate percentage synchronization by all four methods (Table III). There is no correlation between any of the calculations. I recommend the use of the Engelberg method.

TABLE III

QUANTITATIVE ESTIMATION OF THE DEGREE OF SYNCHRONY BY SEVERAL METHODS

Data	Doubling time (T_d) (hours)	Synchrony index (%)			
		Engelberg	Zeuthen	Scherbaum	Blumenthal and Zahler
Figs. 5 and 8	24	62	58	40	42
Fig. 6	22	47	64	43	37
Fig. 7	24	33	62	46	69

VIII. Scaling-Up Procedures

The scale of operations is limited in the procedure described by the cross-sectional area of the separating column. If one wishes to work with more cells, then a larger apparatus is required. This entails mechanical problems, which are best solved by using the STAPUT apparatus (Fig. 9) designed by

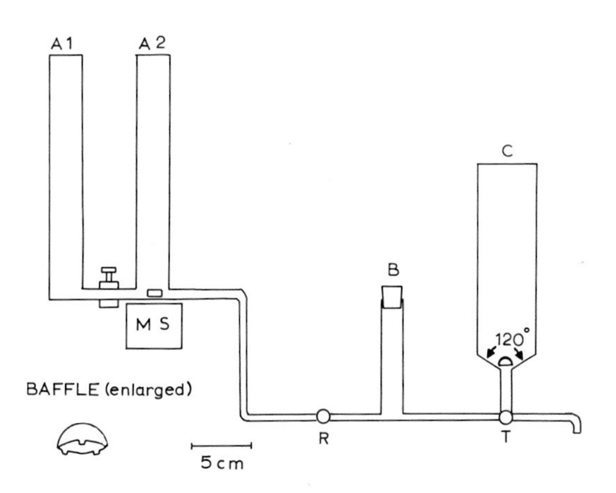

FIG. 9. Apparatus for sedimentation separation on a large scale. Gradient forming vessels A1 and A2 are 27 mm i.d. and 200 mm high. The bottom of the gradient former should be higher than the top of the separating chamber (C). The Rotaflo tap (R) controls the rate of flow from the gradient. The all-glass components are connected by sterilizable silicone tubing. The cell chamber (B) is 18 mm i.d. and 100 mm long and is tightly sealed with a No. 17 silicone bung. The separating chamber (C) is 51 mm i.d. and 150 mm on its straight edge. The baffle (see inset) is made from the base of a 16-mm diameter glass test tube. The bar magnet in cylinder A2 is motivated by the magnetic stirrer (MS). T is a three-way tap, and R is a Rotaflo tap. Separating chamber C is covered to exclude dust. The entire apparatus may be steam sterilized after assembly.

Miller and Phillips (1969) as a development of the STAFLO system of Mel (1964). This is suitable for up to about 10^9 cells.

The 10% sucrose medium (150 ml) is placed in the gradient reservoir A1 (Fig. 9) and the 2.72% solution in the reservoir A2 (Fig. 9). The levels of the solutions in reservoirs A1 and A2 must be the same. Open the clip connecting the two reservoirs. Ten milliliters of medium without serum is placed in the separating column C (Fig. 9). The baffle is positioned after air bubbles are removed, with the concave face down. The baffle ensures that the gradient does not spurt into the separating column, but enters as a thin layer. Then 5 to 10 ml of a concentrated cell suspension containing 10^8 to 10^9 cells, are placed in the cell chamber B (Fig. 9), which is then tightly corked. The cell suspension is driven into the separating column C (Fig. 9) by forcing sterile air into the cell chamber B (Big. 9) with a syringe and needle. The cell suspension should form a narrow band underneath the layer of medium. Without admitting any air into the line, you now allow the gradient to flow into the separating column C (Fig. 9) thus forcing the cells up into the column. It is important that gradient reservoir A2 is always vigorously stirred by the magnetic stirrer. The gradient reservoirs should be kept higher than the separating chamber C to provide enough hydrostatic pressure for the incoming liquid to push the cells up the column.

When the gradient is complete, the separation process may begin. This may be conducted at 37°, 25°, or 4°C. Again, the lower the temperature, the smaller the volume of the cells and the more viscous the medium, and therefore the cells sediment more slowly. The appropriate time for separation must be ascertained for each cell type. For our LS cells with a volume of about 1500 μm^3, about 2 hours at 37°C is necessary. With smaller cells, it is necessary to use 4° or 25°C.

At the end of the separation, the column is fractionated from the bottom into 5- or 10-ml fractions. The cells are gently centrifuged and resuspended. They may then be used either for growth or for analysis. The smallest and youngest cells are at the top, and proceeding down the gradient, the G1, then S, and then G2 cells are encountered. The G1 cells are separated most cleanly.

Instead of sucrose, one can use a 5% to 20% (w/v) gradient of Ficoll, which is a sucrose polymer (Pharmacia Ltd.) (Ayad, Fox, and Winstanley, 1969; Warmsley and Pasternak, 1970). In this case, the Ficoll may be added to the complete medium and the change in osmotic pressure ignored. However, caution is necessary because Ficoll is sometimes toxic to some cell types; we have found it to be detrimental to LS cells and mouse lymphoma cells, L5178Y. Bovine serum albumin or fetal calf serum (Macdonald and Miller, 1970) may be tried as gradient materials, but the first is sometimes toxic and the second expensive and rather variable for this purpose.

This velocity sedimentation selection technique may be performed in small test tubes or in larger tubes. To scale up still further, one can use a zonal centrifuge to sediment the cells (Warmsley and Pasternak, 1970).

REFERENCES

Ayad, S. R., Fox, M., and Winstanley, D. (1969). *Biochem. Biophys. Res. Commun.* **37**, 551.

Blumenthal, L. K., and Zahler, S. A. (1962). *Science* **135**, 724.

Engelberg, J. (1961). *Exp. Cell Res.* **23**, 218.

Engelberg, J. (1964). *In* "Synchrony in Cell Division and Growth" (E. Zeuthen, ed.), pp. 497–508. Wiley, New York.

Macdonald, H. R., and Miller, R. G. (1970). *Biophys. J.* **10**, 834.

Mel, H. C. (1964). *J. Theor. Biol.* **6**, 159, 181, 307.

Miller, R. G., and Phillips R. A. (1969). *J. Cell. Physiol.* **73**, 191.

Mitchison, J. M., and Vincent, W. S., (1965) *Nature (London)* **205**, 987.

Peterson, D. F., and Anderson, E. C., and Tobey, R. A. (1968). *In* "Methods in Cell Physiology" (D. M. Prescott, ed.), Vol. 3, pp. 347–370. Academic Press, New York.

Scherbaum, O. H. (1959). *J. Protozool. Suppl.* **6**, 17.

Schindler, R., Ramseier, L., Schaer, J. C., and Grieder, A. (1970) *Exp. Cell. Res.* **59**, 90.

Shall, S., and McClelland, A. J. (1971). *Nature (London), New Biol.* **229**, 59.

Shortman, K. (1972). *Annu. Rev. Biophys. Bioeng.* **1**, 93.

Sinclair, R., and Bishop, D.H.L. (1965). *Nature (London)* **205**, 1272.

Stubblefield, E. (1968). *In* "Methods in Cell Physiology" (D. M. Prescott, ed.), Vol. 3, pp. 25–43. Academic Press, New York.

Terasima, T., and Tolmach, L. J. (1963). *Exp. Cell Res.* **30**, 344.

Tobey, R. A., and Ley, K. D. (1971). *Cancer Res.* **31**, 46.

Warmsley, A. M. H., and Pasternak, C. A. (1970). *Biochem. J.* **119**, 493.

Zeuthen, E. (1958). *Advan. Biol. Med. Phys.* **6**, 37.

Chapter 15

Methods for Micromanipulation of Human Somatic Cells in Culture[1]

ELAINE G. DIACUMAKOS

Laboratory of Biochemical Genetics, The Rockefeller University, New York, New York

---

[1]Work reported and equipment described have been funded by grant no. T-55, American Cancer Society. Max C. Fleischmann Foundation, and an Institutional grant no. CA-08748, National Cancer Institute, awarded to the Sloan-Kettering Institute for Cancer Research; and grant no. CRBS-248 from The National Foundation–March of Dimes.

I. Introduction

Methods for micromanipulating individual, living cells depend upon: (1) cell type, (2) equipment available to make and control the microtools used to operate on cells, and (3) time required to observe the cells during and after microoperations. The methods to be described here are parts of a methodology designed as the simplest available means of providing the greatest number of options for studying individual, intact and living, human somatic cells for the purpose of isolating and then analyzing operated cells or their progeny.

II. Cell Culture

A. Cell Types

Cells of normal or abnormal origin and of epithelioid or fibroblastic morphology are used. These include HeLa, HeLa clone (HC), a subline of HeLa (ERK), human embryonic lung (HEL), human fetal (HF), Lesch-Nyhan (LN), and mouse L (L) cells.

B. Stock Cultures

Cells are grown in Dulbecco's modification of Eagle's medium (Bablanian *et al.*, 1965) with 10% (v/v) fetal bovine serum (Flow Laboratories, Inc.) that has been preheated at 56°C for 30 minutes. Transfers are done in a sterile tissue culture hood (LabConCo) within a transfer room; and all pipetting, with Pi-Pump attachments (Bel Arts Products). The cells are transferred to sterile, disposable plastic flasks (Falcon No. 3024) that are flushed at a constant flow rate with a mixture of 5% carbon dioxide–95% air (Ohio Medical Products), closed tightly, and kept at 37°C in a Napco water-jacketed, carbon dioxide incubator (Heinicke Co.). Carbon dioxide (bone dry; Matheson Co.) is used. Transfer schedules vary with cell type.

C. Microcultures on Cover Slips

All cell types given in Section I,A can be grown on cover slips in Dulbecco's medium. Stock cultures are used to prepare the microcultures for micromanipulation.

Two sterile, 18 mm square cover slips are inserted in each sterile petri dish (Falcon No. 3002) except that, if cell fusions between two different types of cells are to be done, a triangular cover slip (Fig. 1) is substituted for a square cover slip. A 5 ml aliquot of the cell suspension is used to inoculate each dish, and the dishes are kept in the Napco incubator at 100% humidity. Cell densities are adjusted to provide cultures usable on successive days.

D. Preparation and Sterilization of Materials

1. Glassware

Soiled glassware (immersed in Wescodyne solution [West Chemical Co.]) or new glassware is rinsed in tap water, soaked overnight in 35% nitric acid, and rinsed thoroughly in tap, distilled, and distilled-deionized water, and then dried in an oven. Serological and Pasteur pipettes are removed from Wescodyne solution, rinsed in tap water, and placed in a plastic pipette basket, which is immersed in a plastic pipette jar (Nalgene Co.) containing 35% nitric acid and soaked overnight. The pipettes are removed, rinsed again in tap, distilled, and distilled-deionized water, dried, plugged with non-absorbent cotton, and placed in metal pipette cannisters.

These materials are sterilized for 2 hours at 165°C and left in the sterilizing oven until it has reached room temperature.

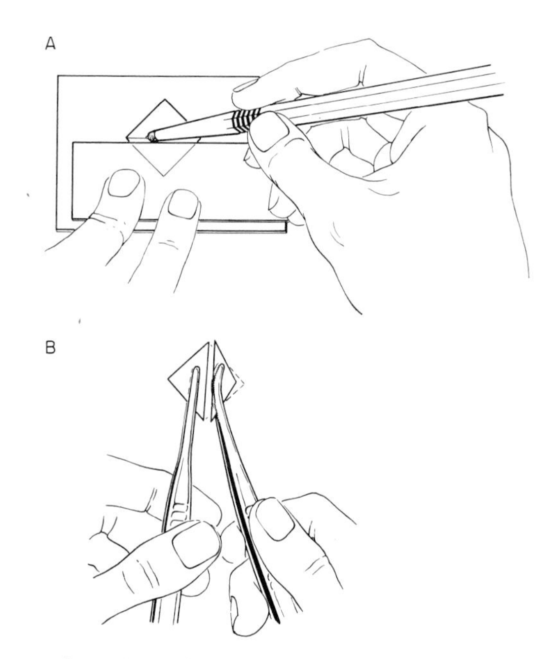

Fig. 1. Diagram of steps in scoring and obtaining two triangular cover slips from an 18 mm square cover slip (Section II,C).

2. Cover Slips

New cover slips are rinsed in tap water and immersed individually in 70% nitric acid to soak overnight. They are rinsed with tap, distilled, and distilled-deionized water and dried in Chen Type A porcelain staining racks in a dust-free container. The cover slips are handled with a stainless steel forceps. Clean, dry cover slips are stored between layers of filter paper, about 10 per layer, in a glass petri dish.

The clean, dry cover slips are cut as shown in Fig. 1, using glass slides to hold them so as not to transfer oils from the fingers. Cutting them after they have been cleaned reduces breakage.

The cover slips are sterilized just before use in the sterile hood. Stacks of six square or triangular cover slips, with edges even, are held in a stainless steel forceps, immersed in ethanol until all the air bubbles between them escape, and, after draining by being touched to the lip of the beaker, are ignited until all the alcohol is burned off.

The sterile cover slips are stored in a sterile, plastic petri dish and then are transferred, one or two at a time, into the dishes used for microcultures. The forceps used to transfer them should be dipped in alcohol and flamed each time. This is also done in the sterile hood.

3. Other Materials

New rubber tubing used with reservoirs is prepared according to the procedure recommended by Parker (1961). After that it is washed in 1% 7X solution (Linbro Co.) and rinsed in tap, distilled, and distilled-deionized water.

Millipore stainless steel filters (Millipore Corp.) are also washed in 7X solution. These are rinsed as above, dried, assembled, and autoclaved according to the manufacturer's instructions.

Syringes and medicine droppers are dried, wrapped in gauze, disassembled, and sterilized by autoclaving in sterilizing bags.

Media and other solutions are sterilized by filtration through HA and GS Millipore filters that are prewashed with phosphate-buffered saline (PBS) solution (Dulbecco and Vogt, 1954).

III. Equipment for Making and Controlling Microtools

A. General

The size and shape of glass microtools depend upon: (1) cell size and shape, (2) microoperations to be performed, (3) diameter and length of microtool holder, (4) adjustability of microtool holder clamp on micropositioner, (5) range of motion of fine controls of micropositioner, (6) distance between micropositioner and optical axis of microscope, (7) microscope and type of optics, (8) size and shape of microsurgical chamber, (9) clearance required for unobstructed movements and replacement of microtools, and (10) limitations that each of the above factors places on any others listed.

The procedures described here take all these factors into account, provide one basic design for all microtools, and take advantage of making them as simple and as short as possible to minimize internal vibration and breakage.

Microneedles are made from 1 mm borosilicate or Pyrex glass rod (Corning Glass Co.), cut in 10 cm lengths, and stored in stoppered tubes. Micropipettes are made from 1 mm diameter, 10 cm long, thin-walled capillaries (Kimble Glass Co.) that have been washed and coated with Desicote (Beckman Instruments Co.), and stored in stoppered tubes.

B. Flameboard

The flameboard, similar to that described by Chambers and Kopac (1950), provides a microburner with an adjustable flame that can be relighted from a second larger burner. Both burners are positioned on a baseboard that is walled on three sides to provide a draft-free work area. Glass rod or capillary tubing is drawn by hand following the steps diagrammed in Fig. 2A-E. The lumen of the capillary is maintained through these steps. The hand-drawn rods or tubes are made in quantity and stored in a vertical position in needle racks that are kept in dust-free containers.

C. Microforge

The instrument shown in Fig. 3 is used to finish the microtool by shaping the microshaft and tip of the needle or pipette. The sequence of steps, as seen through the microscope with the X and Y axes reversed, is diagrammed in Fig. 4, using low magnification (3.5× Leitz objective and 10× ocular) and in Fig. 5, using higher magnification (50× UMK Leitz objective and 20× ocular).

1. Microneedles

Insert the hand-drawn rod as far as possible into the microtool holder and tighten the knurled collar. Insert the holder (M, Fig. 3) in the microforge clamp (C, Fig. 3) at the position shown, with the knurled collar of the holder at the edge of the clamp (Fig. 4A). Rotate the clamp on its axis with Cp (Fig. 3) until the holder is horizontal. Rotate the holder (M, Fig. 3) so that the large glass bend is vertical and the hook is uppermost. This is a critical adjustment. Once adjusted, do not change the position of the holder in the

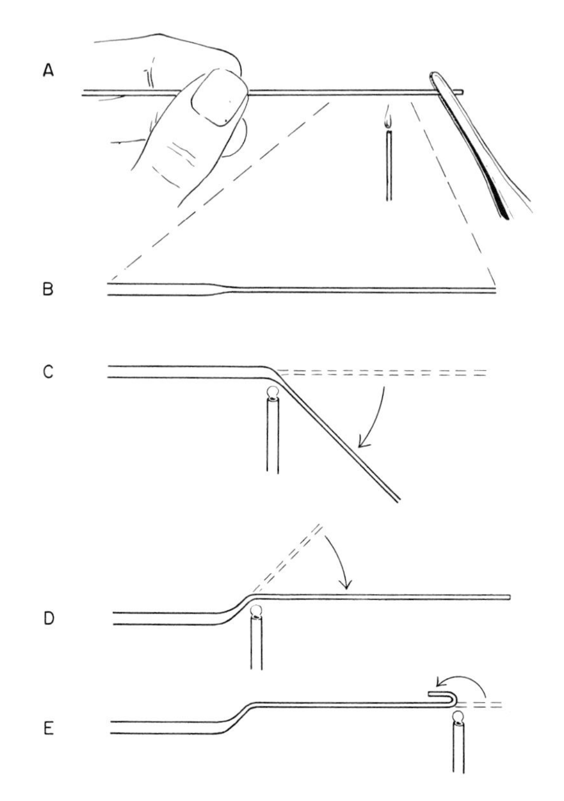

FIG. 2. Diagram of steps in preparing hand-drawn rods and capillaries (Section III,B).

clamp. By rotating the clamp (C, Fig. 3), bring the holder and glass rod to a 45° angle and avoid hitting the platinum heating filament (F, Fig. 3). The glass rod will appear (Fig. 4A). Suspend a 1 gm weight, held in a forceps, from the glass hook of the microtool. The rod may bend but should not break.

Use low magnification. Focus on the rod by moving the microscope body tube. Bring the filament into position so that the rod lies within the smaller U-shaped bend of the filament (Fig. 4A) but does not touch any part of it. Heat the filament, adjusting the heat so that the rod will bend as shown in Fig. 4B. Turn off the filament. Adjust the heat setting, turn on the filament, and apply it just above the bend until the glass begins to taper. Adjust the heat and raise the filament until the gradual taper to produce a blunt needle seen in Fig. 4C is obtained, i.e., when the weight and glass fall free. Turn off the filament. Lower the needle in the field so that the filament position can be shifted to bring the glass bead into position directly above the blunt needle shaft without striking it.

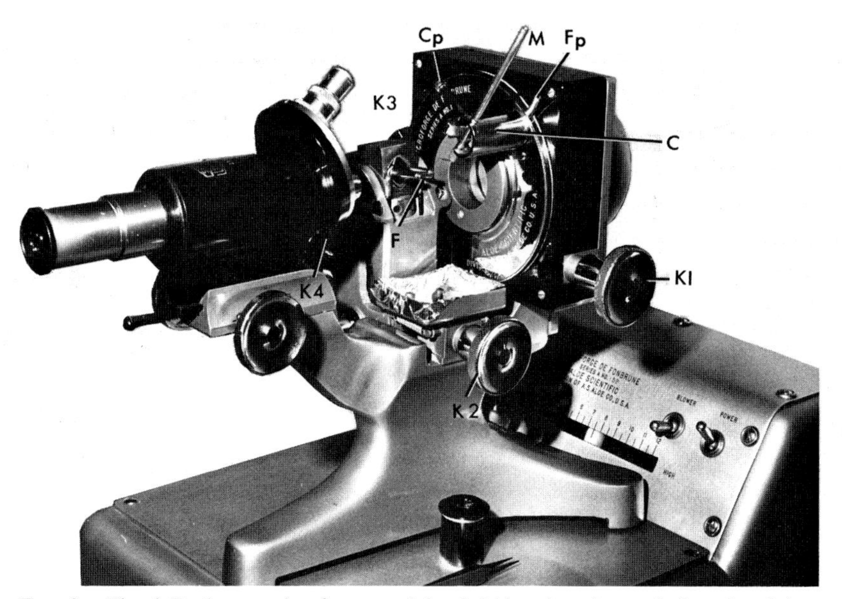

FIG. 3. The deFonbrune microforge used for finishing the microtools from hand-drawn rods or tubes. See Section III,C for the description.

Change to high magnification. Focus on the needle by moving the microscope body tube; the bead and microtip will appear as in Fig. 5A. Bring the blunt tip of the shaft into contact with the heated glass bead (Fig. 5B) and back it off at the rate that produces either an abrupt or a gradual taper (Fig. 5C). If the tip is not satisfactory, break it off against the cold glass bead and repeat the shaping operation. The finished tip is shown in Fig. 5C.

Return to low magnification. At the point where the microshaft begins, apply low heat with the straight part of the filament, between the glass bead and the U-shaped bend, and then bend the shaft until it is almost parallel with the thicker part of the glass rod (Fig. 5D). This produces the finished microneedle shown in side and top view in Fig. 5E and F. Move the filament away for ample clearance. Rotate the clamp (C, Fig. 3) on axis with Cp (Fig. 3) until the holder is horizontal, and remove the holder from the clamp. Loosen the knurled collar of the holder and carefully remove the microneedle. Store it in a vertical position in a needle rack.

2. MICROPIPETTES

a. For Oil Injections. To ensure that there are no air bubbles in the silicone oil syringe reservoir (30 ml capacity) and in the polyethylene tubing connecting the syringe with the micropipette holder, expel oil through the

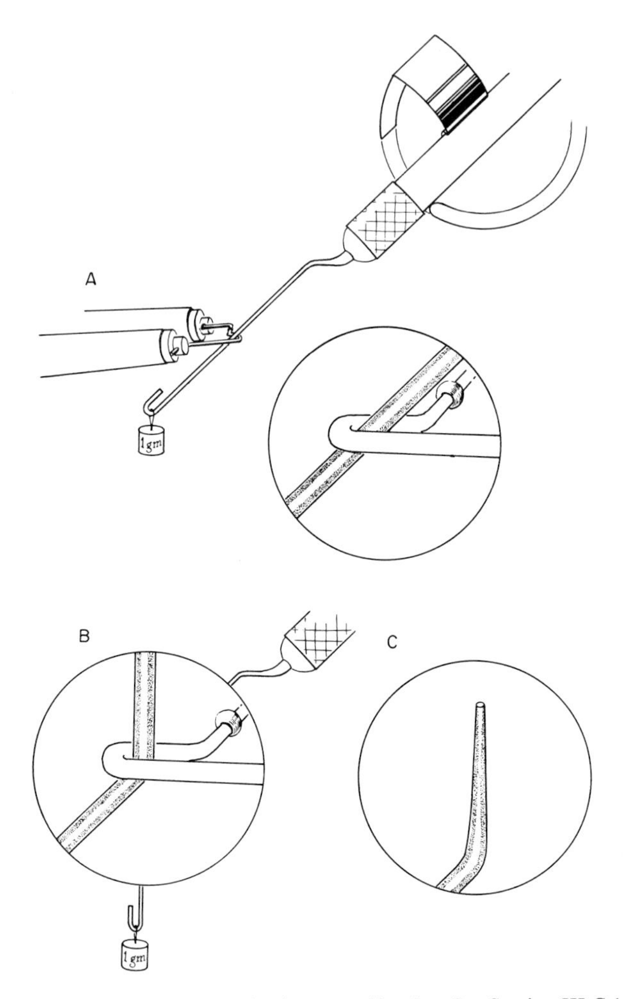

FIG. 4. Diagram of tapering glass rod at low magnification. See Section III,C,1 for details.

opening in the loosened knurled collar of the holder onto a paper towel until no more air bubbles emerge. Insert the capillary just far enough in the holder to bring the oil level midway to the large bend in the capillary. Tighten the collar and retract the oil until it just disappears back into the holder. Loosen the collar and insert the tube deeper in the holder until the oil level again reaches midway to the large bend. Tighten the collar and retract the oil until it just disappears back into the holder. Repeat this procedure as many times as necessary until the capillary is inserted in the holder as far as it will go. With no pressure on the syringe piston, the oil just fills the large bend and remains constant.

Insert the micropipette holder in the clamp (C, Fig. 3) and adjust as in making a microneedle, Section III,C,1. Control the level of oil in the capillary so that it does not reach the region to be heated by the filament, for, should it go beyond this region, and reach the tip, the oil will solidify and seal the pipette. Also control the oil to correct for the expansion and contraction of the air in the capillary as a result of heating and cooling the filament. The tip of the blunt pipette is either polished for use in studying surface reactions or beveled for intracellular microinjections.

With high magnification, the tip of the pipette is polished by touching it to the heated glass bead or beveled by touching the lip at a tangent to the bead and retracting it. The rate of pull determines the length of the bevel at any given temperature. In either case, make the final bend as shown in Fig. 5D, and turn off the filament.

Using low and high magnifications, increase the pressure on the oil until the pipette is filled. The more narrow the pipette, the greater will be the pressure required to force the fluid to the tip.

Move the filament away and remove the micropipette holder, using the

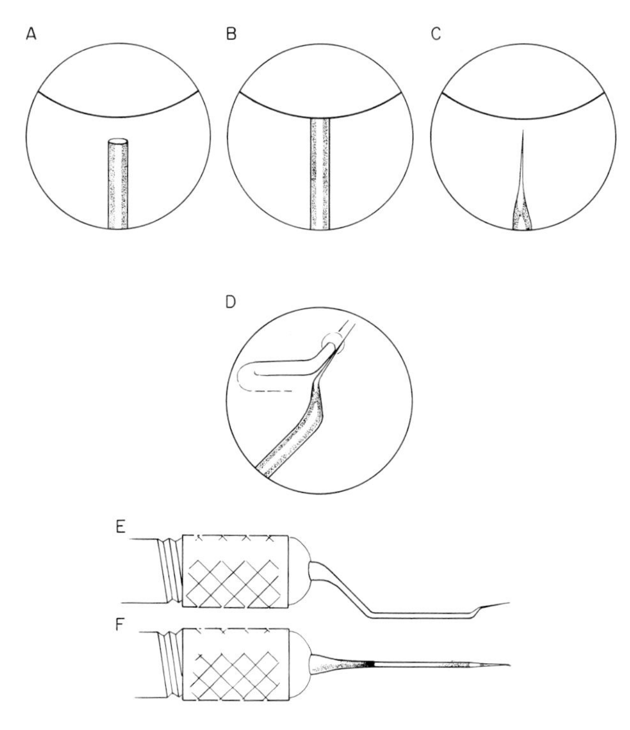

FIG. 5. Diagram of steps in finishing the microtip at high magnification and making the final bend at low magnification. See Section III,C,1 for details.

same procedure as for microneedles (Section III,C,1). Loosen the knurled collar of the pipette holder. By applying pressure to the syringe, hold the pipette in a vertical position and force it up in the holder. Remove the pipette and store it in a vertical position in a micropipette rack or in the holder on the micropositioner.

b. For Aqueous Injections. If an aqueous solution or suspension is to be microinjected, make the capillary, sterilize it with dry heat, and handle it with sterile forceps. Apply heat to the open base of the capillary, creating a partial vacuum, and then insert the base into the solution or suspension which will be drawn into the capillary in sufficient amount for many injections. Carefully wipe the base of the capillary and from there on follow the same procedure throughout as for oil-filled micropipettes, Section III,C,2,a. The only difference is that there will be an aqueous fluid just in front of the silicone oil in the pipette.

The aqueous solutions are less stable than silicone oil, so it is advisable to make the pipettes containing such fluids for microinjection just before they are to be used.

D. Microsurgical Chamber

The microsurgical chamber diagrammed in Fig. 6 is simple in design, inexpensive, and reusable. It is made from a 2 in × 3 in × 1 mm glass microscope slide on which four glass supports, cut from 1 in × 3 in × 1 mm slides, are cemented (Dow Glass Cement). A minimal amount of cement is used, and the supports are weighted to ensure that the cover slip placed on them will be level.

The positions of the supports can be changed to accommodate different sized cover slips or different microscope stages. The 1 inch lengths are not critical but increase the area for adhesion to the base. The positions shown in Fig. 6 are for maximal clearance for the microtools used.

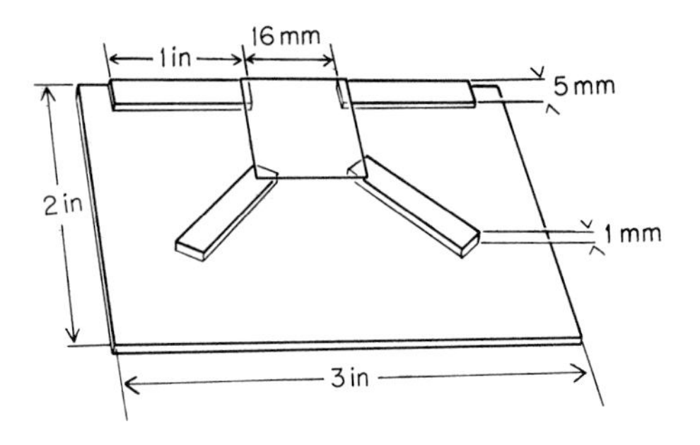

FIG. 6. Diagram with dimensions of microsurgical chamber as described in Section III,D.

The chamber is washed, dried, and autoclaved inverted between two layers of filter paper in the wrapped bottom half of a petri dish; wrap the top half separately. Store the wrapped halves until the chamber is to be used.

E. Micromanipulator

The micromanipulator shown in Fig. 7 consists of a microscope surrounded by six micropositioners of two types and is equipped with two

FIG. 7. The micromanipulator.

microinjection systems. These subunits are attached to a Leitz baseplate (B, Fig. 8) that has been turned around and remachined. The baseplate is shock-mounted to eliminate external vibration.

The micromanipulator is shock-mounted as follows. Place a sheet of dense sponge rubber (28 in × 8 in × 1/4 in) lengthwise on a sturdy table. Place an iron plate (28 in × 18 in × 3/8 in) on the rubber sheet with edges even. Position a 4 inch length of heavy-walled rubber pressure tubing (approximately 1.5 in outer diameter) lengthwise near each corner of the iron plate. The length of tubing near one corner lies perpendicular to the length of tubing at the next corner. After four lengths of tubing are so positioned, rest a second plate, the same size as the first, on the tubing and even with the edges of the lower plate. The micromanipulator rests on the top plate.

The Laborlux II microscope (Leitz) shown in Fig. 7 is equipped with a triocular head with 15× wide-field Galileo oculars and a Leitz photographic attachment (P) with a Leitz camera back (C, Fig. 7). A sliding prism diverts the light for binocular viewing. The revolving nosepiece carries a Leitz diamond scriber, a 10× Leitz, and a 2.5× and 100× Zeiss objective. The set screws that hold the slide clamp on the mechanical stage are replaced with flat-headed screws to provide clearance for the microtools (Fig. 9). A Heine

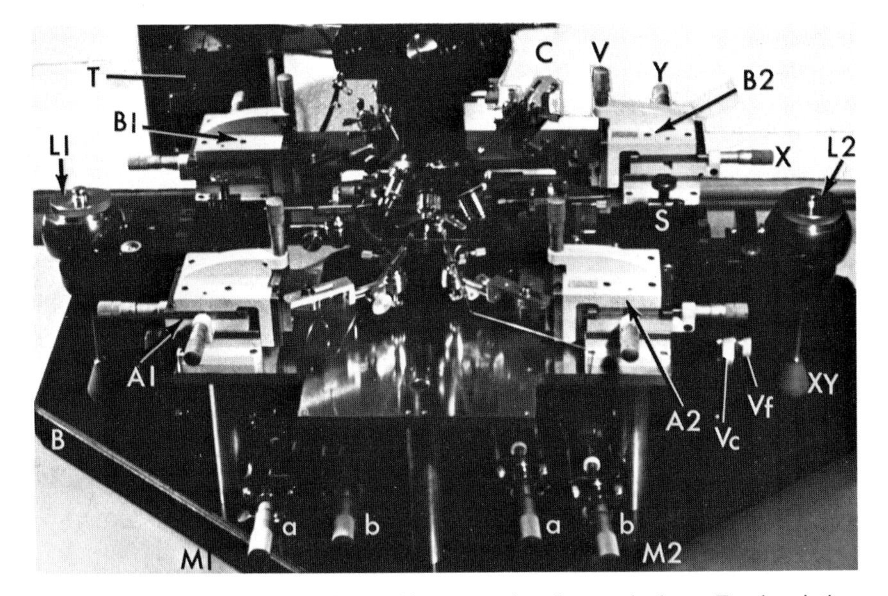

FIG. 8. Closer view of the micropositioners on the micromanipulator. For description, see Section III,E.

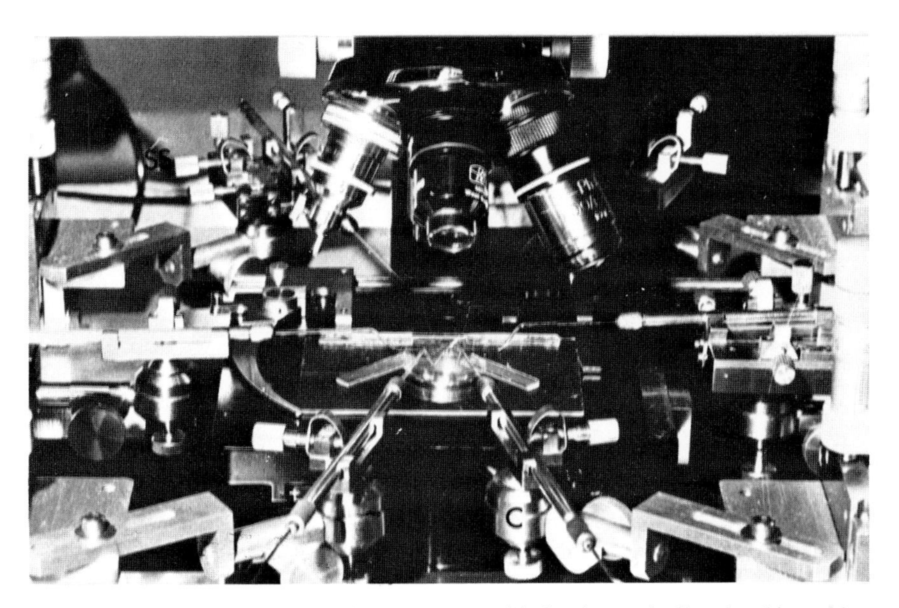

FIG. 9. Close-up view of the microscope stage with six microtools aligned and in position in the chamber with a cover slip roofing it. See Section IV,A.

phase-contrast condenser and a built-in light source (Osram) controlled by a transformer (T, Fig. 7) are used.

On either side of the microscope is a Leitz micropositioner (L1 and L2, Fig. 8) actuated by a single control knob (XY) for X and Y movements in each plane and by a vertical control to change the height of the plane with coaxial coarse (Vc) and fine (Vf) adjustments. In front of the microscope is an aluminum platform supported by three aluminum pillars that screw into the base. On this platform, two Line Tool Co. micropositioners (A1 and A2) are mounted. The horizontal movements and the vertical movement are actuated by the X, Y, and V micrometer heads (Fig. 8) moving against spring-loaded metal slides. Another platform, supported by two aluminum pillars, holds two such micropositioners (B1 and B2) positioned on either side behind the microscope body. All six micropositioners are equipped with adaptors and a Leitz microtool holder clamp (C, Fig. 8), either single or double.

Two microinjection systems (M1 and M2) are mounted on the baseplate (B) as follows. Each Aloe microinjection unit is outfitted with a Hamilton Co. gastight syringe (No. 1002-LT) with a spring-loaded piston actuated by a micrometer head built into the holder. The two syringes are connected in tandem. Fill each syringe with distilled water, making sure that no air bubbles are entrapped. Apply a thin film of silicone grease to the Luer tip of the syringe (M1a, Fig. 8) and attach a Becton-Dickinson three-way valve to it. The outlet at 90° to the right is connected to a Luer-Lok adaptor connected to PE50 polyethylene tubing. The other end of the tubing is fitted to another adaptor attached to the syringe reservoir (M1b, Fig. 8).

Stainless steel capillary tubing connects the injection system (M1a) to the micropipette holder in micropositioner A2. A Luer-Lok adaptor is soldered to the stainless steel tubing and attached to the third outlet of the three-way valve. An adaptor designed for the micropipette holder is soldered to the other end of the tubing. Thus the position of the micropipette and the amount of fluid microinjected can be controlled simultaneously.

Injection system M2 is filled in similar fashion and is used with micropositioner A1 (Fig. 8).

IV. Micromanipulation

A. Alignment Procedure

This procedure is requisite to all operations to be described. It enables the operator to perform one or more of the same or different operations on individual, intact and living, human somatic cells and to retain the option of isolating one or more of these operated cells from neighboring ones for subsequent study. The procedure is as follows. Controls are shown in Fig. 8.

Check the injection systems. Clean the microscope stage and all optical surfaces. Bring the 10× objective into position and raise the microscope body tube to provide clearance for aligning the microtools. Insert a clean 2 in × 3 in microscope slide in the slide clamp to provide a reflecting surface and to protect the condenser. Turn on the microscope light to low intensity.

Align controls of the micropositioner B2 so that the controls X and Y (Fig. 8) are in midpoint of travel (micrometers set at 7). Raise the vertical slide with micrometer (V) to setting 12.

Loosen the knurled collar of the microneedle holder, insert a needle, and tighten the collar. Place the holder in clamp C of micropositioner B2. Slide the holder forward in the clamp and rotate the holder until the large glass bend of the microneedle is perpendicular to the microscope slide, i.e., in a straight line with its reflected image. Advance the holder in this position until the tip of the microtool just intercepts the light from the condenser. Lower the needle (V, Fig. 8) until the thin shaft of the needle, parallel to the microscope slide, is reflected by it. Keeping the large bend perpendicular, use the clamp adjustments (C) to make the thin shaft parallel to its reflection.

Lower the needle with control (V) until the thin shaft just touches the glass slide; then raise it until it just clears the slide. Tighten the set screw (SS, Fig. 9), locking the holder in position.

Lower the 10× objective and, with X and Y (Fig. 8), bring the tip of the needle into position near the center of the field. It appears to enter the field in the lower left quadrant since the optics are reversed. Raise the 10× objective.

With the coarse vertical (Vc) control, raise micropositioner L2 to provide clearance for the needle and move the top slide with control (S, Fig. 8) as far back from the microscope stage as possible. Insert a microneedle in the holder until 1 inch of its thicker shaft extends from the holder. Tighten the collar. Insert the holder in the clamp (C) and advance the holder in the clamp until the microtip of the needle is about 0.5 inch away from the optical axis (center of the condenser). Align the microtool as described above. Tighten the set screw of the clamp.

When these adjustments are complete, lower the body tube until the needle in micropositioner B2 is in focus. Advance the needle in positioner L2 with control S. The needle appears to enter the field from the left. Bring the tip of the needle in L2 to the same focal plane as the tip of the B2 needle. If the L2 needle cannot be lowered sufficiently (detect this as a sudden change in the direction of movement of the needle without changing direction of control movement), raise the needle until it is just free of the slide. Focus on the tip of the L2 needle and bring the tip of the B2 needle into that focal plane with control V. Raise the 10× objective and repeat the above procedures for inserting and aligning the needle in micropositioner B1 and then L1.

With the microinjection systems (M1 and M2) prepared as described in Section III,E, place a small piece of absorbent paper beneath the opening of the pipette holder in positioner A2 (Fig. 8). Loosen the knurled collar of the holder. Draw enough water from syringe M1b into syringe M1a to flush out the stainless steel tubing and holder. Remove the absorbent paper that catches the excess water.

Set the Y control of A2 to micrometer setting 2, and the V control at 12. Leave the X control micrometer at 7. Lift the holder out of the clamp and use micrometer M1a to poise a drop of water on the opening of the holder. Insert the micropipette shaft until it is just past the rubber gasket in the holder and align the micropipette so that the large glass bend will be perpendicular to the glass microscope slide when the holder is reinserted in the clamp. Tighten the collar and replace the holder in the clamp. Slide the holder forward until the tip of the micropipette is close to but higher than the tips of the four needles.

Lower the pipette (V control) and lower the 10× objective until the needles are focused. The X and Y controls will bring the pipette into the upper left quadrant of the 10× field. Position the pipette closer to the microneedles and adjust the pipette height, bringing it to the same focal place as the needles. Make sure that the set screw of the pipette holder clamp (C) is tightened and that there is minimal tension on the rigid tubing when these adjustments are completed. Raise the 10× objective. Repeat this procedure for the other micropipette.

When this part of the procedure is completed, six microtips are arranged in a symmetrical hexagon near the center of the 10× field. All the tips are in one focal plane that is determined by the microtool with the largest microtip angle. Make a final check to see that all set screws on the clamps are tightened. Raise the 10× objective. The microtips and the objective in Fig. 9 are shown in position.

Leave the needles in B1 and B2 (Fig. 8) undisturbed. Raise the others about 2 mm. Hold the microscope slide clamp open and *slide* the glass microscope slide forward and to the right until it can be lifted from the stage. Release the clamp.

Set the stage verniers of the mechanical stage to the coordinates predetermined to center the microsurgical chamber in the optical axis (Fig. 6). Remove the microsurgical chamber from the petri dish and apply a thin film of silicone grease to each of the four supports. Hold the slide clamp open and *slide* the microsurgical chamber to the left and back into the slide holder.

Release the slide clamp while seating the chamber securely. Now lower microtools L1 and L2, then A1 and A2 into the chamber and adjust to the same height (focal plane) as needles B1 and B2 that have been undisturbed.

With a fine-tipped, curved forceps, place a sterile, 18 mm square cover slip on the chamber (Fig. 6). Bring the *lower* surface of the coverslip into focus with the 10× objective, and raise each microneedle to the cover slip and lower it to return it to its original height. This is a test to be sure that each needle tip can reach the cover slip. If any one cannot, remove the cover slip and realign that needle.

Now follow the same procedure to test each pipette. Do not let the fluids to be used for microinjection mix. Should they flow, stop the flow before lowering them. Raise the 10× objective and remove the cover slip. The alignment procedure to this point takes about 20 minutes.

Grasped within their gauze wrappers, assemble two sterile medicine droppers and insert each in a sterile test tube with its rubber bulb resting on the lip of the slanted tube. Fill one dropper with 1–2 ml of sterile silicone oil that has been saturated with growth medium and reinsert in one of the slanted tubes. The oil is saturated by layering sterile oil on sterile medium, shaking it, and leaving it at 37°C for several days prior to use.

Remove the petri dish with the microcultures on cover slips from the incubator. Open the dish just wide enough to insert the other medicine dropper and aspirate 1–2 ml of supernatant growth medium while tilting the dish slightly. Close the dish and reinsert this dropper in the other slanted test tube. With a fine, curved forceps, lift one of the square cover slips with cells on its upper surface from the petri dish and tilt the dish so that there is an adequate layer of medium on the cells. In one quick motion, invert the cover slip so that the cells are now on the lower surface. Holding the cover slip by one edge with forceps, place it on the chamber and align the cover slip, securing it with the silicone grease.

From the oil-filled medicine dropper, place a small drop of silicone oil at the support between needles B2 and L2 and then B1 and L1 at the cover slip corners. From the other medicine dropper, fill the chamber with medium by holding the dropper between the needle L2 and close to the support between that needle and pipette A2. The medium should flow into the chamber with no air bubbles, and the oil drops as placed will coalesce, sealing the back entrance to the chamber. The remaining three entrances are then sealed with a larger drop of oil placed at the center above the microtool at each entrance. The entire chamber is now oil-sealed.

Wipe the upper cell-free surface of the cover slip with a dry, long, cotton-tipped applicator; wash gently with one moistened with lens cleaner; again wipe with a dry applicator; then polish with an applicator wrapped tightly in lens paper. During this process hold the cover slip in position by light pressure at one corner. Check the alignment of the cover slip. The back edge of the cover slip should be even and parallel with the glass supports at the edge of the slide.

Lower the 10× objective, increase the light intensity, and focus on the lower surface of the cover slip. The cells will appear in dark-field illumination because the optics are aligned to produce phase-contrast illumination for the 100× oil immersion objective. To delineate an area for isolation and location of cells to be studied, raise the 10× objective and bring the scriber (offset at the tip to produce a circle about 4 mm diameter) into position. Lower the scriber until the tip of the scriber just depresses the cover slip. By rotating the knurled ring on the scriber, etch a circle on the cover slip. Too much pressure on the cover slip will move or crack it. Too much pressure on the knurled ring may move the scriber from its position on the nosepiece. Scribe the circle once only. After etching the circle, raise the scriber.

Bring the 2.5× objective into position and focus on the circle. The circle should be perfectly scribed, its diameter should be slightly smaller than the diameter of the 2.5× field, and its center mapped by the coordinates of the stage verniers, as previously set (Fig. 9). Raise the 2.5× objective. Place a drop of immersion oil on the cover slip. This part of the procedure takes about 5 minutes.

Slowly lower the 100× objective until contact with the immersion oil is seen by looking from the side. Carefully continue to lower the 100× objective until the cells come into focus. Once in focus, and if the optics are properly aligned, the cells will move in and out of focus without apparent lateral shifting of the image. This is a critical adjustment for precise positioning of the microtools within the cells.

Select a cell in the 100× field that will be recognizable at low (187.5×) magnification and center it in the 100× field. The microtools will not be visible. Switch to the 10× objective to see both the cell selected and the microtools. Focus on the selected cell. Using the X, Y, and XY controls only, bring the microtools in closer to the periphery of the selected cell. The X, Y, and XY controls move the microtools laterally; no controls that move the microtools vertically are used at this point. After this procedure has been completed, the operator can select any cell within the scribed circle on the cover slip by switching back to the 100× objective, center it in the 100× field, and have the microtools ready to perform one or more of the basic microsurgical operations that will be described.

B. Basic Operations

1. Preliminary Observations

Before performing any microsurgery, the following microscopic examination of the living cells in the microsurgical chamber should be made.

Use oil immersion, phase-contrast optics (1875×), and focus on the cells at the level of the lower surface of the cover slip roofing the microsurgical chamber. Thin cell borders of well-spread interphase cells are in focus in this optical plane.

Select an interphase cell, and with the mechanical stage controls, center different parts of that cell in the field. Each time slowly focus down through the cell and then back up to the cover slip. Note the cytoplasmic inclusions at each level of focus and the change in the phase-contrast halo as focus is changed. Note the orientation of the nucleus and cytocenter within the granular cytoplasm. The cytocenter in the living cell appears as a less visibly structured region of the cytoplasm and it lies adjacent to the nucleus, extending partly above or below it as well as out into the granular cytoplasm. Frequently the cytocenter contains small vesicles that are more prominent when the medium becomes more basic.

Repeat this procedure for several different interphase cells. Observe the variation in thickness within the same cell and from cell to cell.

Select a mitotic cell in metaphase and center it in the field at high magnification (1875×). Observe the minute cytoplasmic projections that suspend this cell from the coverslip or from the surfaces of other cells. Focus down through the mitotic cell slowly and back up to the cover slip, noting: (1) cytoplasmic inclusions and how they differ in appearance as compared to interphase cells, (2) chromosome arrangement, (3) cytocenters, and (4) changes in the phase-contrast halo as different levels of the cell are focused.

Compare the shape and thickness of this cell to that of adjacent interphase cells.

Follow one or more cells through cell disjunction (Diacumakos *et al.*, 1972) so that cells at different times in the sequence are easily recognized and distinguished from one another. Observe the changing position of the cytocenters, chromosomes, and cytoplasmic inclusions during the sequence and note the mobilization of the hyaline cell matrix at telophase. Note also the changing shapes and varied orientation of mitotic cells with respect to the cover slip.

By using high magnification with a small depth of focus, the positions of intracellular components can be determined more accurately, thus making the positioning of the microtools more precise.

The working distance of the 100× objective is such that it is not possible to focus on the needles positioned below the mitotic cells without breaking the cover slip. Should it be necessary for any reason to relocate one or more microtools in the chamber, change to lower magnification (187.5× using the 10× objective.

For operations described in the following Sections 2–6, microtools should be aligned as described in Section III, A, below the level of mitotic cells in the chamber.

2. Chromosome Extraction

For chromosome extraction, select any human somatic cell in late prophase or in prometaphase. Center it in the high magnification (1875×) field and record its position by using the vernier stage settings as coordinates.

Focus through the cell and select a pair of chromosomes close to the cell surface and accessible to a microneedle. Raise the needle to the same focal plane (height) beside the cell and then advance the needle into the cell until it touches that pair of chromosomes.

Slowly withdraw the needle, bringing that pair of chromosomes through the cell surface (Fig. 10). Continue moving the needle away, out of the field, and down from the cell. The other chromosome pairs within the cell should follow in chainlike fashion since they are interconnected and still connected to the first pair removed.

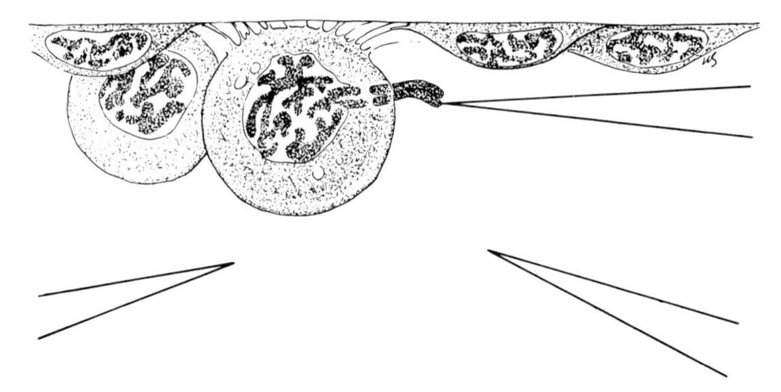

FIG. 10. Artist's concept of chromosomes being extracted from a mitotic cell in prometaphase as it hangs suspended from the cover slip roofing the microsurgical chamber.

The chromosomes will remain attached to the tip of the microneedle that is now out of view. Bring the needle back into the field and up to a clear, cell-free area of the cover slip brought into position by moving the chamber. With an additional needle, lift the chromosomes and attach one or more pairs to the cover slip. Record their position. Lower the needles before moving the chamber to bring other cells into position.

3. Chromosome Expulsion

Since chromosomes can be expelled only by human fibroblasts of normal origin, for chromosome expulsion, select a cell in metaphase, one that is oriented on the cover slip so that one pole of the cell lies in the path of a micropipette, 2–3 μm in diameter, containing silicone oil (50 cs, viscosity). Center the cell and record its position. Raise the pipette to the same focal plane as that of the cell diameter and close to the cell surface near one pole.

Form a fresh drop of silicone oil at the tip of the pipette until the diameter of the drop at the cell surface is about equal to the diameter of the cell. Move the pipette away, leaving the oil drop at the cell surface, and then lower the pipette. Within a minute or two, the chromosomes will move toward the base of the oil drop and will emerge through the cell surface at that site. The chromosomes can be picked up on the tip of a needle that is being raised as they are moving out of the cell. If the oil drop interferes with the visualization, flick it off of the cell with another needle. Lower that needle and the needle with the adherent chromosome aggregate.

Move the chamber bringing a clear area of the cover slip into the field. Raise the needle with the chromosome aggregate and touch the chromosomes to the cover slip. The chromosomes will adhere and the entire complement of the chromosomes of the mitotic cells is available for chromosome dissection or for transplantation. Record the position of the chromosomes. Lower the microneedle.

4. Intracytoplasmic Microinjection

Micropipettes, 1–2 μm in diameter at the orifice, should be used. Silicone oil is the nonaqueous fluid recommended for preliminary injections since the oil is not miscible with the cytoplasm, is not digestible, and can be seen easily.

Select and center an interphase cell. Record its position. Raise the pipette to the level just below the cell surface and near the cytoplasmic region to be injected. Continue to raise the pipette until deformation of the cell surface is seen and advance the pipette horizontally into the cytoplasm. Begin to apply pressure to the syringe by rotating the micrometer head. Maintain the position of the pipette, and, as soon as the oil begins to emerge, rapidly rotate the micrometer head in the opposite direction to decrease the pressure and to control the volume injected. When the desired volume (size) drop has been injected, stop the flow and withdraw the pipette tracing the same path as for its entry until it is out of and below the cell. This procedure leaves the oil drop in the cytoplasm. The diameter of the spherical drop can be measured with an ocular micrometer calibrated for this magnification and used to calculate the volume of oil injected. Different sizes (volumes) of oil drops behave differently. Note the difference in the optical properties of intra- versus extracellular oil drops.

When this operation has been mastered, aqueous fluids can be used. These flow and diffuse locally within the granular cytoplasm but form discrete spheres within the cytocenter. Large amounts of aqueous fluids injected into an interphase cell will disperse throughout the cytoplasm and then disrupt the cell.

A mitotic cell may be injected with nonaqueous or aqueous fluids. Both types of fluids form discrete spheres within the cytoplasm. Aqueous fluids will disappear in time, but nonaqueous fluids will be moved to the periphery of the cells by the kinetics of cell disjunction. Later in this sequence, oil drops injected at the site of apparent constriction move to one cell or the other.

5. Intranuclear Microinjection

An interphase cell of normal or abnormal origin is suitable for this operation if the longer axis of the nucleus is parallel with that of the pipette and if there is no visible cytoplasm between the nucleus and the lower surface of the cell. Center the nucleus in the field. Raise the pipette until it is just beneath the cell surface and in a position such that the tip of the pipette lies beneath but within the nuclear boundary. Raise the tip of the pipette until it just deforms the nuclear membrane, and then advance the pipette in the same plane into the nucleus.

Slowly increase the pressure on the syringe. As it increases, maintain the position of the pipette, and, as soon as the silicone oil begins to flow, rapidly reduce the pressure. When the injection is completed, withdraw the pipette slowly downward and out of the nucleus following the same path as for its entry.

Aqueous injections produce white areas or regions within the nucleoplasm that gradually disappear. Larger volumes of aqueous fluids will first distend and then burst through the nucleus into the cytoplasm. In either case, introduction of the micropipette produces a transitory darkening of the nuclear inclusions such as nucleoli. This is not indicative of cell damage.

6. Cell Fusion

This operation can be performed on telophase cells of the same or different types or species without mixing the cell populations. Locate a pair of cells in telophase and center the pair in the 1875× field. Record the position of this pair. Locate another pair of cells in telophase. Gently detach this pair from the cover slip or from other cells using one or more microneedles without impaling the cells. Lower this pair sufficiently to avoid hitting other mitotic cells.

Center the first pair again by moving the chamber very slowly back to the recorded position so as not to dislodge the second pair on the needle. Raise the second pair adjacent to the first pair and reattach it to the cover slip with gentle pressure. Pivot the second pair into position, using one or more clean microneedles as necessary, so that there is contact between the pairs by only one cell from each.

Raise another, abruptly tapered needle and advance it in the same plane so that it touches lightly the surface of one to the contiguous cells. A small amount of hyaline cell matrix will protrude from the cell surface at this site and will adhere to the tip of the needle. Gently guide the matrix on the microtip over to the surface of the contiguous cell and insert the needle just far enough to fuse the matrix from the two cells, visible by the formation of a hyaline bridge. This is the moment of cell fusion. Carefully withdraw and lower the needle so that the fused cells do not separate.

These cells, undisturbed, will continue fusing and will spread on the cover slip. After about 30 minutes, three individual cells are seen—one hybrid cell formed by the fusion of two individual cells, and two unfused cells, each the presumably identical sister of one of the cells forming the hybrid cell. These three cells can now be isolated.

Fusion of cells of two types or species by microsurgery is accomplished in exactly the same way. To avoid mixing or coculturing two cell populations, grow each type on triangular cover slips (Fig. 1) and use two microcultures on triangular cover slips, one of each type, to roof the microsurgical chamber. Then, with the microtools retracted and raised except for the needles in micropositioners B1 and B2, insert the filled chamber in the microscope slide clamp and reposition the needles.

C. Other Microoperations

All other microoperations, including the mobilization of hyaline cell matrix during cell disjunction, intracellular chromosome displacement and transplantation, or cutting experiments are either modifications or combinations of the basic operations here described.

D. Cell Isolation

Selected, operated somatic cells are isolated as follows. Locate the operated cell by the recorded position and center it in the 1875× field. Examine all the cells that are in contact with that cell. Using the most accessible needle for each, kill the neighboring cells by tearing them, but leave them in position. As the cells die, they retract their connections to the operated cell and after 10–15 minutes the dead cells can be detached from the cover slip without disturbing the operated cell.

Once an area around the operated cell has been cleared, use the needles to sweep from the cover slip the rest of the cells within the area of the etched circle and for a short distance beyond. These cells fall to the floor of the chamber.

Before proceeding with operated mitotic cells, wait until the mitotic cells have spread on the cover slip.

Now remove the lowered microtools from the chamber. Remove the chamber and place it in a sterile hood. Clean the upper surface of the cover slip with ethanol, using cotton-tipped applicators. With a fine-tipped forceps, lift the cover slip from the chamber in one quick motion, turning it cell-side up. Place the cover slip in a sterile, plastic petri dish containing growth medium. Using a stereoscopic dissecting microscope, locate the etched circle on the cover slip. Using a straight-tipped forceps and a diamond scriber, hold the cover slip on the floor of the dish and score from the edge of the circle out to the edges of the cover slip. Slight pressure at the edge of the circle will release the round glass fragment.

With two pairs of forceps lift the glass fragment from this dish and immerse it in another dish containing growth medium. Hold the fragment in one pair of forceps and swirl the medium in the dish to wash the fragment.

Leaving the fragment in this dish temporarily, assemble the Sykes–Moore chamber (Sykes and Moore, 1959, 1960) prepared and autoclaved in the same way as are microsurgical chambers, so that the lower portion of the Sykes–Moore chamber contains a round, 1 inch diameter cover slip floor and the silicone rubber gasket spacer. Lift the fragment and place it cell-side up in the middle of the floor of the chamber. Immediately place the second round cover slip on the rubber gasket, and thread the metal ring, the top portion of the chamber, into position. Tighten the chamber assembly with the wrench provided.

Insert a 26-gauge needle in one port in the chamber wall and through the gasket so that air is displaced as the chamber is filled from a 5 ml syringe via another 26-gauge needle. Through this needle flows the growth medium enriched with 20% (v/v) fetal bovine serum that is refiltered through a Swinnex GS filter assembly (Millipore Corp.) before flowing into the chamber.

Since the fragment is resting on the floor of the chamber, use care in transporting the chamber. The medium in it is changed as necessary following the procedure described by Sykes and Moore (1959, 1960).

V. Photography

Cells may be photographed daily to follow the formation of clonal aggregates and ultimately clones, i.e., clonal aggregates of 50 or more cells all derived from a solitary, operated somatic cell. Useful data are acquired by photographing cells before, during, and after microoperations.

Panatomic-X, 35 mm film (Kodak) is used in the camera (C, Fig. 7) connected to the photographic attachment (P, Fig. 7) shown set at B for 30 seconds exposure, the optimal time for photographing microoperations. The cable releases (Ra and Rb, Fig. 7) control the attachment shutter and prism, and the camera shutter (set at B), normally left open until the exposure is made, is controlled by cable release (Rc, Fig. 7).

Cinematographic records in real time or time-lapse cinematographic records also can be made with appropriate equipment.

VI. Concluding Remarks

Some of the procedures in this communication concerning the micromanipulation of individual cells have been reported heretofore in less detail (Diacumakos *et al.*, 1970, 1971a,b, 1972; Diacumakos and Tatum, 1972); others are reported here for the first time. These procedures provide a means of studying and altering individual, living human somatic cells in a precise and defined manner from a defined starting point.

The microsurgical approach has been evolved as the logical choice for experiments to study an individual living cell in culture from the time of its formation as a single living cell throughout its lifetime; to learn to recognize a cell at different times in its cycle for its similarity to and variation with other cell(s) of the same type as well as different types and examine these similarities and variations; to determine the effects of operations on the cell's means of and role in producing daughter cells and the effect on the daughter cells after one or more divisions; and ultimately to recognize the basis in the individual living cell for the heritable or nonheritable sequelae.

Thus, micromanipulation differs from experiments that deal with large numbers of cells to produce an effect in a minute proportion of the population, a proportion detectable only retrospectively and then considered by criteria or methods devised for a limited situation without a clearly defined base or starting point.

The methods reported here are being applied to study how such processes as biochemical alteration associated with birth defects, malignant transformation, and immunological changes may occur. Extrapolating from these applications, methods for altering these processes may be developed and applied to therapy.

ACKNOWLEDGMENTS

It is a privilege to acknowledge the unique opportunity and inspiration provided by Dr. Frank L. Horsfall, Jr., the late President and Director of the Sloan-Kettering Institute for Cancer Research, to develop and implement the methodology described here. I am grateful for the interest shown by Dr. Leo Wade, Chief of the Division of Special Studies at the Sloan-Kettering Institute. I am grateful to Dr. Pauline Pecora, my colleague in this division, and to my co-workers there, especially Scott Holland for sharing the challenge and enjoyment of this work so competently.

I am most grateful to Professor Edward L. Tatum, Head of the Laboratory of Biochemical Genetics at Rockefeller University, for the opportunity to continue this work based on his classic experiments and applied to human somatic cells and to have the benefit of his close association.

REFERENCES

Bablanian, R., Eggers, H., and Tamm, I. (1965). *Virology* **26**, 100.

Chambers, R., and Kopac, M. J. (1950). *In* "McClung's Handbook of Microscopical Techniques," p. 492. Harper (Hoeber), New York.

Diacumakos, E. G., and Tatum, E. L. (1972). *Proc. Nat. Acad. Sci. U.S.* **69**, 2959.

Diacumakos, E. G., Holland, S., and Pecora, P. (1970). *Proc. Nat. Acad. Sci. U.S.* **65**, 911.

Diacumakos, E. G., Holland, S., and Pecora, P. (1971a). *Nature (London)* **232**, 28.

Diacumakos, E. G., Holland, S., and Pecora, P. (1971b). *Nature (London)* **232**, 33.

Diacumakos, E. G., Holland, S., and Pecora, P. (1972). *Int. Rev. Cytol.* **33**, 27.

Dulbecco, R., and Vogt, M. (1954). *J. Exp. Med.* **99**, 167.

Parker, R. C. (1961). "Methods of Tissue Culture," 3rd ed. Harper, New York.

Sykes, J. A., and Moore, E. B. (1959). *Proc. Soc. Exp. Biol. Med.* **100**, 125.

Sykes, J. A., and Moore, E. B. (1960). *Tex. Rep. Biol. Med.* **18**, 288.

Chapter 16

Tissue Culture of Avian Hematopoietic Cells[1]

C. MOSCOVICI AND M. G. MOSCOVICI

Tumor Virology Laboratory, Veterans Administration Hospital, and Department of Immunology and Medical Microbiology, and Department of Pathology, University of Florida College of Medicine, Gainesville, Florida

I. Introduction

Avian tissues either from embryos or from adult birds have been used extensively in studies of embryology, developmental biology, and virology. In recent years, tissue culture of chicken cells has been a major tool for studying the interaction of several oncogenic viruses with their host. Various methods for growing chick embryo fibroblasts and their response to infection

[1]This work was supported in part by the Veterans Administration and by grant CA10697 from the National Cancer Institute.

to Rous sarcoma virus (RSV) have been published (Temin and Rubin, 1958; Vogt, 1969).

The purpose of this report is to describe a method for growing cells derived from hematopoietic tissues of the chicken embryo. This method has been employed for studying some avian tumor viruses, in particular avian myeloblastosis virus (AMV) (Baluda and Goetz, 1961). This virus has the unique property of inducing morphological changes *in vitro* in certain target cells derived from mesodermal tissues. This paper focuses on a description of how to grow cells from the yolk sac, which is the largest embryonic organ with hematopoietic activity. The method is also applicable to other avian species.

II. Materials

A. Chick Embryos

Fertile eggs can be obtained from various commercial flocks. Practically all flocks have a low incidence of congenital leukosis virus infection varying from 1 to 5%. Embryos can be used from the following breeders in the United States: Kimber, Farm Niles, California; Heisdorf and Nelson, Redmond, Washington; Hy-Line Poultry Farms, Des Moines, Iowa; and SPAFAS, Inc., Norwich, Connecticut. For a complete list of egg distributors in the United States, consult "Hatcheries and Dealers Participating in the National Poultry Improvement Plan," compiled by the United States Department of Agriculture, Beltsville, Maryland.

White Leghorn is the breed most frequently used; approximately half of the embryos from these lines yield cell cultures susceptible to subgroups A, B, and C; the other half are selectively resistant to subgroup B (Hanafusa, 1965; Vogt, 1965).

B. Sera

Calf Serum

Great care should be taken in the choice of the serum to be used for cell growth and focus assays. Not all lots are satisfactory, and pretesting of different samples for toxicity, plating efficiency, and transforming activity is recommended. Calf serum can be obtained from Colorado Serum, Denver, Colorado; Flow Lab, Rockville, Maryland; Microbiological Association, Bethesda, Maryland; GIBCO, Grand Island, New York.

Calf serum can be used inactivated. Storage at −20°C is recommended. It should be used within 2 months after slaughter. Sera should not be frozen and thawed more than once if possible. This treatment is known to be damaging to lipoproteins, some of which may be essential for successful cell culture.

Chicken Serum

Chicken serum can also be purchased from commercial suppliers. It is used in the growth of primary and secondary yolk sac cultures. It should be tested for the presence of RSV and AMV antibodies, which should be absent. Chicken sera must be inactivated at 60°C for 1 hour to eliminate any infectious avian leukosis viruses that may be present. Storage precautions are the same as those recommended for calf serum.

Fetal Calf Serum

Although fetal calf serum promotes attachment and stretching activity of cells, it has been shown to inhibit cellular transformation by AMV (Moscovici, unpublished) and therefore should not be used.

C. Growth Media and Reagent Solutions

Primary yolk sac cultures can be grown with medium 199 (Morgan *et al.*, 1951), F10 (Ham, 1963), or a modified Eagle's medium (Eagle, 1955) as described by Baluda and Goetz (1961). This modified medium has been used in our laboratory and proved to be the most satisfactory for growing hematopoietic cell cultures. For convenience, this medium is called BT-88.

The medium consists of a basal Eagle medium with four times the concentration of amino acids and vitamins, 2.2 gm of bicarbonate per liter, and the additional ingredients listed as follows:

Ingredient	Mg/liter	Ingredient	Mg/liter
Aspartic acid	30	Serine	30
Glutamic acid	60	Adenosine	20
Glutamine	584	Guanosine	1.5
Glycine	50	Phenol Red	16
Hydroxyproline	10	Penicillin	10^5/unit
Proline	40	Streptomycin	25

The pH is adjusted to 7.4 with 10 *N* NaOH. The medium is filtered and stored at 4°C.

BT-88 nutrient medium for growing primary and secondary yolk sac cells is as follows:

BT-88 Basal medium	80 ml
Bacto-tryptose phosphate	10 ml
Calf serum	5 ml
Chicken serum (heat inactivated)	5 ml

Saline Solutions

This buffered saline consists of NaCl, 8 gm/liter; KCl, 0.37 gm/liter; Na_2HPO_4, 0.1 gm/liter; glucose, 1 gm/liter; tris(hydroxymethyl)aminomethane (Tris), 3 gm/liter; 10 *N* HCl, 1.8 ml/liter. Sterilize by filtration.

Trypsin

Trypsin is prepared as a 0.25% solution in Tris-buffered saline from trypsin 1:250 or trypsin 1:300 (Difco Lab, Detroit, Michigan). Sterilize by filtration and store in small aliquots at −20°C.

Bacto-Tryptose Phosphate Broth

This broth is prepared according to the manufacturer's instructions (Difco Lab) and sterilized by autoclaving.

Purified Agar

Purified agar (Difco Lab) is prepared in a 2.5% concentration and autoclaved. It is used for the AMV assay.

Sodium Bicarbonate

Sodium bicarbonate 2.8% solution in distilled water. Sterilize by filtration.

Complex Media

(1) Medium for growing myeloblasts in suspension

Myeloblasts from AMV transformed yolk sac cultures or directly from a leukemic bird can be grown in suspension for several weeks. Myeloblasts are transferred into a small Erlenmeyer flask (50 ml or less) at a concentration of 20 to 40 × 10^6 myeloblasts/ml. The flasks are placed on a shaker rotator Model G-2 (New Brunswick Scientific Co., New Brunswick New Jersey) and set at 50 to 60 cycles per minute. Composition of the medium is as follows:

199 1X	40.0 ml
Chick serum (heat inactivated)	57.8 ml
Glucose	2.0 ml
Folic acid	0.2 ml
	100.0 ml

(2) Freezing medium

Unlike chick embryo fibroblasts, normal yolk sac cells cannot be stored at −70°C. Transformed myeloblasts, however, can be stored at −70°C in small vials for several months. Myeloblasts are distributed in 15.5 × 50 mm vials at a concentration of 10^7 cells per milliliter of freezing medium. The composition of the freezing medium is as follows:

BT-88 Complete	55 ml
Calf serum	25 ml
DMSO	20 ml

(3) Antibiotics

Penicillin and streptomycin are added to cell media and salt solutions at concentration of 10^5 units/liter and 25 mg/liter, respectively.

D. Indispensable Equipment

1. Falcon plastic tissue culture dishes numbered 3003, 3002, 3001 (Falcon Plastics, Los Angeles, California).
2. Falcon plastic centrifuge tubes Nos. 2095, 2070.
3. Stainless steel knives No. 3 with blades No. 15 (Stanisharps, ASR Medical Industries).
4. A clinical centrifuge (Model CL) (International Equipment Co.).
5. A multiple stirring unit for preparing primary cultures from several chicken embryos (Lab-Line Co., Melrose Park, Illinois).
6. Inverted microscope, preferably Leitz with large circular stage, 17.5 cm in diameter. For counting foci, a glass plate with 2 mm^2 grid is necessary. Grid plates can be purchased from Technical Instrument Co., San Francisco, California.
7. If no Coulter Counter is available, cells can be counted with a hemacytometer Spencer, Neubauer 1/10 mm deep.

III. Procedures

A. Primary Yolk Sac Cultures

Twelve- to 14-day-old embryos are used. The shell of the egg is rinsed with a mild solution of 7× detergent and again rinsed with 70% ethanol. The egg is opened on the side of the air sac by punching the shell with the back of a heavy sterile forceps. The same forceps can be used to completely remove the shell above the air sac region and expose the shell membrane. The membrane is peeled off, and the whole embryo can be removed by quickly turning the egg upside down into a large 150-mm petri dish. A gentle tapping on the other end of the egg will facilitate the exit of the embryo. At this point the embryo is removed from the yolk sac, decapitated, eviscerated, and deposited into a beaker containing Tris buffer. The embryo will later be used for determining its phenotype. For further details on the preparation of whole chick embryo cell cultures, see Vogt (1969).

The entire yolk sac is washed several times with Tris-buffered saline to remove as much as possible of its fat content. The washed yolk sac is then placed in a glass petri dish to be cut in small pieces of 1 to 2 mm with dissecting blades No. 15. The minced pieces are collected into a 50-ml beaker containing 5 ml of prewarmed 0.25% trypsin. A 1-inch Teflon-covered stirring bar is added, and the mince is stirred in a magnetic stirrer at a gentle speed for 15 minutes. Care should be taken to maintain a constant temperature of 37°C while the tissue is being trypsinized. This can be done by placing the beaker into the bottom of a plastic dish, where warm water is constantly added. One extraction is sufficient. The trypsinized cells are collected into a plastic centrifuge tube containing chilled growth medium. The suspension is centrifuged at 1000 rpm for 1.5 minutes. The supernatant is discarded, and the pellet is gently resuspended in 10 ml of growth medium. A small sample is diluted 1:10 with growth medium and counted in a hemacytometer. Erythrocytes and cell fragments are not counted. The yield from individual yolk sac varies from 150 to 250 $\times$ 10^6 viable cells. The cells are seeded in 100-mm petri dishes at a concentration of 50 to 60 $\times$ 10^6 cells per dish. Cells are dispersed in 7 ml of growth media and placed in the incubator for 7 to 9 hours. At that time the cell suspension is removed and placed in another 100-mm petri dish, and the original dish is fed again with 10 ml of growth medium. This procedure allows a major part of fibroblast "contaminant" cells to settle down on the bottom of the dish and hence to yield a relatively fibroblast-free culture of yolk sac cells. A relatively homogeneous population of yolk sac cells is then obtained in 4 to 5 days from the 7 to 9 hours of seeding. At that time the supernatant is removed and replaced with 10 ml of fresh growth medium.

After 2 days the cells form a semiconfluent monolayer, at which time the cells are ready to be transferred.

B. Secondary Yolk Sac Cultures

The fluid medium is removed by suction from primary cultures. The cell monolayer is then washed once with Tris-buffered saline and treated for approximately 3 minutes with 2 ml of 0.025% prewarmed trypsin. With a 2-ml pipette (shortie from Bellco), cells are gently detached from the dish. A pool of all the cells derived from a single embryo is made, and the cell suspension is diluted 1:1 with chilled growth medium to stop the action of the trypsin. No centrifugation is necessary. A cell count is obtained with a hemacytometer or an electronic cell counter. If 60 mm Falcon dishes are being used, cells are seeded at a concentration of 3.5 to 4.0 $\times$ 10^5 cells/dish. If 35-mm Falcon dishes are being used, cells are seeded at a concentration of 1.2 to 1.4 $\times$ 10^5 cells/dish. If Lab-Tek Chamber/Slides are being used, cells are seeded at a concentration of 3.5 to 4.0 $\times$ 10^4 cells/chamber. If Falcon No. 3040 Microtest II are being used, cells are seeded at a concentration of 2.0 $\times$ 10^4 cells/well.

The same growth media used for primary cultures is used for secondary yolk sac cultures. Attachment is quite rapid and in 2 to 3 hours is complete. The cells at this point are ready to be infected with the virus.

C. Infection and AMV Assay

Secondary cultures of yolk sac cultures may be inoculated with AMV within 4 hours after seeding. The cultures can be used for viral transforming activity up to 1 week after seeding without reducing the efficiency of the assay.

Transformation of cells by AMV can be assayed by two methods: (a) end-point dilutions of the virus where titers are expressed in 50% transforming doses per milliliter (TD 50/ml) calculated according to the method of Reed and Muench (1938); and (b) scoring transformed foci under an agar overlay. The latter, more accurate method has been described by Moscovici (1967). Briefly, the method consists of infecting the cells with 2-fold virus dilutions. The inoculum is added to 35 mm dishes and adsorbed for 4 hours at 37°C. Maximum adsorption is obtained if infected dishes contain a small volume of fluid, enough to cover the cells. After removing the virus suspension by suction, the plates are overlayered with 2 ml of an agar overlay consisting of the following components:

BT-88 2X	20 ml
BT-88 1X	40 ml
Tryptose phosphate broth	20 ml
Calf Serum	4 ml
Chick serum (heat inactivated)	4 ml
Purified Agar Bacto (2.5%)	23 ml

The overlay is allowed to harden at room temperature for 5 to 10 minutes and the dishes are returned to the incubator. After 6 days, 2 ml of an additional overlay is added to the first and, after hardening, the culture is returned to the incubator for another 4 to 5 days. Avian myeloblastosis virus foci are usually detected at day 11 to day 12 after infection, but counts are preferably made after 2 weeks from infection. Each culture dish is placed on a plain glass plate with a 2 mm^2 grid which sits on the stage of the inverted microscope. The grid can be moved by hand, and the total number of foci counted. Foci are counted with a 2.5× objective and 10× ocular.

Details of focus structure can be distinguished with a higher magnification. An example of a characteristic AMV focus is given in Fig. 7, where individual myeloblasts can be seen tightly packed, forming a large refractile area of transformed cells.

D. Other Recommendations

Several factors can affect the growth of hematopoietic cells and at the same time AMV focus formation. For example, the presence of an excessive number of chick fibroblast cells during the preparation of yolk sac cultures may cause pH changes or cause the cultures during secondary transfer to become excessively crowded, resulting in peeling off of the monolayer. This inconvenience may be overcome by reseeding the cell suspension, as described previously. However, variations occur from egg to egg, and the presence of fibroblast cells remains unavoidable.

For small-scale experiments, bone marrow cultures can be used. Very few cultures, however, can be obtained if individual bone marrows are employed; yolk sac tissue remains the prime source of cells for large-scale experiments. Cells which are genetically resistant to avian leukosis viruses are unsuitable for the AMV assay (Crittenden *et al.*, 1963). Cells congenitally infected with avian leukosis virus do not seem to interfer with AMV transformation, as has been shown with the RSV assay (Rubin, 1960). For focus cloning experiments, leukosis-free cultures are recommended.

Among genetically and physiologically susceptible chick embryo yolk sac cultures one may find a variation in susceptibility to viral transformation. It has been reported that the addition of polycations to the medium has an effect on the plating efficiency of Rous sarcoma virus of subgroups B and C (Toyoshima and Vogt, 1969). In the case of AMV, the addition of polybrene (2 μg/ml of medium) to the cultures at time of infection show a relative enhancement in focus formation, although not as beneficial as reported for RSV.

Appendix

A series of pictures (Fig. 1–8, pp. 321–328) is provided to illustrate the morphological changes in the chicken hematopoietic tissue and in cultures cells after infection with avian myeloblastosis virus.

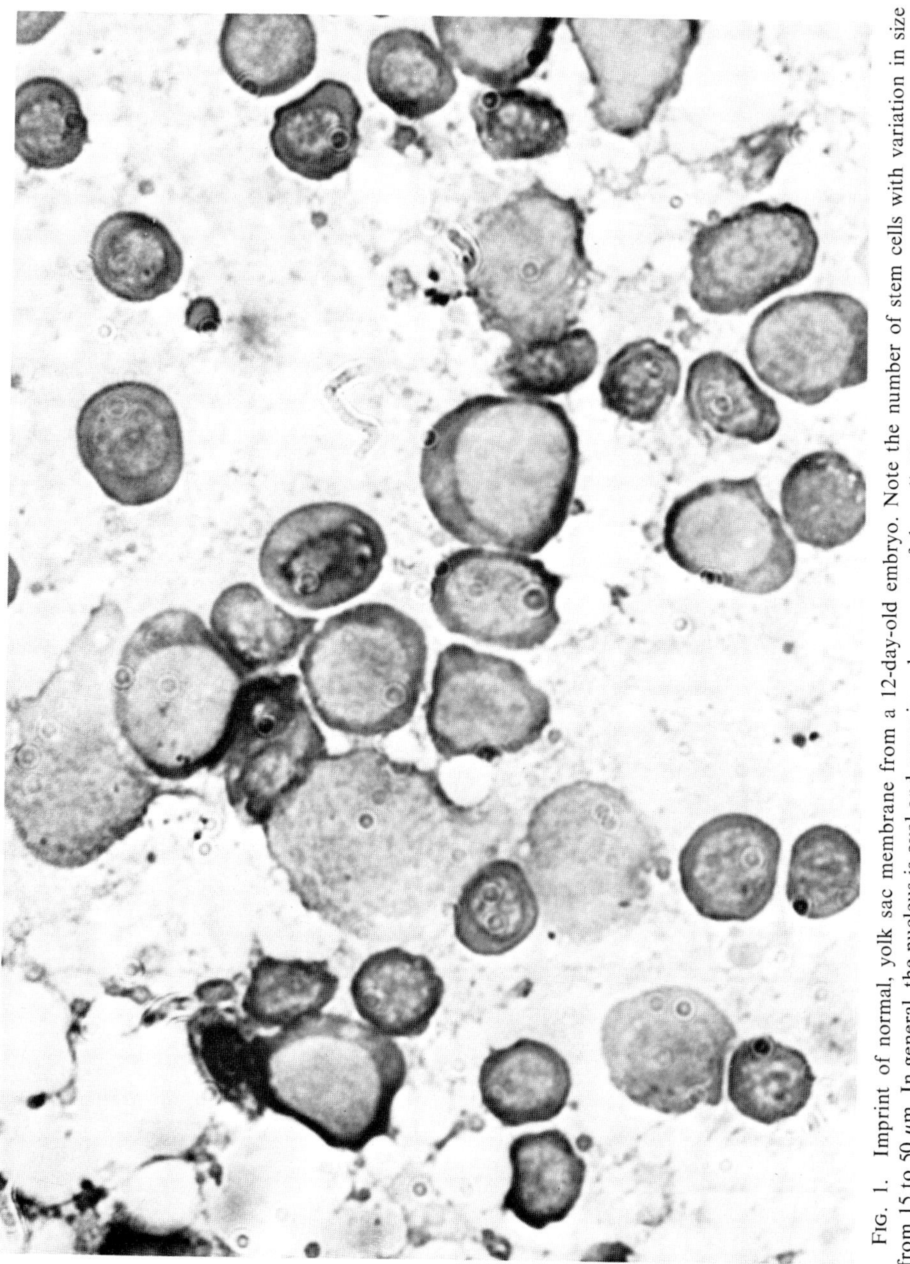

FIG. 1. Imprint of normal, yolk sac membrane from a 12-day-old embryo. Note the number of stem cells with variation in size from 15 to 50 μm. In general, the nucleus is oval and occupies a large part of the cell. The cytoplasm is more basophilic, and the nucleus stains a pale bluish color. Nucleoli are rare. Wright's stain. $\times 1300$.

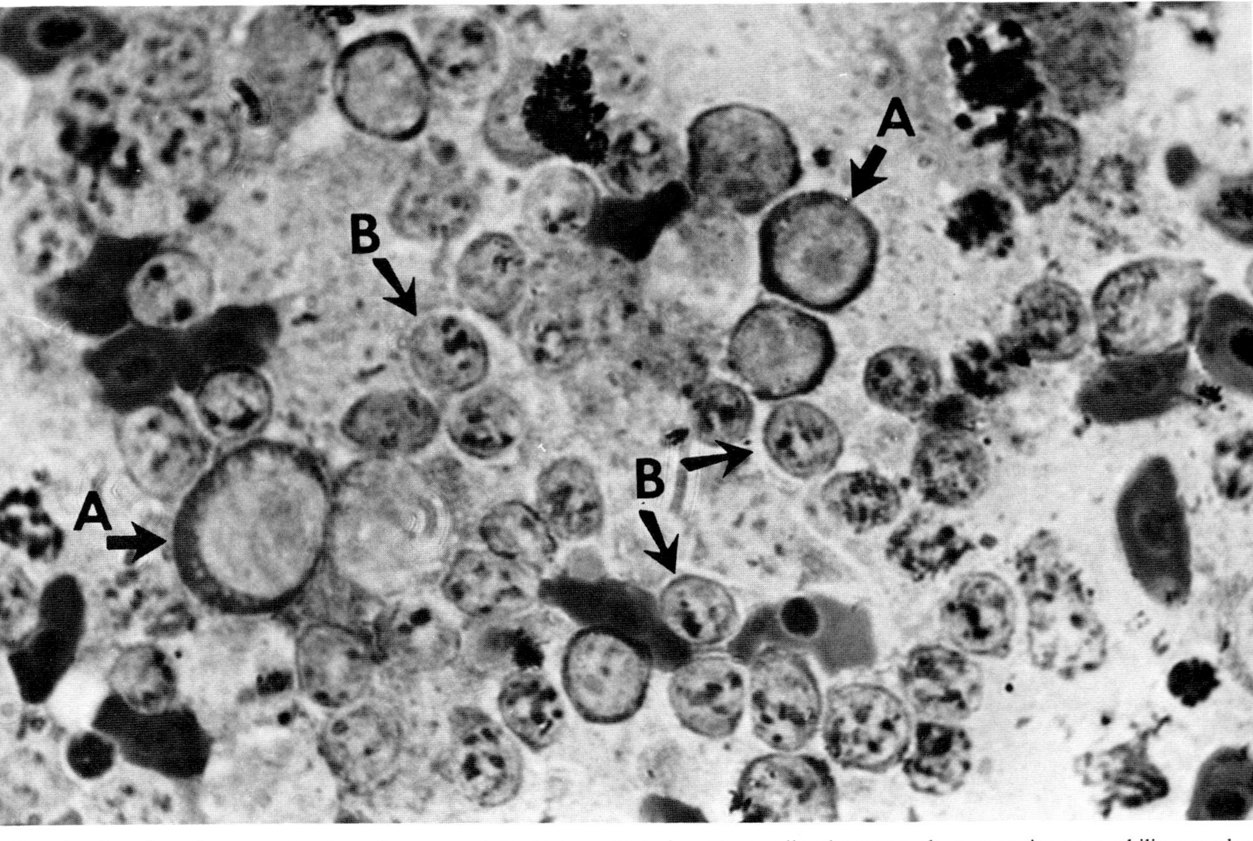

FIG. 2. Imprint of a normal spleen from an adult chicken. Note immature cells whose cytoplasm contains azurophilic granules, probably promyelocytes (arrows A). In some cells, the nucleolus is visible. Few erythrocytes are present. The field contains many small lymphocytes (arrows B). Wright's stain. ×1300.

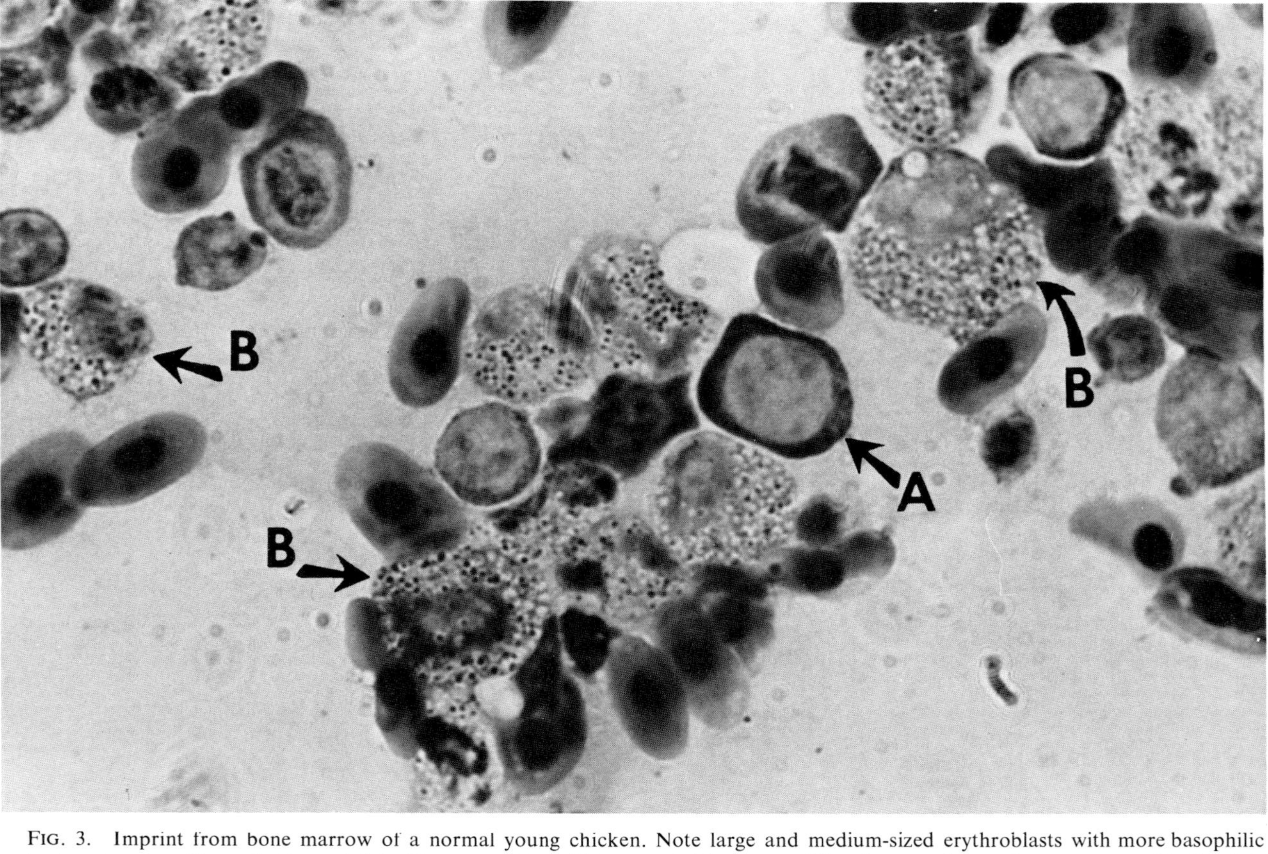

FIG. 3. Imprint from bone marrow of a normal young chicken. Note large and medium-sized erythroblasts with more basophilic cytoplasm (arrows A); nucleoli are visible. The rest of the field is occupied by eosinophil and basophil granulocytes (arrows B). Wright's stain. ×1300.

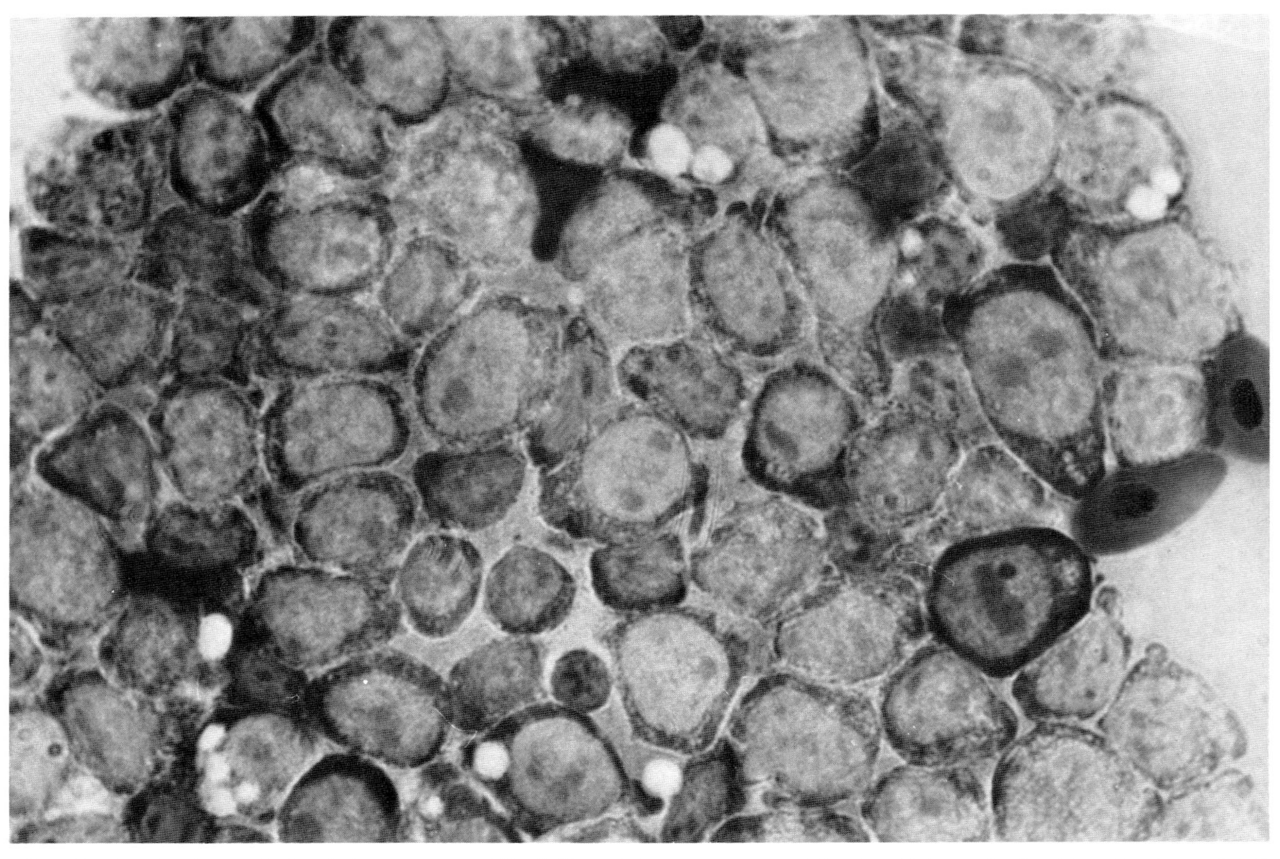

FIG. 4. Imprint from a leukemic bone marrow. The field contains numerous myeloblasts, which vary in size. Granulocytes are totally absent. All myeloblasts contain nucleoli. Wright's stain. $\times 1300$.

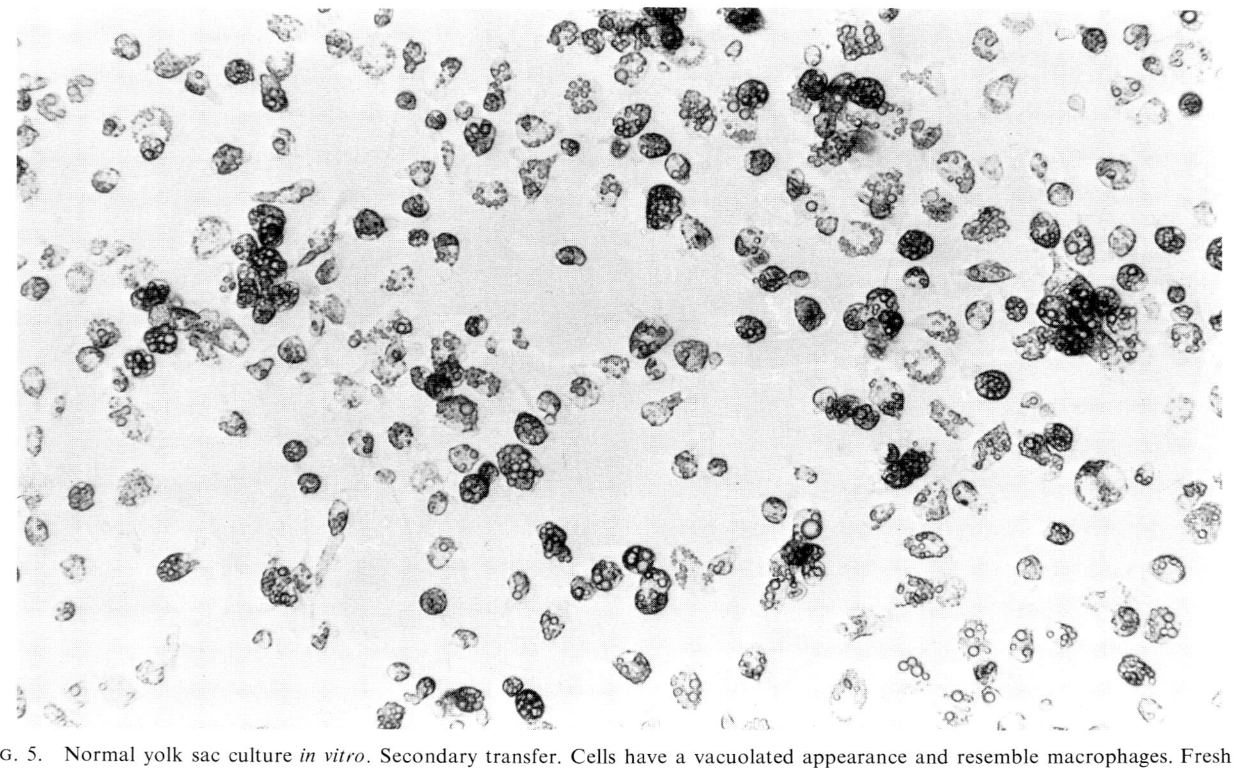

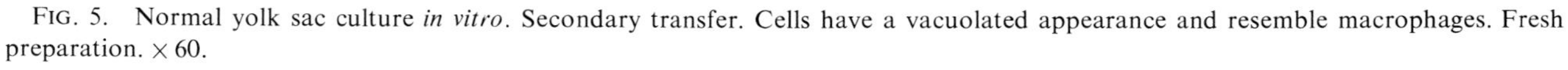

FIG. 5. Normal yolk sac culture *in vitro*. Secondary transfer. Cells have a vacuolated appearance and resemble macrophages. Fresh preparation. $\times 60$.

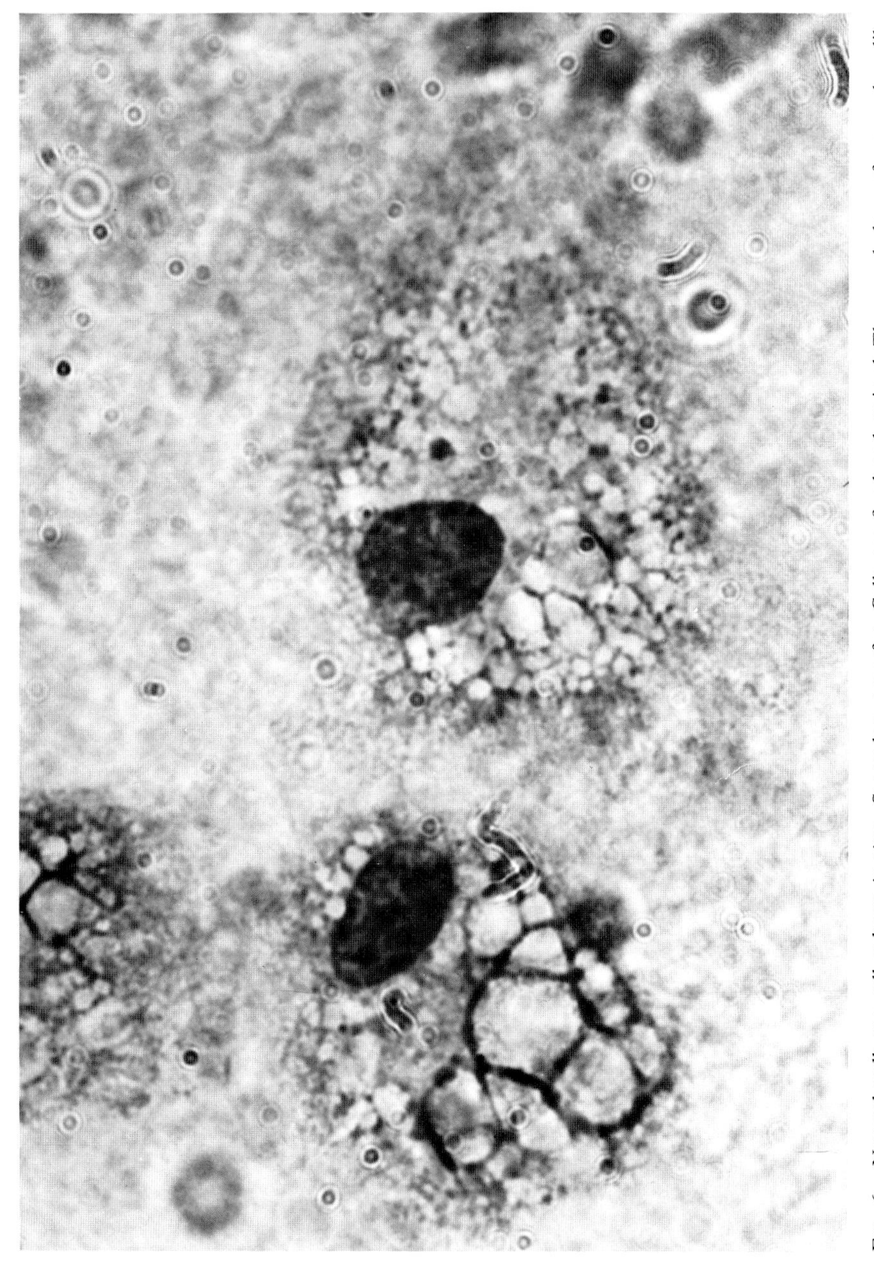

FIG. 6. Normal yolk sac cell culture *in vitro*. Secondary transfer. Cells are fixed and stained. The morphology of macrophagelike cells with extensive vacuolization is clearly shown. May-Grünwald-Giemsa stain. ×1300.

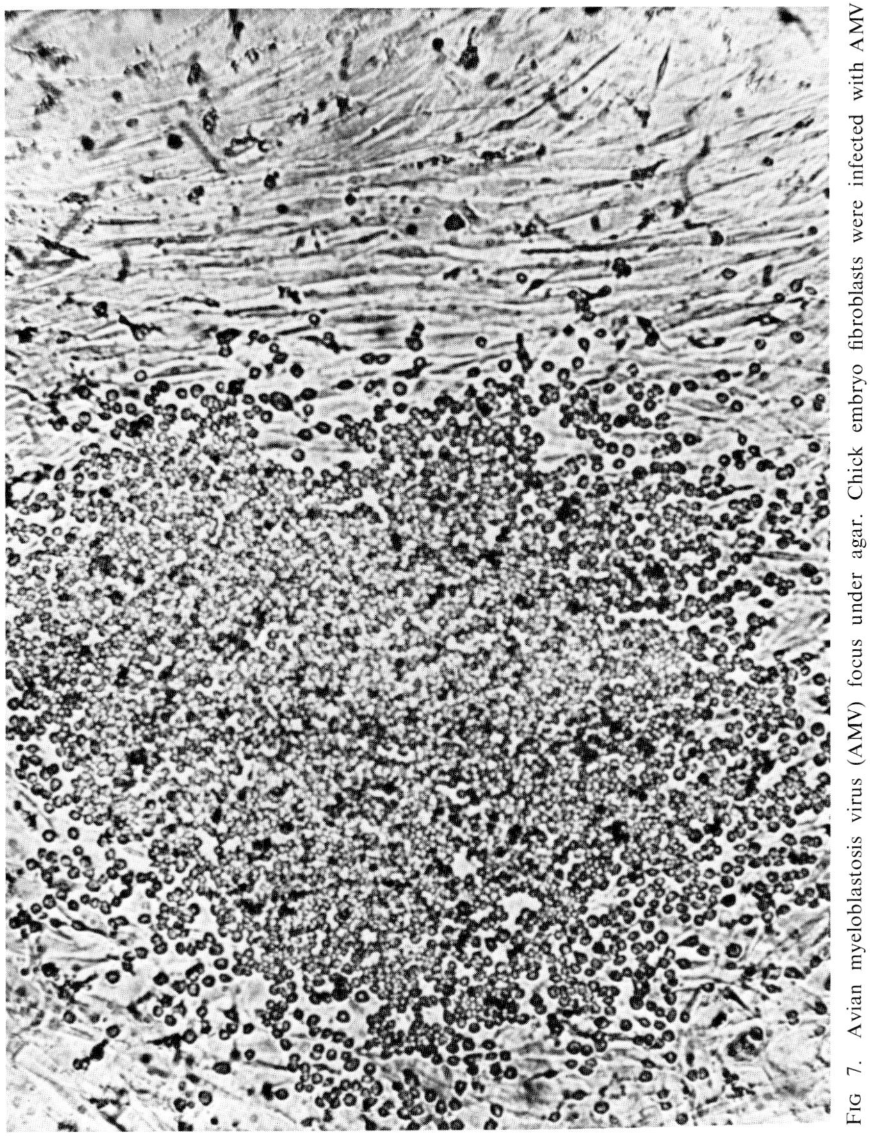

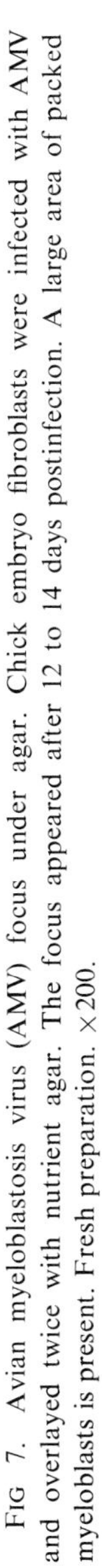

FIG 7. Avian myeloblastosis virus (AMV) focus under agar. Chick embryo fibroblasts were infected with AMV and overlayed twice with nutrient agar. The focus appeared after 12 to 14 days postinfection. A large area of packed myeloblasts is present. Fresh preparation. ×200.

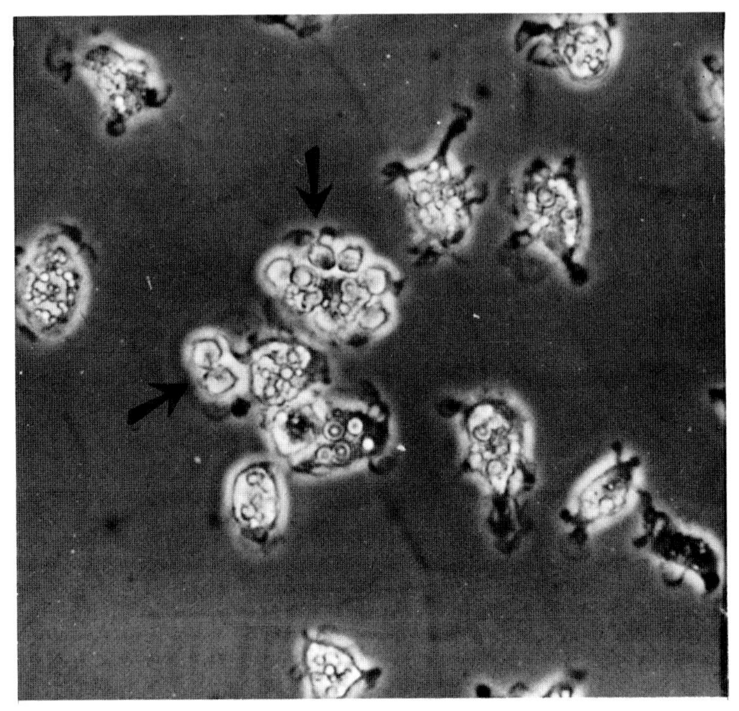

FIG. 8. Early transformation of yolk sac cells by avian myeloblastosis virus. Arrows point to two groups of transformed cells showing small refractile myeloblasts at the periphery of yolk sac cells. Fresh preparation. ×300.

REFERENCES

Baluda, M. A., and Goetz, I. E. (1961). *Virology* **15**, 185.

Crittenden, L. B., Okazaki, W., and Reamer, R. (1963). *Virology* **20**, 541.

Eagle, H. (1955). *J. Exp. Med.* **102**, 37.

Ham, R. G. (1963). *Exp. Cell Res.* **29**, 515.

Hanafusa, H. (1965). *Virology* **25**, 248.

Morgan, J. F., Morton, H. J., Healy, G. M., and Parker, R. C. (1951). *Proc. Soc. Exp. Biol. Med.* **78**, 880.

Moscovici, C. (1967). *Proc. Soc. Exp. Biol. Med.* **125**, 1213.

Reed, L. J., and Muench, H. (1938). *Amer. J. Hyg.* **27**, 493.

Rubin, H. (1960). *Proc. Nat. Acad. Sci. U.S.* **46**, 1105.

Temin, H. M., and Rubin, H. (1958). *Virology* **6**, 669.

Toyoshima, K., and Vogt, P. K. (1969). *Virology* **38**, 414.

Vogt, P. K. (1965). *Virology* **25**, 237.

Vogt, P. K. (1969). *In* "Fundamental Techniques in Virology" (K. Habel and N. P. Salzman, eds.), pp. 198–211. Academic Press, New York.

Chapter 17

Measurement of Growth and Rates of Incorporation of Radioactive Precursors into Macromolecules of Cultured Cells

L. P. EVERHART, P. V. HAUSCHKA, AND D. M. PRESCOTT

Department of Molecular, Cellular and Developmental Biology, University of Colorado, Boulder, Colorado

---

I. Introduction

Experiments with cultured cells often require the measurement of the synthesis of various macromolecules. In this chapter we describe a method for measuring the incorporation of radioactive precursors into the RNA, DNA, and proteins of mammalian cells growing either as monolayers or in suspension. The method could be used to measure the synthesis of other macromolecules and can be applied to other cell types. We have, for example, measured the rates of RNA and DNA synthesis in the protozoan *Tetrahymena*.

The method is simple and accurate and requires only a small number of cells. A most attractive feature is that isotope can be added without perturbing the cells and a large number of samples can be treated under exactly the same conditions. Very short labeling pulses can be applied because the incorporation is terminated by lifting the coverslip from the medium containing the isotope and plunging it into chasing medium or cold washing solution. In effect the beginning and end of a pulse are instantaneous. The method eliminates time-consuming centrifugation, which can lead to loss of material. After counting, the coverslips can be used for other types of analysis such as autoradiography, cytospectrophotometry, and scanning electron microscopy.

The method described here is similar to procedures described by others. Baltimore and Franklin (1962) used a coverslip method to measure the incorporation of uridine-^{3}H and thymidine-^{3}H into L cells and counted the coverslips in a gas-flow counter. Foster and Pardee (1969) used a similar method for measuring the transport of amino acids by 3T3 cells grown on coverslips. Most recently, Ball, Poynter, and van den Berg (1972) described a method for labeling cells grown on the bottom surface of glass scintillation vials (see Ball *et al.*, this volume, p. 349).

II. Culture Growth Conditions

A. Cells

Methods to be described here have been developed for the Chinese hamster ovary (CHO) cell line. The CHO line was originally established by Puck and obtained by us from D. F. Petersen of Los Alamos Scientific Labs. Though we have used this line only, the coverslip method should be adaptable to any cell line which attaches well to glass. A method is also described

(Section IV,F) for measuring incorporation of precursors into suspension cultures of cells.

B. Nutrient Medium

For the experiments described here, cells were grown in Ham's F-12 medium, which was specifically designed for the clonal growth of Chinese hamster cells (Ham, 1965). F-12 is an enriched medium which promotes the growth of individual cells in sparse culture. We supplement the medium with 10% fetal calf serum. Ham's F-12 medium may be made from stocks, as described by Ham (1972), or can be obtained commercially from a number of manufacturers. We omit thymidine from the medium since the cells do not require exogenous nucleosides and because its omission allows labeling at very high specific activities of thymidine-^{3}H.

C. Maintenance of Stock Cultures

Stock cultures are maintained on glass in monolayers. Any type of glass or plastic substrate can be used. We use 200-ml square glass culture bottles (Wheaton) since they are inexpensive, reusable, and provide ample numbers of cells for starting experimental cultures. Cells are subcultured by trypsinization from the substrate, dilution, and replating in fresh medium. Cells are kept in logarithmic growth by subculturing once or twice a week. We use 0.05% trypsin in a solution containing F-12 salts (Ham, 1972). To trypsinize a monolayer of cells the culture medium is decanted from the cells. The cell layer is washed twice with trypsin solution, and a small amount is allowed to remain on the cells for 3 to 5 minutes, or until all cells are detached. Incubation at 37° and gentle rocking both promote detachment. Cells are frequently clumped at this stage and may be dispersed by vigorous shaking or pipetting. An aliquot of the cell suspension is then added to fresh warm medium in a sterile culture bottle.

D. Preparation of Replicate Coverslip Cultures

Cells are grown on 25-mm diameter round glass coverslips in 35 × 10 mm tissue culture dishes (Falcon Plastics). Arthur H. Thomas Red Label coverslips have a lower background radioactivity than other coverslips, an important consideration for counting low amounts of incorporation of isotopes. Coverslips are cleaned by treatment for 10 minutes in 1 *N* HCl, then washed thoroughly with water and passed through several changes of ethyl alcohol. They are then sterilized in porcelain racks (A. H. Thomas) by dry heat. Four racks (48 coverslips) can be sterilized together in covered glass staining dishes. Coverslips are placed aseptically into plastic dishes using forceps which are dipped in ethanol and flamed prior to handling each coverslip. Trypsinized cell suspensions (Section II,C) are diluted to the desired density with fresh complete medium. This suspension of cells is magnetically stirred in a 950-ml bottle to which is fitted an automatic pipetting device. The Repipet® obtained from Lab Industries Inc., Berkeley, California, is accurate and trouble free. The Repipet is fitted with a length of silicone rubber tubing and a tapered glass tip. After freeing the line of air bubbles, 2-ml aliquots are pipetted into each dish. Several aliquots should be taken for direct cell counts. Racks of dishes are swirled gently and then incubated in a 37°C incubator in a humidified atmosphere of 95% air and 5% CO_2. The variation in plating is less than 1%. A similar method for replicate plating of coverslip cultures has been described by Taylor *et al.* (1972).

The plating density is dependent on the objectives of the experiment. With CHO cells, even extremely sparse cultures grow well in the rich F-12 medium. We have plated cells at initial densities as low as 10^4 cells per dish, which allows about 7 doublings of the population before the cells reach confluency at about 10^6 cells per dish. Becuase CHO cells do not show strict density-dependent inhibition of growth, frequent medium changes allow densities as high as 4.0×10^6 cells/dish to be achieved. After planting, cells require about 12 hours to recover from the effects of trypsin treatment, temperature and pH changes, pipetting, etc., and to attach to the substratum. (see Fig. 1). From this point onward it is easy to change the nutrient medium without loss of cells. We generally allow the cells to grow for one to two generations before beginning experimental perturbations.

E. Suspension Cultures

Many established cell lines, including the CHO line, will grow in suspension culture as well as in monlayer. To establish a spinner culture, cells are removed from monolayer stock cultures by trypsinization, as described previously, resuspended, and diluted in fresh medium in a spinner flask or other type of culture vessel which can be stirred and gassed. Attachment of the cells to glass surfaces may be prevented by (1) slow stirring, (2) reducing the concentration of calcium in the growth medium, and (3) coating the culture vessel with Siliclad (Clay Adams).

F. Synchronized Cells

We routinely synchronize cells by the selection technique of mitotic shake-off. The CHO cell line is especially well suited for this procedure because the cells round up during mitosis, retaining a minimal amount of

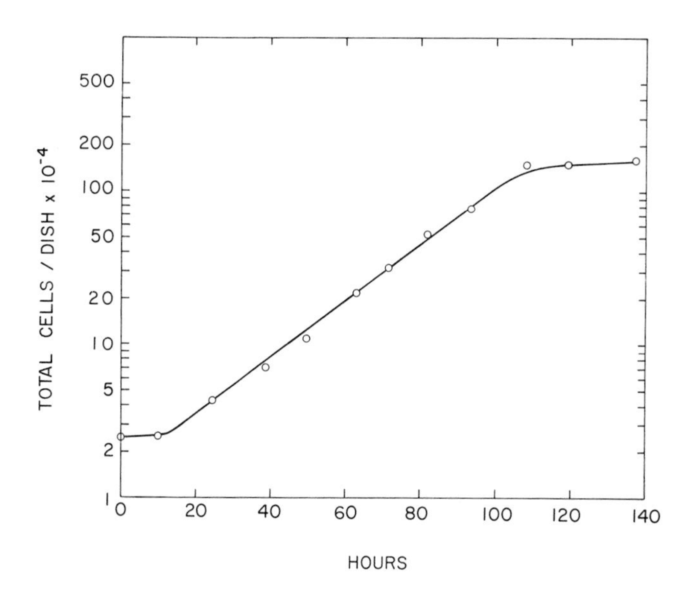

FIG. 1. CHO cell growth at 37°C on replicate coverslips. Cells were grown on round glass coverslips in 35 × 10 mm Falcon Plastics tissue culture dishes containing 2 ml of Ham's F-12 medium plus 10% fetal calf serum. Cells from replicate dishes were removed by trypsin and counted in the Coulter Counter. Counts were corrected for coincidence. The generation time is about 16 hours.

contact with the substrate. The technique has been described in detail by Petersen, Anderson, and Tobey (1968). Aliquots of the suspension of synchronized cells from the final collection are plated out directly into culture dishes containing glass coverslips as described previously. They can then be sampled at intervals to obtain different stages of the cell cycle.

Because only a small fraction of the total number of cells in a monolayer are selected by the mitotic shake-off procedure, the resulting cell density is low. A maximum cell density with the procedure as described by Petersen *et al.* (1968) is about 5000 cells/ml. If more dense cultures are desired, the suspension of cells from the final collection can be concentrated by centrifugation. Because centrifugation may perturb the cells, an alternate procedure in which the cells are concentrated by sedimentation at 1 *g* may be used; 25 ml of cell suspension from the final mitotic shake-off is placed into an 8 dram (25 × 95 mm) flat-bottom shell vial (A. H. Thomas) containing an 18 mm round glass coverslip. The vials are covered with stainless steel culture tube closures (No. 25, Bellco) and placed in a 37°C incubator for sufficient time to allow the cells to settle and attach. After attachment the volume of medium may be reduced or the coverslip transferred to a culture dish to facilitate further manipulations.

III. Methods for Cell Counting

It is usually necessary to measure cell number or some closely related parameter (total cell protein or DNA) in order to express data from the isotopic labeling of cell cultures on a "per cell" basis. Of the several methods discussed below, the electronic particle counter is the simplest and most accurate, especially for small numbers of cells.

A. Electronic Particle Counting

It is feasible to make accurate cell counts on sparsely planted coverslip ($<10^4$ cells/5 cm^2) or on very dilute cell suspensions (10^3 to 10^4 cells/ml). Dilutions to concentrations between 10^3 and 10^5 cells/ml are made in glass scintillation vials to a final volume of 10 ml, which is required by the Coulter Counter (Model F, Coulter Electronics, Hialeah, Florida). The principal obstacle in this procedure is to prevent aggregation of the suspended cells; aggregates of two or more cells register as a single cell in most electronic counting systems. We have found that an isotonic saline-trypsin suspending solution [4 parts sterile 0.9% NaCl plus 1 part trypsin stripping solution (see Section 2a below)] will prevent aggregation of Chinese hamster ovary cells grown in suspension or monolayer. Presumably, in more intractable cases of aggregate formation, other enzymes such as collagenase and neuraminidase may be useful in maintaining monodisperse suspension.

1. Suspension Cultures

Aliquots of the suspension (0.2 ml) are pipetted directly into 9.8 ml of isotonic saline-trypsin at 25°C and mixed vigorously with a Pasteur pipette. The mixing is repeated *immediately* before counting with the electronic counter. Both the pulse-height pattern on the counter and observation with an inverted microscope at 100× will readily reveal the presence of aggregated cells.

2. Monolayer Cultures

Dishes containing monolayer cultures on coverslips may be prepared for counting *in toto* or the coverslip may be removed and counted separately. The latter procedure has the potential disadvantage that suspended or weakly adhering cells are lost during removal of the coverslip.

a. Total Dish Counts. For 35 mm culture dishes, the medium is withdrawn with a siliconized Pasteur pipette and placed in a siliconized graduated 12-ml centrifuge tube. The dish is washed with about 1.5 ml of warm (37°C) trypsin stripping solution (composition in grams per liter: NaCl,

8.0; KCl 0.4; $Na_2HPO_4 \cdot 7H_2O$, 0.045; KH_2PO_4, 0.030; glucose, 1.0; EDTA, 0.1; trypsin, 0.5; adjusted to pH 8 with NaOH), which is withdrawn and placed in the same tube. Another 1.5 ml of stripping solution is added, and the dish is stored (covered) at 37°C for about 10 minutes; a small electric warming plate is convenient for this step. Vigorous agitation with the pipette will break up sheets and clumps of cells adhering to the dish and coverslip into a monodisperse suspension. This is transferred to the graduated tube, followed by two washes of the dish with saline-trypsin. The volume of the total suspension is adjusted to 10.0 ml with saline-trypsin, and after vigorous mixing (avoiding introduction of air bubbles) the suspension is counted in the Coulter Counter. Quantitative recovery of the cells is achieved by using a single pipette for all the operations involved with each culture dish.

b. Coverslip Counts. The coverslip is removed with forceps, rinsed very gently by dipping in isotonic saline, and then placed (cells up) in a clean plastic petri dish (35 mm) containing 2 ml of the trypsin stripping solution. After about 10 minutes at 37° the cells are suspended and diluted to a known volume as described above.

B. Chamber Counting

Following the methods described above for obtaining homogeneous cell suspensions it is possible to count the cells in a ruled glass counting chamber (hemacytometer). The Levy counting chamber (see Everhart, 1972) is suitable for this purpose if the cell titer is 10^5 to 10^6/ml. More dilute suspensions require many replicate counts in order to avoid large statistical sampling errors.

C. Chemical Measurement

1. Protein Determination

Thorough saline washing is essential for removal of interfering materials in the medium, especially serum. Unfortunately, washing removes all loosely adhering and floating cells; these losses can be avoided by collection and washing by centrifugation. The magnitude of interference by the nutrient medium can be very large: a confluent cell layer in a 35 mm petri dish contains about as much protein (~900 μg) as 0.15 ml of nutrient medium supplemented with 10% serum. It is therefore important to wash away the medium remaining underneath the coverslip by gently lifting it with forceps during each wash.

The Lowry method as adapted by Oyama and Eagle (1956) requires about 10 to 200 μg of protein per assay, or approximately 1 to 20 $\times$ 10^4 cells.

Accuracy can be increased by sampling larger volumes of cell suspension or by pooling several monolayer cultures. The washed cells are dissolved in alkaline copper tartrate at a concentration of 10^4 to 10^5 cells/ml. At this point it is convenient to store the alkaline solutions for protein assay at a later time; no loss will be incurred for several days. A visible index of protein content is provided by the purple color of the alkaline solution (λ_{max} = 540 nm). Color development with the Folin-Ciocalteau reagent (λ_{max} = 660 nm) may be carried out with a 5-ml sample volume or at one-fifth or one-tenth of this scale if cell material is limiting. The entire procedure requires only several hours. The color yield should be standardized against a solution prepared with crystalline bovine albumin. Conversion of the values for total cell protein to cell number or cell mass may be done with the reservation that cell size and protein content may vary with the culture conditions.

2. DNA Determination

The colorimetric method of Burton (1968) has been widely used for DNA measurement in cell cultures, but in view of the interference by sialic acid, the modified method of Croft and Lubran (1965) is probably more accurate [see Burton (1968) for details of the methods]. Saline-washed cells are first extracted twice at 0°C for several minutes with 0.5 *M* $HClO_4$ to remove soluble interfering metabolites, and then twice extracted with $HClO_4$ (0.3 to 0.5 *M*) or 5% trichloroacetic acid at 90°C for 30 minutes (Webb and Lindstrom, 1965). The pooled hot acid extracts are diluted with 0.5 *M* $HClO_4$ to a DNA-phosphate concentration of 0.02 to 0.25 μmole/ml before mixing with 2 volumes of fresh diphenylamine-acetaldehyde. Diploid mammalian cells contain about 0.25 μmole of DNA-phosphate/10^7 cells. Fairly large numbers of cells (5 to 20 × 10^5/assay) and the long color development time (17 to 48 hours depending on the temperature) are the obvious disadvantages of this method.

IV. Measurement of Precursor Incorporation

A. Labeling Conditions

The nucleosides, thymidine and uridine, labeled with ^3H or ^{14}C are chosen to study nucleic acid synthesis because these compounds enter most cells more readily than the free pyrimidine bases. Nucleotides do not enter the cell at all. All the amino acids are readily taken up by mammalian cells, but many are involved in metabolic pathways that can lead to labeling of macromolecules other than proteins. ^{3}H- or ^{14}C-labeled leucine is often used because leucine is a significant component of almost all cell proteins and undergoes very few side reactions.

The amount of label that is used will depend on the number of cells, the duration of the labeling period, the concentration of the compound in the medium, and the sensitivity of the counting procedure. The conditions used to obtain the data shown in Section IV, D are representative. To study DNA synthesis at a moderate cell density (10^4 to 2×10^5 cells per coverslip) 1 μCi/ml of thymidine-^{3}H (15 to 20 Ci/mmole) is used for a 15-minute pulse in growth medium from which cold thymidine has been omitted. There is no uridine in Ham's F-12 medium, and 1 μCi/ml of uridine-^{3}H (25 Ci/mmole) is used for a 15-minute pulse with moderate cell densities. To study protein synthesis, 1 μCi/ml leucine-^{3}H (40 Ci/m mole) is used to label moderate density cells in Ham's F-12 medium, which contains 1×10^{-4} *M* leucine. The isotope to be used is diluted if necessary with balanced salt solution or medium so that a readily measured volume may be added. It is convenient to add 10 to 50 μl to 2-ml cultures, since these volumes are easily handled and do not cause significant dilution of the medium. The isotope is added rapidly and with thorough mixing to ensure uniform distribution. These manipulations can be done in a 37°C room or on a warming stage. The culture is then returned to the incubator for the duration of the pulse labeling period. It is very important to remember that fluctuations in both temperature and pH can exert profound effects on rates of incorporation. Where pH must be precisely controlled and cultures cannot be adequately gassed with CO_2, it is necessary to use another buffering system. We use HEPES buffer (*N*-2-hydroxyethylpiperazine-*N*′-2 ethanesulfonic acid) at 30 m*M* as recommended by Ham (1972).

B. Washing, Fixing, and Extracting

The pulse is terminated by lifting the coverslip from the dish with fine tipped forceps and then washing by gently waving the coverslip in a beaker containing ice-cold balanced salt solution. After a second wash in cold saline, the coverslip is passed through a beaker containing 3:1 fixative and then fixed for 10 minutes in a second beaker of 3:1 fixative. The fixative should be prepared the same day by mixing 3 parts of absolute ethanol with 1 part of glacial acetic acid. Coverslips are dehydrated by passing through 2 changes of 95% ethanol and then holding for 5 minutes in 100% ethanol. The coverslips are stored until all the points from an experiment are collected. They are then extracted as a group with acid (1 *N* HCl, 25°C, 5 minutes) to remove counts which are not incorporated into macromolecules. After rinsing in distilled water and dehydrating for 5 minutes in 95% and for

5 minutes in 100% ethanol, the coverslips are dried for 5 to 10 minutes in a 60°C oven.

C. Measurement of Radioactivity

The extracted, dehydrated coverslip cultures are placed (cell side up) in numbered planchets and counted in a windowless, gas-flow planchet counter (Nuclear Chicago). This procedure is simple and inexpensive although other methods of counting may be used. For liquid scintillation measurement the coverslip is fragmented and placed in a scintillation vial by means of a small funnel. It is then digested for 2 hours at 37°C in 1 ml of Nuclear Chicago NCS® solubilizer. Following digestion, 10 ml of Omnifluor® (New England Nuclear) in toluene is added and the sample counted. If the liquid scintillation technique is used, it may be more convenient to grow the cells on the bottom surface of scintillation vials as described by Mizel and Wilson (1972), Ball *et al.* (1972), and Ball *et al.* (this volume).

D. Sample Data

Figures 2, 3, and 4 present data obtained in our laboratory for the incorporation of thymidine-^3H, uridine-^3H, and leucine-^3H into logarithmically

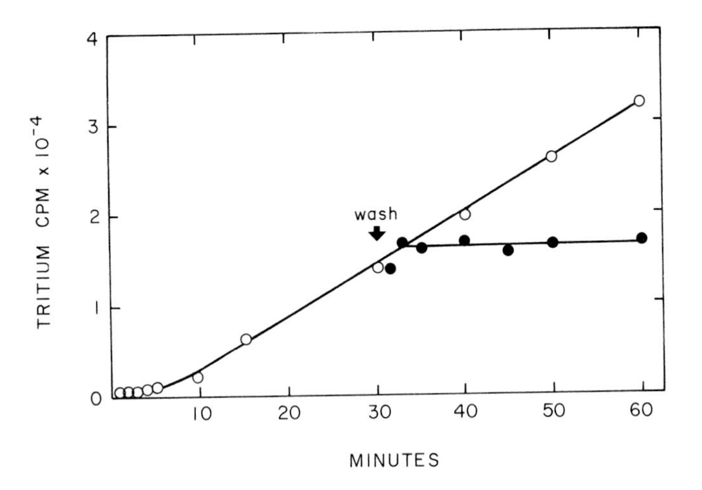

FIG. 2. DNA synthesis. Thymidine-^3H, 5 μCi/ml (49.2 Ci/mmole) was added to log phase cells (about 200,000/dish) at zero time. At the time indicated by the arrow, cold thymidine was added to a final concentration of 2×10^{-4} *M*, resulting in about a 2000-fold dilution of the labeled thymidine. ○, Continuous labeling; ●, chase.

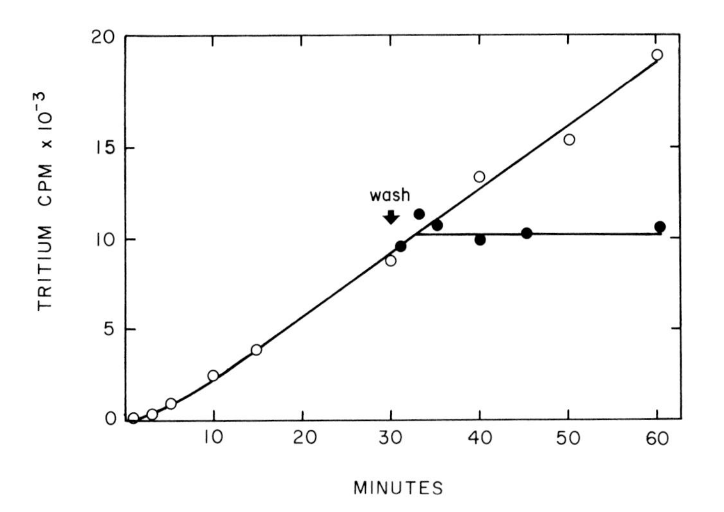

FIG. 3. RNA synthesis. Uridine-^{3}H, 5 μCi/ml (26.4 Ci/mmole) was added to cells in log phase (about 200,000 cells/dish) at zero time. At the time indicated by the arrow, cold uridine was added to a final concentration of 2×10^{-4} M, resulting in about a 1000-fold dilution of the labeled uridine. ○, Continuous labeling; ●, chase.

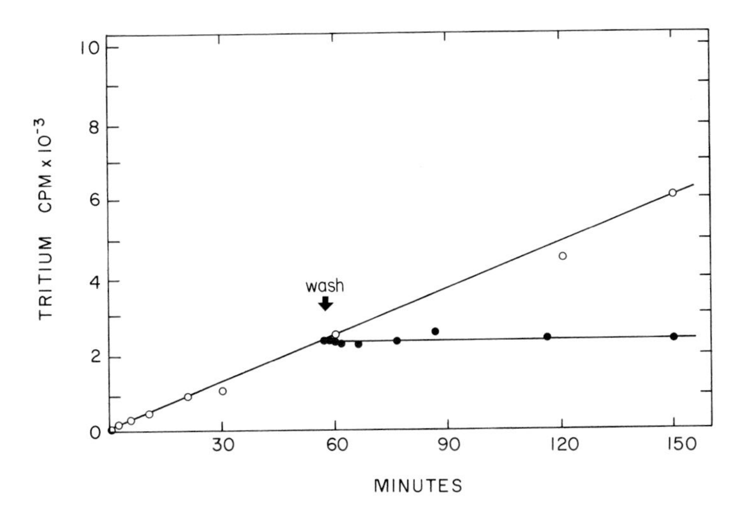

FIG. 4. Protein synthesis. Continuous labeling was carried out in complete Ham's F-12 medium with 1 μCi/ml leucine-^{3}H (31.9 Ci/mmole). At the time indicated by the arrow, the coverslips were washed through a beaker containing complete medium and placed in dishes containing complete conditioned medium. ○, Continous labeling; ●, chase.

growing CHO cells. Specific details concerning the labeling conditions are given in the figure legends.

For thymidine-^{3}H (Fig. 2) the incorporation does not become linear until 5 to 10 minutes (the time at which the nucleotide pool become saturated with labeled thymidine). Pool equilbration time is generally independent of isotope concentration (Gentry *et al.*, 1965; Johnstone and Scholefield, 1965), although it can vary for different precursors. Linear incorporation continues for the course of the experiment, which in this case is 60 minutes. At 60 minutes only 0.7% of the exogenous label has been incorporated. When the exogenous label is diluted, incorporation continues for 3 minutes and then ceases. This indicates that the thymine nucleotide pool is sufficient to sustain DNA synthesis for about 3 minutes under these conditions (see also Cleaver, 1967).

The results obtained for uridine-^{3}H (Fig. 3) are similar to those for thymidine-^{3}H. When exogenous uridine-^{3}H is diluted with unlabeled uridine, incorporation continues for about 3 minutes, and then there is no further increase. Only about 0.5% of the total exogenous label is incorporated in 1 hour under these conditions.

Figure 4 shows that the incorporation of leucine-^{3}H begins without a lag and remains linear for 2.5 hours. Incorporation ceases immediately upon removal of the label by washing because of two factors: (1) the intracellular leucine pool is turning over very rapidly, and (2) unused labeled leucine that has entered the cell is in an unaltered form and may be lost from the cell by washing. Nucleosides are generally phosphorylated upon entry and cannot be washed out of cells.

E. Autoradiography

After the coverslips have been counted they can be prepared for autoradiography. This technique is used to determine the fraction of cells which have incorporated the isotope. For cells which have been labeled with thymidine-^{3}H, this will give an indication of how many cells are in the S phase. Coverslips are attached to glass microscope slides, cells up, with Euparol or other mounting cement. After the cement hardens, the slide and coverslip are dipped into Kodak NTB-2 photographic emulsion at 45°C according to the procedure described by Prescott (1964). Appropriate exposure times range from a few days to a few weeks. For example, cells labeled for 15 minutes with thymidine-^{3}H (1 μCi/ml, 16 Ci/mmole) are conveniently scored after 1 week of exposure. Slides are developed for 2 minutes in Kodak D-11 developer, rinsed in distilled water, and fixed for 7 minutes in Kodak F-5 fixative. They are then washed for 30 to 60 minutes in running water, stained with toluidine blue, and air dried.

F. Suspension Cultures

A small amount of appropriately diluted isotope is placed in the bottom of a 16 × 120 mm sterile plastic screw-cap tube (Falcon Plastics) and before closing, the tube is gassed with 5% CO_2–95% air. At desired times, aliquots of cells are removed from the spinner flask with a sterile pipette, placed into the prepared tubes, and gassed again. The tubes are incubated at 37°C with agitation for the duration of the pulse. Agitation prevents the cells from attaching, and for this purpose we use a wrist-action shaker. The pulse is terminated by placing the tubes into an ice bath; cold saline dilution can also be used. The entire sample or aliquots of it are then collected on filters. We use either glass fiber filters (Whatman GF/A) or Millipore filters (HASP 0.45 μ pore) and a steel filtering chimney obtained from W. A. Shaerr, 356 Gold Street, Brooklyn, New York. Millipore glass chimneys can also be used. The filter is prewashed with balanced salt solution containing an excess of the nonradioactive precursor to prevent adsorption of unincorporated isotope. The cell suspension is sucked onto the filter, washed twice with 5- to 10-ml aliquots of cold balanced salt solution, and then washed twice with cold 5% TCA, allowing at least 1 minute for extraction of unincorporated label. After two washes with 95% ethanol, the filters are dried under an infrared lamp, placed into vials containing scintillation cocktail, and counted. For consideration of some problems involved in counting radioactive materials on solid supports see Bransome and Grower (1970) and Birnboim (1970).

To make autoradiograms of suspension-grown cells a small aliquot of labeled cells (1 to 10 ml) is centrifuged for 2 to 5 minutes at 100 *g* in siliconized, conical-bottom, glass centrifuge tubes. At this point the cells may be washed with balanced salt solution to remove unincorporated counts. It is desirable to carry out the centrifugation steps in the cold to inhibit further incorporation. After washing, most of the supernatant is decanted, and the pellet of cells is resuspended in the remaining small volume of supernatant (about 0.1 ml). A small drop of this suspension of cells is placed with a capillary pipette onto a large drop of 3:1 fixative (see Section IV, B) on a clean glass slide. As the fixative begins to evaporate and pulls away from the affixed cells, the slide is flooded with more 3:1 and turned on edge to dry. The fixed cells are distributed in a circular or crescent-shaped pattern. They can then be washed, extracted with acid, and dehydrated (see

Section IV,B). The slides are then dipped into photographic emulsion, exposed, and developed as described in Section IV,E.

V. Problems Involved in the Interpretation of Results

Three general types of problems are encountered in these experiments.

A. Purity of Precursors

Occurrence of impurities in radioactive compounds is well known. They can arise by radiation decomposition as well as by incomplete or careless purification. Impurities of the latter type have been pointed out by Prescott (1970) and Goldman (1970). Impurities of both types can be spuriously incorporated into macromolecules. Wand *et al* (1967) have described the incorporation of radiation decomposition products present in "old" thymidine-^{3}H preparations into macromolecules other than DNA. A knowledge of the susceptibility of different compounds to radiation decomposition and proper storage conditions, can minimize this type of artifact. Other considerations and appropriate control experiments are discussed by Oldham (1971) in a very useful paper. Two such controls are isotope dilution and enzymatic digestion. In the isotope dilution method a large excess of unlabeled precursor is applied to the cells in the presence

TABLE I

THE EFFECT OF DNASE AND RNASE DIGESTION ON INCORPORATED COUNTS FROM THYMIDINE-^{3}H AND URIDINE-^{3}H[a]

Incorporated precursor	Enzyme	Percent counts removed
Thymidine-5-^{3}H	DNase	91
	RNase	<1
Uridine-5-^{3}H	RNase	96
	RNase	3

[a]Acid-extracted coverslips were treated for 2 hours at 37°C with 0.01% solutions of DNase (Worthington electrophoretically pure DNase) or RNase (Worthington). DNase solutions were prepared in 12.5 m*M* Tris-HCl pH 7.0, 3 m*M* $MgSO_4$. RNase was in 12.5 m*M* Tris-HCl pH 8.0. The RNase solution was boiled for 10 minutes before use to inactivate any contaminating DNase. After digestion the coverslips were washed exhaustively in distilled water, dehydrated through ethyl alcohol, and recounted on the gas-flow planchet counter.

of the labeled precursor. If there is incorporation of impurities from the labeled precursor, the expected decrease in incorporated activity will not result. In our experiments we have identified the product by use of specific enzymatic digestion. The results of DNase and RNase treatment on incorporated uridine and thymidine are shown in Table I. These data indicate that thymidine-^{3}H is incorporated into DNA with very high specificity under the conditions employed. However, there seems to be some incorporation of uridine into material which is digested by DNase. Pathways for the conversion of uridine to thymine nucleotides are known [see Kit (1970) for review].

B. Uptake, Pools, and Metabolism

In choosing a precursor to study the rate of synthesis of some cellular macromolecule, many factors must be considered. Naturally it is desirable to use a precursor that is incorporated with a high degree of specificity. One must consider uptake into the cell, metabolic reactions which activate the precursor so that it can be polymerized into the macromolecule of interest, and metabolic reactions that convert the precursor into inactive compounds or into intermediates that may be incorporated into other macromolecules. Furthermore, all these factors can vary over the cell cycle, the culture cycle, and with experimental conditions. It is not the purpose of this paper to discuss these mechanisms in detail; this is done elsewhere (e.g., Hauschka, this volume). However, it might be helpful to give examples of the various kinds of problems that might be encountered.

1. Uptake

Uptake of nucleosides can vary over the cell cycle and over the culture cycle [see Plagemann (1971)]. Failure to consider these variations can lead to incorrect estimates of the rate of macromolecular synthesis. Stambrook and Sisken (1972) have shown that when uridine-^{3}H is used to measure RNA synthesis in Chinese hamster V79 cells, the apparent rate of synthesis increases 10-fold over a single cell cycle. The true rate of RNA synthesis was shown to increase only 2-fold when measured by incorporation of adenine-^{3}H.

2. Metabolic Conversion

Examples of precursor incorporation into macromolecules other than the desired species are numerous. Albach (1968) demonstrated that tritium from thymidine-5-methyl-^{3}H was incorporated into lipids and RNA of *Tetrahymena*. These results were obtained after very long labeling periods. Several workers have also shown that degradation products of nucleosides can be converted into amino acids and thereby incorporated into proteins [see Oldham (1971)].

Other metabolic activities can reduce the extent of incorporation. For example, some tissues contain high levels of thymidine phosphorylase activity, which degrades thymidine to thymine. This breakdown product is then incorporated into DNA at low rate and after further degradation can also be incorporated into other products.

3. Experimental Alterations

A variety of experimental conditions can alter any of the mechanisms by which precursors become incorporated into macromolecules, including uptake, activation, and polymerization. Care must be taken to establish that an observed reduction in the rate of incorporation actually reflects reduction in rate of synthesis of the macromolecule and is not due to one of these other factors. Some examples follow. When CHO cells are treated with dibutyryl cyclic AMP, the apparent rate of DNA synthesis measured by thymidine-^{3}H incorporation declines 10-fold (Hauschka *et al.*, 1972). This decline results from an inhibition of thymidine uptake, and there is, in fact, no decrease in the actual rate of DNA synthesis. Nakata and Bader (1969) have shown a similar effect of 2-mercapto-1-(β-4-pyridethyl)-benzimidazole (MPB) on the apparent rate of RNA synthesis in cells resulting from an inhibition of exogenous uridine uptake. The rate of RNA synthesis measured by $^{32}PO_4$ incorporation into RNA was unaffected by the presence of MPB. In this case the use of a nonspecific precursor ($^{32}PO_4$) and purification of the product gave much clearer results than the use of a supposedly specific precursor. The use of alternate precursors may often be useful when uptake of other compounds is perturbed.

4. Nonspecific Binding

A final problem results from nonspecific binding of the radioactive precursor to cellular components. Counts which are *not* due to impurities may be bound to the cell and not incorporated into macromolecules. Discussion of this problems for thymidine-^{3}H labeling of hepatocytes appears in a paper by Morley and Kingdon (1972).

C. Counting Efficiency

It is important to know the relationship between the actual amount of radioactivity incorporated (e.g., disintegrations per minute) and the number of counts detected. This relationship is the counting efficiency, and there are primarily two factors to be considered in its determination: (1) the inherent efficiency of the instrument, and (2) quenching in the sample. We have compared counts obtained using the low background planchet counter with liquid scintillation counts obtained from the same samples after solubilization (Fig. 5). The efficiency of counting tritium in the scintillation system was determined to be 36.3% by use of an internal standard. By extrapolation, the efficiency in the low-background gas-flow counter was 18.6%.

It has been reported that for *Tetrahymena* there is a variable amount of self-absorption of tritium counts over the cell cycle (Watanabe, 1971). This could conceivably be a problem for mammalian cells grown on coverslips, because the geometry and surface of the cell change dramatically during progression through the cell cycle (Porter *et al.*, 1973). The efficiency of detection of radioactivity in cells fixed during G1, when the cells are somewhat rounded and blebbed, could be very different from that for S phase cells, which are smooth and flat. To investigate this possibility we used thymidine-^{14}C, for which the emitted β-particle has a much greater mean travel distance than that for tritium. The incorporation of both isotopes

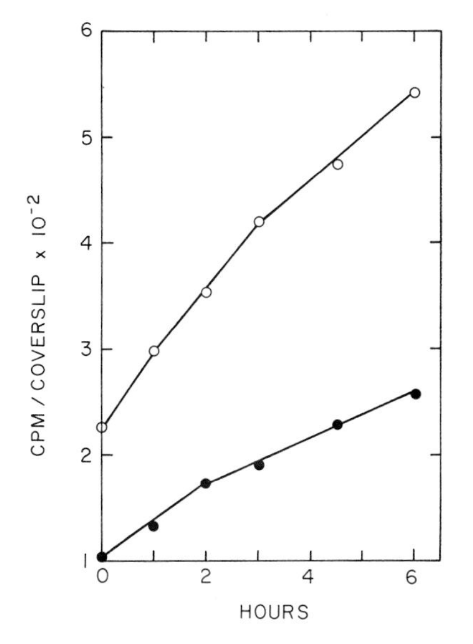

FIG. 5. Comparison of counting by a gas-flow planchet counter (●——●) and liquid scintillation counting (○——○). Log phase cells were pulse-labeled for 15 minutes with thymidine-^3H, washed, fixed, and extracted as described in the text. After counting in the low background planchet counter the same coverslips were fragmented into a scintillation vial and digested for 2 hours at 37° with 1 ml of NCS reagent; 10 ml of Omnifluor solution is added. They were counted in the liquid scintillation counter.

increases in the same fashion as the cells traverse the cell cycle, indicating that gross changes in self-absorption do not occur over the cell cycle (Fig. 6) in CHO cells.

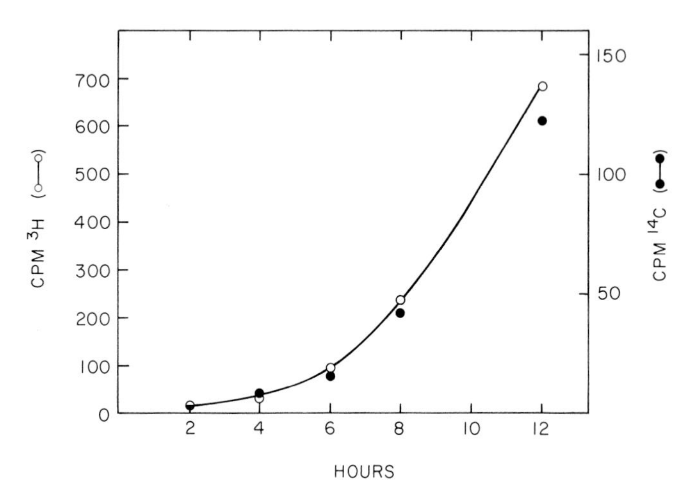

FIG. 6. Comparison of rates of incorporation of thymidine into shake-off synchronized CHO cells using ^{3}H- and ^{14}C-labeled thymidine. Thymidine-^{14}C was added at 10 μCi/ml (54 Ci/mmole). Thymidine-^{3}H was added at 1 μCi/ml (20 Ci/mmole). Cells were pulse labeled for 15 minutes, fixed, and extracted as described in the text. They were then counted in the gas-flow planchet counter. ●, ^{14}C; ○, ^{3}H.

REFERENCES

Albach, R. A. (1968). *J. Protozool.* **14**, 271.

Ball, C. R., Poynter, R. W., and van den Berg, H. W. (1972). *Anal. Biochem.* **46**, 101.

Baltimore, D., and Franklin, R. M. (1962). *Proc. Nat. Acad. Sci. U.S.* **48**, 1383.

Birnboim, H. D. (1970). *Anal. Biochem.* **37**, 178.

Bransome, E. D., Jr., and Grower, M. F. (1970). *Anal. Biochem.* **38**, 401.

Burton, K. (1968). *In* "*Methods in Enzymolology*," Vol. 12: Nucleic Acids (L. Grossman and K. Moldave, eds.), Part B, p. 163. Academic Press, New York.

Cleaver, J. E. (1967). "Thymidine Metabolism and Cell Kinetics." Wiley (Interscience), New York.

Croft, D. N., and Lubran, M. (1965). *Biochem. J.* **95**, 612.

Everhart, L. P. (1972). *In* "Methods in Cell Physiology" (D. M. Prescott, ed.), Vol. V, pp. 219–288. Academic Press, New York.

Foster, D. O., and Pardee, A. B. (1969). *J. Biol. Chem.* **244**, 2675.

Gentry, G. A., Morse, P. A., Jr., Ives, D. H., Gelbert, R., and Potter, V. A. (1965). *Cancer Res.* **25**, 509.

Goldman, I. (1970). *Science* **167**, 237.

Ham, R. G. (1972). *In* "Methods in Cell Physiology" (D. M. Prescott, ed.), Vol. V, pp. 37–74. Academic Press, New York.

Ham, R. G. (1965). *Proc. Nat. Acad. Sci. U.S.* **53**, 288.

Hauschka, P. V., Everhart, L. P., and Rubin, R. W. (1972). *Proc. Nat. Acad. Sci. U.S.* **69**, 3542.
Johnstone, R. M., and Scholefield, P. G. (1965). *Advan. Cancer Res.* **9**, 143.
Kit, S. (1970). *In* "Metabolic Pathways" (D. M. Greenberg, ed.), 3rd ed., Vol. 4, p. 69. Academic Press, New York.
Mizel, S. B., and Wilson, L. (1972). *Biochemistry* **11**, 2573.
Morley, C. G. D., and Kingdom, H. S. (1972). *Anal. Biochem.* **45**, 298.
Nakata, Y., and Bader, J. P. (1969). *Biochim. Biophys. Acta* **190**, 250.
Oldham, K. G., (1971). *Anal. Biochem.* **44**, 143.
Oyama, V. I., and Eagle, H. (1956). *Biochem. J.* **62**, 315.
Petersen, D. F., Anderson, E. C., and Tobey, R. A. (1968). *In* "Methods in Cell Physiology" (D. M. Prescott, ed.), Vol. III, pp. 347–370. Academic Press, New York.
Plagemann, P. G. W. (1971). *J. Cell Physiol.* **77**, 213.
Porter, K. R., Prescott, D., M., and Frye, J. (1973), *J. Cell Biol.* **57**, 815.
Prescott, D. M. (1964). *In* "Methods in Cell Physiology" (D. M. Prescott, ed.), Vol. I, pp. 365–370. Academic Press, New York.
Prescott, D. M. (1970). *Science* **168**, 1285.
Stambrook, P. G., and Sisken, J. E. (1972). *Biochim. Biophys. Acta* **281**, 45.
Taylor, W. G., Price, I. M., Dworkin, R. A., and Evans, V. G. (1972). *In Vitro* **7**, 295.
Wand, M., Zeuthen, E., and Evans, E. A. (1967). *Science* **157**, 436.
Watanabe, Y. (1971). *Exp. Cell. Res.* **69**, 324.
Webb, J. M., and Lindstrom, H. V. (1965). *Arch. Biochem. Biophys.* **112**, 273.

Chapter 18

The Measurement of Radioactive Precursor Incorporation into Small Monolayer Cultures

C. R. BALL, H. W. VAN DEN BERG, AND R. W. POYNTER

Department of Experimental Pathology and Cancer Research, The Medical School, Leeds, England

I. Introduction

The widespread use of radioactively labeled precursors in the measurement of the rate of synthesis of macromolecules has led to a number of methods applicable to cell cultures. In our own work, primarily on DNA and RNA synthesis, we have found a restriction in available methods in that they are not readily applied in experiments where large numbers of treatment conditions are required. The method we have developed for our own use in this type of experiment is the subject of this chapter.

The filter disk method of measuring DNA or RNA synthesis, using an appropriate precursor, as "acid-insoluble radioactivity" on a Millipore or glass fibre disk is very widely used and was recently described by Roberts *et al.* (1971) for cell culture experiments. This method is most convenient for suspension cultures (including ascites cells and lymphocyte cultures) but becomes cumbersome in the type of experiment described above if a stirred suspension has to be maintained for each treatment condition.

For cells in monolayer culture (which will not also grow in suspension), there is a further restriction in available methods. Filter disk methods become intractable owing to the necessity for quantitative transfer of cells from culture to filtration apparatus. A number of methods have been described for processing cells grown on coverslips and attached to the surface throughout processing and counting procedures (Everhart *et al.*, this volume; Rytömaa and Kiviniemi, 1967). Such methods have the advantage of the possibility of autoradiography on the same preparation, but the handling of large numbers of coverslips through the work-up procedures can be tedious.

We describe here a method in which the cells are grown, treated, and processed and radioactivity is recorded while the cells are attached to the bottom surface of glass scintillation vials. We (Ball *et al.*, 1972) have found the method straightforward, rapid, and reproducible. Large numbers of replicates and/or treatment conditions can be used in one experiment, and no specialized culture vessels are required. Replicate samples for autoradiography are readily prepared.

The method as finally used is described in detail in Section II. The way in which it was developed and confirmation of its accuracy are dealt with subsequently, as are further modifications and applications of the technique.

II. Methods

A. Culture Conditions

Glass scintillation vials (Searle Laboratories, Spectrovial III) were used as the culture vessel, and all subsequent operations were carried out in the vial. HeLa cells were used throughout this work and were maintained in spinner culture in spinner modified Minimal Essential Eagle's Medium (MEM) supplemented with 7% fetal calf serum (Biocult Laboratories), penicillin, and streptomycin. When plated into monolayer the medium used was MEM plus 15% fetal calf serum. Monolayer cultures in scintillation vials were seeded by pipetting 1-ml aliquots of a suspension (0.25 to 4×10^5

cells/ml) into a series of sterile vials which were gassed with 5% CO_2–air and sealed. The usual plastic vial caps were replaced by metal caps with silicone rubber liners. This provided a gastight seal and a bottle which could be dry heat sterilized. The vials were incubated at 37°C until used for precursor incorporation studies.

B. DNA and RNA Synthesis

Cultures were allowed to established for at least 16 hours before an experiment was begun (see Section III), and all procedures were carried out in a 37°C room.

Any drug or other treatment required prior to the measurement of nucleic acid synthesis is readily carried out by adding drug-containing medium to the vials. Treatment is most easily terminated by removing the medium by careful suction and replenishing with fresh growth medium. To measure nucleic acid synthesis thymidine-6-^{3}H (> 20 Ci/mmole) or uridine-5-^{3}H (> 15 Ci/mmole) were added by addition of medium (1 ml) containing 1 μCi/ml of the appropriate precursor. At the end of the time period chosen, precursor incorporation was stopped by addition of ice-cold isotonic saline (approximately 10 ml). The vials were then removed from the hot room for further processing. The cold saline solution was poured off and the cells washed once more in saline. At this stage it is essential that washing be done gently so as not to remove cells from the surface of the vial. The vial is now filled with 1.5% v/v perchloric acid, which fixes the cells without cell loss and removes most of the unincorporated readioactive precursor. Two further changes of perchloric acid are applied, and then alcohol is added to the vial to remove lipids. The alcohol is poured off; the vial is then inverted and allowed to drain. Perchloric acid (5% v/v, 1 ml) is then added to each vial, and the samples are heated in an oven at 80°C for 40 minutes to hydrolyze the nucleic acids. After cooling, the samples are ready for scintillation counting.

C. Protein Synthesis

In these model experiments we have used leucine-4,5-^{3}H (> 15 Ci/mmole) as a protein precursor and added it at 1 μci/ml to the cells in leucine-free MEM (Flow Laboratories), having first poured off the growth medium. This procedure increases the incorporated radioactivity about 10-fold over the use of complete MEM. The method is initially identical to that described in Section II,B. The hot perchloric acid extract is discarded, however, and, after brief washing of the vial with isotonic saline to remove residual acid, sodium hydroxide (0.1 *N*, 1.3 ml) is added and the vials are heated again at 80°C for 2 hours to solubilize the protein.

D. Radioactivity Measurements

Radioactivity in the perchloric acid or sodium hydroxide extracts was recorded in a scintillation counter after addition of a Triton X-based scintillator (10 ml), giving a homogeneous solution. The scintillation fluid consisted of Triton X-100 (1000 gm), toluene (2000 ml), PPO (16.5 gm), and dimethyl-POPOP (300 mg). Although it is quite feasible to correct samples for quenching by standard methods, and hence calculate disintegrations per minute, this proved superfluous. Quenching was consistent to within narrow limits between samples, and the method is comparative (i.e., treated vs. control) only in the sense that nucleic acid (or protein) specific activities cannot be obtained.

E. Autoradiography

Commonly in the use of radiolabeled precursors autoradiography is required to locate the site of incorporation. For example, biochemical measurement, as in Section II,A, of the demonstration of inhibition of thymidine-^{3}H incorporation into DNA is a true measure of a decrease in the rate of synthesis only if it is also established that the cell kinetics of the system have not been perturbed by the treatment used. By placing circular coverslips (18 mm) in the bottom of some of the vials before plating the cells, it is possible to obtain cultures for autoradiography which are treated and labeled under identical conditions to those actually attached to the vial and used for biochemical assessment of DNA synthesis. We have used this method to obtain simultaneously the effect of a DNA synthesis inhibitor on incorporation of thymidine-^{3}H into DNA and the proportion of cells in the S phase of the cell cycle.

III. Development of Method

The methods described above are the result of a series of experiments in which the extraction efficiency of the various procedures was checked by recording the radioactivity removed by each process when carried out exhaustively. In this way the regimen described, which removed all non-incorporated radioactivity, was obtained. At the end of this procedure fixed cell ghosts remain attached to the vial surface and contain acid-insoluble radioactivity in nucleic acids or proteins according to the precursor used. It is possible at this point to add scintillation fluid and obtain a measurement of acid-insoluble radioactivity. However, with ^{3}H-labeled precursors the efficiency is low. More reproducible results, with approximately doubled counting efficiency for tritium, were obtained if the radioactivity was solubilized before counting. Nucleic acids were hydrolyzed with hot perchloric acid solution. Repeated extractions showed that one extraction (as used in Section II,B) solubilized 95% of the radioactivity. Any radioactivity remaining in the residue is still recorded, although at lowered efficiency, since it also is in contact with scintillator. Any nonnucleic acid radioactivity in the residue would also be counted, but this was shown to be negligible in this system by agreement with our measurements using the technique of Schmidt and Thannhauser (1945) on bulk HeLa cultures, which showed that thymidine-6-^{3}H and uridine-5-^{3}H are specific precursors for DNA and RNA, respectively, in these cells.

The time course of incorporation of thymidine-^{3}H into DNA and uridine-^{3}H into RNA of HeLa cells as measured by the method are illustrated in Fig. 1. Incorporation was linear over the time interval studied, but with a short time lag before the onset of incorporation. This period is the time required

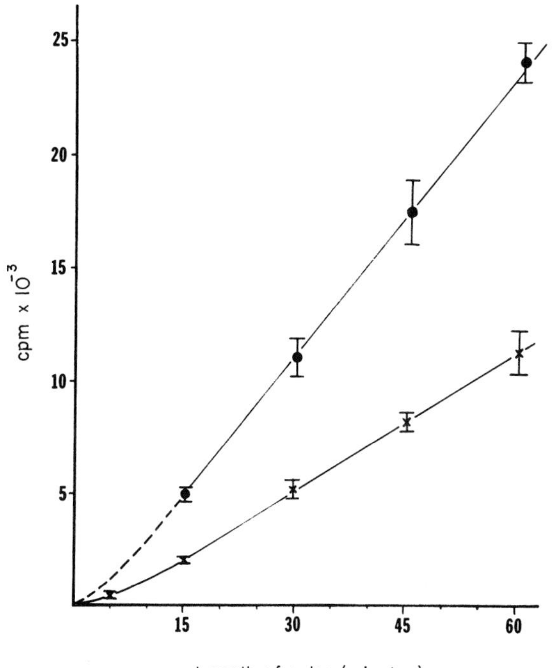

FIG. 1. Incorporation of uridine-^{3}H into RNA (●) and thymidine-^{3}H into DNA (×) as a function of pulse time (mean ±SD).

for the exogeneous precursor to equilibrate with the nucleotide pool. Eight replicates were used per point in these experiments, and on the average the standard deviation of a group was 7% of the mean value.

The method we have used does not measure DNA (or RNA) specific activity and is intended for use in comparative experiments (treated vs. control). Therefore, it was essential to show that precursor incorporation was proportional to cell numbers in individual cultures. This would ensure that between-experiment comparisons were valid even when cell numbers differed. In the course of investigating this point, we found that the time allowed for cultures to "establish" before commencing incorporation studies was critical in determining the resultant radioactivity measured.

When an experiment was commenced 3 hours after plating (Fig. 2) the incorporation of thymidine-^{3}H into DNA was linearly proportional to cell numbers plated. However, 24 hours after plating, when approximately twice as many cells were present, the same pulse of thymidine-^{3}H resulted in approximately three times as much incorporated radioactivity. Similar

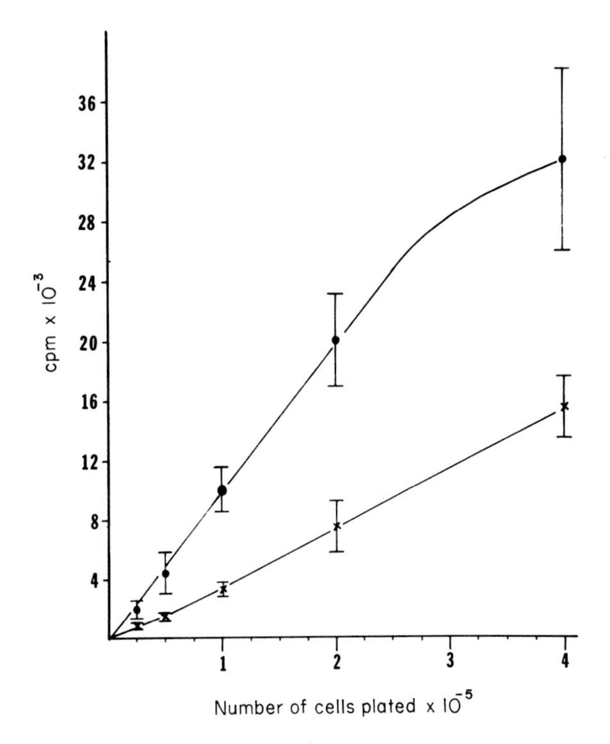

FIG. 2. The effect of cell number per vial on incorporation of thymidine-^{3}H into DNA during a 30-minute pulse commencing 3 hours (×) or 24 hours (●) after plating.

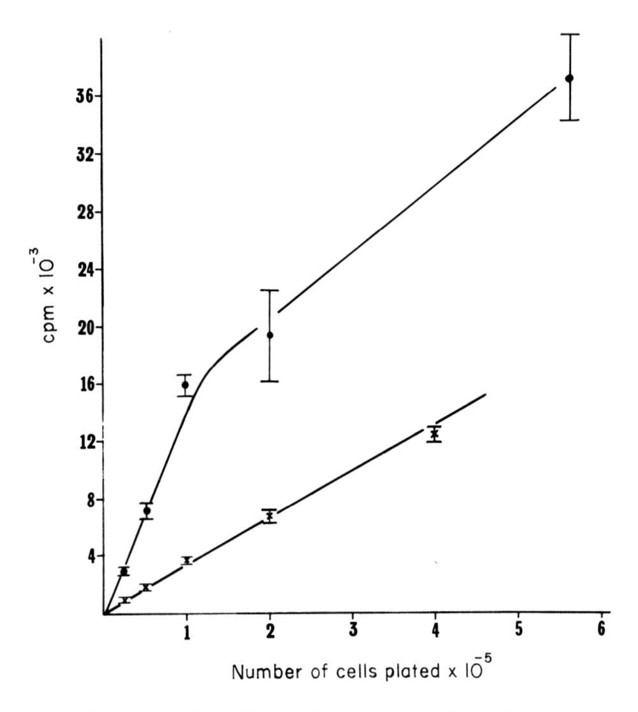

FIG. 3. Incorporation of uridine-^{3}H as a function of cell number 3 hours (×) and 24 hours (●) after plating.

results were obtained for uridine-^{3}H incorporation into RNA (Fig. 3). It is clear that cultures require some time to "establish" following cell attachment. Nucleic acid synthesis returns to normal only after a number of hours, and we have therefore recommended (Section II,A) that cultures be used the day after plating.

The results reported in Fig. 2 and 3 also illustrate that it is important to check that the cell numbers used do not result in density-dependent inhibition of nucleic acid synthesis. With HeLa cells this appears to come into play at around 4×10^5 cells per vial for DNA synthesis (allowing for 1 cell doubling after plating) and 2×10^5 cells per vial for RNA synthesis. Obviously this would have to be checked for any particular cell line. With a cell line such as 3T3, where contact inhibition of growth occurs at very low cell numbers per unit area, it might be necessary to use considerably fewer cells per vial. Using HeLa we have been readily able to make reproducible measurements on 10^3 cells using a precursor concentration on 3 μCi/ml.

The method was intended primarily for measurements of nucleic acid synthesis, but we have briefly looked at its use for measuring gross protein synthesis. The method is described in Section II,C, and the results with incorporation of leucine-^{3}H are illustrated in Figs. 4 and 5.

IV. Applications and Modifications

In view of the various factors described in the preceding section, it would seem that the following conditions should be met before the technique is applied to a cell line for the first time. (1) Incorporation of precursor should increase linearly with time. (2) Incorporation should be proportional to cell numbers in the culture. (3) Sufficient time must be allowed for "establishment" of cultures. (4) Vials must not contain sufficient cells to cause cell density- dependent inhibition of synthesis.

Presently we have used the described method only for cells with high plating efficiency (i.e., greater than 90%). Clearly its use for cells with low

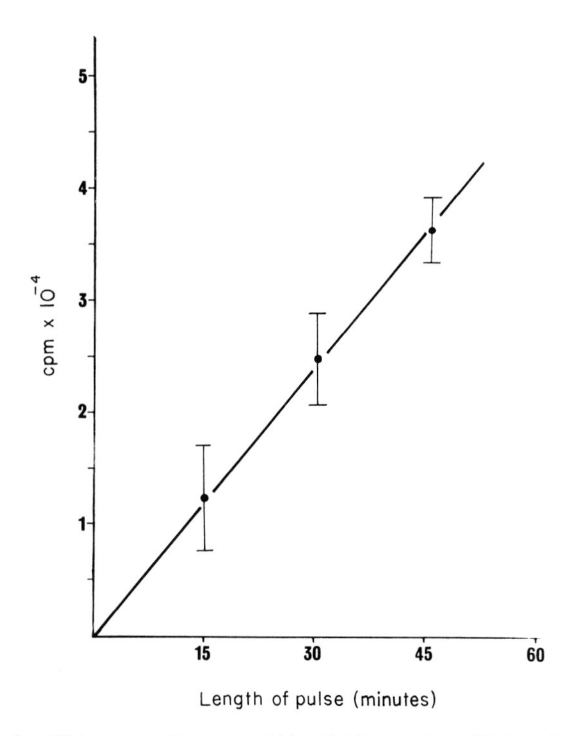

FIG. 4. Leucine-^{3}H incorporation into acid-insoluble proteins of HeLa cells as a function of pulse time 3 hours after plating (4 samples per point; mean ± SD).

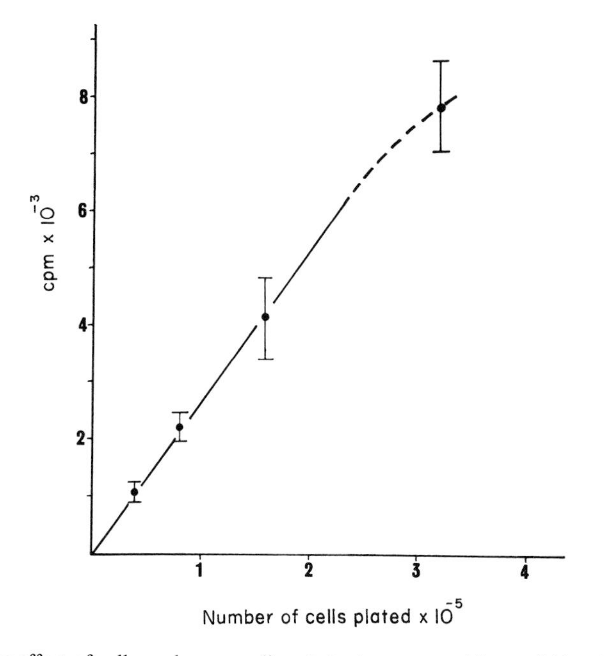

FIG. 5. The effect of cell number on radioactivity incorporated into acid-insoluble protein following a 30-minute pulse with leucine-^{3}H 3 hours after plating (4 samples per point; mean ± SD).

plating efficiency would depend on the reproducibility of cell numbers in scintillation vial cultures grown up over a longer period of time. However, we have found that the method can be applied to much larger cultures equally well. For example, when cultures of 10^7 cells in 16-ounce medical prescription bottles were taken through an identical procedure, equally reproducible results were obtained. In addition, since sufficient nucleic acids were now present for measurement at 260 nm, it was possible to obtain a nucleic acid specific activity measurement. Processing 10^7 cells *in situ* in a bottle in this way was considerably easier than carrying out the conventional Schneider or Schmidt–Tannhauser procedures, since it avoids all the problems of collecting small pellets by centrifugation (Schneider *et al.*, 1950; Schmidt and Thannhauser, 1945). It would seem likely that this latter process would be more suitable for cells of low plating efficiency.

Typical uses of the method in our hands are illustrated in Table I and Figs. 6 and 7. During an investigation of the biochemical effects of the carcinogen methylazoxymethanol acetate (MAMAC) on HeLa cells, were observed (Poynter *et al.*, 1971) that the drug was esterified by serum cholinesterase and that this effect was inhibited by physostigmine (eserine).

TABLE I

EFFECT OF PHYSOSTIGMINE ON INCORPORATION OF THYMIDINE-^{3}H INTO HELA CELLS

	Values (cpm)	Mean	Standard deviation	Student's t	P
Group 1 (control)	2570,2856 2648,2399 2570,2359 2388,2273	2508	176	—	—
Group 2 (10 μg/ml physostigmine)	2678,2606 2577,2610 2481,2445 2439,2734	2571	108	0.86	NS[a]
Group 3 (100 μg/ml physostigmine)	2170,2387 2422,2349 2135,2114 2408,2452	2179	304	2.65	$p < 0.02$[b]

[a]Not significant.
[b]Significant.

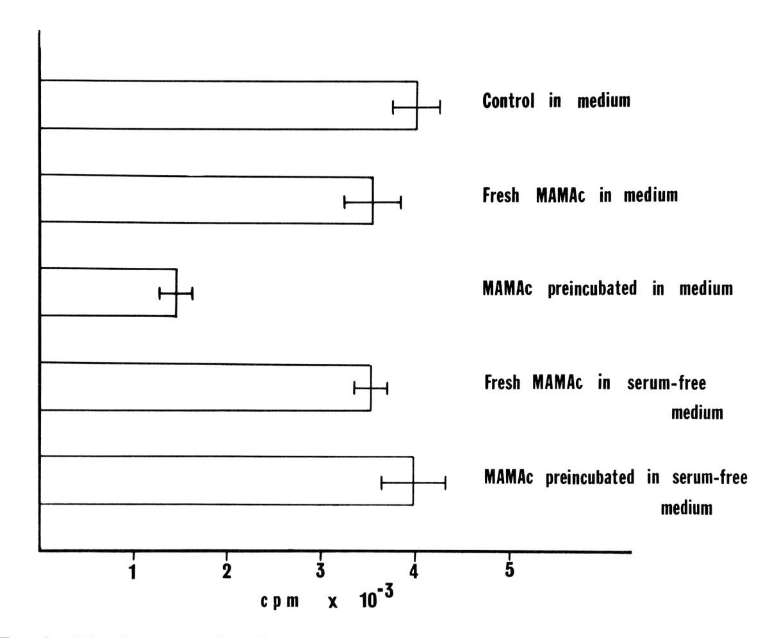

FIG. 6. The demonstration that removal of serum prevents the increased inhibition of DNA synthesis which results when methylazoxymethanol acetate (MAMAC) is incubated in medium for 4 hours prior to use. From Poynter *et al.* (1971) by permission of Elsevier, Amsterdam.

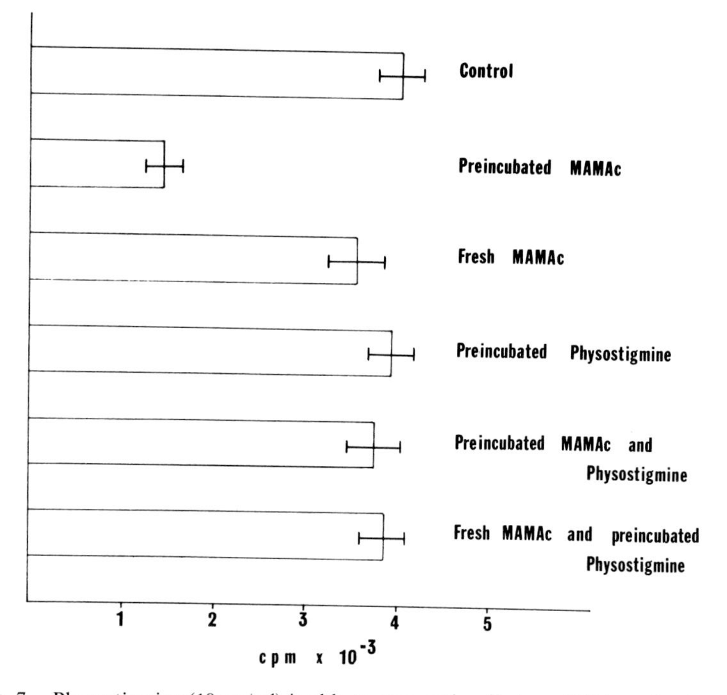

FIG. 7. Physostigmine (10 μg/ml) is able to reverse the effects on DNA synthesis produced when methylazoxymethanol acetate (MAMAC) is incubated in medium containing serum before use. From Poynter *et al.* (1971) by permission of Elsevier, Amsterdam.

The results shown in Table I were obtained in the course of investigating the effects of physostigmine itself in inhibiting DNA synthesis in HeLa cells. Eight replicate vials were used for each treatment condition; after treatment with the drug, DNA synthesis was measured immediately by a 30-minute pulse with thymidine-^{3}H. The actual radioactivity measurements obtained are shown in column 2 (Table I) followed by the mean and standard deviation. Incorporation of thymidine-^{3}H in group 3 was reduced to 87% (2179 cpm) of that in the control group (2508 cpm). However, as shown, the accuracy of the determinations is such that this small difference is significant in a Student t test at $p < 0.02$.

In the course of the same piece of work, the scintillation vial method was extremely useful because of the facility of including numerous treatment conditions within one experiment. We had observed that incubation of MAMAC in the presence of serum enhanced its effects in inhibiting thymidine-^{3}H incorporation into HeLa cells. This was simply illustrated (Fig. 6) by setting up a series of groups of scintillation vials with suitable control treatments. Equally, the observation that this enhanced effect was reversed by physostigmine is illustrated in Fig. 7.

The simplicity of the technique enabled all the relevant "control" treatments to be included in the one experiment.

V. Summary

A straightforward and reproducible technique for measuring incorporation of labeled precursors into nucleic acids and proteins of small monolayer cell cultures is described in detail, together with the background to its development and some illustrations of its use.

Acknowledgment

This work was supported by the Yorkshire Cancer Research Campaign.

References

Ball, C. R., Poynter, R. W., and van den Berg, H. W. (1972). *Anal. Biochem.* **46**, 101–107.

Poynter, R. W., Ball, C. R., Goodban, J., and Thackrah, T. (1971). *Chem.–Biol. Interactions* **4**, 139–143.

Roberts, J. J., Brent, T. P., and Crathorn, A. R. (1971), *Eur. J. Cancer* **7**, 515–524.

Rytömaa, T., and Kiviniemi, K. (1967). *In* "Control of Cellular Growth in Adult Organisms" (H. Tier and T. Rytömaa, eds.). Academic Press, New York.

Schmidt, G., and Thannhauser, S. J. (1945). *J. Biol. Chem.* **161**, 83–89.

Schneider, W. C., Hogeboom, G. H., and Ross, H. E. (1950). *J. Nat. Cancer Inst.* **10**, 977–982.

Chapter 19

Analysis of Nucleotide Pools in Animal Cells

PETER V. HAUSCHKA

Department of Molecular, Cellular and Developmental Biology, University of Colorado, Boulder, Colorado

I. Introduction

Realization of the important functions of nucleotides in a variety of cellular processes has stimulated interest in the measurement of these metabolites in cell extracts. Nucleotides act as carriers and reservoirs of chemical energy, they participate as cofactors, activators, inhibitors, and substrates in myriad enzymatic reactions, and they are the precursors for polymerization of nucleic acids. Most of these processes exhibit some type of dependence of rate on the concentration of participating nucleotides, and in some cases the rate may even be *controlled* by nucleotide concentration. To establish the existence of such control *in vivo*, it is necessary to measure intracellular nucleotide concentrations, or pools.

Control of nucleic acid synthesis by precursor pool concentrations has frequently been proposed in the literature. For instance, Davidson (1962) hypothesized that DNA synthesis takes place at critical concentrations of deoxyribonucleoside triphosphates, and that these compounds are part of a homeostatic control mechanism. Studies on synchronized cell populations, have revealed increased activity during the S period of numerous enzymes involved in the formation of precursors for DNA synthesis (Brent *et al.*, 1965; Littlefield, 1966; Stubblefield and Murphree, 1967; Adams, 1969; Murphree *et al.*, 1969; Brent, 1971). Many of these enzymes are most active in rapidly proliferating cells (Weissman *et al.*, 1960; Bresnick and Thompson, 1965; Maley *et al.*, 1965; Plagemann *et al.*, 1969; Sneider *et al.*, 1969; Nordenskjöld *et al.*, 1970; Weber *et al.*, 1971; Conrad and Ruddle, 1972). The hypothesis of control of nucleic acid synthesis by nucleotide pool size has persisted, even though most of the evidence for it is merely circumstantial—small increases in the pool sizes or in the incorporation of radioactive precursors into pools often accompany conditions of elevated RNA and DNA synthesis, and vice versa (Bucher and Swaffield, 1965, 1969; Yu and Feigelson, 1969; Bosmann, 1971). Recently, data gathered by direct analysis of deoxyribonucleoside triphosphate pools have clearly shown an *absence* of correlation between rates of DNA synthesis and pool size (Lindberg *et al.*, 1969; Colby and Edlin, 1970; Adams *et al.*, 1971; Slaby *et al.*, 1971). In fact, Nordenskjöld *et al* (1970) observed for mouse embryo cells released from growth inhibition that the increase in DNA synthesis actually precedes increases of the dTTP (Refer to Table I for nomenclature) and dATP pools.

The ease with which nucleotide pools can be labeled by a variety of radioactive precursors has been exploited in countless papers for the measurement of rates of nucleic acid synthesis. More often than not, the rate of

TABLE I

NOMENCLATURE

Bases		Nucleosides			
		Ribo-		Deoxyribo-	
A	adenine	rA	adenosine	dA	deoxyadenosine
G	guanine	rG	guanosine	dG	deoxyguanosine
X	xanthine	rX	xanthosine		
H	hypoxanthine	rH	inosine		
C	cytosine	rC	cytidine	dC	deoxycytidine
T	thymine			dT	thymidine
U	uracil	rU	uridine	dU	deoxyuridine
FU	5-fluorouracil			FdU	5-fluorodeoxyuridine
O	orotic acid	rO	orotidine		

Nucleotides		Other	
AMP	adenosine 5′-monophosphate	PRPP	5-phosphoribosyl 1-pyrophosphate
GMP	guanosine 5′-monophosphate	THF	tetrahydrofolic acid
XMP	xanthosine 5′-monophosphate	araC	arabinosyl cytosine
IMP	inosine 5′-monophosphate	araU	arabinosyl uracil
CMP	cytidine 5′-monophosphate	NAD^+	nicotinamide adenine dinucleotide
UMP	uridine 5′-monophosphate	dNMP	deoxyribonucleoside monophosphate
cAMP	adenosine 3′:5′ cyclic monophosphate	dNDP	deoxyribonucleoside diphosphate
cGMP	guanosine 3′:5′ cyclic monophosphate	dNTP	deoxyribonucleoside triphosphate
dAMP	deoxyadenosine 5′-monophosphate	dNXP	deoxyribonucleoside mono-, di-, and triphosphate
dGMP	deoxyguanosine 5′-monophosphate		
dCMP	deoxycytidine 5′-monophosphate	nmole	$= 10^{-9}$ mole
dTMP	thymidine 5′-monophosphate	pmole	$= 10^{-12}$ mole
dUMP	deoxyuridine 5′-monophosphate		

incorporation of label into acid-insoluble nucleic acid has been largely determined by the rate of incorporation of precursor into the nucleotide pool and has not truly reflected the absolute rate of nucleic acid synthesis (e.g., Lindberg *et al.*, 1969; Nakata and Bader, 1969; Plagemann and Roth, 1969; Plagemann, 1971a; Stambrook and Sisken, 1972). Techniques are now available for measuring the specific activity of nucleotide pools in order to obtain the absolute rate of nucleic acid synthesis [e.g., Emerson and Humphreys (1971) and Section VI,B].

Other studies of nucleotide pools concern the metabolism of nucleoside antibiotics. The actions of these compounds are exerted through many different mechanisms, but almost all involve some prior intracellular modification of the nucleoside, such as phosphorylation. Discovery of the mechanism of inhibition usually requires analysis of the nucleotide pools in sensitive and drug-resistant cells.

This is a time when an increasing number of experiments are focused on the behavior of nucleotide pools and nucleoside transport. Knowledge of factors which affect or control these parameters is beginning to accumulate. While this chapter is primarily concerned with methods of pool analysis, several sections are devoted to a review of pertinent information on pool properties, nucleotide metabolism, and precursor transport. The emphasis on animal cell systems is not intended to slight the elegant studies of nucleotide pools in bacteria (e.g., Neuhard, 1966; Neuhard and Munch-Petersen, 1966; Cashel and Gallant, 1968). Many methods developed for bacterial studies have been included for possible adaptation to animal cells. Hopefully, the diversity of methods described will provide the reader with a comfortable background for the design of his own experiments.

II. General Properties of Pools

A. Definitions

Definitions of the term *pool* are fraught with semantic difficulties. In general, a pool of any particular metabolite or family of metabolites is the total intracellular *amount* of the compound which is fed by and available to a set of enzymatic pathways. Rapidly equilibrating metabolites, such as the mono-, di-, and triphosphates of a nucleoside, are generally considered to be in the same pool. Slowly equilibrating metabolites, such as the adenine and guanine nucleotides, are usually considered to be in separate pools (see Section II,D). It is very important to specify which metabolites are components of the pool. For instance, all the following meanings have been used in the literature for what is sometimes vaguely referred to as the "thymidine pool": (1) total free intracellular dT; (2) total intracellular thymine metabolites (i.e., T + dT + dTMP + dTDP + dTTP); (3) total thymine nucleotides available for DNA synthesis; (4) acid-soluble label extractable from cells incubated with radioactive dT, (5) nucleotides derived from radioactive dT.

Normal endogenous pools in cells are expanded to various extents during the process of labeling them with radioactive exogenous precursors (see Section IV,F). Thus, labeling creates a fraction of the pool which was not there before. There is an unfortunate tendency to equate this labeled fraction with the total pool; the endogenous pool has frequently been ignored since it is more difficult to measure.

The intracellular *concentration* (molarity) of various pool components is frequently calculated by dividing the total intracellular amount of the component by the total cell water volume available to it (see Section VII,D). Arguments about enzyme inhibition or activation *in vivo* often utilize these calculated concentrations for comparison with kinetic information (K_m, K_i, V_{max}) gathered from *in vitro* assays (e.g., Adams, 1969). There are certain pitfalls in such arguments. For every pool component, it is likely that a wide spectrum of strong and weak binding sites exists within the cell. Only a fraction of these sites may actually be on enzymes involved in the metabolism of the compound. None of the quantitative methods for pool measurement can distinguish between free and bound intracellular nucleotide. Pool extraction solubilizes all bound and free nucleotides by its acidic, denaturing conditions. In isotopic studies, labeled and unlabeled metabolites are rapidly exchangeable between free and bound states. Hence, the calculated intracellular concentration is merely a *maximum* estimate; the true intracellular concentration of a nucleotide capable of interacting in a regulatory fashion with an enzyme may be vastly lower than this estimate.

B. Four-Factor Model

A model for nucleotide pools was proposed in 1963 by Quastler. There are four basic pathways by which components may enter or leave the pool—hence the name "four-factor model." As shown in Fig. 1, two processes (anabolism and salvage) supply the pool, and two others (catabolism and utilization) drain it. These processes would apply to the thymine nucleotide pool, for example, as described below [see Cleaver (1967) and Section III].

Anabolism: UMP produced by the *de novo* pyrimidine pathway is changed by way of kinases, phosphatases, ribonucleotide reductase, and thymidylate synthetase to dTTP (Figs. 6 and 7). This route, along with some contribution from the deoxycytidylate deaminase pathway, accounts for all endogenous thymine nucleotide synthesis.

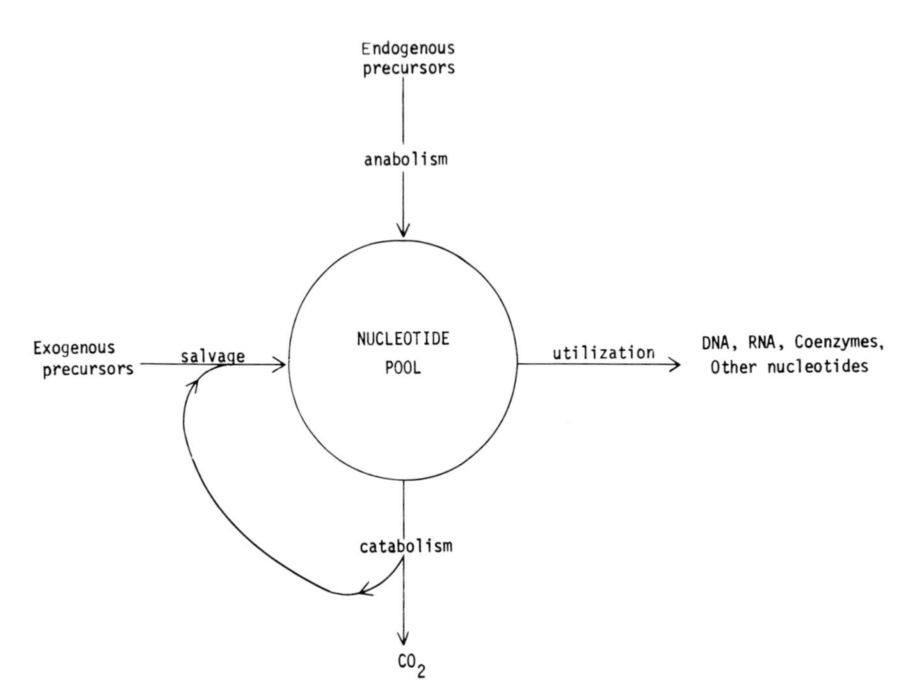

FIG. 1. The four-factor model for nucleotide pools; redrawn from Quastler (1963) and Cleaver (1967), by permission from North-Holland, Amsterdam.

Salvage: The incorporation of exogenous or endogenous thymine or thymidine into the thymine nucleotide pool requires thymidine kinase. When the precursor sources are exogenous, specific uptake systems may also be involved in the salvage process.

Catabolism: Breakdown of thymine nucleotides to thymidine by 5′-nucleotidase and then to thymine by phosphorylase is followed by degradation to CO_2 and other products. The recycling of endogenously formed thymine and thymidine may occur by the salvage pathway.

Utilization: Incorporation of dTTP into DNA is catalyzed by DNA polymerase and is the other major route by which the thymine nucleotide pool is drained. Other pools may have more complex utilization pathways. For instance, ATP enters RNA, DNA (via ADP → dADP), coenzymes, and cyclic AMP, and participates in many enzymatic reactions as an energy or phosphate source.

Several of the enzymes mentioned above are sensitive to feedback inhibition. High levels of dTTP will inhibit ribonucleotide reductase, deoxycytidylate deaminase, and thymidine kinase. Thus, increases in the pool size could tend to shut off pathways which augment the pool.

It is generally assumed that the metabolites of the pool are freely mixing, such that entry and exit are statistically random processes. For experiments involving radioactive precursors of nucleic acids, this assumption dictates that the specific activity of the pool is equal to the specific activity of the precursor in newly synthesized polymer. Deviations from this behavior lead to postulates of multiple pools (Section II,F).

Interrelationships between the four factors or pathways are responsible for such properties of the pool as (1) rate of equilibration with exogenous label, (2) turnover rate, (3) rate of equilibration with other pools, (4) rate of macromolecule labeling, and (5) pool size. The four-factor model has been used specifically for analyzing the kinetics of incorporation of exogenous radioactive thymidine into DNA (Quastler, 1963; Stewart *et al.*, 1965; Cleaver and Holford, 1965). When assumptions and algebraic relationships are appropriately modified, the model can, in principle, predict the fate of virtually any exogenous or endogenous metabolite. Mathematical treatment of the thymine nucleotide pool has yielded some useful relationships (Cleaver, 1967). Unfortunately, essential quantitative information on the kinetics and regulatory properties of the various pertinent pathways is usually not available.

Three different approaches discussed below can yield estimates of the normal rate of flow of endogenously synthesized thymine nucleotides through the pool. The same ideas could be applied to other pools as well.

Theoretical needs of the cell. If the principal reactions involving a nucleotide are known, then the demands which the cell places on the pool can be balanced against known routes of supply. For instance, most of the thymine nucleotide pool is used for DNA synthesis (Cleaver, 1967; Adams, 1969). Catabolism of thymine nucleotides to thymine and other products is apparently small in many rapidly growing cell types, although there may be some exceptions [Cooper *et al.* (1966b) and Section II,C]. Weber *et al*, (1971) estimate for the very rapidly growing 3683-F rat hepatoma that the rate of thymidine incorporation into DNA is about 9200 times greater than the rate of thymidine catabolism to CO_2. Given the cellular DNA complement (about 12 pg/cell; Table X), the base composition of about 30% T, and the duration of the S period (about 8 hours), one can calculate the minimum rate of endogenous dTTP production. This value is about 1.4×10^7 molecules of dTTP/cell per minute or 24 pmoles/10^6 cells per minute for an S period of 8 hours, with proportionally lower rates for longer S periods (see Cleaver, 1967).

Direct measurement. Conrad and Ruddle (1972) have assayed thymidylate synthetase in Chinese hamster cells (Don). In extracts of exponentially growing cells the rate of endogenous dTMP production ranges from 9 to 15 pmoles/10^6 cells per minute. The enzyme activity is 4- to 8-fold reduced in lag phase and stationary phase cultures. While it is obviously possible that the *in vitro* enzyme activity does not equal the *in vivo* level, this rate of dTMP synthesis is in rough agreement with the calculated needs of the cell.

Reversal of specific inhibition. Many drugs can block the normal endogenous route for synthesis of a particular nucleotide. In some instances, the block can be reversed by supplying an appropriate exogenous precursor of the nucleotide. Amethopterin inhibition of thymidylate synthetase will halt cell proliferation, but sufficient amounts of dT will reverse the effect of the inhibitor. The maximum growth rate of Novikoff hepatoma cells in the presence of amethopterin can be preserved by addition of 5×10^{-5} to 10^{-4} *M* dT (Gentry *et al.*, 1965a). In this range of dT concentration, the initial rate of uptake is 24 to 42 pmole dT/10^6 cells per minute for Novikoff rat hepatoma cells (subline NlSl-67) in suspension culture (Plagemann and Erbe, 1972). The average rate of dT uptake over a 3-hour period is lower under the same conditions, amounting to about 5 pmole/10^6 cells per minute (calculated from Plagemann, 1971a). Thus with a fortunately detailed set of information on transport rates, a bracketing estimate of the normal endogenous requirement for thymine nucleotides of about 5 to 40 pmoles/10^6 cells per minute is obtainable.

Studies by Cooper *et al.* (1966a) on human leukemic leukocytes have clarified the degree to which the normal endogenous production of thymine nucleotides can be supplemented by exogenous thymidine. Figure 2 shows that exogenous dT can supply more than 80% of the dTTP for DNA synthesis at high concentrations. *De novo* pathways were apparently not turned off by the precursor in this case, except at very high dT concentrations, where DNA synthesis begins to be inhibited (Fig. 2; see Section III,A for discussion of dT inhibition).

C. Turnover

The turnover rate of a pool is proportional to the fraction of the pool per unit time which is drained by catabolism and utilization. Conditions which change the draining reactions will affect the turnover rate. For instance, hydroxyurea inhibition of DNA synthesis in mouse embryo cells slows the turnover of the dTTP pool (Skoog and Nordenskjöld, 1971). Measurements of turnover usually involve labeling the pool with a radioactive precursor, removing the exogenous label, and then observing the rate of decrease of the radioactivity in the pool. Expansion of the pool by the labeling procedure (see Section IV,F) must be taken into account, because an expanded pool will decay more slowly than a normal one.

Cooper *et al.* (1966b) have developed an accurate procedure for assessing turnover of the thymine nucleotide pool in human leukocytes. Instead of ignoring the pool expansion problem, Cooper *et al.* (1966b) maintained the

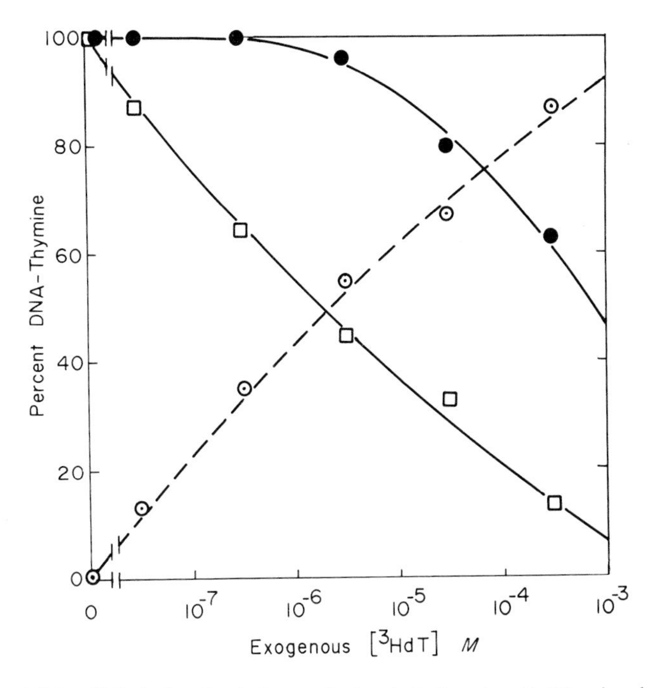

FIG. 2. Origin of DNA-thymine in human leukemic leukocytes. □, Thymine derived from *de novo* synthesis of dTTP; ⊙, thymine derived from exogenous ^{3}HdT; ●, extent of DNA synthesis compared to control cells in the absence of dT. Data from Cooper *et al.* (1966a).

expanded pool at a constant size by changing only the specific activity, not the concentration of the ^{3}HdT for labeling (see Fig. 3). At 14 μM ^{3}HdT, the dTTP pool achieved equilibrium with exogenous label within 10 minutes; 4-fold higher levels of radioactivity were present in the dTTP from cells labeled at 800 μCi/mmole than in those labeled at 200 μCi/mmole (Fig. 3). After 10 minutes, the high specific activity culture was diluted back to 200 μCi/mmole, and the dT concentration was maintained at 14 μM. The amount of radioactivity in dTTP decayed with a half-life of 60 to 90 seconds as it approached the lower equilibrium level. At 14 μM dT, the dTTP pool is expanded to between 12 and 40 times the normal pool level (see Fig. 17), or to about 240 to 800 pmoles/10^6 cells (Table II). Cooper *et al.* (1966b) calculated that only one-third of the dTTP pool was utilized for DNA synthesis, with the rest being drained by phosphatase reactions. Since the expanded pool turned over completely in 3 minutes (Fig. 3), it can be estimated that 1 minute of DNA synthesis utilizes 1 to 4 normal pool complements, or approximately 20 to 80 pmoles of dTTP/10^6 cells per minute. This is in approximate agreement with the calculations made in Section II,B. Ideally,

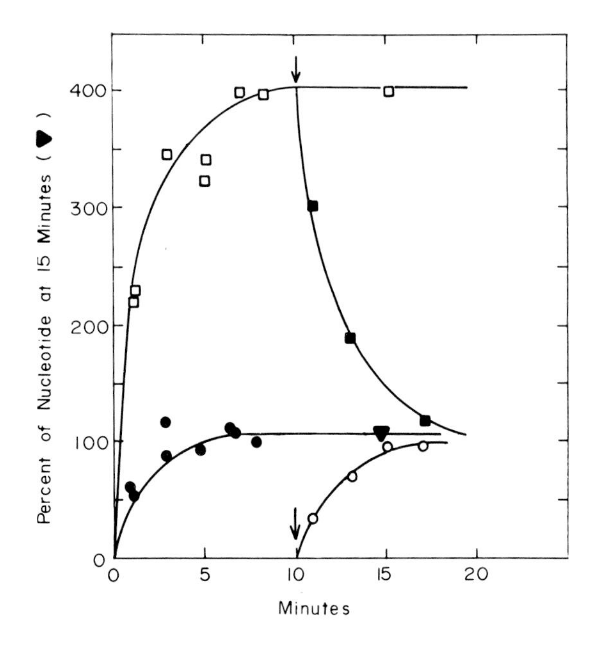

FIG. 3. Turnover of dTTP pool in human leukemic leukocytes at 37°. ●, Control cells incubated from 0 time with 14 μM dT containing 200 μCi of ^{3}HdT/mmole dT. ▼, Value for control cells at 15 minutes; all other points are expressed as a percent of this value. □, Cells incubated from 0 time with 14 μM dT containing 800 μCi of ^{3}HdT/mmole; at 10 minutes, nonradioactive dT was added, maintaining the dT concentration at 14μM but lowering the specific activity of ^{3}HdT to 200 μCi/mmole. ○, Cells incubated with unlabeled 14μM dT from 0 time; at 10 minutes, ^{3}HdT was added, maintaining the dT concentration at 14μM, but raising the specific activity of ^{3}HdT to 200 μCi/mmole. Extracted pools were purified by ion-exchange chromatography and analyzed for radioactivity. Redrawn from Chart 5 of Cooper *et al.* (1966b), by permission of Cancer Research, Inc., Philadelphia, Pennsylvania.

for very accurate turnover studies, the method of Cooper *et al.* (1966b) would be used over a wide range of precursor concentrations (10^{-7} to 10^{-5} M). The true half-life of the *unexpanded* pool could be calculated by extrapolating to a precursor concentration of zero. Other measurements of the turnover rate of the thymine nucleotide pool have been made for animal cells by Cleaver (1967) and Nordenskjöld *et al.* (1970). Generally, the endogenous dTTP pool can support only a few minutes or less of DNA synthesis.

Neuhard and Thomassen (1971) observed rapid turnover of the dCTP and dATP pools in *Escherichia coli* even in the absence of DNA synthesis. As in the human leukocyte system (Cooper *et al.*, 1966b), it was concluded that catabolic pathways play a major role in the turnover phenomenon. Turnover rates may vary with growth conditions (Weber and Edlin, 1971; Neuhard and Thomassen, 1971). The ATP pool in density-inhibited 3T3 cells is labeled by

$^{32}P_i$ more slowly than in exponentially growing cells; reduced phosphate transport and slower pool turnover in the inhibited cultures are the cause of this difference (Weber and Edlin, 1971).

D. Equilibration between Pools

The rate at which intracellular pools equilibrate with an exogenous labeled precursor is determined by transport of the precursor (Section IV) and by metabolic interconversions within the cell (Section III). The unidirectional character of many enzyme catalyzed nucleotide interconversions often causes situations where label from pool A can equilibrate with pool B, but not vice versa. A pool which can only *receive* label from other pools is advantageous for specific labeling of macromolecules. For instance, the dTTP pool (and then DNA) can be labeled with dT or dU with virtually no diversion to other cellular components. Typical patterns of pool equilibration in animal cells are shown in Fig. 4. Solid arrows represent the known directionality of flow of molecules between pools; the broken arrows are slow processes in most cell types. Enzymatic reactions involved in the labeling of individual pools and in the pool equilibration processes are discussed in Section III. If we consider the purine nucleotide pools in Fig. 4, for example, it is a typical observation that guanine nucleotides and adenine nucleotides do not readily equilibrate with one another (McFall and Magasanik, 1960; Emerson and Humphreys, 1971; Crabtree and Henderson, 1971). The conversion of adenine nucleotides to guanine nucleotides is limited by GMP reductase activity, and the reverse process depends on the activity of AMP deaminase (Crabtree and Henderson, 1971). When the pertinent enzyme activity is present, rapid pool equilibration in that direction may be observed. Cook and Vibert (1966) found labeling of the adenine nucleotide pool by guanine and xanthine in mammalian reticulocytes and erythrocytes. Both the adenine and guanine nucleotides can be labeled through the IMP pool when radioactive hypoxanthine is used as a precursor (Brockman *et al.*, 1970).

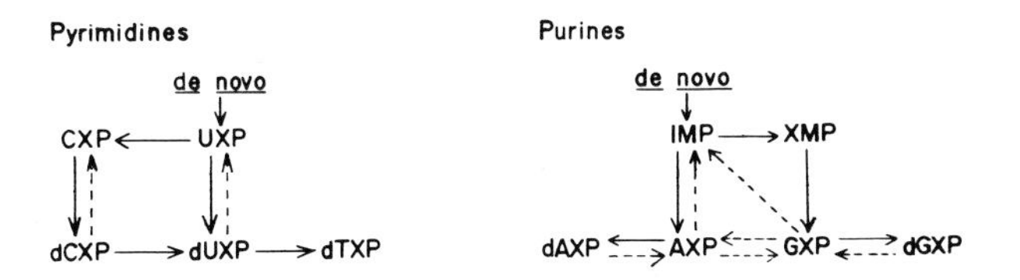

FIG. 4. Typical patterns of pool equilibration in cultured animal cells. Components in a pool are readily converted to other nucleotides in the direction of the solid arrows, and are more slowly converted along the pathways marked by broken arrows.

E. Intracellular Location of Pools

Nuclei isolated from rat liver by organic solvent techniques were found to have high levels of deoxyribonucleotides compared to the bulk tissue (Behki and Schneider, 1962). Adams (1969) reported nuclear localization of the ^{3}HdTTP pool in ^{3}HdT-labeled mouse L929 cells by autoradiography. Presumably, thymidine kinase is also restricted to the nucleus (Adams, 1969). In the presence of hydroxyurea, the ^{3}HdTTP pool appears to be distributed throughout the cytoplasm and nucleus and is less readily available for DNA synthesis (Adams *et al.*, 1971).

In view of the report of a distinct mitochondrial thymidine kinase (Attardi and Attardi, 1972), it is possible that mitochondria will accumulate ^{3}HdTTP under certain conditions of ^{3}HdT labeling.

Ribonucleotide pools are probably distributed equally throughout cells (Ove *et al.*, 1967; Adams *et al.*, 1971). However, Plagemann (1971b, 1972) has proposed the existence of separate, nonequilibrating nuclear and cytoplasmic pools of uracil nucleotides (see Section II,F).

F. Compartmentalization—Multiple Pools

Evidence for multiple pools of the same compound in a single cell derives from precursor incorporation studies in which the labeled precursor does not appear to mix freely with the intracellular pool. For instance, Plagemann (1971b) estimated that only 20 to 25% of the total uracil nucleotides formed from exogenous uridine by Novikoff hepatoma cells were readily available for RNA synthesis. The remaining uracil nucleotides entered RNA at a very slow rate. From this and other evidence, Plagemann (1971b, 1972) has postulated separate nuclear and cytoplasmic pools for the ribonucleotides. The cytoplasmic pool contains about 95% of the total nucleotide, and is expandable by exogenous nucleosides. The small nuclear pool apparently remains small in the presence of exogenous nucleosides (Plagemann, 1972). While these models can explain some of the puzzling RNA labeling kinetics, certain complications are present. Cytoplasmic mengovirus RNA synthesis in the presence of actinomycin D was used to sample the "cytoplasmic" ribonucleotide pools. Several important factors which could affect RNA labeling were not considered: (1) altered enzyme patterns and nucleoside transport in the virus-infected cells (see Sections III and IV); (2) possible effects of actinomycin D on the pool sizes; (3) the rapid turnover of some RNA species. Others have found no evidence for the compartmentalization of the ATP pool (Emerson and Humphreys, 1971; Brandhorst and Humphreys, 1971).

The phenomenon of multiple pools and compartmentalization is related to pool equilibration discussed in Section II,D. For example, one of the major uracil nucleotides is the class of UDP-sugars (see Table II). Rapid formation of these compounds from ^{3}HrU, and then slow conversion back to ^{3}HUTP might cause aberrant RNA labeling kinetics. The observation of preferential use of a precursor (i.e., lack of free mixing with the pool) implies that it is either physically separated from the pool (by membranes) or that it is very strongly bound to some site during the course of several consecutive enzymatic alterations. Werner (1971) has entertained the notion of two separate dTTP pools in *E. coli*.

G. Pool Size

Table II summarizes much of the available information on nucleotide pool sizes in animal cells. For bacterial nucleotide pools, the papers of Neuhard (1966), Bagnara and Finch (1968), and Cashel and Gallant (1968) are useful. Data in Table II were gathered primarily by direct methods (Section VI) or by the ^{32}P method described in Section V. The sizes of expanded pools caused by precursor uptake are considered elsewhere (see Section IV,F). Pool sizes have been expressed as pmoles/10^6 cells to facilitate comparison. For some data this normalization has required certain assumptions regarding cell volume, protein, or DNA content of the studied cells. Hence, the numbers in Table II may vary by as much as a factor of 2 or 3 from the true pool size. Table X contains useful parameters for making these interconversions. Differences between cell volumes should also be considered. Adult rat liver cells are about 3 times as large as cells of the mouse L929 line; erythrocytes are about 2 to 3% of the volume of the adult liver cell. Using 75% for an average cell water content, 10 pmoles of a substance in 10^6 cells would have a maximum concentration (see Section II,A) of about 5 μM thoughout the cell, or about 25 μM if restricted to the nucleus [assuming 2.5 μl/10^6 cells, and 500 μ^3/nucleus; see Table X, and Adams (1969)].

The total ribonucleotide pool in animal cells ranges from about 4 nmoles/10^6 cells in chick embryo fibroblasts (Colby and Edlin, 1970) to almost 40 nmoles/10^6 cells in adult rat liver (Mandel, 1964). Distribution of ribonucleotides within the pool seems to fall into two classes, as pointed out by Mandel (1964). One class is the "metabolic type" such as the liver cell, in which the proportion of nucleotides is roughly: 50 to 60% adenine, 10 to 12% guanine, 5% cytosine, and 10 to 25% uracil. The other class is the "energetic type" such as the erythrocyte where the approximate distribution of nucleotides is 85% adenine, 5% guanine, 2% cytosine, and 8% uracil. The ribonucleotides are generally in the triphosphate form (70 to 90%) except for the uracil nucleotides [see Table II and Mandel (1964)]. Most of the cultured cell lines in Table II seem to belong to the "metabolic type."

TABLE II
POOL SIZES IN ANIMAL CELLS

Cell type	Pool size (pmoles/10^6 cells)					Reference[g]
	Adenine ribonucleotides					
	AMP	ADP	ATP	NAD	Total AXP	
Ehrlich ascites tumor[a]	240	800	3800	140	4900	1
Sarcoma 180	1000	1700	6000		8700	2
Mouse 3T3[b]	70	430	10500		11000	3
Chick embryo fibroblast[b]	80	400	2700		3200	4
Novikoff hepatoma					7900	5
Chinese hamster V79[c]					4000–8000	6
Rat liver[a]			19000		21000	7
Rat skeletal muscle[a]			21000		24000	7
Human erythrocytes[a,d]			2700		3400	7
	Guanine ribonucleotides					
	GMP	GDP	GTP		Total GXP	
Ehrlich ascites tumor[a]			500			1
Mouse 3T3[b]			1900			3
Chick embryo fibroblast[b]			170			4
Novikoff hepatoma[a]	100	100	1400		1600	5
Rat liver[a]			3700		4100	7
Human erythrocytes[a,d]					240	7
	Cytosine ribonucleotides					
	CMP	CDP	CTP		Total CXP	
Mouse 3T3[b]			1500			3
Chick embryo fibroblast	20		70			4
Novikoff hepatoma[a]	60	20	820		900	5
	Uracil ribonucleotides					
	UMP	UDP	UTP	UDP-sugar	Total UXP	
Ehrlich ascites tumor[a]			550			1
Sarcoma 180	460	360	380	390	1600	2
Mouse 3T3[b]	870	75	3100		4100	3
Chick embryo fibroblast[b]	200	30	200		430	4
Novikoff hepatoma					2800	5
Mouse L5178Y	640	550	470	480	2100	9
Rat liver[a]					5400	8
Rat brain[a]					1600	8
Human erythrocytes[a,d]					320	7

TABLE II (*cont'd.*)

Cell type	Pool size (pmoles/10^6 cells)					Reference[g]
	Deoxyribonucleoside triphosphates					
	dATP	dGTP	dCTP	dTTP		
Mouse L929	15	17	17	20		10
Mouse L929[c]	3–15	2–21	5–38	3–80		10
Mouse embryo cells[b]	20			28		10
Mouse embryo cells[e]	5	0.6	11	5		12
Mouse embryo cells[f]	40	7	115	50		12
Mouse 3T3[e]	12	10	9	8		3
Mouse 3T3[f]	96	24	56	77		3
HeLa[c]	3–26	5–22	8–31	10–45		13
Chick embryo fibroblasts[b]	15	6	7	10		4
BHK[a,e]	11	1.0	6	5		14
BHK[a,f]	25	3.8	65	31		14
	Other pools					
	P_i	organic-P	Total P	cyclic AMP	cyclic GMP	
Ehrlich ascites tumor[a]			10000			1
Novikoff hepatoma			62000			5
Mouse 3T3[a]			35000			15
Mouse 3T3[b]	21000	55000	76000			3
Mouse 3T3[a,c]				7–17		16
Mouse 3T3[a,e]				19		17
Mouse 3T3[a,f]				5		17
Chinese hamster CHO[a,c]				11–31		18
Rat liver[a]				1.3	0.11	19
Rat lung[a]				5	2.0	19
Rat cerebellum[a]				25	2.5	19

[a]Data normalized using appropriate values from Table X.

[b]Average of a range of observed values.

[c]Observed range of pool size over the cell cycle in synchronized cell populations.

[d]Erythrocyte pool sizes expressed as pmoles/3×10^7 cells, which is the volume equivalent of about 10^6 cells of the other types listed.

[e]Growth arrested by serum depletion or contact inhibition.

[f]Growth stimulated by fresh serum, or cells in exponential growth.

[g]References: 1, Letnansky (1964); 2, Bekesi and Winzler (1969); 3, Weber and Edlin (1971); 4, Colby and Edlin (1970); 5, Plagemann (1972); 6, Chapman *et al.* (1971); 7, Mandel (1964); 8, Keppler *et al.* (1970); 9, Bosmann (1971); 10, Adams *et al.* (1971); 11, Nordenskjöld *et al.* (1970); 12, Skoog and Nordenskjöld (1971); 13, Bray and Brent (1972); 14, Bjursell *et al.* (1972); 15, Cunningham and Pardee (1969); 16, Burger *et al.* (1972); 17, Otten *et al.* (1971); 18, Sheppard and Prescott (1972); 19, Murad *et al.* (1971).

The total deoxyribonucleotide pool amounts to between 30 and 350 pmoles/10^6 cells, or about 1% of the ribonucleotide pool. Of this, about 70 to 90% is in the triphosphate form (Gentry *et al.*, 1965a; Cleaver, 1967). The smallest component of the pool is usually dGTP, whereas the most abundant, dCTP, can be up to 18 times more plentiful than dGTP. Several cell types have approximately equal amounts of the four dNTPs (see Table II).

H. Cell Cycle and Growth-Rate Dependence of Pools

In studying changes in nucleotide pools over the cell cycle of synchronized cells it is important to realize the limitations of the chosen experimental method. For example, the ability of cells to incorporate a radioactive nucleoside into their nucleotide pool is known to be maximal during the S and G_2 phases of the cell cycle (Miller *et al.*, 1964; Adams, 1969), hence the pool contains more labeled nucleotide during S and G_2. These changes in pool labeling are caused primarily by fluctuation in the activity of nucleoside kinases, however, and may have no relationship to the true pool size. Several studies of nucleotide pool sizes during the cell cycle have been carried out with direct techniques, and thus have avoided the above problems [see Section VI, Adams *et al.* (1971), Chapman *et al.* (1971), Bray and Brent (1972)]. In G_1 arrested mouse L929 cells, the deoxyribonucleoside triphosphate levels are very low. As the cells traverse the S phase the levels increase 2- to 17-fold, and in G_2 they are as much as 27 times larger than the G_1 values. The largest changes are found for dTTP and the smallest for dATP [see Table II and Adams *et al.* (1971)]. Similarly, Bray and Brent (1972) reported that the concentrations of the dNTPs in synchronized HeLa cells changed by 4- to 9-fold over the cell cycle, with minimum levels during G_1.

Many studies have been concerned with the relationship between nucleotide pool sizes and cell growth rate. Ribonucleotide pools are rather constant in mouse 3T3 cells and chick embryo fibroblasts, whether they are rapidly growing or density-inhibited (Colby and Edlin, 1970; Weber and Edlin, 1971). However, the deoxyribonucleoside triphosphate pools are elevated by 2- to 27-fold in several cell types during conditions of rapid growth, serum stimulation, and virus infection (Nordenskjöld *et al.*, 1970; Adams *et al.*, 1971; Skoog and Nordenskjöld, 1971; Weber and Edlin, 1971; Bjursell *et al.*, 1972). Chick embryo fibroblasts showed insignificant changes in dNTP levels with alteration of the growth rate (Colby and Edlin, 1970).

Do the changes in pool size with different growth conditions originate from the known variations in pool size during the cell cycle? Slowly growing cells have a much greater fraction of the population in the G_1 phase than do rapidly growing cells (see Section III,C). As the cell growth rate decreases, and more cells are in G_1, the nucleotide pool sizes approach the low levels characteristic of synchronous G_1 cell populations (e.g., Adams *et al.*, 1971; Bray and Brent, 1972; Bjursell *et al.*, 1972). An analogous argument regarding enzyme levels and growth rate is presented in Section III,C.

Cyclic AMP shows interesting changes during the cell cycle (Burger *et al.*, 1972; Sheppard and Prescott, 1972), and it has been argued that the cyclic AMP pool plays an important role in controlling cell growth rate (Ryan and Heidrick, 1968; Otten *et al.*, 1971; Sheppard, 1971; Burger *et al.*, 1972; Hauschka *et al.*, 1972).

I. Drug and Irradiation Effects

A common mechanism for growth inhibition by drugs is interference with the synthesis of an essential cell metabolite. Numerous compounds that act as growth inhibitors have been tested for this property. The compounds 6-mercaptopurine and 6-methylmercaptopurine ribonucleoside cause a 2- to 3-fold decrease in the adenine nucleotide pool of Sarcoma 180 cells by interference with *de novo* purine synthesis (Scholar *et al.*, 1972). Azaserine reduces both the adenine and thymine nucleotide pools in Sarcoma 180 cells [see Section III and Sartorelli and Booth (1967)]. Aminopterin treatment of mouse L929 cells causes a 2-fold decrease in the dTTP pool, while the other dNTPs increase 2- to 5-fold [see Section III and Adams *et al.* (1971)].

Cytosine arabinoside allows a 2- to 3-fold increase in all dNTPs of mouse embryo cells except for dCTP, which decreases about 2-fold; direct measuring techniques (Section VI) were used (Skoog and Nordenskjöld, 1971). Under similar conditions, the ability of mouse L cells to incorporate ^{3}HdC into the dCTP pool is *increased* about 3-fold by cytosine arabinoside (Graham and Whitmore, 1970). This points out one of the problems with isotopic measurements when information on pool size is desired. Inhibition of DNA synthesis by hydroxyurea can be accompanied by almost no change in the absolute size of the dNTP pools of mouse L929 cells (Adams *et al.*, 1971), or by complete disappearance of the dGTP pool in mouse embryo cells (Skoog and Nordenskjöld, 1971). In the latter study, dATP levels also decreased, while dCTP and dTTP increased 2- to 3-fold, and the turnover of the dTTP pool was abolished. Brockman *et al.* (1970) studied effects of hydroxyurea and guanazole on mouse L1210 cells and found an 8- to 15-fold increase in the ^{3}HdTTP pool (formed from ^{3}HdT) during the drug treatment. The larger dTTP changes in this experiment compared with those of Skoog and Nordenskjöld (1971) may be another manifestation of the radioactive precursor problem mentioned above.

The UTP pool can be diminished 2-fold in Sarcoma 180 cells by D-glucosamine, and there is a simultaneous increase in the pool of UDP-*N*-acetylhexosamine (Bekesi and Winzler, 1969). Under conditions of limiting glucose concentration, the ATP pool is also decreased by D-glucosamine, presumably because of the energy requirement for excess UDP-sugar synthesis (Bekesi and Winzler, 1969). Bosmann (1971) has speculated that neoplastic cells may be more sensitive to growth inhibition by D-glucosamine because of their greater dependence on uridine nucleotides, specifically UTP. However, it is also possible that neoplastic cells are inhibited merely because of their greater ability to take up D-glucosamine (Bekesi *et al.*, 1969).

X-ray effects on DNA synthesis have been examined in relation to nucleotide pool changes. Smets (1969) discussed possible radiation effects on the pool sizes of dCTP and dTTP in human kidney cells. Slaby *et al.* (1971) have reported slight increases in the pools of dTTP and dATP in mouse embryo cells under conditions where DNA synthesis was strongly inhibited by irradiation.

III. Nucleotide Metabolism

While a complete review of nucleotide metabolism is far beyond the scope of this article, an understanding of the subject is essential for the study of nucleotide pools. The ensuing discussion is focused on the metabolic pathways as they operate in *animal* cells and tissues. There is strong temptation to extrapolate from the detailed information in one experimental system to the void surrounding another. Frequently this is not valid even within various tissues of the same rat, let alone from *E. coli* to man. Some of the elegant feedback control systems established for bacterial pathways [e.g., CTP inhibition of aspartate transcarbamylase; Gerhart and Pardee (1962)] do not exist in the animal cells studied to date, and are probably replaced by control at other points (see Blakeley and Vitols, 1968). Enzymatic activities which are absent in one tissue may be abundant in the neighboring tissue (e.g., Cooper *et al.*, 1972).

Isotopic precursors are frequently used for the study of nucleotide pools and nucleic acid synthesis. Special attention is given to the probable fate of various precursors. Three primary factors determine this fate: (1) the enzymatic activities in the system of interest; (2) feedback effects of the precursor and related metabolites on these activities; and (3) the accessibility of the precursor to pertinent enzymes. A knowledge of the expected metabolites which can be derived from a labeled precursor should be valuable in designing experiments and choosing chromatographic methods.

A. Pyrimidines

1. *De Novo* SYNTHESIS

A summary of the pathway in animal cells is shown in Fig. 5. The subject has been reviewed by Reichard (1959), Blakeley and Vitols (1968), Hartman (1970), and Jones (1971).

(1) Carbamyl phosphate synthetase. There are two principal forms of this enzyme (denoted I and II), and it is probably the glutamine-requiring type II enzyme which provides carbamyl phosphate for pyrimidine synthesis (Tatibana and Ito, 1967; Jones, 1971). Carbamyl phosphate synthetase II is a likely control point for the entire *de novo* pathway in animals; it is the rate-limiting reaction for the pathway in several systems (Tatibana and Ito, 1967; Yip and Knox, 1970) and is subject to feedback inhibition by UTP (Tatibana and Ito, 1967). Correlation has been noted between increased enzyme levels and increased growth rates in normal and neoplastic tissues (Tatibana and Ito, 1967; Yip and Knox, 1970).

(2) Aspartate transcarbamylase. The well known feedback inhibition of this enzyme by CTP and UTP in *E. coli* implicates it as the principal control point for pyrimidine biosynthesis in bacteria (Gerhart and Pardee, 1962). Such control probably does not exist in animal cells, however, since the enzyme is not rate-limiting for the pathway and is not sensitive to inhibition by various nucleotides (Tatibana and Ito, 1967; Appel and Pettis, 1967; Yip and Knox, 1970).

(3) OMP pyrophosphorylase and OMP decarboxylase. There is some evi-

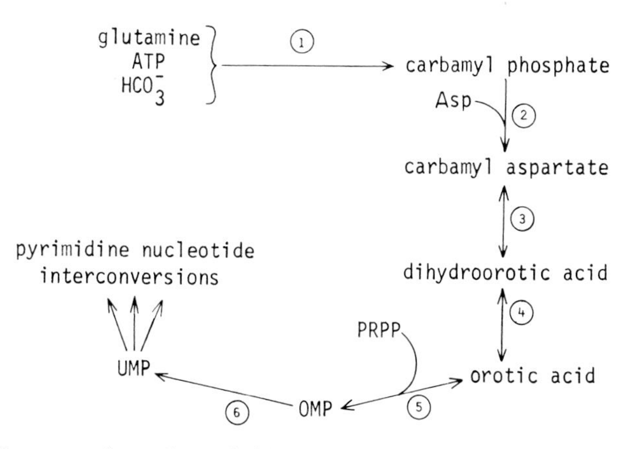

FIG. 5. *De novo* pathway for pyrimidine nucleotide synthesis in animal cells. Enzymes: 1, carbamyl phosphate synthetase; 2, aspartate transcarbamylase; 3 dihydroorotase; 4, dihydroorotic acid dehydrogenase; 5, OMP pyrophosphorylase; 6, OMP decarboxylase.

dence that these enzymes, catalyzing the last two steps of the pathway, share the same protein subunit(s) (Jones, 1971). Levels of both enzymes seem to be controlled by a single mechanism, which is probably substrate induction by the precursor dihydroorotic acid (Pinsky and Krooth, 1967); exceptions to tandem control of the two enzymes have been noted (Jones, 1971). Negative feedback control is also possible as suggested by UMP inhibition of OMP decarboxylase (Blair and Potter, 1961).

(4) Precursors. The first two steps of the *de novo* pathway provide for incorporation of carbon-labeled CO_2 and aspartic acid into pyrimidines. Purines will also be labeled by CO_2, although only the αN of aspartic acid enters purines at the N-1 position. Bresnick (1965) used carbamyl aspartate-2-^{14}C to label both the acid-soluble pool and RNA (mostly as UMP) in rat liver. Orotic acid is the most commonly used precursor which enters through the *de novo* pyrimidine pathway. The permeability of cells to orotic acid is rather high (e.g., Harbers *et al.*, 1959), hence the rate of orotate incorporation into the nucleotide pool is dependent on the activity of two enzymes, OMP pyrophosphorylase and OMP decarboxylase, and the availability of 5-phosphoribosyl 1-pyrophosphate (PRPP). Since PRPP is also required for *de novo* purine synthesis and is possibly rate-limiting for the entire purine pathway (Blakeley and Vitols, 1968; Murray, 1971), the use of excess orotic acid for labeling could actually inhibit purine synthesis by draining off PRPP for the formation of OMP (Rajalakshmi and Handschumacher, 1968). Orotic acid has been used most commonly for *in vivo* RNA labeling. Uridine and cytidine ribonucleotides and deoxyribonucleotides may be labeled. The extent of incorporation can vary widely for different tissues of the same animal. Witschi (1972) found that orotic acid-6-^{14}C entered rat liver RNA 100 times faster than uridine-2-^{14}C, while lung RNA was labeled twice as fast by uridine as by orotic acid. Most of the label (70 to 99%) of both precursors was found as UMP in RNA, with the remainder as CMP (Canellakis, 1957; Witschi, 1972).

(5) Inhibitors. Aside from the feedback inhibition of pyrimidine biosynthesis discussed above, several other compounds are known to inhibit the pathway. The glutamine analog 6-diazo-5-oxo-L-norleucine (DON) inhibits carbamyl phosphate synthetase II (Yip and Knox, 1970); azaserine may also block at this step. Barbituric acid inhibits dihydroorotic acid dehydrogenase; OMP pyrophosphorylase is inhibited by 5-azaorotic acid; and OMP decarboxylase is blocked by the mononucleotide form of 6-azauridine (Rubin *et al.*, 1964; Pinsky and Krooth, 1967; Vesely *et al.*, 1970).

2. Interconversions

Common pathways in animal cells for the interconversion of pyrimidine nucleotides, nucleosides, and bases are shown in Fig. 6. The subject has

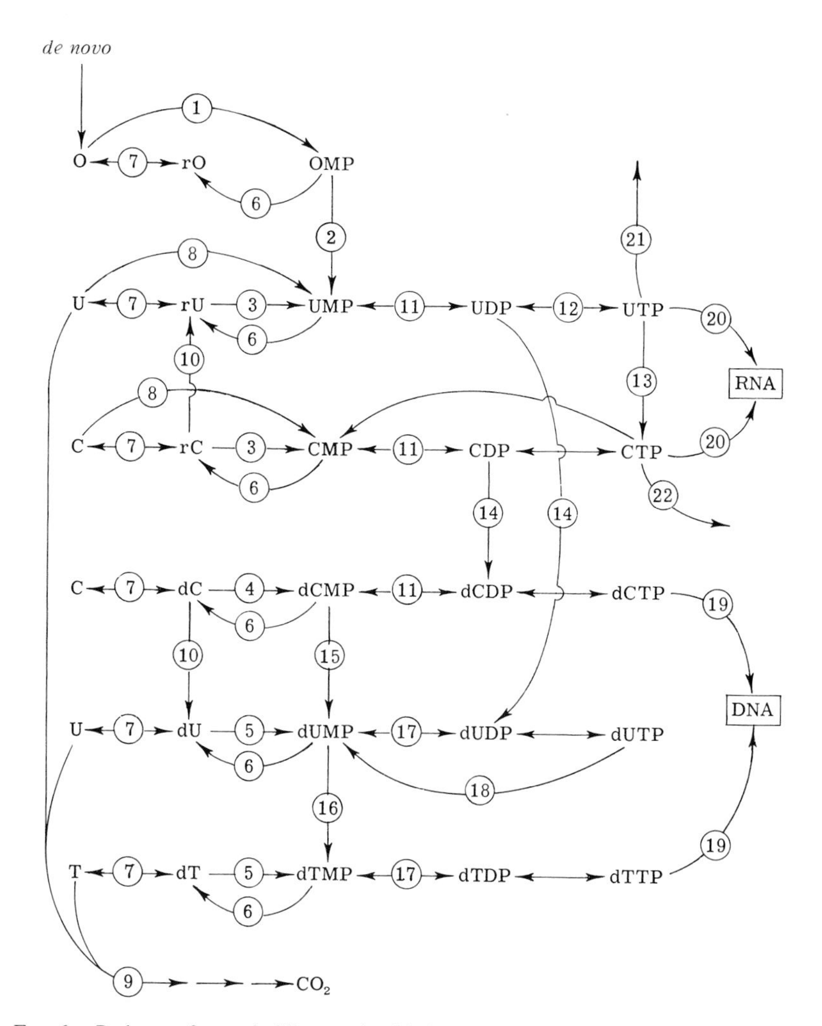

FIG. 6. Pathways for pyrimidine nucleotide interconversion in animal cells. Enzymes: 1, OMP pyrophosphorylase; 2, OMP decarboxylase; 3, uridine kinase; 4, deoxycytidine kinase; 5, thymidine kinase; 6, 5′-nucleotidase; 7, pyrimidine nucleoside phosphorylase; 8, pyrimidine phosphoribosyl transferase; 9, dihydrouracil dehydrogenase; 10, nucleoside deaminase; 11, uridylate kinase; 12, nucleoside diphosphokinase; 13, CTP synthetase; 14, ribonucleotide reductase; 15, deoxycytidylate deaminase; 16, thymidylate synthetase; 17, thymidylate kinase; 18, deoxyuridine triphosphatase; 19, DNA polymerase; 20, RNA polymerase; 21, UDP coenzyme formation; 22, CDP alcohol formation.

been reviewed extensively by Blakeley and Vitols (1968), Hartman (1970), and Kit (1970).

a. Anabolic Reactions. The normal pathways leading to the formation of the four pyrimidine nucleoside triphosphate precursors of nucleic acids

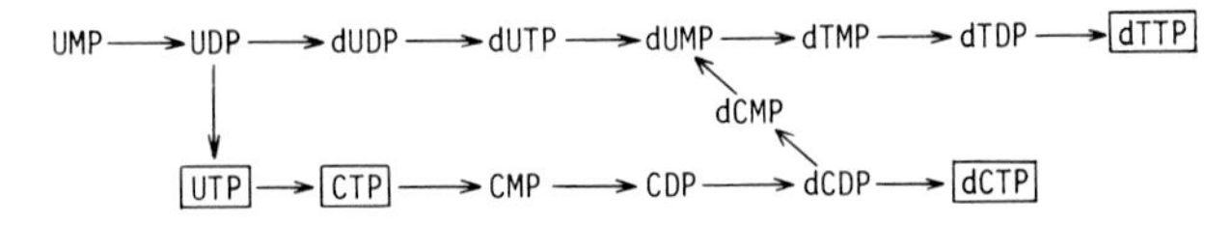

FIG. 7. Normal routes of pyrimidine nucleoside triphosphate formation in animal cells.

from UMP are shown in Fig. 7. The important enzymes catalyzing these reactions are numbered 11 to 17 in Fig. 6.

(1) Uridylate kinase and nucleoside diphosphokinase. Phosphorylation of UMP to form UDP is catalyzed by a nucleoside monophosphokinase which can also utilize CMP and dCMP as substrates (Kit, 1970). Subsequent formation of UTP involves a highly active but relatively nonspecific nucleoside diphosphokinase. The intracellular ratio of UTP:UDP:UMP (see Table II) is indicative of the relative activities of these two kinases. Keeping in mind that the UTP pool is drawn upon heavily for synthesis of UDP-sugars, CTP, and RNA, the preponderance of UTP (Plagemann, 1972) and the low levels of UDP [~6%, Colby and Edlin (1970)] in the uridine nucleotide pool suggests that UDP formation is the rate-limiting step in synthesis of UTP from UMP.

(2) CTP synthetase. This important reaction is the source of all cytidine nucleotides in animal cells. CTP synthesis is irreversible, i.e., deamination of CTP to UTP does not occur, and the enzyme shows some regulatory properties (Hurlbert and Kammen, 1960). GTP stimulates the formation of CTP, and some cytidine nucleotides inhibit the process. Because glutamine is required as an amino donor, inhibition by the analogs DON (strong, irreversible) and azaserine (weak) is observed (Hurlbert and Kammen, 1960).

(3) Ribonucleotide reductase. Reduction of ribonucleoside diphosphates to the corresponding deoxyribonucleotides is the primary mechanism for endogenous production of DNA precursors. Because the pools of deoxyribonucleoside triphosphates are extremely small (Table II) and are sufficient for no more than a few minutes of DNA replication [Cleaver (1967), see also Section II,B], ribonucleotide reductase is thought to occupy a key position in the general control of nucleic acid metabolism (Reichard *et al.*, 1961; Blakeley and Vitols, 1968; Kit, 1970). In common with many enzymes, ribonucleotide reductase increases with increasing cellular growth rate (Elford *et al.*, 1970; Nordenskjöld *et al.*, 1970; Weber *et al.*, 1971), fluctuates during the cell cycle (Murphree *et al.*, 1969) and is possibly a growth rate limiting enzyme in some neoplastic tissues (Cory and Whitford, 1972).

The activity and substrate specificity of mammalian ribonucleotide reductase is controlled in a complex fashion by ATP and several deoxyribonucleoside triphosphates (Moore and Hurlbert, 1966), analogous to the control of the *E. coli* enzyme (Larsson and Reichard, 1966). Reduction of all four substrates (CDP, UDP, ADP, GDP) is apparently catalyzed by the same enzyme in both systems. Strong inhibition of CDP and UDP reduction is possible with low concentrations of dTTP, dUTP, dATP, and dGTP, and the same inhibitors will block ADP and GDP reduction at higher concentrations (see Kit, 1970). This inhibition is probably responsible for the growth-arresting properties of the deoxyribonucleosides dT, dA, and dG (Maley and Maley, 1960; Blakeley and Vitols, 1968). For instance, during the well known "thymidine block" caused by millimolar levels of dT (Ives *et al.*, 1963; Gentry *et al.*, 1965b), high levels of dTTP accumulate intracellularly (Adams *et al.*, 1971), thereby preventing formation of the dCTP required for DNA synthesis. As expected, this inhibition may be reversed by dC which serves as a source of dCTP (Morris and Fischer, 1963; Morse and Potter, 1965; Whittle, 1966). Growth inhibition by dA is reversed only by *simultaneous* application of dC and dG, in which case all four deoxyribonucleoside triphosphates necessary for DNA synthesis can be formed from the three exogenous deoxyribonucleosides [see Figs. 6 and 11 and Klenow, (1962), Blakeley and Vitols (1968)]. The deoxyribonucleoside effects in these systems are probably dependent on the presence of the appropriate nucleoside kinases.

The essential participation of iron and reducing factors in the ribonucleotide reductase reaction (Moore and Reichard, 1964) has prompted the use of some interesting inhibitors of the enzyme. 1-Formylisoquinoline thiosemicarbazone (IQ-1) is a strong iron chelator which apparently interacts with enzyme-bound iron to bring about inhibition at concentrations of 10^{-8} *M* (Moore *et al.*, 1970). Other pyridine carboxaldehyde thiosemicarbazones are also effective (Booth *et al.*, 1971; Moore *et al.*, 1971). The observations on HeLa cells by Robbins and Pederson (1970) and Robbins *et al.* (1972) show that under conditions of iron deprivation by specific chelation with Desferal, DNA synthesis is strongly inhibited while RNA and protein synthesis are not significantly affected. Ribonucleotide reductase inhibition may be the mechanism of this effect.

Hydroxyurea inhibits ribonucleotide reductase in mammalian and bacterial systems, but reversal of the inhibition, either by washing or by addition of deoxyribonucleosides, is not always possible (Turner *et al.*, 1966; Rosenkranz and Carr, 1970). Guanazole has inhibitory properties similar to those of hydroxyurea (Brockman *et al.*, 1970).

(4) Deoxycytidylate deaminase. In addition to supplementing the dUMP pool which is fed primarily by the action of ribonucleotide reductase on UDP (Crone and Itzhaki, 1965), deoxycytidylate deaminase plays a key role in regulating the levels of dTTP and dCTP (Maley and Maley, 1970). Allosteric effects of these two triphosphates have been carefully studied. Feedback inhibition is caused by dTTP, whereas dCTP stimulates the enzyme (Scarano

et al., 1967; Maley and Maley, 1970). Normal intracellular ratios of dCTP to dTTP range from 0.5 to 2 in various types of animal cells (Colby and Edlin, 1970; Weber and Edlin, 1971; Skoog and Nordenskjöld, 1971). Deoxycytidylate deaminase activity is elevated in rapidly growing cells (Maley and Maley, 1960; Scarano *et al.*, 1960; Sneider *et al.*, 1969; Sneider and Potter, 1969a), but because of parallel changes in other enzymes, the dCTP:dTTP ratio is rather insensitive to growth rate (Skoog and Nordenskjöld, 1971; Bjursell *et al.*, 1972).

Purified deoxycytidylate deaminase from mammalian tissues can catalyze the formation of dTMP from 5-methyl dCMP (Geraci *et al.*, 1967). This alternate route for dTTP synthesis may operate in neoplastic mouse cells (Price *et al.*, 1967) and may enable cell populations to resist chemotherapeutic agents which inhibit dTMP synthetase (Sneider and Potter, 1969b). In *E. coli*, as much as 75% of the total dTTP synthesis occurs by deamination of 5-methyl dCMP (Karlstrom and Larsson, 1967). Deoxycytidylate deaminase has been implicated in the conversion of the growth-inhibitory nucleotides of araC to inactive araU nucleotides (Camiener and Smith, 1965). Enhancement of araC effects can be achieved by inhibiting the deamination with tetrahydrouridine (Neil *et al.*, 1970).

(5) Thymidylate synthetase. This enzyme provides the primary route for endogenous dTMP synthesis. Enzyme levels are directly related to growth rate, yet the enzyme is remarkable in that it exhibits no significant activation or feedback inhibition by dTTP or other natural nucleotides (Maley and Maley, 1970; Conrad and Ruddle, 1972). Folic acid and its metabolites are implicated in the control of dTMP synthetase activity, possibly by protecting the enzyme from degradation (Labow *et al.*, 1969; Bertino *et al.*, 1970; Roberts and Loehr, 1971). The primary source of the one-carbon unit for the thymidylate synthetase reaction is serine-C-3; histine-C-2 and exogenous formate can also be utilized (see Fig. 8). Of the formate incorporated into adenine, guanine, and thymine nucleotides in human leukocytes, the fraction in thymine can vary enormously from 3% (chronic lymphocytic leukocytes) to 98% (chronic granulocytic leukocytes) (Wells and Winzler, 1959). The 5-methyl group of dTMP is donated by N^5,N^{10}-methylenetetrahydrofolate to dUMP in a complex reaction in which dihydrofolate is produced (Friedkin, 1963).

Interference with the dTMP synthetase reaction is caused by several folic acid analogs and apparently has several distinct mechanisms: (1) inhibition of dihydrofolate reductase (Hakala, 1971; Ho *et al.*, 1972), thereby preventing regeneration of tetrahydrofolate; (2) competitive inhibition of the binding of N^5,N^{10}-methylenetetrahydrofolate to dTMP synthetase (Friedkin, 1963). The first mechanism, above, is confirmed by studies which show that resistance to folate analogs such as amethopterin is achieved in certain cell types by elevating the level of dihydrofolate reductase (Perkins *et al.*, 1967; Hakala, 1971). In support of the latter mechanism, reduced forms of

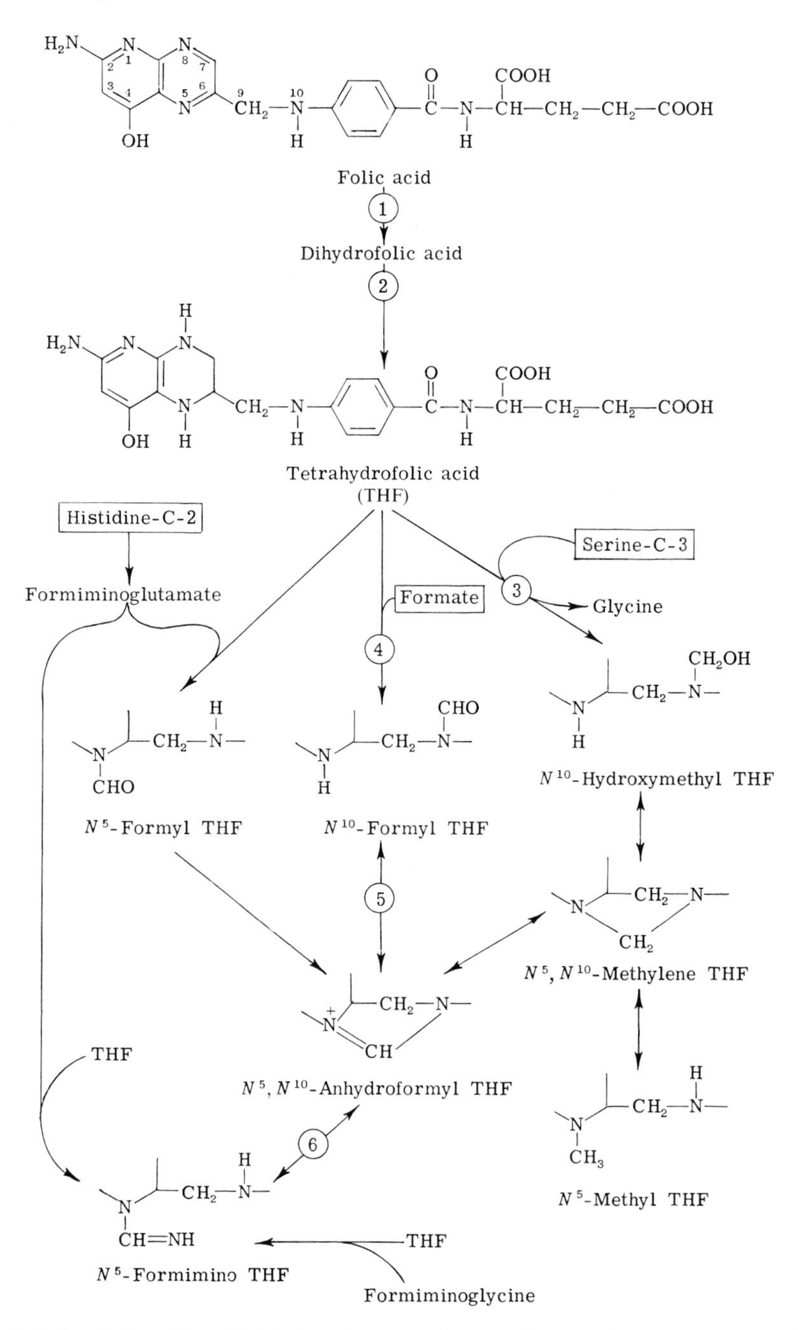

FIG. 8. Folic acid metabolism in animal cells. Donors of one-carbon units are enclosed in boxes. Enzymes: 1, spontaneous chemical reduction; 2, dihydrofolate reductase; 3, serine hydroxymethyl transferase; 4, formate activating enzyme; 5, cyclohydrolase; 6, cyclodeaminase.

aminopterin and amethopterin are much stronger *in vitro* inhibitors of dTMP synthetase than the nonreduced forms because of their greater similarity to the natural reduced substrate (Friedkin, 1963). Also, the differential inhibition of one-carbon transfer reactions by some folate antagonists is known. Homofolate and tetrahydrohomofolate primarily inhibit the introduction of the C-8 carbon in *de novo* purine synthesis and have much weaker effects on dTMP synthetase [see Fig. 10 and Hakala (1971)]. On the other hand, aminopterin can inhibit dTMP synthetase to the extent that incorporation of deoxyuridine into DNA is greatly reduced, while thymidine incorporation can still proceed—implying sufficient pools of purine deoxyribonucleotides in the presence of aminopterin (Friedkin and Roberts, 1956).

Fluorinated pyrimidines (FdUMP, F_3dTMP, and others) are potent inhibitors of dTMP synthetase (Heidelberger, 1965; Reyes and Heidelberger, 1965) and may act in a competitive or noncompetitive manner depending on the conditions. F_3dTMP is capable of alkylating an amino group in the active site of dTMP synthetase (Heidelberger, 1965). Inhibition of dTMP synthetase by FdU is an effective method for achieving very high specific activities of ^{3}HdTTP in the intracellular pool when labeling with ^{3}HdT (D. M. Prescott, personal communication).

(6) Thymidylate kinase. Phosphorylation of dTMP to the triphosphate level is accomplished in two steps (Fig. 6), and available evidence suggests that the thymidylate kinase reaction is the slower of the two (Kit, 1970). The thymine nucleotide pools are mostly composed of dTTP, with only small amounts of dTMP and dTDP (Gentry *et al.*, 1965a; Lindberg *et al.*, 1969). Thymidylate kinase is much less active than the other nucleoside monophosphokinases in many animal systems, and its well characterized fluctuations with the cell cycle and the cell division rate have implicated it in the control of DNA precursor pools (Maley *et al.*, 1965; Kit, 1970).

b. Catabolic Reactions. Four important steps in the degradation of endogenous pyrimidine nucleotides are catalyzed by the enzymes 5′-nucleotidase, pyrimidine nucleoside deaminase, pyrimidine nucleoside phosphorylase, and dihydrouracil dehydrogenase (Fig. 6). The first degradative step is dephosphorylation of the monophosphates by a 5′-nucleotidase of low specificity (Kit, 1970). Deamination of cytidine and deoxycytidine to the corresponding uridine forms is catalyzed by pyrimidine nucleoside deaminase (Ellem, 1968; Kit, 1970).

(1) Pyrimidine nucleoside phosphorylase. Phosphorolysis of nucleosides to yield the free base and ribose- or deoxyribose-1-phosphate is a reversible reaction which is especially important in the breakdown of exogenously supplied nucleosides (Reichard and Sköld, 1958). Flowing in the opposite direction, the reaction provides for the incorporation of free bases into the nucleotide pools (see below). There are large differences in the tissue

distribution of this phosphorylase activity; rabbit intestinal mucosa and liver have several hundred times the activity of skeletal muscle, and brain shows no activity (Friedkin and Roberts, 1954). Nucleoside phosphorylase is elevated in neoplastic tissues (Reichard and Sköld, 1958). Infection of cultured cell lines with mycoplasma (PPLO) is known to increase greatly the phosphorolytic cleavage of thymidine and fluorodeoxyuridine (Hakala *et al.*, 1963). PPLO or L-form infection has been advanced as a possible cause of the unusually active thymidine catabolism (50 to 95% of dT becomes T) in certain sublines of Chinese hamster cells (Dewey *et al.*, 1968). The large effect of nucleoside concentration on the extent of phosphorolysis is an important consideration in the measurement of catabolism (Canellakis, 1957; Dewey *et al.*, 1968).

(2) Dihydrouracil dehydrogenase. Reduction of uracil and thymine by this enzyme is an irreversible reaction (Fritzson, 1957). The degradation products dihydrouracil, β-ureidopropionic acid, and β-alanine cannot be incorporated into the nucleic acids of Ehrlich ascites cells, confirming the irreversibility of the pathway (Lagerkvist *et al.*, 1955). Canellakis (1957) and Queener *et al.* (1971) have discussed the rate-limiting properties of dihydrouracil dehydrogenase for the overall catabolism of pyrimidines. Dihydrouracil dehydrogenase activity and thymidine degradation decrease strongly with increasing growth rates of rat hepatomas (Queener *et al.*, 1971; Ferdinandus *et al.*, 1971; Weber *et al.*, 1971). The tissue distribution of this enzyme is similar to the pyrimidine nucleoside phosphorylase discussed above; highest levels occur in the liver (Potter *et al.*, 1960; Cooper *et al.*, 1972).

Some recent *in vivo* studies concern the systemic effects of the liver on catabolism of free pyrimidines and pyrimidine nucleosides, a factor that severely limits the effective *in vivo* levels of many antineoplastic drugs (Gentry *et al.*, 1971; Cooper *et al.*, 1972). Inhibition of hepatic pyrimidine catabolism at the level of dihydrouracil dehydrogenase by 5-cyanouracil (Gentry *et al.*, 1971) or 5-diazouracil (Cooper *et al.*, 1972) potentiates the growth-controlling effects of fluorouracil by allowing increased incorporation of the drug in all tissues, regardless of the tissue level of dihydrouracil dehydrogenase. Even greater enhancement of pyrimidine incorporation by these inhibitors might be possible with *in vitro* cell lines where hepatic circulation effects would be absent.

c. "Salvage" Reactions. The three reactions that govern the incorporation of free pyrimidines and pyrimidine nucleosides into nucleotide pools are diagrammed in Fig. 9; their relationship to the other pathways is shown in Fig. 6.

(1) Pyrimidine phosphoribosyl transferase (pyrophosphorylase). This reaction is possible only for *ribo*nucleotide formation because of the participation of 5-phosphoribosyl 1-pyrophosphate. It is probably of limited significance for pyrimidine nucleotide formation in most animal tissues (Hartman, 1970), in

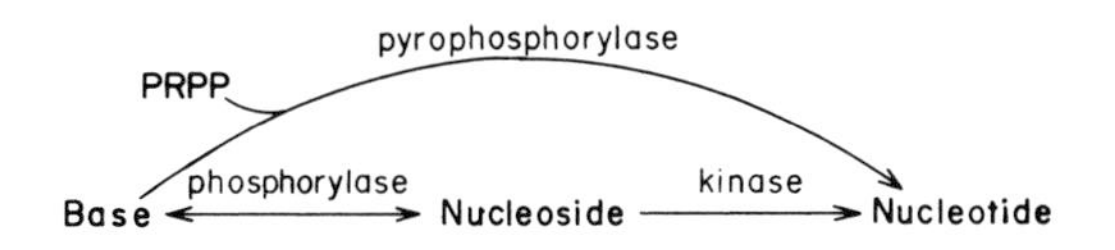

FIG. 9. Alternate routes for salvage of nucleosides and bases.

contrast to the situation for free purine utilization by purine phosphoribosyl transferases (Murray, 1971). Except for orotic acid and fluorouracil [see Section III,A and Kessel *et al.* (1972)], the principal route for pyrimidine base incorporation is through nucleoside intermediates as discussed below.

(2) Pyrimidine nucleoside phosphorylase. The specificity of this enzyme is rather low; ribose-1-P and deoxyribose-1-P are both active. A separate enzyme apparently catalyzes the reaction for purines (Reichard and Sköld, 1958; Kit, 1970). In addition to the formation of nucleoside from free base and pentose phosphate, pentose transfer from a donor nucleoside to a free pyrimidine is catalyzed by nucleoside phosphorylase (Kit, 1970). The overall rate of nucleoside formation is probably limited by availability of pentose phosphate. Strong stimulation (700- to 800-fold) of thymine incorporation into DNA can be effected by dU, dC, dA, and dG (Breitman *et al.*, 1966; Cooper *et al.*, 1972), because these nucleosides can act as deoxyribose donors. Similar findings have been reported for incorporation of uracil, 5-fluorouracil, and 5-bromouracil into nucleic acids of Ehrlich acites cells in the presence of inosine and deoxyinosine (Gotto *et al.*, 1969).

(3) Nucleoside kinase. Phosphorylation of nucleosides by nucleoside kinases completes the two-step formation of nucleotides from free pyrimidines. There are several specific pyrimidine nucleoside kinases in animal cells and all generally utilize ATP for phosphorylation (Hartman, 1970; Kit, 1970). Substrate specificities are fairly well defined. Uridine kinase is active toward rU and rC and their fluorinated derivatives (Hartman, 1970). Deoxycytidine kinase phosphorylates dC, araC, and dG (Schrecker, 1970; Momparler and Fischer, 1968). Thymidine kinase is active toward dT and dU and their halogenated derivatives (Kit, 1970; Hashimoto *et al.*, 1972). Several purine nucleoside kinases are discussed in Section III,B.

Perhaps because of their key role in the incorporation of common nucleoside precursors into nucleic acids, thymidine and uridine kinase have been intensively studied. Thymidine kinase activity increases over the cell cycle from a minimum during G_1 to a maximum in the S, G_2, and M periods (Littlefield, 1966; Stubblefield and Murphree, 1967; Adams, 1969; Bray and Brent, 1972). Enzyme levels are generally elevated in rapidly proliferating cells (Kit, 1970; Weber *et al.*, 1971; Bresnick *et al.*, 1971), and a variety of treatments with materials such as phytohemagglutinin, fresh serum, estrogens, and dibutyryl cyclic AMP can alter kinase activities (Lucas, 1967; Loeb *et al.*, 1970; Nordenskjöld *et al.*, 1970; Garland *et al.*, 1971; Hauschka *et al.*, 1972). Infection by several different types of animal viruses causes increased thymidine kinase levels, even in mutant cell strains which lack the enzyme (Kit, 1970). Multiple forms of thymidine kinase, including a heavy form which is elevated in tumor extracts (Okuda *et al.*, 1972), growth-related variation in the intracellular distribution of the enzyme (Adelstein *et al.*, 1971; Baril *et al.*, 1972), rapid turnover (Bresnick and Burleson, 1970; Adelstein *et al.*, 1971), and substrate protection against turnover (Littlefield, 1966) argue for complex control of the activity. Feedback inhibition of kinases by nucleoside triphosphates is one important aspect of the control. Thymidine kinase is usually inhibited by both potential end products—dTTP (Reichard *et al.*, 1960; Bresnick and Karjala, 1964) and dUTP (Morris and Fischer, 1960). In *Tetrahymena*, thymidine kinase is not inhibited by dTTP (Shoup *et al.*, 1966). dCTP is known to inhibit deoxycytidine kinase (Maley and Maley, 1962).

(4) Pyrimidine base and nucleoside incorporation. Cellular incorporation of free bases is a function of the activities of pertinent "salvage" and catabolic pathways. For example, Reichard and Sköld (1958) found that exogenous uracil was an excellent RNA precursor in Ehrlich ascites cells but was poorly incorporated into rat liver RNA. Active catabolism of uracil to CO_2 in the liver, in contrast to reduced catabolism and elevated uridine kinase activities in the Ehrlich cells explains the original observation. In a series of rat hepatomas, Ferdinandus *et al.* (1971) have found that the ratio of dT catabolism to that which is incorporated into DNA decreases over a 100,000-fold range with increasing growth rate. Under equivalent labeling conditions (10 μmoles/kg, *in vivo*), rat leukemic tissue incorporated thymidine about 90 times more efficiently than thymine (Cooper *et al.*, 1972). The ratio of incorporation dropped to 7 at a precursor concentration of 100 μmoles/kg, presumably because thymidine uptake is less heavely favored over thymine uptake at high concentrations [see Section IV and Cooper *et al.* (1972)]. Other differences in incorporation of nucleosides have similar explanations in terms of transport differences and kinase levels (Plagemann *et al.*, 1969; Plagemann and Roth, 1969; Hauschka *et al.*, 1972), and pool dilution (Lea *et al.*, 1968).

B. Purines

1. *De Novo* SYNTHESIS

A summary of the pathway for IMP synthesis in animal cells is shown in Fig. 10. Because of the important role of folic acid metabolism in *de novo*

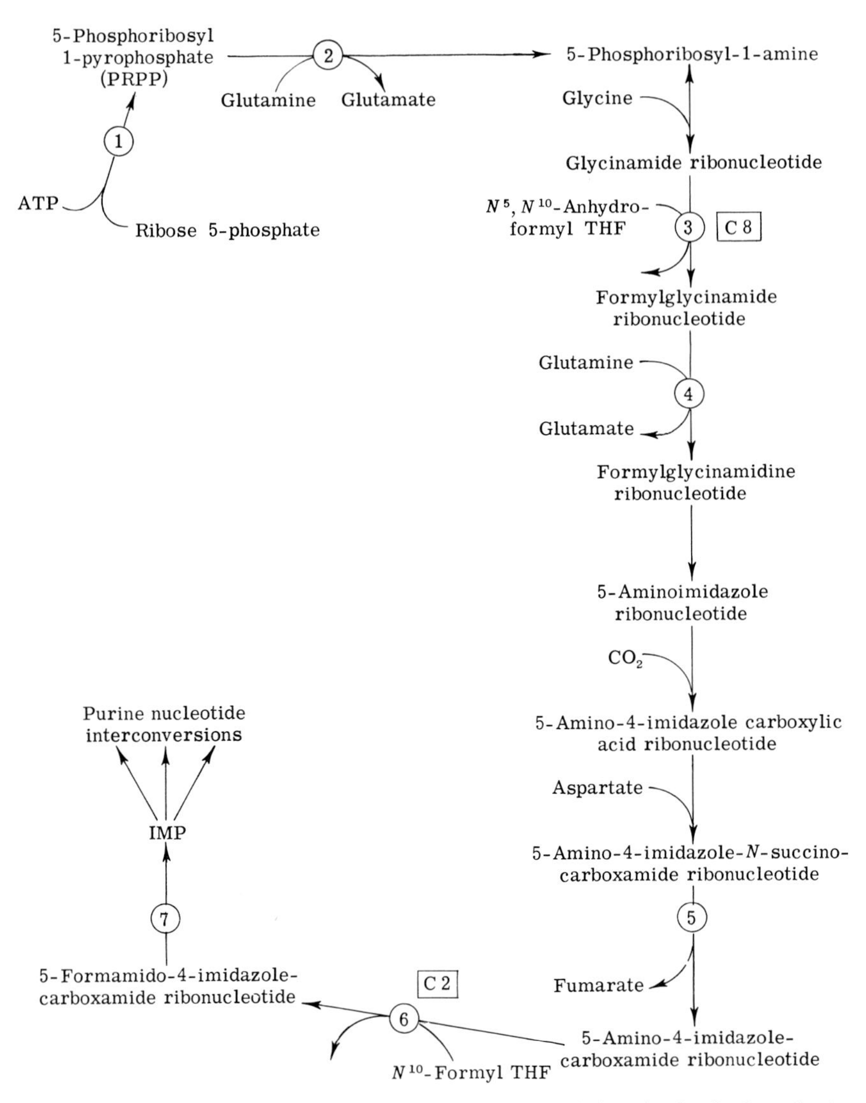

FIG. 10. *De novo* pathway for purine nucleotide synthesis in animal cells. Introduction of the 8-carbon (enzyme 3) and the 2-carbon (enzyme 6) of the purine ring system is indicated by the boxes. Enzymes: 1, PRPP synthetase; 2, PRPP amidotransferase; 3, glycinamide ribonucleotide transformylase; 4, formylglycinamide ribonucleotide amidotransferase; 5, adenylosuccinate lyase; 6, 5-amino-4-imidazolecarboxamide ribonucleotide transformylase; 7, inosinicase.

purine synthesis, the various interconversions of this cofactor are also indicated (Fig. 8). The subject of folate metabolism has been reviewed by Friedkin (1963). Purine pathways are discussed comprehensively by Blakeley and Vitols (1968), Murray *et al.* (1970), Hartman (1970), Kit (1970), and Murray (1971).

(1) PRPP synthetase. In addition to its participation in the *de novo* synthesis of purines, the pool of PRPP in animal cells is utilized for salvage of purines (and, to a limited extent, pyrimidines), formation of orotidylic acid (OMP), and nicotinamide coenzyme production. Formation of PRPP from ATP and ribose-5-P may be limited by deficiency of the latter substrate in certain cases (Murray, 1971). It is known that at least one of the reactions preceding formylglycinamide ribinoculeotide (FGAR) is sensitive to feedback effects of purines (Hartman, 1970). Feedback inhibition of PRPP synthetase by adenosine and inosine nucleotides is possibly the primary regulatory mechanism for the entire pathway (Blakeley and Vitols, 1968; Hartman, 1970). However, McFall and Magasanik (1960) found that feedback inhibition by adenine and guanine nucleotides could reduce the *de novo* purine pathway by only about 50%. L cells grown in the presence of a purine source (adenine or guanosine) apparently control their *de novo* pathway by repression of enzyme synthesis in addition to feedback inhibition of existing enzymes (McFall and Magasanik, 1960).

Levels of PRPP are generally believed to be the rate-limiting factor for the *de novo* purine pathway (Blakeley and Vitols, 1968; Rosenbloom, 1968; Murray, 1971). Depletion of the PRPP pool by exogenous purines (Henderson and Khoo, 1965) or by excess orotic acid (Rajalakshmi and Handschumacher, 1968) can shut down *de novo* purine synthesis. Cells respond to the purine salvage induced draining of PRPP by increased activity of PRPP synthetase (Henderson and Khoo, 1965).

(2) PRPP amidotransferase. As the first "committed" step of purine biosynthesis, this reaction has been considered as a possible point of pathway regulation. While there are some inhibitory effects of purine nucleotides on the amidotransferase (Hartman, 1970), the preferred site of regulation is PRPP synthetase (Blakeley and Vitols, 1968). Strong inhibition of PRPP amidotransferase is caused by 6-methylthiopurine ribonucleotide (Hill and Bennett, 1969; Bennett and Allan, 1971). Glutamine (and possibly ammonia) serves as the nitrogen donor for the formation of phosphoribosylamine; consequently the reaction is inhibited by the glutamine analog DON (Hartman, 1970). The product of the amidotransferase reaction is extremely unstable and reacts immediately with glycine in an enzyme-catalyzed step yielding glycinamide ribonucleotide (GAR).

(3) Glycinamide ribonucleotide (GAR) transformylase. This reaction is the first of two folate-mediated one-carbon transfers to the presumptive

purine ribonucleotide and introduces the C-8 carbon. The transformylase is specific for N^5,N^{10}-anhydroformyltetrahydrofolate (Hartman and Buchanan, 1959). Several inhibitors are believed to selectively inhibit this reaction by interference with some aspect of folate metabolism: homofolate and tetrahydrohomofolate (Hakala, 1971) and 1,3-bis(2-chloroethyl)-1-nitrosourea [BCNU, Groth *et al.* (1971)].

(4) Formylglycinamide ribonucleotide (FGAR) amidotransferase. As in the amidotransferase reaction discussed above, glutamine is the source of the amino group for this reaction. Azaserine and DON are potent inhibitors of FGAR amidotransferase, and the inhibition becomes irreversible with time due to covalent bond formation with the active site of the enzyme (Hartman, 1970). During the block of IMP formation by these inhibitors, the accumulation of large amounts of FGAR and small amounts of GAR suggests that PRPP amidotransferase is inhibited to a lesser extent than FGAR amidotransferase (Hartman, 1970).

(5) 5-Amino-4-imidazolecarboxamide ribonucleotide (AICRP) transformylase. This enzyme is reponsible for the introduction of the C-2 carbon of the purine ribonucleotide and requires N^{10}-formyltetrahydrofolate (Hartman and Buchanan, 1959). There is some inhibition of this reaction by folate analogs, although GAR transformylase and dTMP synthetase seem generally to be more sensitive to these inhibitors.

(6) Precursor incorporation. Of the five carbon atoms in the purine ring system, one comes from CO_2, two are provided by glycine, and two enter through folic acid from formate, serine-C-3, or histidine-C-2 [see Figs. 8 and 10 and Hartman (1970)]. In mouse L1210 ascites cells, formate-^{14}C is incorporated into adenine, guanine, and thymine nucleotides in the proportions 4:1:8 (Groth *et al.*, 1971). Widely varying incorporation ratios were found for different types of human leukocytes by Wells and Winzler (1959). Serine-3-^{14}C is about one-third as effective in labeling these compounds, and histidine-2-^{14}C is less than one-tenth as effective as formate-^{14}C (Groth *et al.*, 1971). Glycine-^{14}C incorporation into adenine and guanine nucleotides *via* the *de novo* purine pathway has been measured by McFall and Magasanik (1960) and Feigelson and Feigelson (1966). Low concentrations of adenine or guanosine will strongly suppress glycine utilization by L cells (McFall and Magasanik, 1960). All of these precursors can appear in a variety of amino acids, proteins, carboxylic acids, and other compounds (Wells and Winzler, 1959; Wheeler and Alexander, 1972; Brockman *et al.*, 1970); chromatographic purification and identification of labeled compounds is essential. Several tissues which do not incorporate common precursors of the *de novo* purine pathway (rabbit bone marrow, human leukocytes and platelets) apparently lack the pathway and derive their purines from the liver (Murray, 1971).

2. INTERCONVERSIONS

Pathways of purine interconversion in animal cells are summarized in Fig. 11.

a. Anabolic Reactions. The four purine nucleoside triphosphate pre-

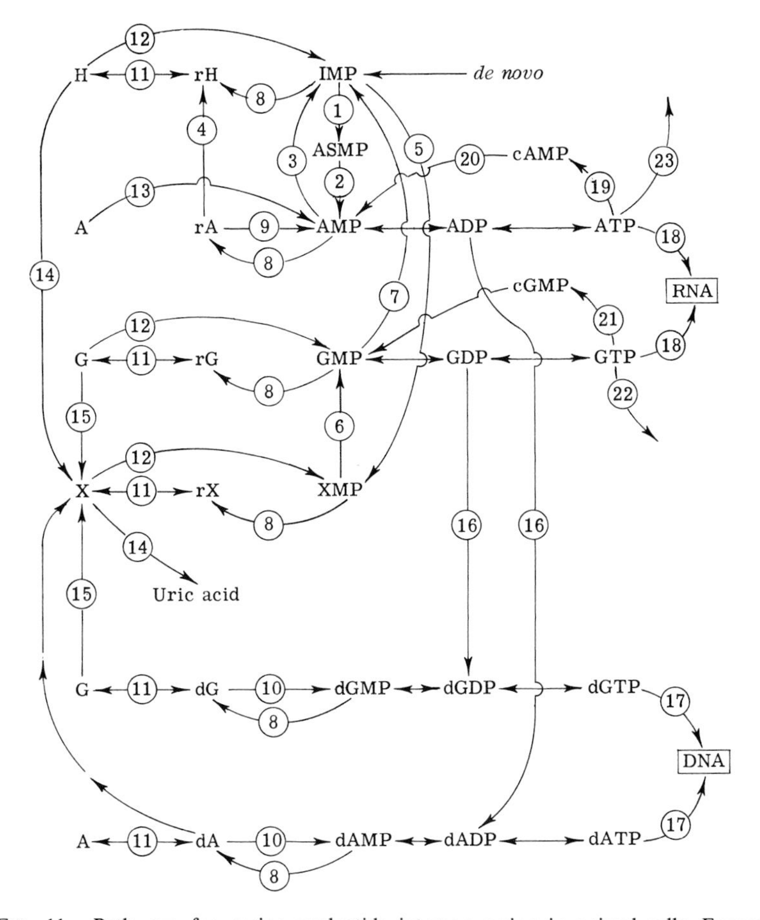

FIG. 11. Pathways for purine nucleotide interconversion in animal cells. Enzymes: 1, adenylosuccinate (ASMP) synthetase; 2, adenylosuccinate lyase; 3, AMP deaminase; 4, adenosine deaminase; 5, IMP dehydrogenase; 6, GMP synthetase; 7, GMP reductase; 8, 5′-nucleotidase; 9, adenosine kinase; 10, deoxyadenosine kinase; 11, purine nucleoside phosphorylase; 12, hypoxanthine–guanine phosphoribosyl transferase; 13, adenine-phosphoribosyl transferase; 14, xanthine oxidase; 15, guanine deaminase; 16, ribonucleotide reductase; 17, DNA polymerase; 18, RNA polymerase; 19, adenyl cyclase; 20, cyclic AMP phosphodiesterase; 21, guanyl cyclase; 22, GDP-sugar pyrophosphorylases; 23, formation of other adenine nucleotides.

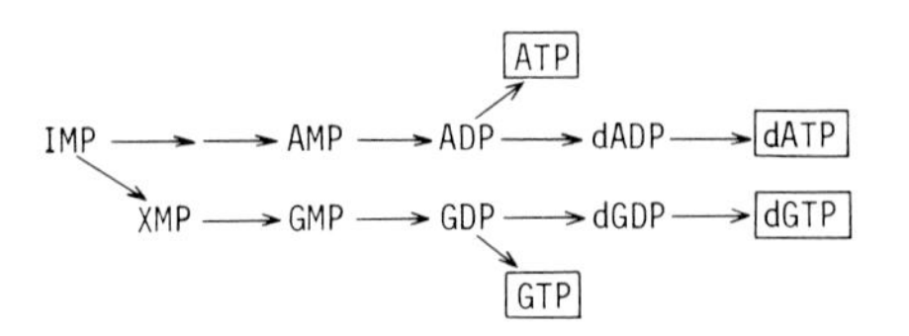

FIG. 12. Normal routes of purine nucleoside triphosphate formation in animal cells.

cursors of nucleic acids are derived from IMP (the terminal product of *de novo* purine synthesis) as shown in Fig. 12.

(1) Adenylosuccinate synthetase and adenylosuccinate lyase. In the absence of exogenous purine sources, all adenine nucleotides are formed from IMP by way of the intermediate, adenylosuccinate. This intermediate is a product of the condensation of L-aspartic acid and IMP, catalyzed by adenylosuccinate synthetase (Hartman, 1970). Crabtree and Henderson (1971) have shown for Ehrlich ascites cells that the availability of aspartic acid limits the rate of this reaction. The aspartate analog, hadacidin, is a strong, reversible inhibitor of adenylosuccinate synthetase (Hartman, 1970). Cleavage of adenylosuccinate to form AMP is catalyzed by adenylosuccinate lyase, an enzyme which also participates in *de novo* purine synthesis. Both steps in the production of AMP are inhibited by 6-thioinosinic acid (Atkinson *et al.*, 1964).

(2) IMP dehydrogenase and GMP synthetase. All guanine nucleotides are normally derived from IMP by intermediate formation of XMP. Dehydrogenation of IMP is inhibited by a variety of 6-chloro- and 6-mercaptopurines (Hartman, 1970). In the second step, amine transfer to XMP by GMP synthetase utilizes glutamine; DON blocks this reaction. Glutamine is rate-limiting for the formation of GMP from IMP in Ehrlich ascites cells (Crabtree and Henderson, 1971).

(3) Nucleotide kinases. The monophosphokinases are quite active and widely distributed. AMP (dAMP) kinase is distinct from GMP (dGMP) kinase (Kit, 1970). XMP and IMP are generally not elevated to the diphosphate and triphosphate levels in animal cells, although it is not clear whether this is caused by low monophosphokinase activities or rapid diversion of XMP and IMP to other products. As is the case for pyrimidines, the purine nucleoside diphosphates are rapidly converted to triphosphates by nucleoside diphosphokinase(s) (Kit, 1970).

(4) Ribonucleotide reductase. This enzyme has been discussed in the context of pyrimidine nucleoside diphosphate reduction (Section III,A); the same enzyme also catalyzes reduction of ADP and GDP. Compared with the pyrimidine substrates (CDP and UDP), reduction of the purine nucleoside diphosphates is relatively insensitive to inhibition by dTTP, dUTP, dATP, and dGTP.

(5) AMP deaminase. This enzyme is found in a wide variety of tissues (Hartman, 1970) and is essential for the conversion of adenine compounds to guanine compounds (AMP⟶IMP⟶XMP⟶GMP). In Ehrlich ascites cells supplied with adequate glutamine, AMP deaminase activity determines the rate at which GMP forms from AMP (Crabtree and Henderson, 1971). AMP deaminase is highly specific for 5′-AMP. The enzyme can be stimulated by ATP, ADP, and AMP, and is inhibited by some guanine nucleotides (Blakeley and Vitols, 1968; Murray *et al.*, 1970).

(6) GMP reductase. The conversion of GMP to AMP via IMP is directly dependent on the activity of GMP reductase in Ehrlich ascites cells (Crabtree and Henderson, 1971). Limited distribution and generally low activity of GMP reductase in animal cells (Hartman, 1970) accounts in part for the poor ability of guanine precursors to enter adenine nucleotide pools (McFall and Magasanik, 1960). In mammalian reticulocytes and erythrocytes, both xanthine and guanine are effective precursors for adenine nucleotides (Cook and Vibert, 1966).

b. Catabolic Reactions. Purine nucleoside monophosphates are hydrolyzed to their respective nucleosides by 5′-nucleotidase(s) (Kit, 1970). The low rate of nucleoside formation from 5′-nucleotides in human kidneys has been exploited by LePage *et al.* (1972) to achieve long-term systemic release of nucleoside antimetabolites from injected 5′-nucleotide derivatives. Deoxyribonucleotides are dephosphorylated at a lower rate than ribonucleotides, and there is some evidence for inhibition of 5′-nucleotidase by various nucleoside triphosphates (Murray *et al.*, 1970). Further degradation of nucleosides to free bases occurs by phosphorolysis, with the exception of adenosine and deoxyadenosine. These nucleosides are converted to inosine and deoxyinosine by adenosine deaminase, an enzyme distinct from the AMP deaminase discussed above. Adenosine deaminase is apparently not subject to regulatory effects of nucleotides and is specific for the deamination of adenosine, deoxyadenosine, and some analogs of these compounds; N^6-methyladenosine is a powerful inhibitor of the enzyme (Murray *et al.*, 1970).

(1) Purine nucleoside phosphorylase. Phosphorolytic cleavage of purine nucleosides yields pentose 1-phosphate and free purine. The reaction is reversible, but it is not clear whether the equilibrium favors nucleoside formation (Murray *et al.*, 1970) or breakdown (Hartman, 1970). Both ribonucleosides and deoxyribonucleosides are substrates for the phosphorylase, although adenosine nucleosides are first converted to their inosine equivalents by deamination (Gotto *et al.*, 1964; Murray *et al.*, 1970). In addition to nucleoside cleavage, the phosphorylase catalyzes slow base exchange between free purines and nucleosides of purines or pyrimidines (Reichard and Sköld, 1958; Paterson and Simpson, 1965; Murray *et al.*, 1970).

(2) Guanine deaminase and xanthine oxidase. Deamination of guanine to

xanthine is the principal route for degradation of guanine and some of its analogs such as 6-thioguanine, 1-methylguanine, 8-azaguanine, and 1-methyl-6-thioguanine (Murray *et al.*, 1970). Deamination at the nucleoside level (rG⟶rX) is slight and is probably catalyzed by a different enzyme (Hartman, 1970). Xanthine oxidase fosters the conversion of hypoxanthine to xanthine and finally to uric acid. Inhibition of xanthine oxidase by allopurinol has been used to increase the effectiveness of xanthine as a nucleic acid precursor in the mouse (Pomales *et al.*, 1965).

c. "Salvage" Reactions. As discussed for pyrimidine salvage (Fig. 9), there are two alternate routes for incorporation of free bases into nucleotide pools: (1) base⟶nucleoside⟶nucleotide (phosphorylase—kinase) and (2) base⟶nucleotide (pyrophosphorylase). The second pathway is generally favored for purine bases.

(1) Purine phosphoribosyl transferase (pyrophosphorylase). Two enzymes of this type have been studied extensively. One is specific for adenine and 5-amino-4-imidazolecarboxamide, while the other acts on hypoxanthine, guanine, and slowly on xanthine (Hartman, 1970; Murray *et al.*, 1970; Murray, 1971). The equilibrium for adenine phosphoribosyl transferase lies heavily in favor of AMP formation from adenine and PRPP (Flaks *et al.*, 1957). The distribution of these enzymes is broader than that of the *de novo* purine pathway (Hartman, 1970), and in at least one case the adenine phosphoribosyl transferase exhibits large changes in activity related to the growth rate of the tissue (Murray, 1966). Inhibition of the transferases by their reaction products is known to occur (Murray *et al.*, 1970). Formycin A inhibits the adenine enzyme (Henderson *et al.*, 1967). As discussed for pyrimidines (Section III,A), the supply of PRPP is a very important factor in determining the rate of conversion of free purine bases to nucleotides.

(2) Purine nucleoside kinases. Adenosine kinase is the only purine ribonucleoside kinase generally found in animal cells. The enzyme is specific for adenosine and many of its nucleoside analogs (Kit, 1970), although it apparently does not phosphorylate deoxyadenosine (Schrecker, 1970; Krygier and Momparler, 1971). There are reports of low activities of guanosine kinase (Schrecker, 1970) and inosine kinase (Pierre and LePage, 1968). Rigorous proof of these activities is difficult because the nucleosides inosine and guanosine are readily cleaved by purine nucleoside phosphorylase, and the resulting free bases can form IMP and GMP by the hypoxanthine-guanine phosphoribosyl transferase reaction. In both of the above studies, however, this possible artifact was eliminated.

Specificities of mammalian deoxyribonucleoside kinases have not been established unambiguously because of difficulties in purification. While apparently distinct from the deoxyadenosine kinase, the kinase for deoxyguanosine seems to be identical to that for deoxycytidine and araC (Schrecker,

1970). Krygier and Momparler (1971) reported a relatively pure kinase from calf thymus which phosphorylates deoxyguanosine and cytidine in addition to deoxyadenosine. Analogous to the inhibition of thymidine kinase by dTTP, deoxyguanosine (dC-araC) kinase is strongly inhibited by dCTP (Schrecker, 1970). Kinase activity toward deoxyadenosine and deoxyguanosine does not vary significantly over the cell cycle in 3T3 mouse fibroblasts, in contrast to the variations of deoxycytidine kinase (4-fold) and thymidine kinase (10-fold) (Bernard and Brent, 1971).

(3) Precursor incorporation. Much of the literature concerning the incorporation of preformed purines into nucleotides by mammalian cells has been reviewed (Murray *et al.*, 1970). The ability of various cell types to incorporate free purines is dependent on (1) purine transport, (2) the availability of PRPP, (3) activity of the pertinent purine phosphoribosyl transferase, and (4) loss of the purine by catabolic routes. Large differences in the incorporation of adenine-8-^{14}C by various mouse tissues *in vivo* (Wheeler and Alexander, 1972) reflect the magnitude of variation which the above factors can exhibit.

Hypoxanthine and adenine are the more commonly used purine precursors because their conversion to nucleotides is generally more efficient than for the other purines. Guanine and xanthine can be rapidly degraded by guanine deaminase and xanthine oxidase without significant incorporation into nucleotide pools (Pomales *et al.*, 1965; Crabtree and Henderson, 1971). Undoubtedly, some lines of cultured cells will be found to have aberrant behavior with respect to these patterns. Hypoxanthine-8-^{14}C enters the adenine and guanine nucleotide pools in equivalent amounts (Brockman *et al.*, 1970), and its incorporation can be made very efficient by methotrexate (amethopterin) induction of the "purineless" state (Hryniuk, 1972). In this condition, exogenous hypoxanthine satisfies all the purine requirements of cultured cells (Hakala, 1957). Adenine is incorporated into the adenine nucleotide pool rather selectively both *in vivo* (Brockman *et al.*, 1970; Wheeler and Alexander, 1972) and *in vitro* (Crabtree and Henderson, 1971). Labeled, acid-soluble metabolites derived from adenine-^{14}C in the above studies are composed of 90 to 95% adenine nucleotides, with the remainder composed of IMP, rH, H, XMP, rX, X, GMP, rG, and G. Limited activity of AMP deaminase is responsible for the sluggish conversion of adenine nucleotides to other purine nucleotides (Crabtree and Henderson, 1971). Of the free guanine which is not catabolized by Ehrlich ascites cells *in vitro*, about 95% remains in guanine nucleotides with little conversion to adenine nucleotides; GMP reductase activity limits this conversion (Crabtree and Henderson, 1971). Purine incorporation may be stimulated by exogenous nucleosides which serve as a source of pentose phosphate for the purine nucleoside phosphorylase reaction (Paterson and Simpson, 1965). Low activity of the phosphorylase toward adenine (Kim *et al.*, 1968) would probably eliminate this type of stimulation for adenine incorporation.

A rough proportionality between the efficiency of incorporation of several precursors into total nucleic acid (RNA plus DNA) can be calculated from the data of Brockman *et al.* (1970) for L1210 murine leukemia cells *in vivo*. Normalizing to equivalent isotopic dose, the ratio of incorporation is approximately: formate-^{14}C:thymidine-2-^{14}C:hypoxanthine-8-^{14}C:adenine-8-^{14}C:uridine-5-^{3}H::1:35:50:70:130.

With the possible exception of adenosine, labeling with purine nucleosides does not have the advantages found for pyrimidine nucleoside labeling. Rapid degradation to free bases by purine nucleoside phosphorylase and low activities of purine nucleoside kinases is the cause of this problem. Consider, for example, the attempts of Lieberman *et al.* (1971) to incorporate deoxyadenosine and deoxyguanosine into DNA during the repair process in lymphocytes exposed to alkylating agents. Although the major sites of alkylation damage are the purine bases, the low incorporation of dA and dG compared to that for dC and dT does not accurately reflect the extensive repair which is occurring at these purine sites. Lieberman *et al.* (1971) found that most of the dA and dG were utilized for RNA synthesis, presumably because of preferential incorporation into the ribonucleotide pools by sequential action of purine nucleoside phosphorylase and purine phosphoribosyl transferase.

C. Variable Aspects of Metabolic Pathways

Knowledge of the target enzymes for feedback inhibition of nucleotide pathways is accumulating rapidly. In addition to these specific control processes, there is some evidence of what might be termed "higher order" mechanisms for the control of the relationships between anabolic, catabolic, and salvage pathways. Study of such control in animal systems has had its locus within the discipline of cancer research since 1930 when Warburg advanced his hypothesis of the glycolytic origins of neoplasia (Warburg, 1930, 1956). Subsequent interest by others, and gradual shifting of attention from energetics to nitrogen metabolism to nucleic acids has produced other biochemically oriented postulates of carcinogenesis (Potter, 1944; Potter and LePage, 1949; Greenstein, 1954). A more recent hypothesis is one which Weber and colleagues have called the Molecular Correlation Concept (Weber, 1961; Weber *et al.*, 1971). This essentially states that there is a meaningful, nonrandom relationship between the growth rate of neoplastic cells and the enzymatic activities for their various metabolic pathways (see Fig. 13). It is a difficult task to prove that these relationships are tightly linked with the mechanism of carcinogenesis. A basic criterion which must be satisfied by such a proof is that growth of a tumor must be inhibited when its enzyme levels are restored to those of the homologous normal tissue (Weber and Cantero, 1957). In addition, we must come to understand the mechanisms by which enzyme levels fluctuate with variation in the physiological state of *normal* tissues (Knox, 1967). Are there relatively simple mechanisms for control of pathway interrelationships, or are these "higher order" control processes rather a consequence of the total spectrum of feedback inhibition, activation, and competition of all metabolites with all enzymes?

1. Growth-Related Effects

Many examples of growth-rate related enzyme activities have been given in the sections on purine and pyrimidine metabolism. Figure 13, from Weber *et al.* (1971), summarizes some of these changes. It should be remembered that tissues and cell lines of nonhepatic origin may deviate from this pattern because of an intrinsically different spectrum of normal enzyme levels. The possibility that at least some of these changes are related to the growth-rate dependent pool of cyclic AMP has been pointed out (Hauschka *et al.*, 1972).

The fluctuation of enzyme levels with the growth rate of cultured animal cells is well known for thymidine kinase (Weissman *et al.*, 1960; Sneider *et al.*, 1969; Hare, 1970; Cory and Whitford, 1972), uridine kinase (Kit *et al.*, 1964; Plagemann *et al.*, 1969), choline kinase (Plagemann, 1969), thymidylate synthetase (Sneider *et al.*, 1969; Conrad and Ruddle, 1972), thymidylate kinase and deoxycytidylate deaminase (Sneider *et al.*, 1969), and ribonucleotide reductase (Cory and Whitford, 1972). It is likely that the pattern of enzyme activity over the *cell cycle* is important in determining the way in which the activity changes with the *cell growth rate.* The duration of the $S + G_2 + M$ phases of the cell cycle is relatively constant (about 8 hours) for mammalian cells. Changes in the cell generation time are almost always caused by increases in the G_1-phase. Hence, the total fraction of cells existing in the $S + G_2 + M$ phases of the cell cycle at any particular time is directly related to the growth rate of the population. A cell line doubling every 10 hours would have about 5 times as many cells in $S + G_2 + M$ ($\sim 80\%$) as a population of cells doubling every 48 hours. Thymidine kinase is known to be most active during the S and G_2 phases of the cell cycle (Stubblefield and Murphree, 1967; Adams, 1969). Compared with slowly growing cells, a rapidly growing cell population would therefore be expected to exhibit higher thymidine kinase activity because of its greater proportion of cells posessing high enzyme activity. All enzymes which increase with increasing growth rate should, by this argument, be maximally active during the period $S + G_2 + M$. Enzymes which are most active during G_1 should decrease in activity as the population proliferates more rapidly.

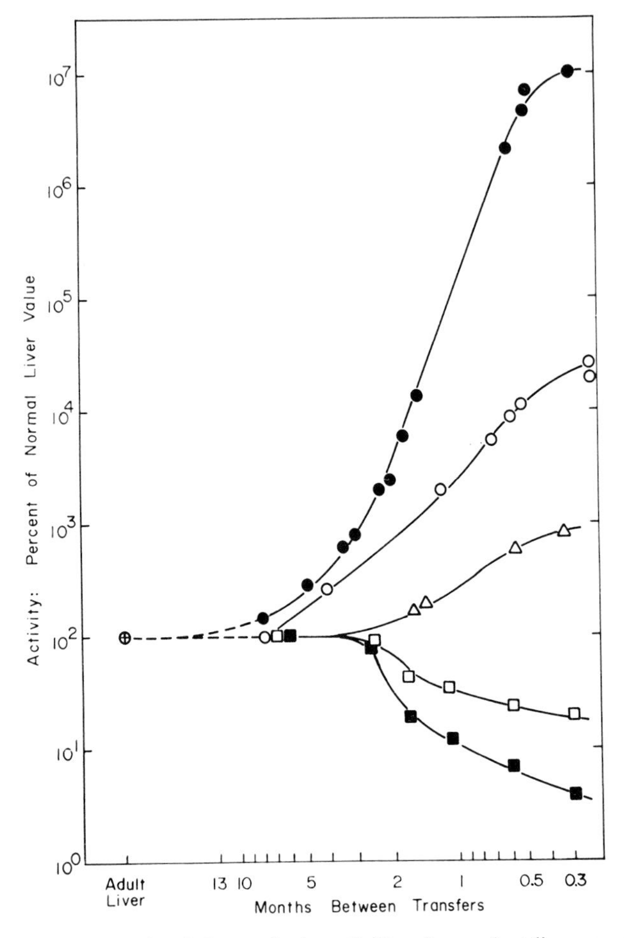

FIG. 13. Growth rate-related changes in the activities of several rat liver enzymes involved in pyrimidine metabolism. The data points represent enzyme activities in homogenates of normal liver (⊕) and hepatomas with various growth rates. Increasing growth rate is indicated by decreasing time between transfers (abscissa) ●, Ratio of pathway for dT ⟶ DNA to pathway for dT ⟶ CO; ○, ribonucleotide reductase; △ aspartate transcarbamylase; □, adenylate kinase; ■, thymine catabolic pathway (T ⟶ CO_2). Redrawn from Fig. 10 and 11 of Weber *et al.* (1971), by permission of Pergamon, Oxford and New York.

2. DRUG EFFECTS AND MECHANISMS OF DRUG RESISTANCE

The use of drugs for the artificial control of cell growth requires some awareness of the mechanisms by which inhibition may be achieved. Through perturbations of the machinery of the inhibitory process, cells and tissues can develop resistance to even the most potent antimetabolites. Frequently the mechanism of resistance can be discovered by analysis of the intracellular forms of the drug by methods detailed in Sections V and VI of this chapter. Brockman (1963) and Law (1956) have reviewed the effects of many anticancer agents and some of the mechanisms of cellular resistance to these compounds: (a) decreased permeability to the inhibitor; (b) decreased conversion of the inhibitor to an active form; (c) increased degradation of the inhibitor (d) increased synthesis of the inhibited enzyme, or increased overall activity of the target pathway; (e) decreased sensitivity of the target enzyme to the inhibitor (altered enzyme); (f) decreased dependence on the inhibited pathway (alternate routes of anabolism).

There are numerous examples of the above-listed resistance mechanisms. The most frequently encountered form of resistance is the loss of kinase, phosphorylase, or pyrophosphorylase activity which is essential for conversion of free base and nucleoside antimetabolites to the active nucleotide forms (e.g., Brockman, 1963; Sköld, 1963; Chu and Fischer, 1965; Schrecker and Urshel, 1968). Multiple application of inhibitory drugs can be used to overcome cellular resistance in certain cases (see Sartorelli, 1969).

IV. Transport of Precursors

A. General Properties

The mechanisms by which cells take up nutrients and other matabolites from their environment are complex. Because transport processes play a central role in the kinetic behavior of intracellular pools, they are discussed below in some detail. Numerous factors are known to alter the rate of nucleoside transport: serum (Cunningham and Pardee, 1969; Hare, 1972a,b), cell density (Weber and Rubin, 1971), pH (Ceccarini and Eagle, 1971), phytohemagglutinin (Kay and Handmaker, 1970; Peters and Hausen, 1971a,b), and dibutyryl cyclic AMP (Hauschka *et al.*, 1972), to name a few. Transport of other precursors, such as phosphate (Nakata and Bader, 1969) and adenine (Stambrook and Sisken, 1972), may be less sensitive to such alterations. Knowledge of transport rates or intracellular specific activity is essential for the interpretation of most precursor incorporation studies. Methods outlined in Section V provide for the direct measurement of transport rates in animal cells. The frequently made assumption that transport of exogenous labeled metabolites (precursors) is a constant, unperturbable process is usually unwarranted.

Several distinct mechanisms for transport of small molecules are known. Table III indicates several characteristic properties of each. The reader is

TABLE III

PROPERTIES OF TRANSPORT PROCESSES

Process	Cellular energy requirement	Catalytic component (permease)	Concentrative uptake	V_{max} and saturation	Chemical alteration of permeant
Passive					
Simple diffusion	No	No	No	No	No
Facilitated diffusion	No	Yes	No	Yes	No
Active					
Coupled transport	Yes	Yes	Yes	Yes	No
Group translocation	Yes	Yes	Yes	Yes	Yes

referred to Heilbrunn (1952), Stein (1967), Pardee (1968), Roseman (1969), Christensen (1969), Berlin (1970), and Oxender (1972) for further information. At high concentrations, many types of compounds can enter cells at significant rates by *simple diffusion*. However, this type of transport is generally not believed to be physiologically important for most cell types (Plagemann and Erbe, 1972). *Facilitated diffusion* is the mechanism which probably applies to purine, pyrimidine, and nucleoside transport in animal cells. *Coupled transport* is characteristic of amino acid and sugar uptake by animal cells, where the sodium ion may be cotransported with other permeants. *Group translocation* occurs in the phosphotransferase-mediated uptake of sugars by bacterial cells. Because of the confusing role played by nucleoside kinases, group translocation remains a possible mechanism for nucleoside uptake in some types of animal cells (Plagemann and Erbe, 1972).

Transport of nucleosides has been studied quantitatively in many vertebrate cell types and can be accounted for by two passive transport mechanisms—simple diffusion and facilitated diffusion. The facilitated process is by far the more important of the two under typical physiological conditions (i.e., low exogenous nucleoside concentrations). Facilitated diffusion operates down a concentration gradient and is a nonconcentrative, carrier-mediated process with a saturable rate (Table III and Fig. 14). Evidence for the operation of this mechanism is necessarily of a kinetic nature. Despite the abundance of kinetic data in the literature (Table IV), *rigorous* demonstration of the mechanism has been possible only with those cell types that do not metabolize the permeant [e.g., human erythrocytes (Oliver and Paterson, 1971; Lieu *et al.*, 1971; Cass and Paterson, 1972) and murine leukemia cells (Kessel and Shurin, 1968; Kessel and Dodd, 1972)].

B. Kinetics of Simple and Facilitated Diffusion

Typical dependence of uptake velocity on permeant concentration is shown in Fig. 14 [redrawn from Plagemann and Erbe (1972)]. Above a certain saturating concentration of nucleoside (in this case, about 2 μM dT) the initial velocity of uptake increases linearly because of simple diffusion. The saturable, facilitated portion of the uptake may be clearly seen by subtracting the contribution of simple diffusion (see Fig. 14) (Schuster and Hare, 1971; Hare, 1970; Plagemann, 1970; Plagemann and Erbe, 1972; Roos and Pfleger, 1972).

Kinetic data are generally analyzed according to the Lineweaver-Burk form of the Michaelis-Menten relationship for enzymatic reactions (Eq. 1), where V is the initial rate of uptake and S is the exogenous nucleoside concentration.

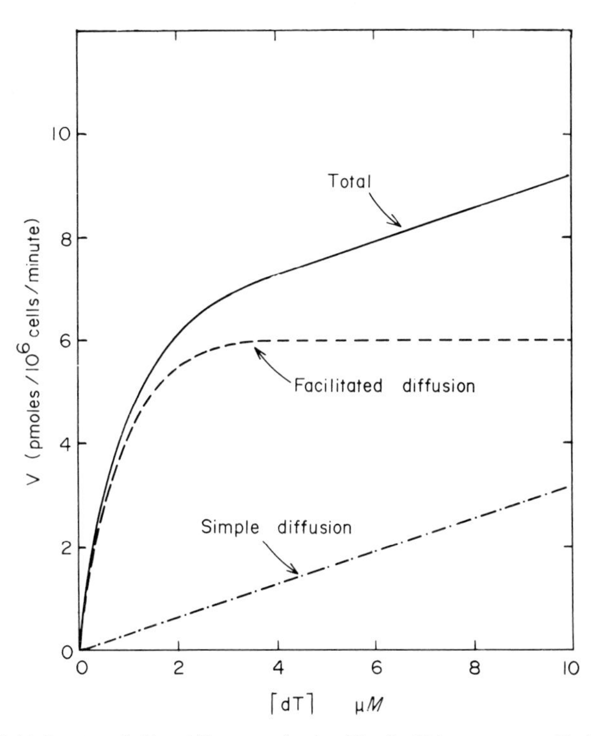

FIG. 14. Initial rate of thymidine uptake by Novikoff hepatoma cells in suspension. The total observed uptake (——) is a sum of two separate processes: simple diffusion (-·-·-) and facilitated diffusion (-----). The facilitated process is saturated above about 2 to 3 μM dT and has a K_m of about 0.5 μM. Redrawn from Plagemann and Erbe (1972), by permission of the Rockefeller University Press, New York.

TABLE IV

FACILITATED UPTAKE OF NUCLEOSIDES AND PURINES

Compound	V_{max}^{a}	K_m^b	T(°C)	Cell type	Reference[f]
Nucleoside					
rU	110	10	37	Novikoff rat hepatoma	1
	100–200	14	37	Novikoff rat hepatoma	2
	125	13	37	Novikoff rat hepatoma	3
	200	8.3	37	HeLa	4
	300[c]	10	37	Chick embryo fibroblasts	5
	10–40[c]	15	25	Mouse embryo cells	6
	—	8	39	Chick embryo cells	7
	—	10	37	Rat embryo cells	7
	50[c]	710[d]	37	Human erythrocytes	8
	710[c]	700[d]	37	Human erythrocytes	9
	0.1[c]	200[c]	—	Bovine lymphocytes	10
rC	100–200	23	37	Novikoff rat hepatoma	2
	300[c]	8.4	37	Chick embryo fibroblasts	5
	1180[c]	3500[d]	37	Human erythrocytes	9
dC	4200[c]	7500[d]	0	L1210 murine leukemia	11
araC	4200[c]	7500[d]	0	L1210 murine leukemia	11
	490[c]	1700[d]	37	Human erythrocytes	9
dT	8–12	0.5	37	Novikoff rat hepatoma	12
	45[c]	—	22	Hamster tumor cells	13
	1.3	1.0	0	HeLa	4
	8.3	0.24	37	HeLa	4
	16	40	37	Rabbit leukocytes	14
	340[c]	1300[d]	37	Human erythrocytes	9
rA	250	12–20	37	Novikoff rat hepatoma	1
	110	6	37	Novikoff rat hepatoma	3
	100–200	8.8	37	Novikoff rat hepatoma	2
	67	20	37	HeLa	4
	13	10	37	Rabbit leukocytes	14
	2500[c]	10	37	Chick embryo fibroblasts	5
rG	140	12	37	Novikoff rat hepatoma	2
	400[c]	10	37	Chick embryo fibroblasts	5
	670	3.6	37	HeLa	4
rH	140	11	37	Novikoff rat hepatoma	2
	5	21	37	Rabbit leukocytes	14
Purine					
A	7.6	7	38.5	Rabbit leukocytes	15
	18,300	100,000[e]	38.5	Rabbit leukocytes	15
X	170	2300	38.5	Rabbit leukocytes	15

[a] V_{max} in units of pmole/10^6 cells/minute.

[b] K_m, μM.

[c] Estimated from available data, using parameters in Table X, if necessary. V_{max} values for erythrocytes should not be directly compared with data for other cell types, since the volume of 10^6 erythrocytes is about 1/30 of that for the same number of "typical" animal cells and the surface area may be 1/5 to 1/20 as great.

[d] Cells lack enzymes for metabolism of the nucleoside.

[e] Nonspecific, low affinity system for adenine uptake.

[f] References: 1, Plagemann (1970); 2, Plagemann (1971c); 3, Plagemann (1971a); 4, Mizel and Wilson (1972); 5, Scholtissek (1968); 6, Hare (1972b); 7, Steck *et al.* (1969); 8, Oliver

$$\frac{1}{V} = \frac{K_m}{V_{\max}} \cdot \frac{1}{[S]} + \frac{1}{V_{\max}} \tag{1}$$

Although linear behavior of the data is observed in most instances (after subtraction of the contribution of simple diffusion), the true meaning of the deduced values for K_m and $V_{\max}$ is not clear. The growing tendency to assign these affinity and velocity constants to presumed membrane-bound carrier proteins (permeases) is not necessarily valid. Table IV contains kinetic constants for nucleoside uptake in many different systems. The absolute initial rate of facilitated uptake (V_f) may be simply estimated for any nucleoside concentration [S] using the Michaelis-Menten equation (Eq. 2) and the pertinent K_m and V_{max} values from Table IV.

$$V_f = \frac{V_{\max}\,[S]}{(K_m + [S])} \tag{2}$$

Given a $V_{\max}$ of 100 pmole/10^6 cells/minute and a K_m of 10 μM, the initial rate of facilitated uptake would be about 9 pmoles/10^6 cells per minute at 1 μM nucleoside concentration and 50 pmole/10^6 cells/minute at 10 μM nucleoside.

Some caution must be exercised in the comparison of $V_{\max}$ values. Experimental methods vary in their ability to yield absolute initial rates. Since the instantaneous rate of uptake is constantly decreasing after zero time, sluggish measurements (5- to 10-minute labeling) using centrifugation techniques on suspended cells will give lower estimates of $V_{\max}$ than rapid measurements (<1 minute of labeling) with cells on coverslips.

Simple diffusional transport has an initial rate (V_d) which is a linear function of nucleoside concentration (Eq. 3). Values for the parameter v_{dif}, which is the slope of the simple diffusional process in Fig. 14,

$$V_d = v_{\text{dif}} \cdot [S] \tag{3}$$

are found in Table V. For nucleosides and bases the values of v_{dif} are apparently dictated by hydrophobic character ("lipoid solubility") rather than by size. Hence thymidine, with its extra methyl group, diffuses somewhat more readily than uridine (Table V). Roos and Pfleger (1972) have found excellent correlation between the octanol/water partition coefficient and the rate of diffusion of purine ribonucleosides into guinea pig erythrocytes. Diffusion rates were in the proportion inosine, 1 : adenosine, 15 : N^6-methyladenosine, 87 : N^6-dimethyladenosine, 790.

It is simple to calculate the contribution of simple diffusion to the total uptake. For example, at a concentration of 1 mM dT, the initial rate of dT

---

and Paterson (1971); 9, Lieu *et al.* (1971); 10, Peters and Hausen (1971a); 11, Kessel and Shurin (1968); 12, Plagemann and Erbe (1972); 13, Hare (1970); 14, Taube and Berlin (1972); 15, Hawkins and Berlin (1969).

TABLE V
SIMPLE DIFFUSIONAL UPTAKE

Cell type	Nucleoside	T(°C)	v_{dif}[a]	V_{max}[b]	Reference[c]
Hamster embryo cells	dT	25	0.70	140	1
Hamster embryo cells	rU	25	0.50	110	1
Hamster tumor cells	dT	25	1.45	200	1
Hamster tumor cells	dT	5	0.54	0	1
Hamster tumor cells	rU	25	0.68	100	1
BHK 21/13 (normal)	dT	25	1.05	70	1
BHK 21/13 (normal)	dT	5	0.54	0	1
BHK 21/13 (TK^-)[d]	dT	25	0.84	0	1
BHK 21/13 (TK^-)[d]	rU	25	0.35	130	1
Novikoff hepatoma	dT	22	0.34	4	2
Novikoff hepatoma	dT	37	0.34	10	2
Novikoff hepatoma	rU	6	0.14	20	3
Novikoff hepatoma	rU	16	0.19	34	3
Novikoff hepatoma	rU	37	0.23	110	3
Novikoff hepatoma	rU	37	0.32	60	4
Novikoff hepatoma	rA	6	0.25	40	3
Novikoff hepatoma	rA	16	0.27	90	3
Novikoff hepatoma	rA	26	0.30	143	3
Novikoff hepatoma	rA	37	0.32	250	3

[a] v_{dif} in units of pmole/10^6 cells per minute per micromole (see Eq. 3).
[b] V_{max} in units of pmole/10^6 cells per minute for the saturated facilitated diffusion process.
[c] 1, Schuster and Hare (1971); 2, Plagemann and Erbe (1972); 3, Plagemann (1970); 4, Plagemann (1971c).
[d] TK^-: cells lack thymidine kinase.

uptake by hamster embryo cells (Table V) would be 700 pmoles/10^6 cells per minute by simple diffusion (i.e., 0.70×1000) plus 140 pmoles/10^6 cells per minute by the saturated facilitated diffusion process. By combining Eqs. (2), (3), and (4), the fractional contribution of facilitated diffusion to total uptake can be simply expressed (Eq. 5).

$$V_{total} = V_f + V_d \tag{4}$$

$$\frac{V_f}{V_{total}} = \frac{V_{max}}{V_{max} + v_{dif}(K_m + [S])} \tag{5}$$

Figure 15 illustrates the decreasing contribution of V_f to V_{total} for several typical sets of parameters. Use of Eq. (5) is integral to the proper design of nucleoside uptake experiments. Apparent changes in uptake rates caused by cell density, medium composition, inhibitory or stimulatory substances, genetic alteration, or other factors should be assayed in the most sensitive manner possible.

Suppose one is studying a "typical" vertebrate cell type which possesses facilitated nucleoside uptake systems and nucleoside kinase activities. Assume that a particular transport change in these cells is entirely due to *nonselective* changes in plasma membrane permeability (i.e., simple diffusion rates, v_{dif}, have changed) and that the facilitated uptake systems remain essentially unchanged. Uptake studies at low nucleoside concentrations (0.1 to 10 μM) would probably reflect only marginal transport changes because of the generally small contribution of simple diffusion to the total uptake process in this concentration range (see Figs. 14 and 15). Changes in the rate of simple diffusion are best measured at high nucleoside concentrations (0.1 to 10 mM), where simple diffusion dominates the uptake process. Alternatively, if some treatment of these cells alters only the facilitated uptake process, then this is most accurately measured at very low nucleoside concentrations (< 1 μM, see Fig. 14) where facilitated uptake is dominant [see also Roos and Pfleger (1972)].

The fundamental cause of transport changes may be assessed by comparing initial rates of nucleoside uptake as a function of nucleoside concentration (see Schuster and Hare, 1971; Plagemann and Erbe, 1972). Figure

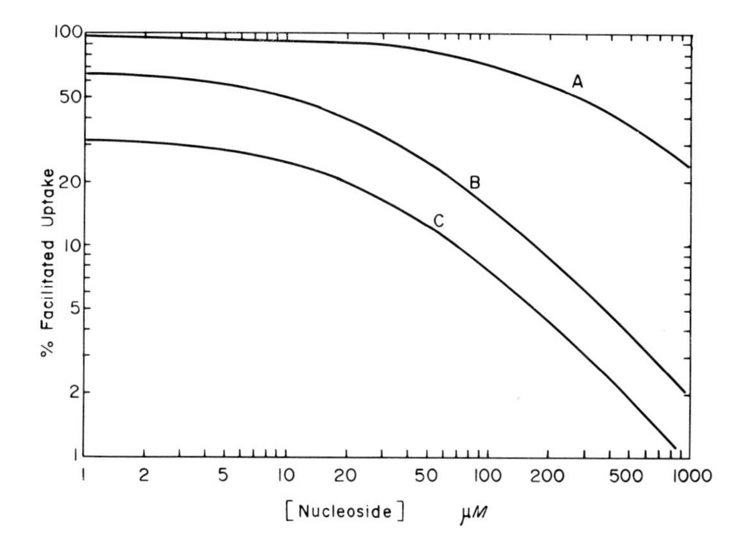

FIG. 15. Contribution of facilitated diffusion to the total initial rate of nucleoside uptake. Equation (5) has been plotted for three sets of values for V_{max} and K_m, with $v_{dif} = 0.5$ pmole/10^6 cells per minute per micromole per liter: All uptake which is not by facilitated diffusion is assumed to result from simple diffusion. Curve A, $V_{max} = 150$ pmole/10^6 cells per minute, $K_m = 10$ μM; curve B, $V_{max} = 10$, $K_m = 10$ μM; curve C, $V_{max} = 5$, $K_m = 20$ μM.

16 indicates the various characteristic relationships that would be expected for alterations in simple or facilitated diffusion. Because the temperature dependence of facilitated diffusion ($Q_{10} \sim 1.8$ to 2.2) is usually greater than that of simple diffusion ($Q_{10} \sim 1.1$ to 1.4) [see Table V and Plagemann and Erbe (1972), Craig and Chen (1972)], the contribution of simple diffusion is increased at low temperatures. For thymidine uptake by Novikoff hepatoma cells (Plagemann and Erbe, 1972), the calculated contribution of simple diffusion to the total uptake process (at 0.1 μM dT) increases from 2% at 37°C to 15 to 40% at 1°C.

Because uptake studies invariably involve radioactive nucleosides, the choice of isotope becomes important. At radioactive concentrations of 1 to 10 μCi/ml, uptake rates are conveniently measured with samples of about 10^5 cells. If high specific activity ^{3}H nucleosides (~10 Ci/mmole) are used, the nucleoside concentration will be 1 μM or less, while with ^{14}C nucleosides (~50 mCi/mmole) the nucleoside concentration would be 20 to 200 μM. Hence, tritiated nucleosides are suitable for measuring both facilitated uptake and simple diffusion if diluted appropriately. ^{14}C nucleosides generally are not suitable for facilitated transport studies.

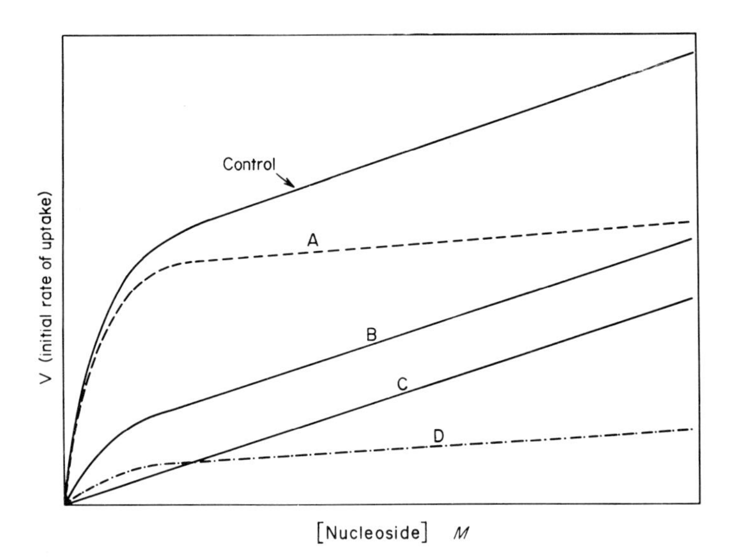

FIG. 16. Expected changes in the concentration dependence of nucleoside uptake caused by various cellular changes. Control (see Fig. 14); curve A, unchanged facilitated diffusion and decreased simple diffusion; curve B, decreased facilitated diffusion and unchanged simple diffusion; curve C, complete loss of facilitated diffusion and unchanged simple diffusion (see Schuster and Hare, 1971); curve D, decreased facilitated diffusion and decreased simple diffusion.

C. Role of Nucleoside Kinases

Considerable evidence implicates nucleoside kinases in the uptake of nucleosides by animal cells. Nucleoside kinase substrate specificities are very similar to the specificities of the transport systems (see Table VII and Section III). Under many conditions where nucleoside uptake is reduced, kinase activities are also low; stimulation of uptake is often accompanied by increased kinase activities. Cells in crowded or confluent cultures have decreased permeability to exogenous nucleosides as compared to sparse, rapidly growing cultures (Weissman *et al.*, 1960; Cunningham and Pardee, 1969; Weber and Rubin, 1971; Cass, 1972). Transformation by viruses (Hare, 1970; Nordenskjöld *et al.*, 1970), addition of fresh serum (Cunningham and Pardee, 1969; Wiebel and Baserga, 1969; Nordenskjöld *et al.*, 1970), dilution to lower cell densities (Weissman *et al.*, 1960; Plagemann *et al.*, 1969), and stimulation by phytohemagglutinin (Kay and Handmaker, 1970) or hormonal changes such as those following partial hepatectomy (Bollum and Potter, 1959) rapidly stimulate nucleoside uptake with concomitant increase in nucleoside kinase activities.

In a mutant hamster cell line lacking thymidine kinase activity, dT is taken up exclusively by simple diffusion (Fig. 16) and accumulates in the nucleoside form; facilitated uptake of dT is observed only in the normal parental line which has thymidine kinase (Schuster and Hare, 1971). Both cell lines have uridine kinase activity and exhibit facilitated uptake of uridine (see Table V).

In cells possessing the appropriate kinase activities, the equilibrium for nucleoside phosphorylation is heavily in favor of nucleoside triphosphate formation. Usually at least 70 to 95% of a nucleoside is phosphorylated to the triphosphate level upon uptake (Gentry *et al.*, 1965a; Lindberg *et al.*, 1969; Adams, 1969). However, the portion of intracellular nucleoside that is not phosphorylated is known to increase as the contribution of simple diffusional transport increases at high nucleoside concentrations (Lindberg *et al.*, 1969). Apparently, kinase activities are insufficient to completely phosphorylate such large amounts of nucleoside entering by simple diffusion. Thus, there is a distinct possibility that *all* nucleoside which has been transported by the specific facilitated mechanism is phosphorylated. At issue is whether or not phosphorylation of the nucleoside occurs concomitantly with uptake, in which case the uptake should be termed group translocation (Table III), or whether the nucleoside is phosphorylated *after* being released by the transport system inside the cell. This issue obviously does not exist for facilitated uptake in the absence of nucleoside kinases. Also, it is possible that under some conditions the free nucleoside detected inside labeled cells is the result of 5′-nucleotidase action on nucleoside monophosphates, rather than merely being nucleoside which has entered and has not yet been phosphorylated.

Evidence against the involvement of nucleoside kinases in nucleoside transport takes several forms. Quantitative changes in the *in vitro* specific activity of kinases are often not proportional to the observed change in nucleoside uptake rate (Breslow and Goldsby, 1969; Kay and Handmaker, 1970; Divekar *et al.*, 1972; Hare, 1972b; Hauschka *et al.*, 1972), although proportionality is exhibited in other systems (Plagemann *et al.*, 1969; Hare, 1970). Nucleoside uptake is inhibited by many compounds which fail to inhibit nucleoside kinase activity *in vitro*. Among these compounds are heterologous nucleosides (Steck *et al.*, 1969), puromycin (Plagemann and Erbe, 1972), *p*-chloromercuribenzoate (Schuster and Hare, 1970; Plagemann and Erbe, 1972), $HgCl_2$ (Schuster and Hare, 1970), persantin (Roos and Pfleger, 1972), 2-mercapto-l-(β-4-pyridethyl)benzimidazole (Nakata and Bader, 1969), colchicine (Mizel and Wilson, 1972), cytochalasin B (Plagemann and Estensen, 1972), and dibutyryl cyclic AMP (Hauschka *et al.*, 1972).

Comparison of K_m and V_{max} values for transport and kinase activities has also been used as evidence for (physical) independence of the two processes (Plagemann and Erbe, 1972). Such comparisons may not be rigorously justified. Problems of preserving and assaying kinase activity *in vitro* are the same as for other enzymes, and in addition there is the possibility that alternate forms of nucleoside kinase with different specific activity can be generated or lost during extraction procedures (Adelstein *et al.*, 1971; Okuda *et al.*, 1972; Baril *et al.*, 1972).

In some bacteria there is direct participation of enzymes in the uptake process. For example, purine nucleoside uptake in *E. coli* is a two-step event in which the nucleoside is first cleaved by membrane-bound nucleoside phosphorylase, and the free bases are then taken up by a group translocation reaction involving PRPP and adenine phosphoribosyl transferase (Hochstadt-Ozer and Stadtman, 1971a, b; Hochstadt-Ozer, 1972).

D. The Steady State

Most of the kinetic data for nucleoside transport concern *initial* rates of uptake. These are essentially the maximum transport rates, and with increasing time of labeling, the instantaneous net rate of uptake into the acid soluble pool decreases to zero, at which time the steady state condition has been achieved. Cells usually require a relatively invariant time of 3 to 15 minutes to reach equilibrium with an enormously wide range of exogenous nucleoside concentrations [at least 10^{-7} to 10^{-3} *M* (Gentry *et al.*, 1965a; Cleaver, 1967)]. Presumably this rapid equilibration is a consequence of simple diffusion, the rate of which increases linearly with increasing precursor concentration (Eq. 3). The rate of simple diffusion at any concentration is such that only a few minutes of operation at its initial rate (see Table V) would bring in enough nucleoside to raise the intracellular/extracellular concentration ratio (R) to 1.0.

Net increase of the acid-soluble pool probably continues until $R = 1.0$ for free nucleoside (simple diffusional equilibrium) and until the phosphorylated products of the nucleoside begin to inhibit the kinase and/or the facilitated transport system.

E. Concentrative Uptake

Frequent observations of the ability of animal cells to accumulate extracellular nucleosides to high intracellular levels (see Table VI) have generated some confusion regarding the role of active transport processes in nucleoside uptake. The passive transport processes (simple and facilitated diffusion) are not capable of concentrative uptake (Table III), yet the uptake of thymidine can result in a 50-fold ratio (R) of intracellular to extracellular *isotope* concentration (Hare, 1970). Note that the intracellular concentration used for calculating R includes, in addition to free nucleoside, all the nucleotide derivatives of the precursor. All known cases of concentrative uptake of nucleosides (i.e., where $R > 1.0$) involve systems where the nucleoside is rapidly phosphorylated intracellularly by the appropriate nucleoside kinase

TABLE VI

CONCENTRATION OF NUCLEOSIDES BY CELLS

Cell type	Nucleoside	Ratio[a]	Kinase[b]	Reference
Novikoff hepatoma	dT	4–5	Yes	Gentry *et al.* (1965a)
Hamster tumor cells	dT	30–50	Yes	Hare (1970)
Chinese hamster (CHO)	dT	45	Yes	Hauschka (unpublished obs.)
BHK (normal)	dT	1.1–6.6	Yes	Schuster and Hare (1971)
BHK (TK^-)	dT	1.0	No	Schuster and Hare (1971)
BHK (normal)	rU	0.6–4.4	Yes	Schuster and Hare (1971)
Mouse L1210	rU	2–6	Yes	Kessel and Dodd (1972)
Mouse L1210	rA	2–5	Yes	Kessel and Dodd (1972)
Mouse L1210	dC	2	Yes	Kessel and Dodd (1972)
Mouse L1210	dC	1.0	No	Kessel and Shurin (1968)

[a]Ratio of intracellular to extracellular concentration of labeled compounds, denoted R in the text.

[b]Cells with the appropriate kinase activity accumulate nucleosides as phosphorylated derivatives (nucleotides).

(Table VI). Because cells are impermeable to nucleotides, free equilibration between intracellular nucleotides and extracellular nucleosides cannot take place. The trapped nucleotide causes a net accumulation of precursor derivatives. Certain compounds may induce concentrative uptake of nucleosides even in the absence of nucleoside kinases by interfering with the nucleoside exit process or by forming intracellular complexes with the nucleoside. Kessel and Shurin (1968) observed such effects of the uranyl ion (UO_2^{2+}) on the uptake of araC and dC by murine leukemia cells.

For thymidine, the most extensively studied of the nucleosides, the extent of concentration by cells depends strongly on the exogenous thymidine concentration. R decreases from about 50 at 10^{-6} M dT to about 1 to 3 at 10^{-3} M dT (see Fig. 17). Also, at higher dT concentrations, there is a decreasing

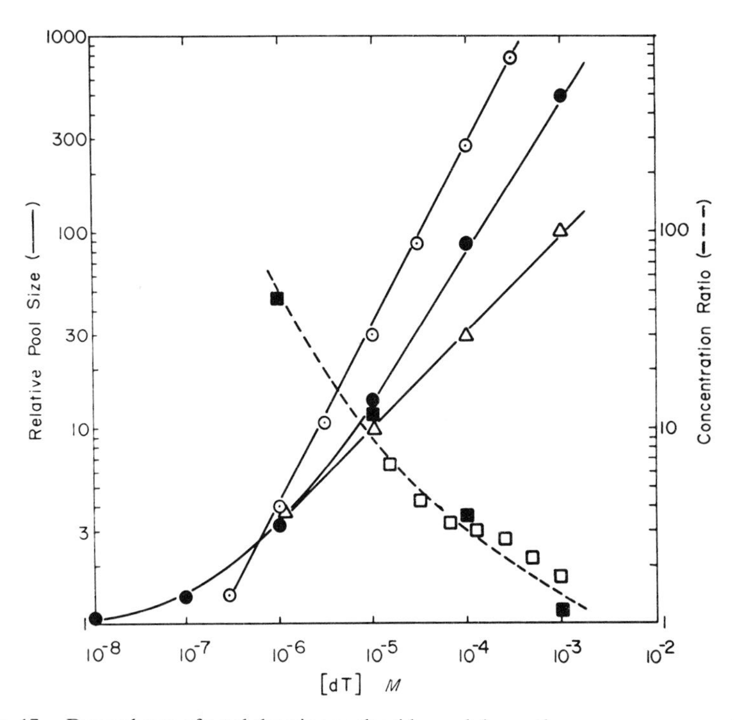

FIG. 17. Dependence of total thymine nucleotide pool size on the exogenous concentration of thymidine. ⊙, Data of Cooper *et al.* (1966b) for human leukemic leukocytes normalized to a relative pool size of 4 at 1 μM dT. ●, Data of Adams (1971) for mouse L929 cells normalized to absolute thymine nucleotide pool size of 20 pmole/10^6 cells at dT concentrations less than 10^{-8} M. Δ, Data of Gentry *et al.* (1965a) for Novikoff hepatoma cells normalized as for ●. (----) Ratio (right ordinate) between the intracellular and extracellular concentrations of total thymine compounds. ■, Data of Gentry *et al.* (1965a) for Novikoff hepatoma cells □. Data of Schuster and Hare (1971) for baby hamster kidney cells.

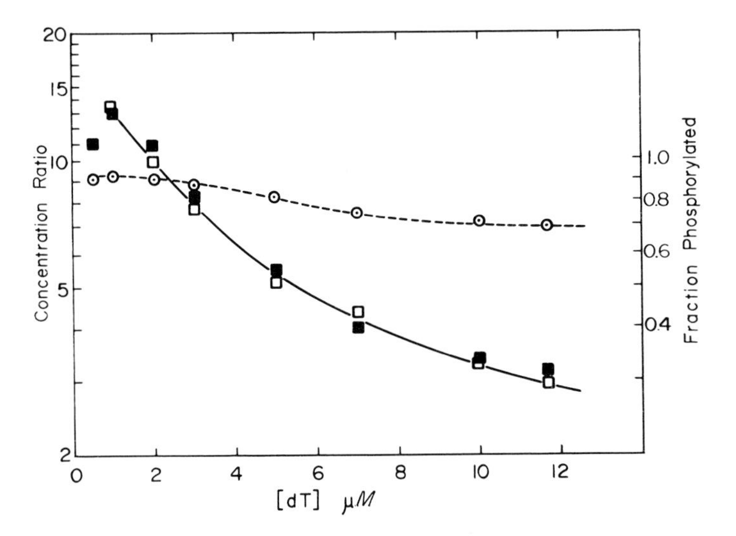

FIG. 18. Relationship between the concentration ratio for thymine compounds and the extent of phosphorylation. Data for mouse embryo cells from Lindberg *et al.* (1969). □, Relative observed concentration ratio of intracellular to extracellular thymine compounds. ⊙, fraction of intracellular thymidine in the phosphorylated form; ■, calculated concentration ratio using Eq. (6) and the observed values for the fraction of thymidine in the phosphorylated form.

fraction of intracellular label in the phosphorylated form [Fig. 18 and Lindberg *et al.* (1969)]. The reason for the large concentration factor ($R \sim 50$) at low dT concentration may be related to thymidine kinase. In this low range the kinase reaction is appreciable in comparison with dT influx. At higher dT concentrations, the absolute rate of dT phosphorylation may be slightly increased because of saturation at higher substrate concentrations, but it is minute in comparison with the dT influx by simple diffusion. Apparently, most of the increase in pool size at high dT concentrations is due to free intracellular dT (Lindberg *et al.*, 1969).

If we make the hypothesis for equilibrated cells that free dT is always at equal concentrations inside and outside the cells, regardless of the exogenous concentration, and that all phosphorylated dT (dTXP) is unable to leave the cell, then a useful relationship can be derived (Eq. 6). P is the fraction of intracellular dT which is phosphorylated.

$$\frac{[\text{intracellular dT + dTXP}]}{[\text{extracellular dT}]} = R = \frac{1}{(1 - \text{P})} \tag{6}$$

Data from the work of Lindberg *et al.* (1969) have been plotted according to Eq. (6) in Fig. 18. The agreement between the observed relative concentration factor and the R calculated from Eq. (6) is remarkably close. It is not known whether such a model will apply to other cell systems and over a wider concentration range, as no other data are presently available for testing.

The known feedback inhibition of thymidine kinase by dTTP (Ives *et al.*, 1963; Bresnick and Karjala, 1964) probably accounts for the lack of extensive phosphorylation at high dT concentrations. Only a small amount of dTTP can raise the intracellular concentration to a level at which thymidine kinase would be strongly inhibited *in vitro*. For instance, the normal intracellular dTTP pool of 15 to 20 pmoles/10^6 cells [see Adams *et al.* (1971) and Table II] is equivalent to an intracellular dTTP concentration of about 10^{-5} *M* (see Section VII, D). This level could increase to about 10^{-4} *M* after accumulation of several hundred pmoles of dTTP. Thymidine kinase is competitively inhibited by dTTP with a K_i of about 1.5 to 7 μM (Bresnick and Karjala, 1964; Hauschka, unpublished observations). Hence, ample inhibitory concentrations of dTTP can be achieved (see Section II,A), although the actual extent of inhibition might depend on other factors. For instance, if all the thymidine kinase and dTTP are restricted to the cell nucleus (Adams, 1969) then less dTTP would be required to reach inhibitory concentrations in the smaller nuclear volume.

F. Expansion of the Pool

Figure 17 shows the behavior of the thymine nucleotide pool as a function of exogenous dT concentration. Increases as great at 750-fold over the basal level (15 to 20 pmole/10^6 cells; see Table II) have been observed at 1m*M* dT (Cooper *et al.*, 1966b; Adams, 1969). Other studies of pool expansion are reported by Cleaver and Holford (1965), Cleaver (1967), Lindberg *et al.* (1969), and Adams *et al.* (1971). As mentioned in the preceding section, most of the increase in the pool at high concentration is probably due to the influx of free dT [see also Lindberg *et al.* (1969)]. However, there are possibly some exceptions where dTTP accumulates to high levels, especially during lengthy labeling periods (e.g., Adams *et al.*, 1971; Gentry *et al.*, 1965a).

Other nucleotide pools are also known to expand in the presence of exogenous nucleoside, although the increases are not so dramatic as for thymidine. Ribonucleoside triphosphate levels of Novikoff hepatoma cells are increased 50 to 300% by 1 m*M* levels of the homologous nucleoside (Plagemann, 1972). Adenosine at 10 μM concentration can increase ATP levels in mouse L cells by 3-fold (Brandhorst, personal communication). Murphy and Attardi (1973) have observed a 2.5-fold increase in the UTP pool of HeLa cells during incubation in 10 m*M* rU; however, 5 m*M* rC caused no increase in the CTP pool. The wide variation in pool expansion properties for different pools and different cells types is presumably related to differences in nucleoside kinase activities. Kinases seem to be essential for precursor accumulation, especially at low concentrations (see Section IV,E).

Expansion of the pool permits the choice of two fundamentally different conditions for labeling nucleic acids with radioactive precursors. These are termed "trace" and "flooding" conditions, and are related to the concentration of labeled nucleoside (see Cleaver, 1967). At very low exogenous nucleoside concentrations ("trace") the endogenous anabolic pathways contributing to the pool of a particular nucleotide will dilute the specific activity of the labeled precursor by a significant amount (see Figs. 1 and 2). In this situation, the apparent rate of nucleic acid synthesis is dependent on the extent of dilution, which in turn is dependent on precursor uptake and nucleoside kinase activity. "Trace" labeling conditions must be used with serious reservations when rates of nucleic acid synthesis are being compared. "Flooding" conditions are those which expand the pool with exogenous precursor to the extent that anabolic routes make an insignificant contribution to the pool. This situation causes the specific activity of the pool to be very nearly equal to that of the exogenous labeled precursor. Unfortunately, nucleoside precursors often inhibit nucleic acid synthesis at high concentrations (Fig. 2 and Section III, A), so the true rates of synthesis may again be perturbed by the measuring process.

Figure 2 (see Section II,B) indicates the amount of dilution of exogenous labeled precursor (^{3}HdT) which is caused by the *de novo* synthesis of thymine nucleotides in human leukemic leukocytes (Cooper *et al.*, 1966a). The reduction in the specific activity of ^{3}HdT as it enters the pool would be approximately equal to the percent DNA-thymine derived from the *de novo* pathway. This is about 88% at 10^{-8} M ^{3}HdT, and decreases to less than a 10% reduction in specific activity at 10^{-3} M ^{3}HdT. Cell types with greater transport and kinase activities should show a higher percentage of incorporation of ^{3}HdT into DNA at any particular concentration than those with very low transport and kinase activities. Low activity cells would not be subject to "flooding" conditions of labeling even at the highest nucleoside concentrations.

G. Specificity of Transport

Competitive inhibition and substrate specificity studies of nucleoside uptake have yielded information on the selectivity of the various transport systems. In systems where nucleosides are not metabolized, accelerative exchange diffusion studies have also been fruitful. Cells that have been preloaded with one nucleoside will leak this nucleoside into the extracellular medium at a rate that is dependent on the extracellular concentration of *other* nucleosides sharing the same transport system (Kessel and Shurin, 1968; Oliver and Paterson, 1971; Lieu *et al.*, 1971; Cass and Paterson, 1972). Winkler and Wilson (1966) have argued that the accelerative exchange effect of exogenous compounds is exercised by blocking the recapture of the leaving metabolite; the recapture event is normally 80 to 90% efficient.

Some data for transport specificity are presented in Table VII; for comparison, known specificities for the nucleoside kinases (see Sections III,A and III,B) are found in the top line of Table VII. In many cell types a rather strong parallel exists between transport and kinase specificities. This may be taken as further evidence for kinase-mediated transport, or it may merely reflect the fundamental classes of nucleoside stereochemical structure. The last four entries in Table VII indicate unusually nonselective substrate requirements. In these few cases, pertinent kinase activities were not detectable in the cells, and K_m values for the facilitated uptake systems are in the millimolar range instead of the usual micromolar range (see Table IV). Some type of coupling of kinases with the facilitated uptake system may confer greater specificity on the uptake process.

Studies of the transport of free purines and pyrimidines have demonstrated that this process is probably independent of the nucleoside transport systems (Oliver and Paterson, 1971; Stambrook and Sisken, 1972; Kessel and Dodd, 1972). Separate specific transport mechanisms for adenine and xanthine are present in rabbit leukocytes (Hawkins and Berlin, 1969).

In contrast, the uptake of free bases (A, T, U, FU) by Ehrlich ascites cells has been attributed solely to simple diffusion (Jacquez and Ginsberg, 1960; Jacquez, 1962a). In the same system, nucleoside uptake is clearly a combination of facilitated and simple diffusion (Jacquez, 1962b). The rates of purine and pyrimidine uptake are rapid; essentially complete equilibration between intracellular and extracellular concentrations of the bases is achieved in 0.5 minute at 39°C and in 9 minutes at 5°C. Diffusion constants (v_{dif}; Eq. 3) calculated from Jacquez' data are roughly the same for bases and nucleosides (ranging from 0.7 to 5 pmoles/10^6 cells per minute per micromole per liter), and are somewhat larger for Ehrlich ascites cells than for the cell types in Table V.

H. Inhibitors of Transport

Competitive inhibition of nucleoside transport by heterologous nucleosides has been studied by most of the authors listed in Table VII. Many other inhibitors of nucleoside transport are known: persantin (Scholtissek, 1968; Plagemann and Roth, 1969; Plagemann, 1971c; Kessel and Dodd, 1972; Roos and Pfleger, 1972); tubericidin, puromycin, formycin B, and

TABLE VII

SUBSTRATE SPECIFICITIES OF NUCLEOSIDE KINASES AND TRANSPORT SYSTEMS[a]

Nucleosides											
dA	dG	rH	rG	rA	dC	araC	dU	dT	rU	rC	Ref.[d]
											1–7[b]
											8
											9
											10
											11
											12–14
											15
											16
											17
											18[c]
											19[c]
											20[c]
											21[c]

[a]Each separate horizontal bar represents a distinct class of specificity for phosphorylation or transport of the indicated nucleosides. Dotted lines connect various substrates belonging to the *same* specificity class.

[b]Nucleoside kinase specificities shown in the first line *only*. All other data are for transport systems.

[c]Nucleoside phosphorylation is not observed in these systems because nucleoside kinases are absent.

[d]References: 1, Bresnick and Thompson (1965); 2, Schrecker (1970); 3, Kozai and Sugino (1971); 4, Hashimoto *et al.* (1972); 5, Murray (1968, 1971); 6, Momparler and Fischer (1968); 7. Krygier and Momparler (1971); 8, Scholtissek (1968); 9, Steck *et al.* (1969); 10, Mizel and Wilson (1972); 11, Hare (1970); 12, Stambrook and Sisken (1972); 13, Schuster and Hare (1971); 14, Plagemann and Erbe (1972); 15, Plagemann (1971c); 16, Taube and Berlin (1972); 17, Cass and Paterson (1972); 18, Lieu *et al.* (1971); 19, Oliver and Paterson (1971); 20, Kessel and Dodd (1972); 21, Kessel and Shurin (1968).

other unusual nucleosides and analogs (Krenitsky *et al.*, 1968; Nakata and Bader, 1969; Oliver and Paterson, 1971; Paterson and Oliver, 1971; Warnick *et al.*, 1972; Taube and Berlin, 1972); phenylethyl alcohol (Plagemann and Roth, 1969); colchicine (Mizel and Wilson, 1972); $HgCl_2$ (Schuster and Hare, 1970; Lieu *et al.*, 1971); *p*-chloromercuribenzoate (Schuster and Hare, 1970; Plagemann and Erbe, 1972); oligomycin, phlorizin, and fluorodinitrobenzene (Hare, 1972a); and borate (Taube and Berlin, 1972). The specificity and extent of transport inhibition varies for each compound, so the individual references should be consulted for details.

V. Isotopic Methods

Isotopic methods for pool analysis consist basically of four steps; labeling, washing, extraction, and analysis. The details of each step are related to the experimental purpose, whether it be measurement of the absolute pool size of a particular nucleotide, or the transport kinetics of a nucleoside, or the intracellular conversion of an anticancer drug to its active form.

A. Labeling

The metabolic pathways available to a precursor and its derivatives directs the choice of a particular compound for pool labeling. While thymidine-^3H can label the thymine nucleotides very selectively, other substances such as $^{32}P_i$ and $^{14}CO_2$ will generally label all nucleotide pool components. A discussion of metabolic pathways in Section III should be helpful in the selection of precursors. Also, it is important to be aware of the labeling and equilibration properties of pools and how they may change with environmental conditions (see Section II).

1. Duration

Two basic schedules for labeling cells are followed in most experiments, pulse labeling and continuous labeling. Short-term labeling is useful for measuring transport kinetics, intracellular enzyme activities, and approximate rates of precursor incorporation into nucleic acids. Continuous labeling is necessary for quantitative pool measurements and accurate determination of rates of nucleic acid synthesis and turnover, since these require constant specific activity of the intracellular label. The time required for exogenous isotope to equilibrate with intracellular pools of animal cells ranges from a few minutes for thymidine (Gentry *et al.*, 1965a; Cleaver, 1967) to as much as 48 hours for phosphate (Weber and Edlin, 1971). True equilibrium is achieved when the specific radioactivity of every acid-soluble metabolite containing the precursor reaches a constant level. The levels need not all be the same, because pool dilution effects by *de novo* synthesis may differ for each metabolite. A false equilibrium may be obtained in many cases where cells become readily "saturated" with label, but the label is not freely distributed among its eventual metabolites. For instance, cellular uptake of radioactive phosphate into the inorganic phosphate pool may be several times more rapid than its rate of incorporation into slowly turning over ribonucleotides (Weber and Edlin, 1971).

Typical kinetics of uptake of a nucleoside are shown in Fig. 19 for ^{3}HdT labeling of Chinese hamster ovary cells (Hauschka, unpublished observations). The incorporation of ^{3}H into acid-insoluble material achieves a constant rate at the same time that pool saturation occurs (13 minutes), therefore the saturation point probably corresponds to a true equilibrium (constant specific activity of dTTP). In mouse L cells, equilibration with 2.5 μM ^{3}HdT occurs within 3 to 5 minutes (Cleaver, 1967). Gentry *et al.* (1965a) observed identical kinetics of saturation (within 10 minutes) of the thymine nucleotide pool of Novikoff hepatoma cells over a 1000-fold range of ^{3}HdT concentration. The pool size was strongly dependent on the exogenous precursor concentration [Gentry *et al.* (1965a), see also Section II,G]. However, once saturated, the size of the pool remained constant for 56 hours, provided that exogenous label was not depleted (Gentry *et al.*, 1965a).

The speed of pool equilibration is affected by the following factors: (1) rate of precursor entry; (2) rate of metabolism of the precursor (e.g., phosphorylation of a nucleoside); (3) rate of turnover of the pool; (4) pool size; (5) expansion of pool by labeling. Variation in all these factors is possible (see Sections II and III). Large pools, such as ATP and phosphate, usually equilibrate rather slowly, but it is possible to overcome this difficulty. Brandhorst (personal communication) has been able to reduce the time required for steady-state labeling of the ATP pool in L cells from about 60 minutes to 3 minutes. After labeling with high specific activity ^{3}HrA ($\sim$ 25 Ci/mmole, 5×10^{-8} M) for 2 minutes, a 200-fold excess of unlabeled rA (10 μM) is added to the culture. The absolute intracellular specific activity of ATP [measured by the method of Emerson and Humphreys (1971)] overshoots slightly and then remains at a constant level from 3 minutes to more than 12 hours.

Continuous labeling of cells invites the possibility of cell damage or death due to ionizing radiations. Sensitivity of animal cells to ^{3}HdT incorporation has been discussed by Cleaver (1967). The specific activity (Ci/mmole) of the labeled precursor seems to be more important for killing than the radioactive

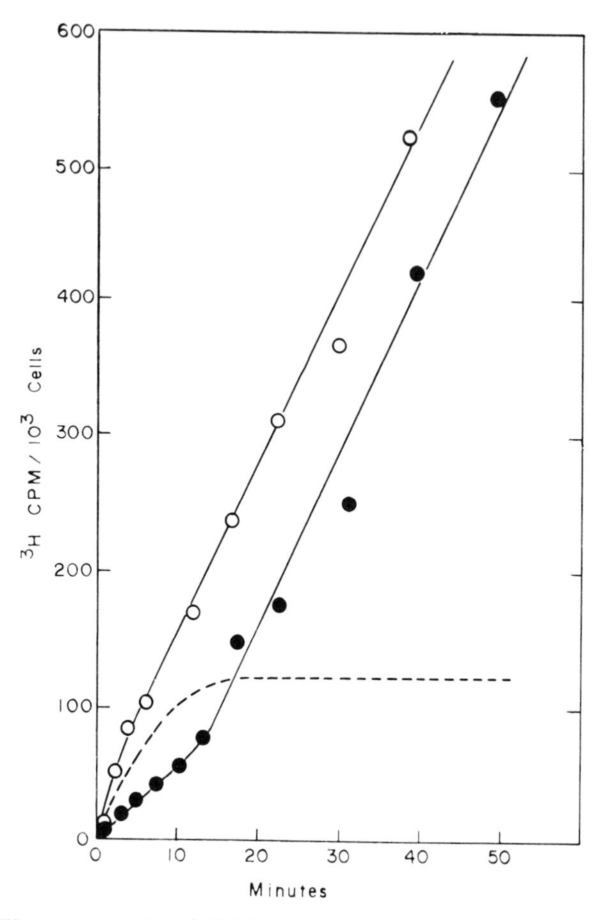

FIG. 19. Kinetics of uptake of ^{3}HdT by Chinese hamster ovary cells (CHO) growing on glass coverslips. ○, Total uptake; ●, acid-insoluble incorporation; (-----), acid-soluble uptake (total minus acid-insoluble). Pairs of coverslips, each containing about 10^5 cells in exponential growth, were placed for various intervals in conditioned medium at 37°C containing ^{3}HdT (2 μCi/ml; 0.125 μM). One coverslip was washed in cold saline, dried, and counted for total uptake of label (○). The other coverslip (●) was washed in cold saline briefly and then acid extracted before drying and counting (see Section V,B for details).

concentration (μCi/ml). A 15-minute pulse of ^{3}HdT at 1 μ Ci/ml (15 Ci/mmole) can kill more than 90% of some cultured cell lines; ^{3}HrU can have similarly strong killing effects (Prescott, personal communication).

2. Phosphate Uptake and the Phosphate Pool

Labeling cells with radioactive orthophosphate ($^{32}P_i$ or $^{33}P_i$) has become a popular method for quantitative measurement of nucleotide pools in bacteria (Neuhard, 1966; Cashel and Gallant, 1968; Bagnara and Finch, 1968; Yegian, 1973) and in animal cells (Colby and Edlin, 1970; Weber and Edlin, 1971; Plagemann, 1972). Phosphate labeling has also been advantageous in certain cases for studying the rate of nucleic acid synthesis. For instance, Nakata and Bader (1969) found for several animal cell lines that pulse labeling of RNA and DNA with various nucleoside precursors was a poor reflection of the actual rate of synthesis in the presence of the inhibitor 2-mercapto-1-(β-4-pyridethyl) benzimidazole. Pulse labeling with $^{32}P_i$ and purification of the RNA and DNA showed that the true rate of synthesis had not been depressed; rather, the transport of nucleosides (but not phosphate) had been inhibited by the drug.

As in all tracer studies, the prerequisite for quantitative results is knowledge of the extent of equilibration of exogenous label with the intracellular pool (i.e., the intracellular specific activity). Since direct measurement of phosphate specific activity is difficult with small amounts of cell material, most workers have resorted to indirect estimation of this value. The *total* radioactivity in particular nucleotides separated by chromatography is measured at various times after labeling the culture. Attainment of constant levels of radioactivity is assumed to represent an equilibrium situation where intracellular and exogenous specific activities of phosphate are equal. This type of study must be done for each growth condition where nucleotide pool sizes are to be measured: exponential growth, contact inhibited growth, drug altered growth, etc. In each case, phosphate uptake has a characteristic rate which, in conjunction with the turnover rates of the phosphate and nucleotide pools, determines the time of equilibration. Times for equilibration of exogenous $^{32}P_i$ with the intracellular pool range from 4 to 48 hours, depending on the cell type and growth rate. Rapidly growing cells will generally reach steady state within 12 hours (Jeanteur *et al.*, 1968; Colby and Edlin, 1970; Plagemann, 1972). The reduced rate of phosphate uptake in contact inhibited chick embryo fibroblasts and mouse 3T3 fibroblasts is presumably responsible for the slower equilibration exhibited by these cells (Colby and Edlin, 1970; Weber and Edlin, 1971). Once an equilibrium state is reached, the absolute pool sizes can be calculated from the known specific activity and the amount of label in each pool.

The total pool of acid-soluble phosphorus in vertebrate cells amounts to about 60 to 80 nmoles of P per 10^6 cells (Weber and Edlin, 1971; Plagemann, 1972). About one-third of the phosphorous occurs as orthophosphate (P_i), and the remainder is organic phosphate (Weber and Edlin, 1971). Acid-soluble organic phosphate consists primarily of ribonucleoside triphosphates ($\sim$45 to 95%), diphosphates (2 to 13%), and monophosphates (2 to 4%), and deoxyribonucleoside triphosphates (1 to 2%) (Colby and Edlin, 1970; Weber and Edlin, 1971; Plagemann, 1972). Phosphorylcholine (30%) and phosphorylated sugars (2 to 3%) can also be significant components of the acid-soluble organic phosphorus (see Plagemann, 1972). The acid-insoluble reserves of phosphorus are somewhat larger than the acid-soluble pool. In addition to phosphoproteins and phospholipids, DNA (~38 nmoles of P per 10^6 cells) and RNA (~ 100 to 200 nmoles of P per 10^6 cells) are the principal species.

Uptake of phosphate into the acid-soluble pool proceeds at rates ranging from 0.04 to 2 nmoles/10^6 cells per minute (Cunningham and Pardee, 1969; Weber and Edlin, 1971; Plagemann, 1972). Acid-insoluble material is labeled at about 10 to 25% of the above rates. The dependence of the rate of phosphate uptake on the exogenous concentration has not been studied in detail. However, a relationship can be gleaned from the literature as shown in Fig. 20. The plotted values are *initial* rates of uptake, and it is not known whether the size of the acid-soluble phosphorus pool is actually proportional to the exogenous P_i concentration, as is true for nucleosides (Section IV,F, Fig. 17).

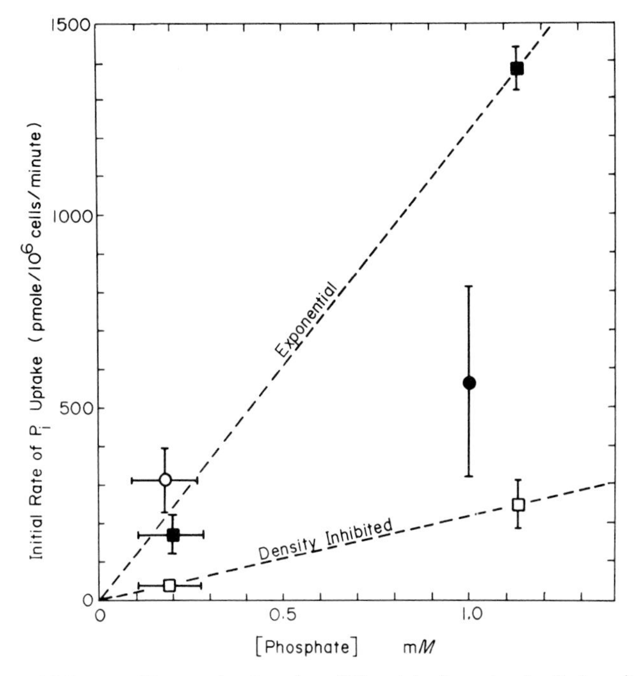

FIG. 20. Initial rate of inorganic phosphate (P_i) uptake by animal cells in culture. Uptake by mouse 3T3 cells in exponential growth (■) and under conditions of density inhibition (□). ○, Polyoma transformed 3T3 cells in exponential growth [data for 3T3 cells from Weber and Edlin (1971)]. ●, Novikoff hepatoma cells growing exponentially in suspension culture [data of Plagemann (1972)]. Error bars represent uncertainty in estimating initial rates of uptake from the published data. Uncertainty in the phosphate concentration is caused by use of undialyzed serum in "low phosphate" medium.

The choice of P_i concentration for labeling nucleotide pools is dependent on several factors. Phosphate becomes limiting for vertebrate cell growth at concentrations ranging from 30 μM for low density monolayers (R. G. Ham, personal communication) to 0.6 mM for heavy suspension cultures (Birch and Pirt, 1971). Most nutrient media contain 1 to 4 mM P_i, and the phosphate content of nondialyzed serum (1 to 3 mM) should be taken into consideration when using serum-supplemented media. Several advantages are gained by labeling with $^{32}P_i$ at the lowest phosphate concentration where growth is not limited. First is the economy of isotope: at a cell density of 2×10^6 cells/ml in medium containing 1 mM phosphate, about 30% of the total P_i is incorporated by the cells within 24 hours (Plagemann, 1972). Higher concentrations of P_i will give correspondingly lower incorporation of isotope. Second, the lethal effects of ^{32}P radiation on cultured cells may be reduced by lowering the total radioactive concentration in the medium. At 20 μCi of $^{32}P_i$ per milliliter, death of cultured 3T3 mouse cells is appreciable after 72 hours (Weber and Edlin, 1971). $^{32}P_i$ concentrations as high as 200 μCi/ml have been used (Weber and Edlin, 1971; Yegian, 1973) but are reliable only for short-term experiments. Most $^{32}P_i$ labeling of pools in vertebrate cells and bacteria is carried out at 3 to 20 μCi/ml (10 to 50 Ci/mole) for several cell generations. With this procedure, sufficient samples of labeled nucleotides may be obtained from about 10^6 to 10^7 vertebrate cells (Weber and Edlin, 1971; Plagemann, 1972) or 10^8 to 10^9 bacteria (Cashel and Gallant, 1968). Phosphorus-$^{33}P_i$ has not been extensively used for pool studies. The low emission energy of this isotope (0.26 MeV compared to 1.72 MeV for ^{32}P) would favor it for long-term labeling. Nazar *et al.* (1970) have reported $^{33}P_i$ labeling of *E. coli* nucleotide pools.

3. Measurement of ^{32}P Specific Activity

Most growth media are sufficiently defined that the total phosphate concentration is known with reasonable accuracy. Trace impurities in reagents generally contribute some "blackground" phosphate, equivalent to 3 to 6 μM P_i in one carefully studied system (Birch and Pirt, 1971). Serum, proteose peptone, milk, and other additives to culture media can contribute variable amounts of phosphorus. Large errors in specific activity are invited when it is assumed that the only phosphate in "low-phosphate" media is derived from added inorganic phosphate. Measurement of the phosphate concentration in media is straightforward, and should be done whenever the value is not accurately known. Several methods may be used. Bartlett's (1959) modification of the Fiske-SubbaRow procedure for total phosphorus requires about 0.1 μmole of phosphorus. The method of Martin and Doty (1949) is accurate for similar amounts of inorganic phosphate. Inorganic and organic phosphate can also be distinguished by selective precipitation

(Sugino and Miyoshi, 1964). The total phosphorus value is probably the most meaningful for specific activity calculations. Measurements can be made with nonradioactive media since the contribution of carrier-free $^{32}P_i$ (1×10^{-13} mole/μCi) to total phosphate concentration is infinitesimal. After counting aliquots of the labeled culture medium (including cells) by liquid scintillation or gas flow techniques, the resulting value for counts per minute per picomole of phosphorus is used to calculate the sizes of individual nucleotide pools, keeping in mind the molar content of phosphate in each separate compound. Other methods for measuring specific activity of nucleotide pools are given in Section VI.

B. Washing Procedures

Before pool extraction can proceed, cells and tissues must be thoroughly washed free of contaminating nutrient media and interfering extracellular metabolites. In experiments where pools are to be measured with radioisotopes, washing is one of the most important yet generally neglected aspects of the study. Large excesses of extracellular isotope can lead to spurious results. It is a general rule that as the specificity of the method for nucleotide pool analysis increases, the importance of washing decreases. Hence a study of the total radioactivity in the acid-soluble pool of ^{3}HdT labeled cells would be critically sensitive to washing, whereas the analysis of specific nucleoside triphosphate levels by either of two methods (thin-layer chromatography of $^{32}P_i$ labeled cells, or DNA polymerase assay of the limiting triphosphate) usually removes possible contaminants concomitantly, or the analysis does not respond to them, and washing can be minimal.

1. Thorough Washing: Isotopically Labeled Cells

Three to six buffered isotonic saline washes of labeled cells are generally used to remove extracellular label. Uniform exposure of all cells to the washing solution is critical. Monolayer cultures are relatively easy to wash by bathing in saline, but thorough resuspension of centrifuged cells is required for each wash of suspension cultures.

Before developing a heavy dependence on a particular washing regimen, the effectiveness should be checked by analysis of the radioactivity in each wash (Fig. 21). Variables that should be considered in washing are considered below.

a. Washing Medium. Washing solutions should be isotonic to prevent leakage by osmotic swelling or shrinking. For most cell culture studies, a suitable washing solution is the buffered inorganic salt portion of the nutrient medium employed for growth. Phosphate or Tris·HCl buffers are frequently added to iosotonic saline at a final concentration of 2 to 10 mM in order to

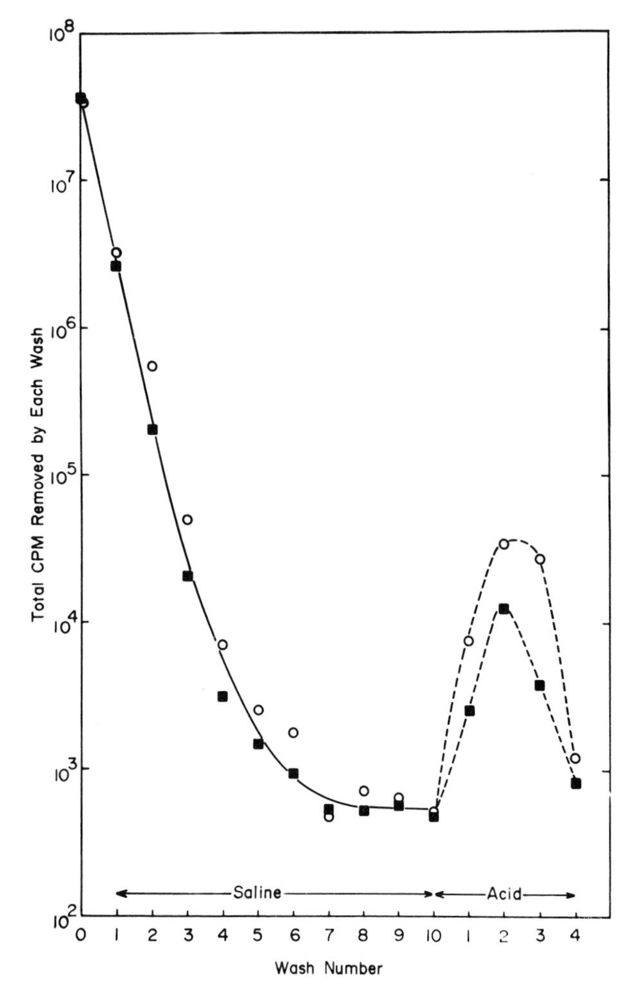

FIG. 21. Removal of ³HdT from Chinese hamster (CHO) cells by washing at 0°C. Cells growing in 75 cm^2 plastic bottles (Falcon Plastics) were labeled for 40 minutes with ³HdT (2 μCi/ml, 0.125 μM, 0°C); ○, control cultures; ■, cells pretreated for 16 hours with 0.3 m*M* dibutyryl cyclic AMP. Ten saline washes (10 ml each, 1 minute, 0°C) were followed by 4 extractions with 0.5 *N* HCl (1 ml each, 2 minutes, 0°C). More than 3 times as much acid-soluble thymidine compounds were extracted from the control cells than from the cells treated with dibutyryl cyclic AMP. Similar results were obtained with extraction by 0.5 *M* $HClO_4$. Data from Hauschka (unpublished observations); see also, Hauschka *et al.* (1972).

maintain constant pH during operations outside the CO_2 incubator, where bicarbonate–CO_2 buffer systems rapidly become alkaline. Table IX lists the ingredients of some commonly used washing solutions.

It is a common practice to include excess amounts (0.1 to 1 m*M*) of homologous unlabeled ("cold") metabolite in the washing solution when cells have been labeled with a radioactive metabolite (e.g., Cunningham and Pardee, 1969). While this procedure may decrease the amount of spuriously bound extracellular label by competing for binding sites, it could introduce serious artifacts in transport studies. Specifically, in systems where a nucleoside is not metabolized (usually by phosphorylation) upon uptake, loss of the nucleoside can be greatly accelerated by exogenous cold nucleoside (Kessel and Shurin, 1968; Oliver and Paterson, 1971; Kessel and Dodd, 1972; Cass and Paterson, 1972).

b. Duration of Wash. Dilution of extracellular radioactivity occurs almost immediately upon addition of the washing solution. Each wash requires adequate mixing and removal but should last no more than 1 minute. Most pool components of interest are highly charged and will not ordinarily diffuse out of cells during a wash (see Schuster and Hare, 1971). The danger of exhaustive washing is that metabolism of acid-soluble pool components may occur during the washing. Triphosphates may be cleaved to di- and monophosphates and nucleosides, and phosphorolysis of nucleosides may allow leakage of free bases from the cell. In this regard, monolayer cultures are advantageous because no lengthy centrifugation steps are needed to collect the cells after each wash.

c. Number of Washes. Large amounts of unincorporated label, in the form of droplets or films on the walls of the culture vessel, can persist through many washes if care is not exercised. It is suggested that the washing method be practiced in a spare culture vessel with ink and water, where the most effective method for rapid removal of the color becomes immediately obvious. Three careful washes are better than ten sloppy ones. We have examined the removal of unincorporated ^{14}CdT from monolayers of CHO cells grown in 75 cm^2 plastic bottles (Falcon Plastics). As shown in Fig. 21, there is a two-stage character to the washing. First is the rapid dilution of the original labeling medium. The dilution factor is about 10$\times$ per wash during the first four 10-ml washes, each of which is rocked gently for 1 minute and then poured off. This dilution can be increased to about 40$\times$ per wash by thorough removal of each wash with a fine-tipped suction tube. From the sixth to tenth wash, a relatively constant amount of total label is removed in each step. This probably represents slow leakage of nucleoside from the cells. Extraction of the acid-soluble pool from washed cells is rapid (Fig. 21), and the baseline radioactivity in the final saline washes represents only 0.8 to 4.7% of the total acid-soluble pool.

Schuster and Hare (1971) have investigated the washing of ^{3}HdT labeled hamster cell monolayers with Tris-buffered saline at 5°C. They find that the equivalent of only 1% of the acid-soluble fraction is present in the third or fourth wash. The rate of efflux of pool components from prelabeled cells is slow; rinsing for 0.5 or 5 minutes removes about the same amount of radioactivity (Schuster and Hare, 1971). Adenine-^{14}C leaches from rabbit leukocytes with first-order kinetics such that 3 to 4% is lost during a 25-second wash; xanthine is not lost during a short wash (Hawkins and Berlin, 1969). Losses of adenosine and thymidine from prelabeled rabbit leukocytes are about 9% and 6%, respectively, during a 25- to 120-second wash in cold saline (Taube and Berlin, 1972).

Sample: 10^6 to 10^7 cells in monolayer culture (petri dish or bottle)

(1) Remove labeled medium by suction. A blunt 20-gauge syringe needle attached by tubing to a vacuum trap is convenient.

(2) Immediately cover the cell layer gently with ice cold buffered isotonic saline (see Table IX), using about 0.2 ml/cm^2 of surface, and swirl the liquid, taking care not to dislodge any cells. After 10 to 60 seconds, tilt the vessel and remove the wash by suction. Repeat the washing step 2 to 5 more times.

(3) Extract the acid-soluble pool immediately (Section V,C).

(4) References: Adams *et al.* (1971), Schuster and Hare (1971), Weber and Rubin (1971), and Hauschka *et al.* (1972).

Sample: 10^4 to 10^6 cells in monolayer culture on a glass coverslip (recommended method for transport kinetic studies)

(1) Fill at least six 150-ml beakers with cold buffered isotonic saline and place on ice.

(2) Remove a coverslip culture from the labeling medium with fine-tipped stainless forceps and gently dip and wave in successive beakers of saline, spending about 5 to 10 seconds in each beaker. After three coverslips have been carried through this procedure, discard the first beaker and place a fresh one in the last position.

(3) The washed coverslips can be drained on edge on absorbant paper for subsequent counting of total radioactivity [see (5c) below].

(4) References: Hawkins and Berlin (1969) and Hauschka *et al.* (1972).

(5) Extraction of the acid-soluble pool can be carried out in two ways:

(a) If the pool is to be analyzed, place the coverslip face down in a 35 mm petri dish (Falcon Plastics) containing at least 0.2 ml of ice cold extracting solution such as 0.4 M $HClO_4$ (see Section V,C). The coverslip can be removed after 15 minutes and washed with several additional drops of acid. Neutralization and analysis of the extract is described in Sections V,C and V,D.

(b) If the pool is merely to be extracted so that acid-insoluble counts can be measured, place the coverslip on edge in a small rack *immediately* after the saline wash, and immerse for 10 minutes in about 50 ml of fresh, 3:1

ethanol/acetic acid (see Section V,C and Everhart *et al.*, this volume). The acid wash is followed by 5 minutes in 95% ethanol, 5 minutes in 100% ethanol, drying, and counting (see below).

(c) To count radioactivity in the cells either before or after acid extraction as described above, the coverslips may be placed face-up in planchets for gas-flow counting. This is useful if autoradiograms are to be prepared. Alternatively, the coverslips can be crushed directly into scintillation vials using a small wide-mouth plastic funnel and a sturdy glass rod. The radioactivity should be solubilized to minimize self-absorption [particularly problematic with tritium, Hodes *et al.* (1971)]. Digestion can be effected by 0.2 ml 0.5 *N* KOH (60 minutes at ~40°C) followed by neutralization with $HClO_4$ using phenol red as an indicator (Hawkins and Berlin, 1969). The commercial organic solubilizer NCS (Nuclear Chicago) is also effective (1 ml/vial, 2 hours, 37°C) (Everhart *et al.*, this volume).

Sample: Suspension culture.

(1) Pour a sufficient aliquot of the culture (usually 10^6 to 10^7 cells) into a cold conical centrifuge tube. Swirl the tube in a bath of dry ice/ethanol or ice/NaCl until the contents reach about 4°C.

(2) Sediment the cells at 500 *g* for 2 to 5 minutes at 0° to 4°C, using the minimum time required to form a soft, fluffy pellet; do not pack.

(3) Gently resuspend the cells in 10 ml of ice cold buffered isotonic saline and centrifuge immediately. Carefully withdraw and discard the supernatant, resuspend the cells in 10 ml of washing solution and repeat the centrifugation. The pellet should be acid extracted immediately.

(4) References: Bray and Brent (1972) and Plagemann (1972).

(5) In cases where labeled cells are centrifuged directly out of the labeling medium and counted without washing, corrections must be made for radioactivity trapped in the extracellular space by preparing control tubes with inulin-^{14}C and ^3H-labeled water as described by Kessel and Dodd (1972).

(6) A rapid centrifugation method for separating suspended erythrocytes from the labeling medium has been used by Oliver and Paterson (1971). Because of their high specific gravity, erythrocytes will pellet in di-1-butylphthalate (DBP), leaving the aqueous medium at the top of the tube (Danon and Marikovsky, 1964). Samples of erythrocytes suspended in aqueous labeling medium (~1 ml) are layered over 5 ml of DBP in 13 × 100 mm tubes at 25°C and centrifuged 1.5 minutes at 1700 *g*. Both influx and efflux of nucleosides can be measured in this fashion (Oliver and Paterson, 1971; Cass and Paterson, 1972), although rather high concentrations of cells are usually required.

(7) An alternative method to centrifugation for collecting and washing

suspension cultures is filtration. This has been used in studies with bacteria (Nazar *et al.*, 1970). We have found that CHO cells labeled in suspension culture can be readily collected on glass fiber filters (GF/A) and washed free of acid-soluble radioactivity on a Millipore suction device. We have not yet investigated the possibility of making accurate nucleotide pool measurements with filter-collected cells. Problems may be caused by lysis during suction or loss of cells from the filter. Roberts and Loehr (1972) have used filtration on 0.45 μm MF-Millipore cellulose ester filters for measurement of nucleoside uptake and incorporation by a human lymphoblastic cell line (CCRF-CEM) grown in spinner culture. After incubation with labeled precursors under the desired conditions, 2-ml aliquots of the cell suspension were filtered and washed with 0.9% NaCl only (for total uptake) or with 0.9% NaCl followed by cold 5% trichloroacetic acid (for acid insoluble incorporation). Filters were dried and counted by liquid scintillation. The acid-soluble pool was equal to the difference between the above two quantities.

2. Minimal Washing: ^{32}P-Labeled Cells

The purpose of washing in this type of experiment is to remove unincorporated $^{32}P_i$. Since chromatography of the acid soluble extract is also designed to separate $^{32}P_i$ from the nucleotides of interest, the requirement for washing is directly related to the chromatographic system. Plagemann (1972) washed Novikoff hepatoma cells twice by centrifugation to reduce problems caused by tailing of $^{32}P_i$ in his two-dimensional electrophoresis and chromatography system. The first dimension of chromatography in Yegian's system runs $^{32}P_i$ right off the thin layer onto an absorbant wick, hence no washing of the cells is required before extraction (Yegian, 1973).

Sample: 10^6 animal cells in monolayer culture (50-mm petri dish)

(1) Remove radioactive medium by suction.

(2) Cover cells once with 5 ml of ice cold 0.15 *M* NaCl–0.1 *M* Tris HCl (pH 7.4) and quickly remove the washing solution by suction. Other buffered washing solutions listed in Table IX could also be used.

(3) Extract the acid-soluble pools immediately with 0.4 *M* $HClO_4$ at 0°C for 30 minutes.

(4) Reference: Colby and Edlin (1970).

Sample: Suspension culture of animal cells; see previous section and Plagemann (1972)

Sample: E. coli in suspension culture

(1) The culture is directly extracted with no washing (see Section V,C).

(2) References: Neuhard and Munch-Petersen (1966), Bagnara and Finch (1968), Cashel and Gallant (1968), and Yegian (1973).

3. Minimal Washing: Unlabeled Cells

Analysis of pools by the direct chemical and enzymatic methods described in Section VI usually requires little or no washing because of the specificity of the assays and the lack of extracellular interfering compounds.

Sample: Suspension culture of animal cells

(1) Collect cells by centrifugation at 0° to 4°C (see Section V,B above).

(2) Extract immediately without washing (Solter and Handschumacher, 1969), or wash twice with cold buffered isotonic saline and then extract with $HClO_4$ (Bray and Brent, 1972).

Sample: Monolayer culture of animal cells

(1) Remove medium by suction.

(2) Extract immediately without washing (Skoog, 1970), or wash twice with cold buffered isotonic saline and then extract with $HClO_4$.

C. Extraction of Intracellular Pools

1. Principles

The method chosen for pool extraction should be tailored to the requirements of the experiment. Simple studies involving short-term incorporation of a tritiated nucleoside into the acid-soluble pool need not be concerned, necessarily, with the chemical form of the extracted label. On the other hand, analysis of deoxyribonucleoside triphosphate pools by sensitive chromatographic or enzymatic methods requires careful extraction by nondegradative procedures. Most extraction techniques are variations of a single theme: solubilization of free and bound cellular nucleosides and nucleotides under conditions where polynucleotides and proteins are insoluble and enzymes are inactive. As outlined below, many methods have been developed. Some are cumbersome, or nonquantitative, or chemically destructive, but all generally have merits, at least for their original application. Because all methods of pool analysis involve extraction, the accuracy of the results ultimately depends on the extraction procedure. To date, the most widely applied and versatile technique is perchloric acid extraction followed by neutralization. This method is recommended for all to whom pool analysis is a passing fancy. New methods which exhibit higher extraction efficiencies and different rates of hydrolysis of nucleoside triphosphates (Cashel and Gallant, 1968; Nazar *et al.*, 1970) have not yet been applied to a sufficient number of vertebrate cell systems to prove their versatility. Figure 22 shows the efficiency of perchloric acid extraction of nucleotide pools in *E. coli* (Bagnara and Finch, 1972). The same $HClO_4$ method has been effective for vertebrate cells in culture (Colby and Edlin, 1970). As pointed out by Cleaver (1967), some cell types are resistant to standard extraction procedures, so any method should first be tested on the system of interest.

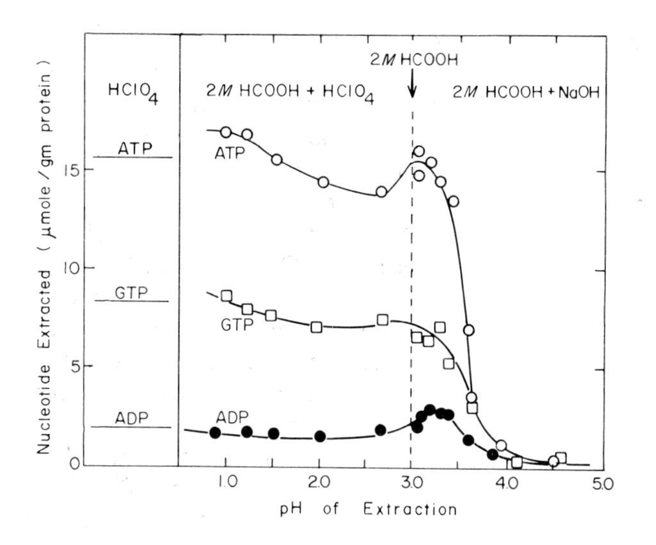

FIG. 22. Effect of pH on extraction of nucleotides from *Escherichia coli* [redrawn from Fig. 4 of Bagnara and Finch (1972), by permission of Academic Press, New York]. The standard $HClO_4$ extraction method (Section V,C) gives a reproducibly high yield of nucleotides compared to other extraction solvents [see also, Bagnara and Finch (1968)].

Jonsen and Laland (1960) outlined three general types of extraction procedures—acid, neutral, and alkaline. Each has its advantages for preservation of certain chemical linkages. Acid conditions may rupture certain N-glycoside, glycosyl phosphate, or sugar pyrophosphate linkages, but generally will not destroy nucleoside phosphates at 0° to 4°C during 1 hour of exposure (Bagnara and Finch, 1972). Neutral conditions, while allowing some extremely labile substances to be studied, will generally not provide quantitative extraction. Alkaline conditions are essential for the extraction of reduced pyridine nucleotides (Estabrook *et al.*, 1967) but can destroy amino acyl adenylates and esters, nucleoside diphosphate esters of sugars, and will solubilize ribopolynucleotides.

Often of greater importance than the sensitivity of particular compounds to hydrolysis by extraction solvents is their enzymatic sensitivity prior to and during extraction. Instantaneous disruption of the normal balance of enzymatic activities and pool sizes within cells and tissues may be assumed to occur when pool extraction begins. In rat brain there is a substantial change in ATP level (70% decrease) during the 20-second dissection procedure prior to freezing in liquid air (Mandel and Harth, 1961); *in situ* freezing and extraction of only the outer layer of frozen liver is the only feasible method for obtaining accurate data on nucleotide levels in this system (Mandel, 1964). Muscle offers similar problems because of the rapid deamination of adenine nucleotides occurring during isolation (Szentkiralyi, 1957). Scholtissek (1967) observed extensive enzymatic degradation of UTP and GTP during several days of storage of washed cells at $-20°C$. It is therefore important to extract nucleotide pools immediately and store the extracts, rather than to store the frozen cell samples for extraction at a later time.

Rapid inactivation of all enzyme activities can best be achieved by instantaneous freezing or exposure to inhibitors. Perchloric acid and trichloroacetic acid, because of their large anions (ClO_4^- and $CCl_3CO_2^-$) and relatively strong acidity, are potent precipitating agents for proteins and nucleic acids. With very few exceptions (Wood, 1971), enzymes are completely inactivated by the combination of precipitation, low pH, and low temperature which characterizes $HClO_4$ and TCA extraction methods. Virtually instantaneous attainment of these inhibitory conditions is possible in monolayer and suspension cultures of vertebrate cells. Assuming no permeability barriers under the hypertonic conditions of acid extraction, diffusion of the large anions into the interior of all cells should be complete within less than one second after exposure. Tissue chunks present geometric difficulties for rapid fixation. Quick freezing, pulverization, and acid extraction of the tissue powder appears to be the best procedure (Mandel, 1964).

Other extraction solvents described below may be effective for solubilizing nucleotides in certain systems, but unlike $HClO_4$ and TCA, they may allow residual enzyme activity to alter pool sizes during the extraction procedure.

2. Perchloric Acid

Perchloric acid was introduced by E. J. King in 1932 as a versatile substitute for H_2SO_4–H_2O_2 in the oxidative ashing of organic phosphates—a prerequisite for phosphate analysis. The new reagent rapidly gained favor in biochemistry because it facilitated analysis of many organic phosphate esters which were commonly isolated as barium salts. Tedious removal of barium ions was no longer required for phosphate analysis, since barium perchlorate was soluble and noninterfering, unlike barium sulfate.

a. Advantages of $HClO_4$. (1) $HClO_4$ extraction usually gives the highest yields of nucleotides (see Fig. 22). (2) $HClO_4$ solutions have negligible UV absorption, enabling optical quantitation of dissolved nucleotides. (3) The low solubility of $KClO_4$ (about 0.054 *M*, saturated, 0°C) permits simple removal of ClO_4^- by neutralization of $HClO_4$ solutions with KOH. Neutralized supernatants are sufficiently salt-free to be used directly for chromatography (Colby and Edlin, 1970; Plagemann, 1972) or for DNA polymerase analysis of deoxyribonucleoside triphosphates (Solter and Handschumacher, 1969; Lindberg and Skoog, 1970) (see Sections V,D and VI,A). (4) Nucleic acids and proteins remain insoluble in 0.5 *M* $HClO_4$ at 0°C. RNA can be selectively removed from an $HClO_4$-extracted pellet by alkaline hydrolysis under the conditions of Davidson and Smellie (1952); 0.3 *N* KOH, 16 hours, 30°C. DNA is still quantitatively precipitable by 0.5 *M* $HClO_4$ after this treatment (Schmidt, 1957). By heating the DNA-protein precipitate in 0.5 *M* $HClO_4$ for 15 minutes at 80°C (Skoog and Nordenskjöld, 1971) or 30 minutes at 70°C (Plagemann, 1972), it can be solubilized for scintillation counting. Eberhard (1972) has described a procedure for solubilizing DNA in the presence of large excesses of protein (0.3 *M* $HClO_4$, 90°C, 150 minutes). Colorimetric determination of DNA content, even without alkaline hydrolysis of the RNA, is also a useful procedure (Burton, 1968; Bray and Brent, 1972). Hence, initial use of $HClO_4$ for nucleotide extraction is compatible with DNA and RNA determinations on the same cell material.

b. Preparation of Perchloric Acid Solutions. Perchloric acid (MW = 100.5) is generally available in aqueous solutions at 60% or 70% concentration by weight. The high specific gravity of these stock solutions (1.54 and 1.67, respectively) must obviously be taken into account when preparing working solutions. A 0.4 *M* solution (~ 4%) is prepared by diluting 34.4 ml of 70% $HClO_4$ or 43. 5 ml of 60% $HClO_4$ to 1 liter with distilled H_2O. A 0.5 *M* (~ 5%) solution is prepared by diluting 43 ml of 70% or 54.3 ml of 60% acid to 1 liter. Bagnara and Finch (1972) studied the extraction of nucleotides from *E. coli* by various concentrations of $HClO_4$ and found 0.4 *M* $HClO_4$ to be optimal.

c. Extraction Method. This is essentially the method of Bagnara and Finch (1968, 1972) as used by Colby and Edlin (1970) for chicken fibroblast monolayers. For a 5-cm petri dish containing about 10^6 cells:

(1) Remove medium and wash appropriately, as described in Section V,B.

(2) Cover with 1.5 ml 0.4 *M* $HClO_4$ at 0° to 4°C, swirl gently, and let stand in cold for 30 minutes.

(3) Mix gently with a cold pipette and transfer 1.0 ml of the extraction solution, including suspended cell debris, to a small, cold centrifuge tube.

(4) Neutralize by adding 0.5 ml 0.72 *N* KOH–0.6 *M* $KHCO_3$ at 0°C. This should be done with adequate mixing (vortex) to prevent alkaline conditions from developing locally in the solution. Large amounts of RNA in the cell debris are sensitive to alkali and could affect the results if hydrolyzed. Other neutralization methods involving K_2CO_3 and triethanolamine (Estabrook *et al.*, 1967) or KOH and Tris·HCl (Plagemann, 1972) have been applied successfully. In order to avoid hydrolysis of triphosphates, the neutralization procedure should follow within an hour after acid extraction (Bagnara and Finch, 1972).

(5) After about 5 minutes on ice, the $KClO_4$ (heavy white precipitate) should have completely formed. Centrifuge at 7000 rpm for 10 minutes and transfer the supernatant to a clean tube.

(6) The supernatant contains the free bases, nucleosides, and nucleotides

of interest and can be stored frozen at −20°C for several weeks without appreciable loss of nucleoside triphosphates (Nordenskjöld *et al.*, 1970). Remember that this solution contains only two-thirds of the cellular pools because of the aliquot taken in step 3. Analysis of the extract should be carried out by the methods in Section V,D or VI as soon as possible.

(7) If it is desirable to measure the DNA content of the total cell material extracted, then the procedure of Nordenskjöld *et al.* (1970) may be substituted for steps 2 and 3. To the washed monolayer is added 0.3 ml 0.5 *M* $HClO_4$ at 0°C and the cells are scraped off the dish into a cold centrifuge tube with a rubber policeman. Quantitative recovery of the cells is important, so two additional washes of the dish and rubber policeman with 0.6 ml of 0.5 *M* $HClO_4$ are advisable. The tube is centrifuged after 30 minutes at 0°C, the supernatant is removed, 1.0 ml is neutralized as in step 4, and the cell debris saved for DNA analysis by the Burton (1968) method. Adams *et al.* (1971) observed that about 50% of the dTTP pool in hydroxyurea-treated mouse L929 fibroblasts was lost if a rubber policeman was used to collect the cells *before* extracting with 5% TCA. Similarly, Piez and Eagle (1958) found rapid proteolysis of cell proteins and loss of amino acid pools when scraping of monolayers preceded addition of TCA. This should not be a problem in the above procedure where $HClO_4$ is added, and thus most enzyme activities are destroyed, before the scraping step.

(8) Cells grown in suspension culture can be prepared for $HClO_4$ extraction by centrifugation and rapid washing [see Section V,B and Bray and Brent (1972)]. All procedures are carried out in centrifuge tubes. The time period between harvesting and extraction should be minimized because of the continuous degradation of nucleotides described above.

3. Trichloroacetic Acid

Kerr and Blish first used TCA in 1932 for the extraction of purine nucleotides from muscle and blood. Their quantitation of purines by Kjeldahl nitrogen analysis was dependent on total removal of interfering nitrogenous compounds, the bulk of which was protein and was completely precipitated by TCA.

a. Advantages of TCA.

(1) TCA at 5 to 10% concentration effectively precipitates polynucleotides, proteins, and inactivates most enzymes. Because it solubilizes most small phosphate-containing compounds, TCA is used for cleaning up DNA and RNA prior to phosphate analysis (Schmidt, 1957).

(2) DNA is hydrolyzed under the conditions: 10% TCA, 70°C, 1 hour (see Weber and Rubin, 1971). Hydrolysis of RNA does not occur.

(3) Separation of TCA from nucleotides can be achieved by several methods: (a) Evacuation or lyophilization of the solution removes TCA as a

vapor (see Schuster and Hare, 1971). (b) Extraction of the solution 4 to 8 times with equal volumes of diethyl ether quantitatively removes TCA in the ether phase (Adams *et al.*, 1971; Bagnara and Finch, 1972). (c) Adsorption of the nucleotides to acid washed charcoal (Norit A), washing with water, elution with ammoniacal 50% aqueous ethanol, and evaporating to dryness will effectively concentrate the nucleotides free of TCA [see Sugino and Miyoshi (1964) and Section V,D].

b. Disadvantages of TCA.

(1) Strong UV absorption of TCA solutions precludes its use as a solvent where optical determination of nucleotides is desired.

(2) Procedures for removing TCA are more cumbersome and time consuming than for $HClO_4$ removal. Some aqueous phase dissolves in the ether, thereby removing small amounts of nucleotides. Losses are also noted with the evaporation method. Bagnara and Finch (1972) estimate that 10 to 30% of the total acid-soluble radioactivity is lost from the TCA sample during either extraction and evaporation to dryness.

(3) TCA extraction of nucleotide pools is sometimes not reproducible (Cashel and Gallant, 1968). Bagnara and Finch (1968) have compared the effectiveness of extraction by 5% TCA with that for 0.4 *M* $HClO_4$. Values for ribonucleoside triphosphate pools in a stringent strain of *E. coli* were very reproducible using 0.4 *M* $HClO_4$. Extraction with 5% TCA yielded smaller and more variable pool sizes, possibly because TCA does not efficiently extract nucleotides which are bound to intracellular polycations (Bagnara and Finch, 1968). Polyamines accumulate in bacteria during amino acid starvation (Raina and Cohen, 1966), and the binding of purine nucleotides, especially those of guanine, to polylysine and polyarginine is rather strong (Wagner and Arav, 1968). The greater ability of $HClO_4$ to disrupt these interactions could explain its observed advantages over TCA for nucleotide extraction.

c. Preparation of TCA Solutions. TCA (MW = 163.4) is extremely deliquescent, and even the crystals in unopened bottles are usually wet. We generally prepare stock solutions at a concentration of 50% (w/w) using an entire 1-lb bottle of TCA. It is unwise to store solutions at concentrations below 30% because of the accumulation of decomposition products ($CHCl_3$, HCl, CO_2, and CO). Working solutions of 5% and 10% (about 0.51 *M*) are prepared weekly and stored at 4°C.

d. Extraction Method. This is essentially the method of Adams *et al.* (1971) as applied to monolayers of mouse L929 cells. For a glass bottle containing about 10^7 to 10^8 cells:

(1) Remove the medium and wash appropriately (see Section V,B).

(2) Bathe the monolayer in 10 ml 5% TCA at 0°C, gently rocking several times.

(3) After 30 minutes, transfer the extraction solution to a cold centrifuge tube. If any cell debris is present, remove by sedimentation at 0°C and place the supernatant in a clean tube.

(4) Extract the supernatant four times with an equal volume of cold ether, mixing well and discarding the ether layer each time.

(5) Lyophilize the aqueous solution and dissolve the residue (nucleotides) in water or an appropriate buffer for chromatography, enzymatic analysis, or scintillation counting.

4. Acetic Acid

Nazar *et al.* (1970) have developed an extraction procedure for *E. coli* which is particularly useful for chromatography of nucleotide pools because the solvent is completely volatile, thereby eliminating salt effects. Pool samples can be concentrated *in situ* on the chromatography sheets using a drying stream of air. As applied to *E. coli*, the method is as follows: (a) Collect cells from suspension on prewarmed 25 or 47 mm Millipore filters. (b) Quickly immerse the filter in 5.5 ml of 1 *N* acetic acid at 0°C in a small beaker; agitate gently for 5 minutes. (c) Freeze and thaw three times in succession, with 5 minutes of gentle agitation between freezes. (d) Lyophilize 5 ml of the extract, redissolve in 0.6 ml cold water, and centrifuge 5 minutes at 12,000*g*. Up to 0.5 ml of the supernatant can be slowly applied to a single origin for thin-layer chromatography.

5. Formic Acid and Sodium Formate

Because it is volatile, formic acid is also useful for pool extraction where the samples are to be chromatographed. Formic acid solutions are sometimes not sufficiently acidic to extract nucleotides quantitatively as indicated in Fig. 22 (Bagnara and Finch, 1972). Nazar *et al.* (1970) reported that 0.33 *N* or 1 *N* formic acid could be substituted for 1 *N* acetic acid in their extraction procedure for *E. coli*. Freezing and thawing was still required, and the yields were slightly lower. Cashel and Gallant (1968) have used sodium formate (pH 3.4) at a final concentration of 0.33 *M* for complete extraction of *E. coli* nucleotides in 30 minutes at 0°C. The centrifuged extract was chromatographed in a "step formate" system, so the salt in the nucleotide sample was not problematic. Yegian (1973) developed a formic acid extraction procedure for *E. coli* which was compatible with his periodate oxidation step for destruction of ribonucleotides. Cell suspensions are mixed directly with 0.5 volumes of ice cold 2 *N* formic acid and centrifuged after 15 minutes to remove cell debris.

6. Methanol

Skoog (1970) extracted mouse embryo cells with 60% methanol in order to rapidly concentrate the extract for enzymatic analysis of dGTP. The monolayer culture (60 mm petri dish) was drained of medium, placed on ice, and covered with 1 ml 60% methanol (0°C). The cells and methanol were immediately scraped into a cold centrifuge tube with a rubber policeman. After 16 hours at -20°C the tube was mixed and then centrifuged to sediment insoluble material. The supernatant was evaporated to dryness on a Buchler Rotary Evapo Mix at 30°C and then dissolved in 0.2 ml of buffer for use.

7. ETHANOL

Mandel (1964) reported improved recovery of nucleotides from certain tissues such as brain by using 70% ethanol. Extraction with hot 80% ethanol has been effective for several types of mouse tumor cells (Brockman *et al.*, 1970; Wheeler and Alexander, 1972). As with acetic acid and methanol, the concentration of ethanolic pool extracts is very simple. Hydrolysis of nucleoside triphosphates is a possible problem with this method (see paragraph on hot water extraction).

8. ETHANOL-ACETIC ACID

The common fixative 3:1 (v/v) ethanol–glacial acetic acid effectively removes acid soluble nucleotide pools (Cleaver, 1967). We use this solvent to fix pulse labeled mammalian cells on glass coverslips for autoradiography (Everhart *et al.*, this volume). Exposure for 10 minutes to the 3:1 solution at 20°C removes all nonincorporated ^{3}HdT; further hydrolysis of the cells for 10 minutes in 1 *N* HCl at 20°C releases no more label. Nucleotides extracted in this way have not yet been used for analysis of pool composition.

9. HOT WATER

Rapid extraction of nucleotides can be achieved with hot water. Bagnara and Finch (1972) collected *E. coli* on 23-mm Millipore filters which were extracted by 1.0 ml H_2O at 90°C for 1 minute. Removal of 0.7 ml of the extract to a cold tube for a 10-minute centrifugation at 8000 *g* produced the nucleotide sample. The hydrolysis of ribonucleotides was extensive in this procedure, with more than 50% of the ATP being degraded to ADP. In a study of the thymine nucleotide pools in human leukocytes, Cooper *et al.* (1966a) used a hot water extraction procedure after five freeze–thaw cycles. Only 2 to 35% of the pool was found to be dTTP. When compared with other results which show 80 to 90% dTTP in the thymine nucleotide pool (Gentry *et al.*, 1965a), it seems likely that hot water extraction has allowed extensive hydrolysis of pool components and should generally not be used.

10. KINETICS OF POOL EXTRACTION

Extraction times reported in the literature for acid-soluble pools vary from 5 minutes to 2 hours, with most workers using a 30 minute extraction. A compromise must be made between quantitative removal of the intracellular nucleotides and minimal hydrolysis by prolonged exposure to low pH (Bagnara and Finch, 1972) and residual hydrolytic enzyme activities (Wood, 1971). The kinetics of nucleotide extraction from *E. coli* using 0.4 *M* $HClO_4$ were studied by Bagnara and Finch (1972); plateau levels are reached within 20 minutes, and the authors have pointed out that extraction with 5% TCA takes about 10 minutes longer. As is true of most extraction methods, nucleoside diphosphates are released more rapidly than triphosphates.

In our experience with monolayers of Chinese hamster ovary cells, the extraction of acid-soluble ^{3}HdT and its phosphorylated derivatives is quite rapid. Using three consecutive 0.5 *N* HCl washes, over 95% of the acid-soluble label is released within 6 minutes. Similar results have been obtained for 0.5 *M* $HClO_4$ and ethanol–glacial acetic acid. Other types of extraction can be considerably slower. Using 1 *N* acetic acid, Nazar *et al.* (1970) found that less than half of the total nucleoside triphosphates could be extracted from *E. coli* in 1 hour. However, two freeze–thaw cycles in 1 *N* acetic acid removed virtually all of the nucleotides. In an investigation of triphosphate pools in mouse embryo cells, Skoog (1970) found that extraction with 60% methanol facilitated the concentration and measurement of dGTP, but the extraction required 16 hours at $-20°C$.

Adaptation to other cell systems of any of the extraction methods described above is certainly feasible. Several tests must be made. (1) Does the method give nucleotide yields comparable to those obtainable with 0.4 M to 0.5 *M* $HClO_4$ in a reasonably short period of time (<1 hour)? The $HClO_4$ method is chosen as a standard because it is known to be the most versatile and effective extraction method for most systems. (2) If nucleoside triphosphates are being measured, does the method give 100% recovery of known amounts of triphosphates added to the washed cells and carried through the entire procedure?

D. Analysis of Extracts of Labeled Cells

Quantitative measurement of the radioactivity extracted from labeled cells may be done without regard for the distribution of label among various metabolites, or it may be carried out after various degrees of resolution by chromatographic and electrophoretic methods. Such resolution generally permits more meaningful interpretation of the results. Techniques for complete separation of virutally all pyrimidine and purine metabolites by thin-layer chromatography have been thoroughly developed by the Randeraths and their colleagues (see Randerath and Randerath, 1967). Because thin-layer techniques are convenient and give excellent resolution with extremely small samples, they have been widely adopted for pool studies. Articles by Neuhard *et al.* (1965) and Randerath and Randerath (1967) contain much useful information regarding the practice of thin-layer techniques. Other separation methods involving paper chromatography, electrophoresis, ion exchange column chromatography, and various combinations of these techniques are advantageous for certain experiments. Table VIII contains a list of references that should be useful in choosing an appropriate separation method [see also Nickless (1971)].

1. Resolution of Complex Mixtures of Ribonucleotides and Deoxyribonucleotides

Labeling of nucleotide pools with nonspecific precursors such as $^{32}P_i$ or $^{32}P_i$ places the greatest demand on chromatographic resolution of pool components. Ribonucleotide pools are much larger than the corresponding deoxyribonucleotide pools (see Table II) and often obscure the latter compounds in chromatographic systems. In several studies of ribonucleotide pools, the small contribution of the deoxyribonucleotide pools has been ignored, and chromatographic methods have been employed which do not

TABLE VIII

Methods for Resolution of Pool Components[a]

Compounds to be resolved	Contaminating compounds						
	rNTP	dNTP	rNXP	dNXP	rN	dN	Pu and Py
rNTP	1, 4, 5, 8[b]	—	—	—	—	—	—
dNTP	4, 5, 14	1, 4, 5, 14	—	—	—	—	—
rNXP	1, 5	5	1, 3, 5, 6, 9, 11	—	2, 5, 7	2, 5, 7	5, 7
dNXP	5	1, 5	3, 5, 11	1, 3, 5, 6, 11	2, 5, 7	2, 5, 7	5, 7
rN	5	5	3, 5, 7, 11	3, 5, 7, 11	3, 5, 10, 11	—	—
dN	5	5	3, 5, 7, 11	3, 5, 7, 11	3, 5, 11 12	3, 5, 11, 12, 13	—
Pu and Py	5	5	3, 5, 7, 11	3, 5, 7, 11	3, 5, 11	3, 5, 11, 13	3, 5, 11, 13

[a] Abbreviations: r, ribo; d, deoxyribo; N, nucleoside; TP, triphosphate; XP, mono-, di- and triphosphate; Pu, purines; Py, pyrimidines.

[b] References: 1, Randerath and Randerath (1964a,b); 2, Bresnick and Karjala (1964) and Roberts and Tovey (1970); 3, Grippo *et al.* (1965); 4, Neuhard *et al.* (1965); 5, Randerath and Randerath (1967); 6, Smith (1967); 7, Raaen and Kraus (1968); 8, Cashel *et al.* (1969); 9, Silver *et al.* (1970) and Serlupi-Crescenzi *et al.* (1968); 10, van den Bos *et al.* (1970); 11, Schwarz-Mann (Orangeburg, New York) Radiochemical Catalog (1971); 12, Al-Arif and Sporn (1972); 13, Holguin-Hueso and Cardinaud (1972); 14, Yegian (1973).

resolve compounds according to their sugar moiety (e.g., Colby and Edlin, 1970; Plagemann, 1972). The "step LiCl" and "step formate" methods (Randerath and Randerath, 1964a, b) and the "step acetic acid-LiCl" method (Neuhard *et al.*, 1965) are suitable for this purpose. Adequate resolution of deoxyribonucleotides from ribonucleotides can be achieved either by chromatography in the presence of borate, or by periodate oxidation of the ribonucleotides followed by chromatography.

a. Borate Methods. Complexes formed in solution between borate ions and the *cis* hydroxyl groups of the ribose moiety of ribonucleotides have lower mobilities in thin-layer systems than the corresponding deoxyribonucleotides which do not form borate complexes. Various borate-containing solvent systems have been developed and are frequently used in combination with other solvents for two-dimensional resolution of complex mixtures of nucleotides on thin layers of cellulose or poly(ethylene)imine (PEI)-cellulose (Neuhard *et al.*, 1965; Grippo *et al.*, 1965; Randerath and Randerath, 1967).

b. Periodate Oxidation of Ribonucleotides. Yegian (1973) developed an ingenious method for eliminating the large interfering pools of ribonucleotides by periodate oxidation of the ribose moiety to the 2′, 3′-dialdehyde derivative. The oxidized ribonucleotides are easily separated from the deoxyribonucleotides. Although the procedure has been used only for the study of deoxyribonucleoside triphosphate pools in ^{32}P labeled *E. coli*, it should be applicable to animal cell systems. The acid extract (0.67 *N* HCOOH, in this case) is mixed with $NaIO_4$ to a final concentration of 0.027 *M* and incubated for 30 minutes before addition of stoichiometric amounts of ethylene glycol to consume the excess periodate. Chromatography is carried out on PEI-cellulose thin layers as follows (Yegian, 1973):

(1) Spot the oxidized acid extracts of ^{32}P labeled cells and marker dNTPs on the origin line 16.25 cm above the bottom edge of a 15 × 20 cm PEI-cellulose sheet. Many separate samples can be analyzed in parallel lanes on the same sheet. An absorbant wick (10 layers of Whatman 3 MM paper) is stapled 2.5 to 3.0 cm above the origin.

(2) Immerse the sheet in 0.08 *M* H_3PO_4 at 37°C to a depth of 15 cm and allow it to stand for 8 to 12 hours (with change of wick if it becomes saturated) until the fastest of the marker dNTPs has migrated about 1.5 cm or more toward the wick.

(3) Remove and discard the wick by cutting off the top edge of the sheet about 0.5 cm above the leading edge of the marker dNTPs. The wick contains virtually all the normally problematic interfering ^{32}P labeled compounds: ^{32}Pi, dNMPs, dNDPs, and the periodate oxidized ribonucleotides.

(4) Attach a new wick to the opposite end of the sheet and without drying, place the end from which the wick was just cut barely below the solvent meniscus in a second tank containing 1 cm of 0.45 *M* $(NH_4)_2SO_4$ (adjusted

at 37°C to pH 8.0 with NH_4OH). After 15 minutes, the dNTPs have migrated sufficiently to lower the sheet to the bottom of the tank.

(5) Chromatography is continued at 37° until the fastest marker dNTP (dTTP) has moved up to within 2 cm of the wick. The wick is discarded immediately and, after drying, the sheet is analyzed for radioactivity by autoradiography and counting.

c. Trailing of $^{32}P_i$. The trailing of only a small portion of the radioactivity associated with orthophosphate in ^{32}P labeled cells can readily obscure nucleotide spots on chromatograms. Charcoal adsorption methods and Yegian's wick procedure described above may alleviate $^{32}P_i$ interference. In addition, Cashel *et al.* (1969) have developed a procedure involving 0.85 *M* KH_2PO_4 (pH 3.4) as the second dimension solvent for PEI-cellulose thin-layer chromatography; trailing of $^{32}P_i$ was eliminated by this solvent.

2. Problems with the Sample

Two general difficulties attend the chromatographic analysis of extracted pools of labeled cells: (1) the nucleotide concentrations in the extracts may be very low, and (2) salts and other compounds may cause trailing on the chromatograms. The second problem frequently arises when, for reasons of inadequate concentration, large volumes of extract containing nonvolatile salts are applied to a single origin. Optimum sample size is determined by the opposing factors of detectability and overloading. The limits of detection of nucleotides on thin layers of cellulose by ultraviolet light range from 0.05 to 0.3 nmole for one-dimensional systems to 0.2 to 5 nmoles for two-dimensional systems where the spots are larger (Grippo *et al.*, 1965; Randerath and Randerath, 1967). Chromatography papers usually require 20 nmoles or more per spot because of their greater thickness. Overloading causes trailing of spots. In thin layer systems, unmodified cellulose is somewhat more sensitive to overloading than PEI-cellulose; Randerath and Randerath (1967) suggest sample sizes of less than 20 nmoles per compound to achieve the best resolution. Frequently, these factors apply only to the use of marker compounds for cochromatography with the radioactive sample, since the labeled material may be an insignificant portion of the total and yet be easily detectable by autoradiographic or counting methods.

a. The Concentration Problem—Charcoal Adsorption. Aside from the trivial solution of increasing the cell sample size and decreasing the volume of extraction solvent, several methods can be used to concentrate pool components. Nazar *et al.* (1970) developed an acetic acid pool extraction procedure to allow concentration of the extracts directly on the chromatogram. Others have used extraction solvents in which the salt does not interfere with the chosen chromatography system (Cashel and Gallant, 1968).

Adsorption of nucleotides, nucleosides, and free bases to activated charcoal is frequently used to concentrate these compounds and to remove interfering salts and metabolites (see Hurlbert, 1957; Sugino and Miyoshi, 1964; Cashel *et al.*, 1969). In general, 5 to 20 mg of acid-washed activated powdered charcoal (Norit or Darco) per milliliter of acidic or neutral solution gives nearly complete adsorption of nucleotides and their derivatives (Hurlbert, 1957). Inorganic phosphate, sugar phosphates, salts, and proteins are not strongly adsorbed. The charcoal is collected by centrifugation or filtration after 5 minutes of incubation, and is then mixed with 4 ml of ethanol: water: concentrated NH_4OH (65:35:0.3) for elution of the nucleotides (Cashel *et al.*, 1969). Filtration through Celite to remove the charcoal is followed by concentration of the eluate on a warming plate with a stream of warm air or nitrogen (Fleming and Bessman, 1967; Cashel *et al.*, 1969). Recovery of nucleotides from the charcoal usually ranges from 80 to 90% (Hurlbert, 1957); there may be much lower yields for guanine compounds (see Randerath and Randerath, 1964a) and thus some caution is required in the use of this method.

b. The Salt Problem. Concentration by charcoal or use of volatile extracting solutions may eliminate the interference by excess salts. Alternatively, Randerath and Randerath (1964a) have washed thin-layer plates in anhydrous methanol after spotting the sample in order to remove unwanted salt from the origin. Although resolution is improved, and less than 10% of the nucleotides are lost by washing, this method is not to be used for studies involving nucleosides and bases. The latter compounds, with the exception of xanthosine, are completely removed from PEI-cellulose by washing with methanol (Raaen and Kraus, 1968). Foster *et al.* (1966) have made other suggestions regarding salts and thin-layer chromatography.

3. Recovery and Quantitation of Resolved Compounds

After chromatographic separation, compounds may be located by one of several techniques: (1) ultraviolet detection of cochromatographed marker compounds, (2) autoradiography, or (3) strip scanning (relatively insensitive).

a. Autoradiography. Although best suited to the detection of ^{32}P, autoradiography can be used for ^{14}C and ^{3}H detection also. X-ray film (Kodak No-Screen or Royal Blue) is placed in close contact with the chromatogram and exposed in total darkness for the required period of time before developing. Realignment of the film and chromatogram after development are facilitated by the use of a hole punch or radioactive ink. Exposure for 24 hours should reveal ^{32}P spots containing about 100 dpm ($\sim$ 50 pCi) or more and ^{14}C spots of the order of 2000 dpm or more. ^{3}H labeled spots containing about 10^4 dpm/cm^2 can be visualized in 1 day if the chromatogram is first dipped in diethyl ether containing 7% PPO and then dried before exposure to Royal Blue RB54 medical X-ray film at $-78°C$ (Randerath, 1970).

b. Elution and Conventional Counting. Nucleotides may be quantitatively eluted from the thin layer with a small paper wick using 0.7 *M* $MgCl_2$:2.0 *M* Tris·HCl, pH 7.4 (100:1, v/v); bases and nucleosides are eluted with 0.1 *N* HCl (Randerath and Randerath, 1967). The compounds may be quantitated by UV absorption if more than a few nmoles are present and no carrier was cochromatographed (Randerath and Randerath, 1967). Radioactivity can be measured by liquid scintillation techniques using aliquots of the eluted compounds (Randerath *et al.*, 1972). Counting with a gas-flow planchet counter is also convenient. The circumscribed spots of labeled compounds are cut out and glued to the planchets, or else moistened with a minute amount of water and transferred with a small spatula (Neuhard *et al.*, 1965). Problems of quenching by the adsorbant may be appreciable for ^{14}C and ^{3}H.

c. Cherenkov Counting of ^{32}P. A counting method which may prove useful for ^{32}P labeled pool materials is based on Cherenkov radiation (see Hash, 1972). In aqueous media, β particles with energies greater than about 0.26 MeV will cause emission of photons in the *absence* of scintillation fluors. ^{32}P can be counted in conventional liquid scintillation counters with an efficiency of 25 to 30%, and because no fluor or special organic solvents are necessary, up to 20 ml of aqueous solution can be placed in a single vial. There is virtually no chemical quenching ($< 1\%$) by 0 to 6.3 *M* $HClO_4$, so acid-soluble pools could be counted with minimal manipulation. Elution of purified pool components from chromatograms can be carried out with convenient volumes of eluent (10 to 20 ml) in a centrifuge tube, and an aliquot of the sample could be treated with charcoal (see above) before Cherenkov counting to confirm the proportion of the total eluted counts which are due to nucleotide.

4. Other Separation Methods

The strong binding of nucleotides, but not nucleosides or bases, to DEAE-cellulose (e.g., Bresnick and Karjala, 1964) has provided a convenient method for simple pool analysis. Uptake of radioactive nucleosides and bases is frequently accompanied by conversion of these compounds to the nucleotide form, and many cases of reduced uptake or drug resistance are caused by change in the rate of this conversion (see Sections III and IV). The ratio of labeled nucleotide to labeled precursor in the acid-soluble pool can be determined by placing 25- to 50-μl aliquots of an acid extract of labeled cells on 24 mm disks of DEAE paper (Whatman DE-81). After the liquid has been totally absorbed (but *not* evaporated), the disks are soaked for 15 minutes in several changes of 1 m*M* ammonium formate and then ethanol to quantitatively remove nonphosphorylated compounds (Bresnick and Karjala, 1964). An identical disk is prepared for each sample and not washed. Both disks are dried and counted by liquid scintillation in omnifluor-toluene. Because of quenching of tritium by the filter disks, ^{14}C-labeled compounds are recommended for this procedure (Hauschka *et al.*, 1972). The filter disk method should be used with caution, since it is known that excess salt in the sample can inhibit quantitative binding of nucleotides (Roberts and Tovey, 1970). Furthermore, Schrecker (1970) found that only about 90% of the phosphorylated derivatives of deoxycytidine and thymidine were retained by DEAE-cellulose disks, and that guanine could not be completely washed off the disks by the conventional procedure.

Ion exchange column techniques for separation of nucleotides are not usually advantageous for pool analysis when the quantity of cell material is limited. Elegant measurements of ribonucleotide and even deoxyribonucleotide pools in gram quantities of animal tissues have been carried out by Potter *et al.* (1957), Mandel (1964), and others using Dowex columns as developed by Cohn (1950) and later by Hurlbert *et al.* (1954). These studies have generally depended on UV absorption for quantitation of the separated compounds and were limited, therefore, by the extent of dilution on the column. It may be worthwhile to apply the column methods to ^{32}P labeled nucleotide pools, since Cherenkov counting (Section V,D) of dilute column fractions would be extremely simple and accurate.

VI. Direct Chemical Methods for Pool Analysis

Despite the sensitivity and simplicity of many of the isotopic techniques discussed in the preceding section, such methods are haunted by questions of equilibration with exogenous label, pool expansion, multiple pools, and radiation damage caused by the isotope. The labeling period required for attainment of true equilibrium can be as long as two cell generations for ^{32}P labeling (Colby and Edlin, 1970; Quigley *et al.*, 1972). In addition to this inconvenience, there is no guarantee that the equilibrium with exogenous isotope will be maintained during subsequent experimental manipulation (drugs, fresh serum, etc.). Direct measurement of nucleotides in unlabeled cells can be extremely sensitive and yet avoids these ambiguities. Furthermore, direct techniques are better suited to the analysis of pools in nongrowing or synchronized cells, serum-stimulated cells, and drug-treated cells where transport of exogenous precursors is frequently perturbed (see Section IV).

All the early studies of nucleic acids and free nucleotides in cells and tissues utilized direct methods of analysis. Through the decades, sensitivity of the various assays was limited by the method for measuring nitrogen (e.g., Jackson, 1923; Kerr and Blish, 1932), and later, pentose or phosphorus (Berenblum *et al.*, 1939; Schmidt and Thannhauser, 1945), and more recently UV absorption (Schmitz *et al.*, 1954; LePage, 1957; Mandel, 1964). Even microbiological assays were used for quantitative measurement of deoxyribonucleotides (Schneider, 1955; Rotherham and Schneider, 1958). It is of historical interest that reasonably accurate analyses of nucleotides in tissues were made as early as 40 to 50 years ago (Thannhauser and Czoniczer, 1920; Jackson, 1923; Buell and Perkins, 1928; Kerr and Blish, 1932). Although rather crude by today's standards, the early methods for analysis were ingenious.

The procedure for adenine nucleotide in human blood as developed by Buell and Perkins (1928) was: (1) Precipitation of proteins with tungstic acid. (2) Precipitation of adenine nucleotide with uranyl nitrate. (3) Hydrolysis of the precipitate with dilute H_2SO_4 to liberate free adenine. (4) Removal of uranium by ammonia. (5) Precipitation of adenine as the silver complex by ammoniacal silver nitrate in the presence of the "protective colloid" gelatin. (6) Nephelometric (nitrogen) comparison of silver-adenine with standard adenine sulfate solutions.

Until about 1970, the most successful of the direct methods has involved quantitation by UV absorption of chromatographically purified nucleotides. Because of the extremely small amounts of free deoxyribonucleotides in tissues this method was principally applied to ribonucleotide pool studies (Hurlbert *et al.*, 1954; Schmitz *et al.*, 1954; Mandel, 1964). For instance, LePage (1957) required 75 gm of a rat tumor ($\sim 3 \times 10^{10}$ cells) for the quantitative ($\pm$ 15%) determination of dATP. In contrast, the newer analytical techniques described below can be carried out on samples of the order of 10^6 cells, at the same time providing great accuracy.

A. Enzymatic Determination of Deoxyribonucleoside Triphosphates

Solter and Handschumacher (1969) developed an important assay for all four deoxyribonucleoside triphosphates (dATP, dGTP, dCTP, and dTTP) involving *in vitro* synthesis of DNA by *E. coli* DNA polymerase. The method is based on the observation of Bessman *et al.* (1958) that all four dNTPs are required for polymerization. Mixtures containing calf thymus DNA as a template, *E. coli* DNA polymerase, and excess quantities of three of the four dNTPs (one of which is radioactive) will form an amount of labeled DNA directly proportional to the added quantity of the fourth (limiting) triphosphate. The reaction mixture is precipitated with acid, collected, and washed on a filter; the labeled DNA is counted by liquid scintillation techniques. Standard curves are prepared by adding known amounts of the limiting dNTP to the reaction mixture. Aliquots of neutralized, acid-soluble pools containing 0.1 to 1 nmole of each dNTP in a volume of about 0.1 ml can be assayed by comparison with the standard curves. The method of Solter and Handschumacher (1969) has been applied to animal cells by Adams *et al.*

(1971) and Bray and Brent (1972). In both of these studies, the useful range of the assay was lowered to between 5 and 500 pmoles of dNTP by the substitution of high specific activity ^{3}HdTTP and ^{3}HdATP for the lower specific activity ^{14}CdNTPs used by Solter and Handschumacher (1969).

Lindberg and Skoog (1970) and Skoog (1970) have markedly improved the dNTP assay by the use of synthetic DNA templates and ^{3}HdNTPs. Synthetic poly[d(A-T)] is a strictly alternating sequence which provides a system for measurement of either dATP or dTTP. Lindberg and Skoog (1970) found that the assay worked optimally in the range of 0.5 to 7 pmoles of the limiting dNTP. Using synthetic poly[d(I-C)] for the DNA polymerase template, Skoog (1970) developed a similar procedure for determining dGTP and dCTP in the range of 0.2 to 4 pmoles. These extremely sensitive methods have proved very useful for animal cell pool studies because only 10^5 to 10^7 cells are required for measurement of all four dNTPs (Nordenskjöld *et al.*, 1970; Skoog and Nordenskjöld, 1971; Slaby *et al.*, 1971; Bjursell *et al.*, 1972).

Problems and Controls

These enzymatic methods require a very clean *in vitro* system. Nuclease activities in the DNA polymerase preparation will digest the radioactive DNA product as it is formed, and phosphatases and kinases can alter the amounts of dNTPs added to the assay mixture. The major source of enzymatic contamination is the polymerase sample; acid extracts of cells do not have appreciable enzymatic activity. Appropriate kinetic controls for nuclease contamination are described by Solter and Handschumacher (1969), Lindberg and Skoog (1970), and Skoog (1970). The time course of radioactive DNA formation is studied for various concentrations of DNA polymerase. Optimal enzyme concentrations allow DNA synthesis to reach a maximum yield at 10 to 15 minutes, followed by a very slow decline in the amount of product. At high polymerase concentrations (or at low concentrations of a "dirty" enzyme preparation), endogenous nuclease activity rapidly degrades whatever radioactive DNA is being formed.

Skoog (1970) studied the problem of phosphatase contamination in the polymerase preparation. The rate of degradation of added dNTPs by the reaction mixture varied for each triphosphate and was affected by the concentration of the other dNTPs present. Preincubation of the polymerase preparation with Hg^{2+} was found to abolish all troublesome phosphatase activity (Skoog, 1970).

It is important to check for the presence of inhibitors or activators in the cell extracts which are being assayed for dNTP content. Most studies have involved the assay of KOH-neutralized $HClO_4$ extracts. Saturated $KClO_4$ solutions added at one-tenth to one-fourth of the volume of the total assay mixture do not affect the procedure, and no inhibitors have been detected in leukemia cell or mouse embryo cell extracts (Solter and Handschumacher, 1969; Lindberg and Skoog, 1970). Ribonucleoside triphosphates do not interfere, even when present in 8000-fold excess compared to the dNTP being measured (Skoog, 1970). The rate of labeled DNA formation in the assay is strongly dependent on pH and the Mg^{2+} concentration (Skoog, 1970), so it is important to neutralize carefully the acid extracts before analysis (see Section V,C) and to control the ion content.

One final possible source of error in the polymerase assay system is the state of the template material. Single-strand breaks in the structure can increase the amount of DNA which is synthesized, although typical templates generally have enough initiation sites so that nicking is not required (see Lindberg and Skoog, 1970). Residual nuclease activity in the template preparation can have two effects: (1) the template becomes more active with time, and (2) the nucleotides produced by digestion of the template can lead to high background incorporation, i.e., DNA synthesis in the absence of addition of the limiting dNTP (Lindberg and Skoog, 1970).

Extension of the polymerase method to measurement of nucleotides other than the dNTPs may be feasible by conversion of monophosphates or diphosphates to the triphosphate level with purified kinases. However, the appeal of the method as it is now applied is that (1) no prior purification or treatment of the cell extract is necessary and (2) most of the deoxyribonucleotides are in the triphosphate form anyway, so that it is perhaps less important to measure monophosphate and diphosphate pools. Using RNA polymerase and synthetic templates, it may also be possible to assay ribonucleotide pools.

B. Enzymatic Analysis of Ribonucleotides

1. Uracil Nucleotides

Keppler *et al.* (1970) have developed a method applicable to complex mixtures and tissue extracts for determination of 3 to 600 nmoles of uracil nucleotides and UDP-sugars by sequential conversion *in vitro* with purified enzymes. After forming UDP-glucose from uracil nucleotides in the mixture, UDP-glucose dehydrogenase is added and NAD^+ reduction is followed spectrophotometrically.

2. Adenine Nucleotides

Adam (1962) has described a method for 5′AMP measurement using adenylate kinase. In addition, all adenine 5′-nucleotides (ADP, ATP, etc.) can be converted to 5′AMP with snake venom phosphodiesterase for determination of *total* adenine nucleotide content by the adenylate kinase assay.

Keppler *et al.* (1970) used this same procedure to measure total uracil nucleotides (see above). The principal adenine nucleotide in most cells and tissues is ATP (Table II), and very sensitive techniques are available for measuring ATP by coupled enzymatic reactions (e.g., Aurbach and Houston, 1968; Goldberg *et al.*, 1969a). However, these methods frequently require some sort of purification of the ATP before assay. Firefly luciferase requires ATP for light emission, and it is this reaction that provides the most convenient method for determination of ATP in complex mixtures. Originally conceived by McElroy (1947), the procedure has been adapted to liquid scintillation counters by many workers, including Cole *et al.* (1967) and Stanley and Williams (1969). The useful range is between 0.1 and 1000 pmoles of ATP, depending on the particulars of the technique (Cole *et al.*, 1967; Lin and Cohen, 1968; Stanley and Williams, 1969; Johnson *et al.*, 1970).

Emerson and Humphreys (1971) measured ATP in the range 10 to 30 pmoles with 5% error. Hence, samples of animal cells or tissue as small as 10^3 cells could provide adequate material for the determination of ATP. Samples of ATP are assayed one at a time by injection into vials containing the luciferase–luciferin reaction mixture. After a predetermined interval (usually about 10 seconds), the vial is counted in the tritium channel with the coincidence circuit shut off, since single photons cannot be registered by more than one phototube (Stanley and Williams, 1969). Because the light emission decays rather quickly, the counts during the first minute or less are the most useful. The method is calibrated with aliquots of standard ATP solutions, and it is important that these are properly prepared. Stanley and Williams (1969) noted that very dilute ATP solutions (10^{-8} *M* or less) would "degenerate" within 2 hours at 0°C in phosphate buffer. Presumably some type of adsorption to the vessel surface is involved in this degeneration. Stock solutions can be prepared in 0.4 *M* $HClO_4$ if desired, since the assay system will work (although at lower sensitivity) for nonneutralized $HClO_4$ extracts (Stanley and Williams, 1969). There is slow degradation of ATP in 0.4 *M* $HClO_4$ at 0° (about 5% loss in 16 hours). The firefly luciferase ATP assay has been used for pool measurements in *E. coli* (Cole *et al.*, 1967) and chick embryo fibroblasts (Emerson and Humphreys, 1971). Because the method is highly specific for ATP, and because ATP is usually the predominant nucleotide in cells and tissues, purification of samples before assay is generally unnecessary.

The specific radioactivity of ribonucleoside triphosphate pools of labeled cells has been measured directly by several groups. Kijima and Wilt (1969) determined the specific activity of GTP in sea urchin embryos by ion exchange chromatography, UV absorption, and liquid scintillation counting. A loss of tritium label from the guanosine-8-^{3}H precursor by base catalyzed exchange with water was noted in this system; this may be a problem with other tritiated purines (Wilt, 1969; Brandhorst and Humphreys, 1971).

Sensitive techniques have also been applied to measurement of the specific activity of the UTP pool [thin-layer chromatography, UV absorption, and scintillation counting, Roeder and Rutter (1970)] and the ATP pool [luciferase assay, chromatography, and scintillation counting of the ATP spot (Emerson and Humphreys, 1971; Brandhorst and Humphreys, 1971)].

C. Cyclic Nucleotides

Numerous methods have been developed for assay of cyclic nucleotides, particularly adenosine 3′:5′ cyclic monophosphate (cyclic AMP) and guanosine 3′:5′ cyclic monophosphate (cyclic GMP); Breckenridge (1971) has recently reviewed the principles of these techniques. Many of the procedures are hampered by the requirement that the cyclic nucleotide be purified before assay. For instance, the method of Goldberg *et al.* (1969a) involves enzymatic conversion of purified cyclic AMP to 5′ AMP and then to ATP. A further series of coupled enzymatic reactions requiring ATP serve to amplify the sensitivity. As little as 0.05 pmole of cyclic AMP (Goldberg *et al.*, 1969a) or cyclic GMP (Goldberg *et al.*, 1969b) is detectable. Johnson *et al.* (1970) developed an analogous assay for cyclic AMP where, after purification and conversion to ATP, the compound is measured by the firefly luciferase technique.

Kuo and Greengard (1970) have assayed cyclic AMP with a cyclic AMP-dependent protein kinase. The amount of phosphorylation of a protein substrate, such as histone, is proportional to the amount of cyclic AMP added. ATP labeled with ^{32}P in the terminal phosphate provides a sensitive method for detecting phosphoprotein. Although picomole amounts of cyclic AMP can be measured, the sample must first be purified. The ATP present in a typical acid extract would be 10^2 to 10^4 times as plentiful as cyclic AMP (Table II). This ATP would cause a drastic reduction in the specific activity of ATP-γ ^{32}P in the reaction mixture, thereby nullifying the assay.

Cyclic nucleotides can be assayed in crude mixtures by the protein binding methods of Gilman (1970) for cyclic AMP and Murad *et al.* (1971) for cyclic GMP. Certain protein components present in protein kinase preparations will bind cyclic nucleotides with a high degree of specificity. Competition for the protein *in vitro* between a constant amount of ^3H cyclic nucleotide and a variable amount of unlabeled cyclic nucleotide is followed by adsorption of the protein-nucleotide complex to a Millipore filter. The amount of adsorbed label is inversely proportional to the quantity of cold cyclic nucleotide (Gilman, 1970; Murad *et al.*, 1971). As little as 0.5 pmole of cyclic AMP or cyclic GMP can be measured by these methods. Levels of interference by other nucleotides are so low that in most cases it is not necessary to purify the cyclic nucleotide before the assay (Gilman, 1970).

Steiner *et al.* (1969) have developed a radioimmunoassay for cyclic AMP involving competition with ^3H cyclic AMP for binding to a specific purified immunoglobulin.

D. Other Methods

Randerath *et al.* (1972) reported an interesting method for tritium labeling of ribonucleosides *after* their isolation from biological materials. Periodate oxidation is followed by reduction with NaB^3H_4 to form the labeled trialcohol derivatives. Chromatographic separation, localization by autoradiography, and quantitation by liquid scintillation has permitted accurate compositional analysis of picomole quantities of individual bases from transfer RNA. Although not applicable to deoxyribonucleotides, the method may have some use in ribonucleotide pool studies when limited quantities of materials are available. Specific enzymes may be used to convert ribonucleotides to ribonucleosides (see Randerath *et al.*, 1972). It is necessary to measure the extent of oxidation, reduction, and recovery of all ribose moieties if quantitative results are desired.

VII. Miscellaneous

A. Purity of Radioactive Compounds

Obvious problems can result from radioactive impurities in labeled compounds. Economic factors generally encourage long-term storage of ^3H- and ^{14}C-labeled compounds, and the radiolytic decomposition products which invariably accumulate are frequently disregarded. In the usual case where no more than one radioactive atom is present in each molecule, the products of decay from *intra*molecular radiolysis are no longer radioactive and present few problems because their chemical concentration is usually insignificant. However, decomposition also proceeds by *inter*molecular radiolysis and simple chemical breakdown, both of which yield labeled impurities. Some of the general problems arising from isotopic impurities have been discussed along with several possible remedies (Wand *et al.*, 1967; Oldham, 1971). Extensive studies concern the decomposition of various nucleosides and nucleotides (Evans and Stanford, 1963a; Sheppard, 1972), especially thymidine (Evans and Stanford, 1963b; Cleaver, 1967).

The rate and products of decomposition are dependent on the type and location of the isotopic atom. For instance, dT-2-^{14}C is more stable in aqueous solution (about 1% decomposition per year) than any of the ^3H-labeled compounds: dT-5-methyl-^3H, 6 to 12% per year; dT-6-^3H and generally labeled ^3HdT, 12 to 24% per year (Evans and Stanford, 1963b; Evans, 1966). Thymine and 2-deoxyribose are common decomposition products of ^{3}HdT; in addition, dT-5-methyl-^{3}H and dT-6-^{3}H are known to form peroxides and glycols of dT (Apelgot and Ekart, 1963). Free-radical reactions are important in the decomposition of all labeled nucleosides, and these can be minimized by inclusion of ethanol, benzyl alcohol, sodium formate, or cysteamine in the solution (Evans, 1966; Apelgot *et al.*, 1964). Because radiolytic decomposition is often faster at $-20°$ than at $0°$ or $-70°$, it is probably most useful to store solutions in the refrigerator at about 2°C (Evans and Stanford, 1963b; Evans, 1966; Apelgot *et al.*, 1964). Other storage conditions are described by Sheppard (1972).

Degradation of labeled compounds is often a consequence of bacterial or fungal growth in the stock solution. Inclusion of 50% ethanol in the stock solution largely overcomes this problem. Sterile preparation of isotope dilutions is advisable, and the possibility of freezing aqueous stocks to prevent biological contamination should be weighed against the disadvantages of more rapid radiolysis in the $-20°$C range. Rinsing syringes or pipets in absolute ethanol before withdrawing the isotope also helps to prevent contamination.

1. Detection and Removal of Impurities from Labeled Nucleosides

Chromatography of radioactive compounds on ion exchange columns, paper, or thin-layer plates of silica gel, cellulose, or PEI-cellulose allows simple detection of impurities; in most cases purification is possible by the same method. Some useful systems for various separations are outlined in Table VIII. As an example, we found that a commercial lot of ^{3}H-deoxycytidine (8 Ci/mmole) contained about 20% ^{3}HdCMP (probably of biological origin) when resolved on silica gel thin layers. Impurities were detected by spotting 1 μCi amounts ($\sim$1 μl) of the isotope along with known marker compounds on a 5×20 cm silica gel plate and developing with water. Using a small, square-ended spatula, 3 mm-wide bands of the silica gel were scraped directly into vials containing 10 ml of a toluene-based scintillation fluid for counting. Identical chromatographic patterns were obtained with and without cold carrier dC. High specific activity material (20 μCi/spot) was run without carrier dC for purification. The 1 cm^2 spot of desired pure material on the silica gel plate was scraped directly into 10 ml of fresh Ham's F-12 saline (Table IX), and after 15 minutes at 37° (> 90% elution) the suspension was filtered through a disposable sterile membrane filter (Nalge) and refrigerated until needed for culture labeling studies.

2. Problems with Radioactive Phosphate

Inorganic ^{32}P is generally available as carrier-free $H_3{}^{32}PO_4$ in 0.02 *N* HCl. The material often contains polyphosphates and phosphosilicates which can be extremely problematic in thin-layer chromatography because of similar mobility to the ^{32}P-labeled nucleotides. Yegian (1973) detected polyphosphates and phosphosilicates in $^{32}P_i$ as radioactivity which chromatographed well behind the main $^{32}P_i$ peak on PEI-cellulose. Isotope which contained more than about 0.02% of these impurities was unsuitable for use. Bagnara and Finch (1972) used charcoal adsorption of ^{32}P isolated from thin-layer chromatograms as a secondary criterion for its presence in nucleotides. Non-adsorbable radioactivity would presumably be in polyphosphates, phosphosilicates, sugar phosphates, or other nonnucleotide compounds.

Method of Sugino and Miyoshi (1964) for removal of polyphosphates from $^{32}P_i$: (a) Add carrier P_i (KH_2PO_4) to carrier-free $^{32}P_i$ such that the desired specific activity is obtained. (b) Adjust the solution to a final concentration of 1 *N* HCl and heat for 15 minutes at 100°C to hydrolyze the polyphosphates. (c) Treat with acid washed Norit A and neutralize to pH 8.0 with NaOH. (d) Purify the $^{32}P_i$ by column chromatography on Dowex 1 (X8, 200 to 400 mesh, formate form) using formic acid as a developer. (e) Remove the formic acid by evaporation under reduced pressure and then lyophilize.

B. Washing Solutions for Cells

TABLE IX

COMMON ISOTONIC WASHING SOLUTIONS[a] (MG/LITER)

Components	F12[b] Ham's	Gey's	Earle's	Hank's	PBS (Dulbecco)	TBS[c]	Puck's G
NaCl	7600	8000	6800	8000	8000	8000	8000
KCl	224	370	400	400	200	380	400
$MgSO_4 \cdot 7H_2O$	—	70	200	100	—	—	154
$MgCl_2 \cdot 6H_2O$	122	210	—	100	100	214	—
$CaCl_2 \cdot 2H_2O$	44	225	265	185	132	132	16
KH_2PO_4	—	30	—	60	200	—	150
$NaH_2PO_4 \cdot H_2O$	—	—	140	—	—	—	—
$Na_2HPO_4 \cdot 7H_2O$	268	225	—	90	2170	189	290
$NaHCO_3$	1176	227	2200	350	—	—	—
Tris[d]	—	—	—	—	—	3000	—
Glucose	1802	1000	1000	1000	—	1000	1100
Phenol Red	1.2	—	10	10	—	—	5

[a]Corrected for use of salts of different hydration states. Original reference: Grand Island Biological Company Catalog. (Grand Island, New York), 1972–1973. All solutions are adjusted to pH 7.4 with 1 *N* NaOH or 1 *N* HCl, filtered through Whatman No. 2 paper, and stored at 4°C.

[b]Ham (1972).

[c]Smith *et al.* (1960).

[d]Tris (hydroxymethyl) aminomethane ("Sigma 7–9").

C. Useful Cell Parameters

TABLE X
USEFUL CELL PARAMETERS

Cell type	Parameter	Reference
	Volume ($\mu l/10^6$ cells)	
Rat liver (adult)	4.5	Weber *et al.* (1971)
Rat liver (newborn)	1.5	Weber *et al.* (1971)
Novikoff hepatoma	2.4	Gentry *et al.* (1965a)
Pig kidney	3.3–4.0	Cass (1972)
BHK (low density)	4.5	McDonald *et al.* (1972)
BHK (high density)	2.5	McDonald *et al.* (1972)
Mouse L929	1.6	Adams (1969)
Chinese hamster CHO	1.4–2.2	Hauschka (unpubl. obs.)
Human erythrocyte	0.08–0.1	Calculated
	Protein content ($\mu g/10^6$ cells)	
Rat liver (adult)	910	Weber *et al.* (1971)
Rat liver (newborn)	310	Weber *et al.* (1971)
Pig kidney	640–720	Cass (1972)
Eight mammalian cell lines	560–830	Oyama and Eagle (1956)
	DNA content ($\mu g/10^6$ cells)	
Rat liver (adult)	11.4	Weber *et al.* (1971)
HeLa	4–10	Bray and Brent (1972)
Mouse L strain	12	Cleaver (1967)
Mouse embryo cells	12.5	Nordenskjöld *et al.* (1970)
Mouse cells	7	Riegler (1966)

D. Determination of Cell Volume and Intracellular Concentration of Pools

Calculation of intracellular concentrations of small molecules requires a knowledge of the cell volume accessible to these molecules. Measurement of this volume is most easily carried out with concentrated cell suspensions. Total cell volume is determined by the microhematocrit method, where a known number of cells are sedimented in a calibrated centrifuge or capillary tube [McDonald *et al.* (1972), see Table X for some typical cell volumes].

The cell water is assumed to be the difference between wet weight and dry weight of a cell pellet, after correction for extracellular water (Winkler and Wilson, 1966). Total intracellular plus extracellular water (W_t) in cell pellets is determined by measuring the dilution of tritiated water (Kessel and Shurin, 1968), or by drying cell pellets at 110°C to constant weight (Hawkins and Berlin, 1969; Oliver and Paterson, 1971; McDonald *et al.*, 1972). Extracellular water (W_e) in cell pellets (usually 7 to 11%) is estimated by dilution of inulin-^{14}C (Kessel and Shurin, 1968; Hawkins and Berlin, 1969; Oliver and Paterson, 1971) or dilution of sucrose-^{14}C (Jacquez, 1962a; McDonald *et al.*, 1972). The intracellular concentration (C_i) of a labeled substance is calculated from the formula $C_i = (B - C_e W_e)/(W_t - W_e)$, where B is the total amount of label in the cell pellet, and C_e is the extracellular concentration of label.

Acknowledgments

I wish to thank Dr. D. M. Prescott for his critical reading of the text and for providing the laboratory facilities necessary for the writing of this chapter; Drs. B. P. Brandhorst, L. P. Everhart, S. Follansbee, and R. W. Rubin for many helpful discussions; my wife, Barbara, for her patient assistance; Gayle Prescott for her flawless typing; and the Jane Coffin Childs Memorial Fund for Medical Research for its generous support.

References

Adam, H. (1962). *In* "Methoden der enzymatischen Analyse" (H.-U. Bergmeyer, ed.), p. 573. Verlag Chemie, Weinheim.
Adams, R. L. P. (1969). *Exp. Cell Res.* **56**, 49.
Adams, R. L. P., Berryman, S., and Thomson A. (1971). *Biochim. Biophys. Acta* **240**, 455.
Adelstein, S. J., Baldwin, C., and Kohn, H. I. (1971). *Develop. Biol.* **26**, 537.
Al-Arif, A., and Sporn, M. B. (1972). *Anal. Biochem.* **48**, 386.
Apelgot, S., and Ekart, B. (1963). *J. Chim. Phys. Physiochim. Biol.* **60**, 505.
Apelgot, S., Ekart, B., and Tisne, M. R. (1964). *In* "Conference on Methods of Preparing and Storing Marked Molecules," p. 939, Euratom.
Appel, S. H., and Pettis, P. (1967). *Fed. Proc., Fed. Amer. Soc. Exp. Biol.* **26**, 292.
Atkinson, M. R., Morton, R. K., and Murray, A. W. (1964). *Biochem. J.* **92**, 398.
Attardi, B., and Attardi, G. (1972). *Proc. Nat. Acad. Sci. U.S.* **69**, 2874.
Aurbach, G. D., and Houston, B. A. (1968). *J. Biol. Chem.* **243**, 5935.
Bagnara, A. S., and Finch, L. R. (1968). *Biochem. Biophys. Res. Commun.* **33**, 15.
Bagnara, A. S., and Finch, L. R. (1972). *Anal. Biochem.* **45**, 24.
Baril, E., Baril, B., and Elford, H. (1972). *Proc. Amer. Ass. Cancer Res.* **13**, 84.
Bartlett, G. R. (1959). *J. Biol. Chem.* **234**, 466.
Behki, R., and Schneider, W. C. (1962). *Biochim. Biophys. Acta* **61**, 663.
Bekesi, J. G., and Winzler, R. J. (1969). *J. Biol. Chem.* **244**, 5663.
Bekesi, J. G., Bekesi, E., and Winzler, R. J. (1969). *J. Biol. Chem.* **244**, 3766.
Bennett, L. L., Jr., and Allan, P. W. (1971). *Cancer Res.* **31**, 152.
Berenblum, I., Chain, E., and Heatley, N. G. (1939). *Biochem. J.* **33**, 68.
Berlin, R. D. (1970). *Science* **168**, 1539.
Bernard, O., and Brent, T. P. (1971). Unpublished data quoted by Krygier, V., and Momparler, R. L. (1971). *J. Biol. Chem.* **246**, 2745.
Bertino, J. R., Chasmore, A. R., and Hillcoat, B. L. (1970). *Cancer Res.* **30**, 2372.
Bessman, M. J., Lehman, I. R., Simms, E. S., and Kornberg, A. (1958). *J. Biol. Chem.* **233**, 171.
Birch, J. R., and Pirt, S. J. (1971). *J. Cell Sci.* **8**, 693.
Bjursell, K. G., Reichard, P. A., and Skoog, K. L. (1972). *Eur. J. Biochem.* **29**, 348.
Blair, D. G. R., and Potter, V. R. (1961). *J. Biol. Chem.* **236**, 2503.

Blakeley, R. L., and Vitols, E. (1968). *Annu. Rev. Biochem.* **37**, 201.
Bollum, F. J., and Potter, V. R. (1959). *Cancer Res.* **19**, 561.
Booth, B. A., Moore, E. C., and Sartorelli, A. C. (1971). *Cancer Res.* **31**, 228.
Bosmann, H. B. (1971). *Biochim. Biophys. Acta* **240**, 74.
Brandhorst, B. P., and Humphreys, T. (1971). *Biochemistry* **10**, 877.
Bray, G., and Brent, T. P. (1972). *Biochim. Biophys. Acta* **269**, 184.
Breckenridge, B. M. (1971). *Ann. N.Y. Acad. Sci.* **185**, 10.
Brent, T. P. (1971). *Cell Tissue Kinet.* **4**, 297.
Brent, T. P., Butler, J. A. V., and Crathorn, A. R. (1965). *Nature (London)* **207**, 176.
Breitman, T. R., Perry, S., and Cooper, R. A. (1966). *Cancer Res.* **26**, 2282.
Breslow, R. E., and Goldsby, R. A. (1969). *Exp. Cell Res.* **55**, 339.
Bresnick, E., (1965). *J. Biol. Chem.* **240**, 2550.
Bresnick, E., and Burleson, S. S. (1970). *Cancer Res.* **30**, 1060.
Bresnick, E., and Karjala, R. J. (1964). *Cancer Res.* **24**, 841.
Bresnick, E., and Thompson, U. B. (1965). *J. Biol. Chem.* **240**, 3967.
Bresnick, E., Mayfield, E. D., Jr., Liebelt, A. G., and Liebelt, R. A. (1971). *Cancer Res.* **31**, 743.
Brockman, R. W. (1963). *Advan. Cancer Res.* **7**, 129.
Brockman, R. W., Shaddix, S., Laster, W. R., Jr., and Schabel, F. M., Jr. (1970) *Cancer Res.* **30**, 2358.
Bucher, N. L. R., and Swaffield, M. N. (1965). *Biochim. Biophys. Acta* **108**, 551.
Bucher, N. L. R., and Swaffield, M. N. (1969). *Biochim. Biophys. Acta* **174**, 491.
Buell, M. V., and Perkins, M. E. (1928). *J. Biol. Chem.* **76**, 95.
Burger, M. M., Bombik, B. M., Breckenridge, B. M., and Sheppard, J. R. (1972). *Nature (London), New Biol.* **239**, 161.
Burton, K. (1968). *In* "Methods in Enzymology", Vol. 12: Nucleic Acids (L. Grossman and K. Moldave, eds.), Part B, p. 163. Academic Press, New York.
Camiener, G. W., and Smith, C. G. (1965). *Biochem. Pharmacol.* **14**, 1405.
Canellakis, E. S. (1957). *J. Biol. Chem.* **227**, 701.
Cashel, M., and Gallant, J. (1968). *J. Mol. Biol.* **34**, 317.
Cashel, M., Lazzarini, R. A., and Kalbacher, B. (1969). *J. Chromatogr.* **40**, 103.
Cass, C. E. (1972). *Exp. Cell Res.* **73**, 140.
Cass, C. E., and Paterson, A. R. P. (1972). *J. Biol. Chem.* **247**, 3314.
Ceccarini, C., and Eagle, H. (1971). *Proc. Nat. Acad. Sci. U.S.* **68**, 229.
Chapman, J. D., Webb, R. G., and Borsa, J. (1971). *J. Cell Biol.* **49**, 229.
Christensen, H. N. (1969). *Advan. Enzymol. Relat. Areas Mol. Biol.* **32**, 1.
Chu, M. Y., and Fischer, G. A. (1965). *Biochem. Pharmacol.* **14**, 333.
Cleaver, J. E. (1967). *In* "Thymidine Metabolism and Cell Kinetics," Frontiers of Biology (A. Neuberger and E. L. Tatum, eds.), Vol. 6. Wiley (Interscience), New York.
Cleaver, J. E., and Holford, R. M. (1965). *Biochim. Biophys. Acta* **103**, 654.
Cohn, W. E. (1950). *J. Amer. Chem. Soc.* **72**, 1471.
Colby, C., and Edlin, G. (1970). *Biochemistry* **9**, 917.
Cole, H. A., Wimpenny, J. W. T., and Hughes, D. E. (1967). *Biochim. Biophys. Acta* **143**, 445.
Conrad, A. H., and Ruddle, F. H. (1972). *J. Cell Sci.* **10**, 471.
Cook, J. L., and Vibert, M. (1966). *J. Biol. Chem.* **241**, 158.
Cooper, G. M., Dunning, W. F., and Greer, S. (1972). *Cancer Res.* **32**, 390.
Cooper, R. A., Perry, S., and Breitman, T. R. (1966a). *Cancer Res.* **26**, 2267.
Cooper, R. A., Perry, S., and Breitman, T. R. (1966b). *Cancer Res.* **26**, 2276.
Cory, J. G. and Whitford, T. W., Jr. (1972). *Cancer Res.* **32**, 1301.

Crabtree, G. W., and Henderson, J. F. (1971). *Cancer Res.* **31**, 985.
Craig, L. C., and Chen, H.-C. (1972). *Proc. Nat. Acad. Sci. U.S.* **69**, 702.
Crone, M., and Itzhaki, S. (1965). *Biochim. Biophys. Acta* **95**, 7.
Cunningham, D. D., and Pardee, A. B. (1969). *Proc. Nat. Acad. Sci. U.S.* **64**, 1049.
Danon, D., and Marikovsky, Y. (1964). *Lab. Clin. Med.* **64**, 668.
Davidson, J. N. (1962). *In* "The Molecular Basis of Neoplasia," The Univ. of Texas, M. D. Anderson Hospital and Tumor Institute, p. 420. Univ. of Texas Press, Austin, Texas.
Davidson, J. N., and Smellie, R. N. (1952). *Biochem. J.* **52**, 594.
Dewey, W. C., Humphrey, R. M., and Sedita, B. A. (1968). *Exp. Cell Res.* **50**, 349.
Divekar, A. Y., Fleysher, M. H., Slocum, H. K., Kenny, L. N., and Hakala, M. T. (1972). *Cancer Res.* **32**, 2530.
Eberhard, A. (1972). *Anal. Biochem.* **46**, 660.
Elford, H. L., Freese, M., Passamani, E., and Morris, H. P. (1970). *J. Biol. Chem.* **245**, 5228.
Ellem, K. A. O. (1968). *J. Cell. Physiol.* **71**, 17.
Emerson, C. P., Jr., and Humphreys, T. (1971). *Anal. Biochem.* **40**, 254.
Estabrook, R. W., Williamson, J. R., Frenkel, R., and Maitra, P. K. (1967). *In* "Methods in Enzymology," Vol. 10: Oxidation and Phosphorylation (R. W. Estabrook and M. E. Pullman, eds.), p. 474. Academic Press, New York.
Evans, E. A. (1966). *Nature (London)* **209**, 169.
Evans, E. A., and Stanford, F. G. (1963a). *Nature (London)* **197**, 551.
Evans, E. A. and Stanford, F. G. (1963b). *Nature (London)* **199**, 762.
Feigelson, M., and Feigelson, P. (1966). *J. Biol. Chem.* **241**, 5819.
Ferdinandus, J. A., Morris, H. P., and Weber, G. (1971). *Cancer Res.* **31**, 550.
Flaks, J. G., Erwin, M. J., and Buchanan, J. M. (1957). *J. Biol. Chem.* **228**, 201.
Fleming, W. H., and Bessman, M. J. (1967). *J. Biol. Chem.* **242**, 363.
Foster, J. M., Abbott, H., and Terry, M. L. (1966). *Anal. Biochem.* **16**, 149.
Friedkin, M. (1963). *Annu. Rev. Biochem.* **32**, 185.
Friedkin, M., and Roberts, D. (1954). *J. Biol. Chem.* **207**, 245.
Friedkin, M., and Roberts, D. (1956). *J. Biol. Chem.* **220**, 653.
Fritzson, P. (1957). *J. Biol. Chem.* **226**, 223.
Garland, M. R., Ng, T., and Richards, J. F. (1971). *Cancer Res.* **31**, 1348.
Gentry, G. A., Morse, P. A., Jr., Ives, D. H., Gebert, R., and Potter, V. A. (1965a). *Cancer Res.* **25**, 509.
Gentry, G. A., Morse, P. A., Jr., and Potter, V. R. (1965b). *Cancer Res.* **25**, 517.
Gentry, G. A., Morse, P. A., Jr., and Dorsett, M. T. (1971). *Cancer Res.* **31**, 909.
Geraci, G., Rossi, M., and Scarano, E. (1967). *Biochemistry* **6**, 183.
Gerhart, J. C., and Pardee, A. B. (1962). *J. Biol. Chem.* **237**, 891.
Gilman, A. G. (1970). *Proc. Nat. Acad. Sci. U.S.* **67**, 305.
Goldberg, N. D., Larner, J., Sasko, H., and O'Toole, A. G. (1969a). *Anal Biochem.* **28**, 523.
Goldberg, N. D., Dietz, S. B., and O'Toole, A. G. (1969b). *J. Biol. Chem.* **244**, 4458.
Gotto, A. M., Meikle, A. W., and Touster, O. (1964). *Biochim. Biophys. Acta* **80**, 552.
Gotto, A. M., Belkhode, M. L., and Touster, O. (1969). *Cancer Res.* **29**, 807.
Graham, F. L., and Whitmore, G. F. (1970). *Cancer Res.* **30**, 2627.
Greenstein, J. P. (1954). "Biochemistry of Cancer," 2nd ed. Academic Press, New York.
Grippo, P., Iaccarino, M., Rossi, M., and Scarano, E. (1965). *Biochim. Biophys. Acta* **95**, 1.
Groth, D. P., D'Angelo, J. M., Vogler, W. R., Mingioli, E. S., and Betz, B. (1971). *Cancer Res.* **31**, 332.
Hakala, M. T. (1957). *Science* **126**, 255.
Hakala, M. T. (1971). *Cancer Res.* **31**, 813.

Hakala, M. T., Holland, J. F., and Horoszewica, J. S. (1963). *Biochem. Biophys. Res. Commun.* **11**, 466.

Ham, R. G. (1972). *In* "Methods in Cell Physiology" (D. M. Prescott, ed.), Vol. V, pp. 37–74. Academic Press, New York.

Harbers, E., Chaudhuri, N. K., and Heidelberger, C. (1959). *J. Biol. Chem.* **234**, 1255.

Hare, J. D. (1970). *Cancer Res.* **30**, 684.

Hare, J. D. (1972a). *Biochim. Biophys. Acta* **255**, 905.

Hare, J. D. (1972b). *Biochim. Biophys. Acta* **282**, 401.

Hartman, S. C. (1970). *In* "Metabolic Pathways" (D. M. Greenberg, ed.), 3rd. ed., Vol. 4, pp. 1–68. Academic Press, New York.

Hartman, S. C., and Buchanan, J. M. (1959). *J. Biol. Chem.* **234**, 1812.

Hash, J. H. (1972). *In* "Methods in Microbiology" (J. R. Norris and D. W. Robbins, eds.), Vol. 6B, p. 110. Academic Press, New York.

Hashimoto, T., Arima, T., Okuda, H., and Fujii, S. (1972). *Cancer Res.* **32**, 67.

Hauschka, P. V., Everhart, L. P., and Rubin, R. W. (1972). *Proc. Nat. Acad. Sci. U.S.* **69**, 3542.

Hawkins, R. A., and Berlin, R. D. (1969). *Biochim. Biophys. Acta* **173**, 324.

Heidelberger, C. (1965). *Progr. Nucl. Acid Res. Mol. Biol.* **4**, 1.

Heilbrunn, L. V. (1952). "An Outline of General Physiology," 3rd ed., Chapters 12 and 13. Saunders, Philadelphia, Pennsylvania.

Henderson, J. F., and Khoo, M. K. Y. (1965). *J. Biol. Chem.* **240**, 2358.

Henderson, J. F., Paterson, A. R. P., Caldwell, I. C., and Hori, M. (1967). *Cancer Res.* **27**, 715.

Hill, D. L., and Bennett, L. L., Jr. (1969). *Biochemistry* **8**, 122.

Ho, Y. K., Hakala, M. T., and Zakrzewski, S. F. (1972). *Cancer Res.* **32**, 1023.

Hochstadt-Ozer, J. (1972). *J. Biol. Chem.* **247**, 2419.

Hochstadt-Ozer, J., and Stadtman, E. R. (1971a). *J. Biol. Chem.* **246**, 5304.

Hochstadt-Ozer, J., and Stadtman, E. R. (1971b). *J. Biol. Chem.* **246**, 5312.

Hodes, M. E., Kaplan, L. A., and Yu, F. L. (1971). *Anal. Biochem.* **43**, 644.

Holguin-Hueso, J., and Cardinaud, R. (1972). *J. Chromatogr.* **66**, 388.

Hryniuk, W. M. (1972). *Cancer Res.* **32**, 1506.

Hurlbert, R. B. (1957). *In* "Methods in Enzymology," Vol. 3: Preparation and Assay of Substrates (S. P. Colowick and N. O. Kaplan, eds.), p. 785. Academic Press, New York.

Hurlbert, R. B., and Kammen, H. O. (1960). *J. Biol. Chem.* **235**, 443.

Hurlbert, R. B., Schmitz, H., Brumm, A. F., and Potter, V. R. (1954). *J. Biol. Chem.* **209**, 23.

Ives, D. H., Morse, P. A., Jr., and Potter, V. R. (1963). *J. Biol. Chem.* **238**, 1467.

Jackson, H., Jr. (1923). *J. Biol. Chem.* **57**, 121.

Jacquez. J. A. (1962a). *Proc. Soc. Exp. Biol. Med.* **109**, 132.

Jacquez, J. A. (1962b). *Biochim. Biophys. Acta* **61**, 265.

Jacquez, J. A., and Ginsberg, F. (1960). *Proc. Soc. Exp. Biol. Med.* **105**, 478.

Jeanteur, P., Amaldi, F. and Attardi, G. (1968). *J. Mol. Biol.* **33**, 757.

Johnson, R. A., Hardman, J. G., Broadus, A. E., and Sutherland, E. W. (1970). *Anal. Biochem.* **35**, 91.

Jones, M. E. (1971). *Advan. Enzyme Regul.* **9**, 19.

Jonsen, J., and Laland, S. (1960). *Advan. Carbohyd. Chem.* **15**, 201.

Karlstrom, O., and Larsson, A. (1967). *Eur. J. Biochem.* **3**, 164.

Kay, J. E., and Handmaker, S. D. (1970). *Exp. Cell Res.* **63**, 411.

Keppler, D., Rudigier, J., and Decker, K. (1970). *Anal. Biochem.* **38**, 105.

Kerr, S. E., and Blish, M. E. (1932). *J. Biol. Chem.* **98**, 193.

Kessel, D., and Dodd, D. C. (1972). *Biochim. Biophys. Acta* **288**, 190.

Kessel, D., and Shurin, S. G. (1968). *Biochim. Biophys. Acta* **163**, 179.

Kessel, D., Deacon, J., Coffey, B., Bakamjian, A. (1972). *Mol. Pharmacol.* **8**, 731.
Kijima, S., and Wilt, F. H. (1969). *J. Mol. Biol.* **40**, 235.
Kim, B. K., Cha, S., and Parks, R. E. (1968). *J. Biol. Chem.* **243**, 1763.
King, E. J. (1932). *Biochem. J.* **26**, 292.
Kit, S. (1970). *In* "Metabolic Pathways" (D. M. Greenberg, ed.), 3rd ed., Vol. 4, pp. 69–275. Academic Press, New York.
Kit, S., Valladares, Y., and Dubbs, D. R. (1964). *Exp. Cell Res.* **34**, 257.
Klenow, H. (1962). *Biochim. Biophys. Acta* **61**, 885.
Knox, W. E. (1967). *Advan. Cancer Res.* **10**, 117.
Kozai, Y., and Sugino, Y. (1971). *Cancer Res.* **31**, 1376.
Krenitsky, T. A., Elion, G. B., Henderson, A. M., and Hitchings, G. H. (1968). *J. Biol. Chem.* **243**, 2876.
Krygier, V., and Momparler, R. L. (1971). *J. Biol. Chem.* **246**, 2745.
Kuo, J. F., and Greengard, P. (1970). *J. Biol. Chem.* **245**, 4067.
Labow, R., Maley, G. F., and Maley, F. (1969). *Cancer Res.* **29**, 366.
Lagerkvist, U., Reichard, P., Carlsson, B., and Grabosz, J. (1955). *Cancer Res.* **15**, 164.
Larsson, A., and Reichard, P. (1966). *J. Biol. Chem.* **241**, 2540.
Law, L. W. (1956). *Cancer Res.* **16**, 698.
Lea, M. A., Morris, H. P., and Weber, G. (1968). *Cancer Res.* **28**, 71.
LePage, G. A. (1957). *J. Biol. Chem.* **226**, 135.
LePage, G. A., Lin, Y.-T., Orth, R. E., and Gottlieb, J. A. (1972). *Cancer Res.* **32**, 2441.
Letnansky, K. (1964). *Biochim. Biophys. Acta* **87**, 1.
Lieberman, M. W., Sell, S., and Farber, E. (1971). *Cancer Res.* **31**, 1307.
Lieu, T.-S., Hudson, R. A., Brown, R. K., and White, B. C. (1971). *Biochim. Biophys. Acta* **241**, 884.
Lin, S., and Cohen, H. P. (1968). *Anal. Biochem.* **24**, 531.
Lindberg, U., and Skoog, L. (1970). *Anal. Biochem.* **34**, 152.
Lindberg, U., Nordenskjöld, B. A., Reichard, P., and Skoog, L. (1969). *Cancer Res.* **29**, 1498.
Littlefield, J. W. (1966). *Biochim. Biophys. Acta* **114**, 398.
Loeb, L. A., Ewald, J. L., and Agarwal, S. S. (1970). *Cancer Res.* **30**, 2514.
Lucas, Z. J. (1967). *Science* **156**, 1237.
McDonald, T., Sachs, H. G., Orr, C. W., and Ebert, J. D. (1972). *Develop. Biol.* **28**, 290.
McElroy, W. D. (1947). *Proc. Nat. Acad. Sci. U.S.* **33**, 342.
McFall, E., and Magasanik, B. (1960). *J. Biol. Chem.* **235**, 2103.
Maley, F., and Maley, G. F. (1962). *Biochemistry* **1**, 847.
Maley, F., and Maley, G. F. (1970). *Advan. Enzyme Regul.* **8**, 55.
Maley, G. F., and Maley, F. (1960). *J. Biol. Chem.* **235**, 2964.
Maley, G. F., Lorenson, N. G., and Maley, F. (1965). *Biochem. Biophys. Res. Commun.* **18**, 364.
Mandel, P. (1964). *Progr. Nucl. Acid Res. Mol. Biol.* **3**, 299.
Mandel, P. and Harth, S. (1961). *J. Neurochem.* **8**, 116.
Martin, J. B., and Doty, D. M. (1949). *Anal. Chem.* **21**, 965.
Miller, O. L., Jr., Stone, G. E., and Prescott, D. M. (1964). *J. Cell Biol.* **23**, 654.
Mizel, S. B., and Wilson, L. (1972). *Biochemistry* **11**, 2573.
Momparler, R. L., and Fischer, G. A. (1968). *J. Biol. Chem.* **243**, 4298.
Moore, E. C., and Hurlbert, R. B. (1966). *J. Biol. Chem.* **241**, 4802.
Moore, E. C., and Reichard, P. (1964). *J. Biol. Chem.* **239**, 3453.
Moore, E. C., Zedeck, M. S., Agrawal, K. C., and Sartorelli, A. C. (1970). *Biochemistry* **9**, 4492.
Moore, E. C., Booth, B. A., and Sartorelli, A, C. (1971). *Cancer Res.* **31**, 235.
Morris, N. R., and Fischer, G. A. (1960). *Biochim. Biophys. Acta* **42**, 184.
Morris, N. R., and Fischer, G. A. (1963). *Biochim. Biophys. Acta* **68**, 84.

Morse, P. A., Jr., and Potter, V. R. (1965). *Cancer Res.* **25**, 499.
Murad, F., Manganiello, V., and Vaughan, M. (1971). *Proc. Nat. Acad. Sci. U.S.* **68**, 736.
Murphree, S., Stubblefield, E., and Moore, E. C. (1969). *Exp. Cell Res.* **58**, 118.
Murphy, W., and Attardi, G. (1973). *Proc. Nat. Acad. Sci. U.S.* **70**, 115.
Murray, A. W. (1966). *Biochem. J.* **100**, 664.
Murray, A. W. (1968). *Biochem. J.* **106**, 549.
Murray, A. W. (1971). *Annu. Rev. Biochem.* **40**, 811.
Murray, A. W., Elliott, D. C., and Atkinson, M. R. (1970). *Progr. Nucl. Acid Res. Mol. Biol.* **10**, 87.
Nakata, Y., and Bader, J. P. (1969). *Biochim. Biophys. Acta* **190**, 250.
Nazar, R. N., Lawford, H. G., and Wong, J. T. (1970). *Anal. Biochem.* **35**, 305.
Neil, G. L., Moxley, T. E., and Manak, R. C. (1970). *Cancer Res.* **30**, 2166.
Neuhard, J. (1966). *Biochim. Biophys. Acta* **129**, 104.
Neuhard, J., and Munch-Petersen, A. (1966). *Biochim. Biophys. Acta* **114**, 61.
Neuhard, J., and Thomassen, E. (1971). *Eur. J. Biochem.* **20**, 36.
Neuhard, J., Randerath, E., and Randerath, K. (1965). *Anal. Biochem.* **13**, 211.
Nickless, G. (1971). *J. Chromatogr.* **62**, 173.
Nordenskjöld, B. A., Skoog, L., Brown, N. C., and Reichard, P. (1970). *J. Biol. Chem.* **245**, 5360.
Okuda, H., Arima, T., Hashimoto, T., and Fujii, S. (1972). *Cancer Res.* **32**, 791.
Oldham, K. G. (1971). *Anal. Biochem.* **44**, 143.
Oliver, J. M., and Paterson, A. R. P. (1971). *Can. J. Biochem.* **49**, 262.
Otten, J., Johnson, G. S., and Pastan, I. (1971). *Biochem. Biophys. Res. Commun.* **44**, 1192.
Ove, P., Takai, S. I., Umeda, T., and Lieberman, I. (1967). *J. Biol. Chem.* **242**, 4963.
Oxender, D. L. (1972). *Ann. Rev. Biochem.* **41**, 777.
Oyama, V. I., and Eagle, H. (1956). *Proc. Soc. Exp. Biol. Med.* **91**, 305.
Pardee, A. B. (1968). *Science* **162**, 632.
Paterson, A. R. P., and Oliver, J. M. (1971). *Can. J. Biochem.* **49**, 271.
Paterson, A. R. P., and Simpson, A. I. (1965). *Can. J. Biochem.* **43**, 1701.
Perkins, J. P., Hillcoat, B. L., and Bertino, J. R. (1967). *J. Biol. Chem.* **242**, 4771.
Peters, J. H., and Hausen, P. (1971a). *Eur. J. Biochem.* **19**, 502.
Peters, J. H., and Hausen, P. (1971b). *Eur. J. Biochem.* **19**, 509.
Pierre, K. J., and LePage, G. A. (1968). *Proc. Soc. Exp. Biol. Med.* **127**, 432.
Piez, K. A., and Eagle, H. (1958). *J. Biol. Chem.* **231**, 533.
Pinsky, L., and Krooth, R. S. (1967). *Proc. Nat. Acad. Sci. U.S.* **57**, 1267.
Plagemann, P. G. W. (1969). *J. Cell Biol.* **42**, 766.
Plagemann, P. G. W. (1970). *Arch. Biochem. Biophys.* **140**, 223.
Plagemann, P. G. W. (1971a). *J. Cell. Physiol.* **77**, 213.
Plagemann, P. G. W. (1971b). *J. Cell. Physiol.* **77**, 241.
Plagemann, P. G. W. (1971c). *Biochim. Biophys. Acta* **233**, 688.
Plagemann, P. G. W. (1972). *J. Cell Biol.* **52**, 131.
Plagemann, P. G. W., and Erbe, J. (1972). *J. Cell Biol.* **55**, 161.
Plagemann, P. G. W., and Estensen, R. D. (1972). *J. Cell Biol.* **55**, 179.
Plagemann, P. G. W., and Roth, M. F. (1969). *Biochemistry* **8**, 4782.
Plagemann, P. G. W., Ward, G. A., Mahy, B. W. J., and Korbecki, M. (1969). *J. Cell. Physiol.* **73**, 233.
Pomales, R., Elion, G. B., and Hitchings, G. H. (1965). *Biochim. Biophys. Acta* **95**, 505.
Potter, R. L., Schlesinger, S., Buettner-Janusch, V., and Thompson, L. (1957). *J. Biol. Chem.* **226**, 381.
Potter, V. R., (1944). *Advan. Enzymol.* **4**, 201.
Potter, V. R., and LePage, G. A. (1949). *J. Biol. Chem.* **177**, 237.
Potter, V. R., Pitot, H. C., Ono, T., and Morris, H. P. (1960). *Cancer Res.* **20**, 1255.

Price, F. M., Rotherham, J., and Evans, V. (1967). *J. Nat. Cancer Inst.* **39**, 529.
Quastler, H. (1963). *In* "Actions chemiques et biologiques des radiations" (M. Haissinsky, ed.), p. 147. Masson, Paris.
Queener, S. F., Morris, H. P., and Weber, G. (1971). *Cancer Res.* **31**, 1004.
Quigley, J. P., Rifkin, D. B., and Einhorn, M. H. (1972). *Anal. Biochem.* **47**, 614.
Raaen, H. P., and Kraus, F. E. (1968). *J. Chromatogr.* **35**, 531.
Raina, A., and Cohen, S. S. (1966). *Proc. Nat. Acad. Sci. U.S.* **55**, 1587.
Rajalakshmi, S., and Handschumacher, R. E. (1968). *Biochim. Biophys. Acta* **155**, 317.
Randerath, K. (1970). *Anal. Biochem.* **34**, 188.
Randerath, K., and Randerath, E. (1964a). *J. Chromatogr.* **16**, 111.
Randerath, E., and Randerath, K. (1964b). *J. Chromatogr.* **16**, 126.
Randerath, K., and Randerath, E. (1967). *In* "Methods in Enzymology," Vol. 12: Nucleic Acids (L. Grossman and K. Moldave, eds.), Part. A, p. 323. Academic Press. New York.
Randerath, E., Yu, C.-T., and Randerath, K. (1972). *Anal. Biochem.* **48**, 172.
Reichard, P. (1959). *Advan. Enzymol. Relat. Subj. Biochem.* **21**, 263.
Reichard, P., and Sköld, O. (1958). *Biochim. Biophys. Acta* **28**, 376.
Reichard, P., Canellakis, Z. N., and Canellakis, E. S. (1960). *Biochim. Biophys. Acta* **41**, 558.
Reichard, P., Canellakis, Z. N., and Canellakis, E. S. (1961). *J. Biol. Chem.* **236**, 2514.
Reyes, P., and Heidelberger, C. (1965). *Mol Pharmacol.* **1**, 14.
Riegler, R. (1966). *Acta Physiol. Scand.* **67**, Suppl. 267, 18.
Robbins, E., and Pederson, T. (1970). *Proc. Nat. Acad. Sci. U.S.* **66**, 1244.
Robbins, E., Fant, J., and Norton, N. (1972). *Proc. Nat. Acad. Sci. U.S.* **69**, 3708.
Roberts, D., and Loehr, E. V. (1971). *Cancer Res.* **31**, 1181.
Roberts, D., and Loehr, E. V. (1972). *Cancer Res.* **32**, 1160.
Roberts, R. M., and Tovey, K. C. (1970). *Anal. Biochem.* **34**, 582.
Roeder, R. G., and Rutter, W. J. (1970). *Biochemistry* **9**, 2543.
Roos, H., and Pfleger, K. (1972). *Mol. Pharmacol.* **8**, 417.
Roseman, S. (1969). *J. Gen. Physiol.* **54**, 138s.
Rosenbloom, F. M. (1968). *Fed. Proc., Fed. Amer. Soc. Exp. Biol.* **27**, 1063.
Rosenkranz, H. S., and Carr, H. S. (1970). *Cancer Res.* **30**, 1926.
Rotherham, J., and Schneider, W. C. (1958). *J. Biol. Chem.* **232**, 853.
Rubin, R. J., Reynard, A., and Handschumacher, R. E. (1964). *Cancer Res.* **24**, 1002.
Ryan, W. L., and Heidrick, M. L. (1968). *Science* **162**, 1484.
Sartorelli, A. C. (1969). *Cancer Res.* **29**, 2292.
Sartorelli, A. C., and Booth, B. (1967). *Mol. Pharmacol.* **3**, 71.
Scarano, E., Talarico, M., Bonaduce, L., and dePetrocellis, B. (1960). *Nature (London)* **186**, 237.
Scarano, E., Geraci, G., and Rossi, M. (1967). *Biochemistry* **6**, 192.
Schmidt, G. (1957). *In* "Methods in Enzymology," Vol. 3: Preparation and Assay of Substrates (S. P. Colowick and N. O. Kaplan, eds.), p. 671. Academic Press, New York.
Schmidt, G., and Thannhauser, S. J. (1945). *J. Biol. Chem.* **161**, 83.
Schmitz, H., Potter, V. R., and Hurlbert, R. B. (1954). *Cancer Res.* **14**, 66.
Schneider, W. C. (1955). *J. Biol. Chem.* **216**, 287.
Scholar, E. M., Brown, P. R., and Parks, R. E., Jr. (1972). *Cancer Res.* **32**, 259.
Scholtissek, C. (1967). *Biochim. Biophys. Acta* **145**, 228.
Scholtissek, C. (1968). *Biochim. Biophys. Acta* **158**, 435.
Schrecker, A. W. (1970). *Cancer Res.* **30**, 632.
Schrecker, A. W., and Urshel, M. J. (1968). *Cancer Res.* **28**, 793.
Schuster, G. S., and Hare, J. D. (1970). *Exp. Cell Res.* **59**, 163.
Schuster, G. S., and Hare, J. D. (1971). *In Vitro* **6**, 427.
Serlupi-Crescenzi, G., Paolini, C., and Leggio, T. (1968). *Anal. Biochem.* **23**, 263.

Sheppard, G. (1972). *At. Energy Rev.* **10**, 3.
Sheppard, J. R. (1971). *Proc. Nat. Acad. Sci. U.S.* **68**, 1316.
Sheppard, J. R., and Prescott, D. M. (1972). *Exp. Cell Res.* **75**, 293.
Shoup, G. D., Prescott, D. M., and Wykes, J. R. (1966). *J. Cell Biol.* **31**, 295.
Silver, M. J., Rodalewicz, I., Duglas, Y., and Park, D. (1970). *Anal. Biochem.* **36**, 525.
Sköld, O. (1963). *Biochim. Biophys. Acta* **76**, 160.
Skoog, L. (1970). *Eur. J. Biochem.* **17**, 202.
Skoog, L., and Nordenskjöld, B. (1971). *Eur. J. Biochem.* **19**, 81.
Slaby, I., Skoog, L., and Thelander, L. (1971). *Eur. J. Biochem.* **21**, 279.
Smets, L. A. (1969). *J. Cell. Physiol.* **74**, 63.
Smith, J. D. (1967). *In* "Methods in Enzymology," Vol. 12: Nucleic Acids (L. Grossman and K. Moldave, eds.), Part. A, p. 350. Academic Press, New York.
Smith, J. D., Freeman, G., Vogt, M., and Dulbecco, R. (1960). *Virology* **12**, 185.
Sneider, T. W., and Potter, V. R. (1969*a*). *Advan. Enzyme Regul.* **7**, 375.
Sneider, T. W., and Potter, V. R. (1969*b*). *Cancer Res.* **29**, 2398.
Sneider, T. W., Potter, V. R., and Morris, H. P. (1969). *Cancer Res.* **29**, 40.
Solter, A. W., and Handschumacher, R. E. (1969). *Biochim. Biophys. Acta* **174**, 585.
Stambrook, P. J., and Sisken, J. E. (1972). *Biochim. Biophys. Acta* **281**, 45.
Stanley, P. E., and Williams, S. G. (1969). *Anal. Biochem.* **29**, 381.
Steck, T. L., Nakata, Y., and Bader, J. P. (1969). *Biochim. Biophys. Acta* **90**, 237.
Stein, W. D. (1967). "The Movement of Molecules Across Cell Membranes." Academic Press, New York.
Steiner, A. L., Kipnis, D. M., Utiger, R., and Parker, C. (1969). *Proc. Nat. Acad. Sci. U.S.* **64**, 367.
Stewart, P. A., Quastler, H., Skougaard, M. R., Wimber, D. R., Wolfsberg, M. F., Perotta, C. A., Ferbel, B., and Carlough, M. (1965). *Radiat. Res.* **24**, 521.
Stubblefield, E., and Murphree, S. (1967). *Exp. Cell Res.* **48**, 652.
Sugino, Y., and Miyoshi, Y. (1964). *J. Biol. Chem.* **239**, 2360.
Szentkiralyi, E. M. (1957). *Arch. Biochem. Biophys.* **67**, 298.
Tatibana, M., and Ito, K. (1967). *Biochem. Biophys. Res. Commun.* **26**, 221.
Taube, R. A., and Berlin, R. D. (1972). *Biochim. Biophys. Acta* **255**, 6.
Thannhauser, S. J., and Czoniczer, G. (1920). *Z. Physiol. Chem.* **110**, 307.
Turner, M. K., Abrams, R., and Lieberman, I. (1966). *J. Biol. Chem.* **241**, 5777.
van den Bos, R. C., van Kamp, G. J., and Planta, R. J. (1970). *Anal. Biochem.* **35**, 32.
Vesely, J., Cihak, A., and Sorm, F. (1970). *Cancer Res.* **30**, 2180.
Wagner, K., and Arav, R. (1968). *Biochemistry* **7**, 1771.
Wand, M., Zeuthen, E., and Evans, E. A. (1967). *Science* **157**, 436.
Warburg, O. (1930). "The Metabolism of Tumors." Constable, Press, London.
Warburg, O. (1956). *Science* **123**, 309.
Warnick, C. T., Muzik, H., and Paterson, A. R. P. (1972). *Cancer Res.* **32**, 2017.
Weber, G. (1961). *Advan. Cancer Res.* **6**, 403.
Weber, G., and Cantero, A. (1957). *Cancer Res.* **17**, 995.
Weber, G., Queener, S. F., and Ferdinandus, J. A. (1971). *Advan. Enzyme Regul.* **9**, 63.
Weber, M. J., and Edlin, G. (1971). *J. Biol. Chem.* **246**, 1828.
Weber, M. J., and Rubin, H. (1971). *J. Cell. Physiol.* **77**, 157.
Weissman, S. M., Smellie, R. M. S., and Paul, J. (1960). *Biochim. Biophys. Acta* **45**, 101.
Wells, W., and Winzler, R. J. (1959). *Cancer Res.* **19**, 1086.
Werner, R. (1971). *Nature (London), New Biol.* **233**, 99.
Wheeler, G. P., and Alexander, J. A. (1972). *Cancer Res.* **32**, 1761.
Whittle, E. D. (1966). *Biochim. Biophys. Acta* **114**, 44.

Wiebel, F., and Baserga, R. (1969). *J. Cell. Physiol.* **74**, 191.
Wilt, F. H. (1969). *Anal. Biochem.* **27**, 186.
Winkler, H. H. and Wilson, T. H. (1966). *J. Biol. Chem.* **241**, 2200.
Witschi, H. (1972). *Cancer Res.* **32**, 1686.
Wood, T. (1971). *Anal. Biochem.* **43**, 107.
Yegian, C. (1973). *Anal. Biochem.* In press.
Yip, M. C. M., and Knox, W. E. (1970). *J. Biol. Chem.* **245**, 2199.
Yu, F. L., and Feigelson, P. (1969). *Arch. Biochem. Biophys.* **129**, 152.

Author Index

Numbers in italics refer to the pages on which the complete references are listed.

D

L

Y

Z

Subject Index

E

I

J

K

L

R

S

X

Y